P9-AFZ-610

Methods in Cell Biology

VOLUME 46
Cell Death

Series Editors

Leslie Wilson
Department of Biological Sciences
University of California, Santa Barbara
Santa Barbara, California

Paul Matsudaira
Whitehead Institute for Biomedical Research and
Department of Biology
Massachusetts Institute of Technology
Cambridge, Massachusetts

Methods in Cell Biology

Prepared under the Auspices of the American Society for Cell Biology

VOLUME 46
Cell Death

Edited by

Lawrence M. Schwartz

Morrill Science Center
University of Massachusetts
Amherst, Massachusetts

and

Barbara A. Osborne

Department of Veterinary and Animal Science
University of Massachusetts
Amherst, Massachusetts

ACADEMIC PRESS

San Diego   New York   Boston   London   Sydney   Tokyo   Toronto

Cover photograph (paperback edition only): Dying cells in a stage 12 *Drosophila* embryo revealed by acridine orange staining and confocal microscopy. Staged embryos were dechorionated in 50% bleach, shaken in heptane and the vital dye acridine orange (5 g/ml). Whole embryos were then examined by confocal microscopy and computer-generated false color were then examined by confocal microscopy and computer-generated false color was added to facilitate discrimination of the dying cells. While dying cells can be observed in several tissues, the major loss at this stage occurs along the ventral mid-line, where cell death allows the nervous system to delaminate from the underlying epidermis. Photo by Donna Saatman, Randall Phillis, and Lawrence M. Schwartz.

This book is printed on acid-free paper.

Copyright © 1995 by ACADEMIC PRESS, INC.

All Rights Reserved.
No part of this publication may be reproduced or transmitted in any form or by any means, electronic or mechanical, including photocopy, recording, or any information storage and retrieval system, without permission in writing from the publisher.

Academic Press, Inc.
A Division of Harcourt Brace & Company
525 B Street, Suite 1900, San Diego, California 92101-4495

United Kingdom Edition published by
Academic Press Limited
24-28 Oval Road, London NW1 7DX

International Standard Serial Number: 0091-679X

International Standard Book Number: 0-12-564147-8 (case)

International Standard Book Number: 0-12-632445-X (comb)

PRINTED IN THE UNITED STATES OF AMERICA
95   96   97   98   99   00   EB   9   8   7   6   5   4   3   2   1

CONTENTS

12. Methods for Studying Cell Death and Viability in Primary
Neuronal Cultures

James E. Johnson

13. Neuron Death in Vertebrate Development: *In Vivo* Methods

Peter G. H. Clarke and Ronald W. Oppenheim

14. Methods for the Study of Cell Death in the Nematode
Caenorhabditis elegans

Monica Driscoll

15. Programmed Cell Death during Mammary Gland Involution

Robert Strange, Robert R. Friis, Lynne T. Bemis, and F. Jon Geske

CONTRIBUTORS

Numbers in parentheses indicate the pages on which the authors' contributions begin.

Steven W. Barger (187), Department of Anatomy and Neurobiology, Sanders-Brown Research Center on Aging, University of Kentucky, Lexington, Kentucky 40536

James G. Begley (187), Department of Anatomy and Neurobiology, Sanders-Brown Research Center on Aging, University of Kentucky, Lexington, Kentucky 40536

Lynne T. Bemis (139,355), Division of Laboratory Research, AMC Cancer Research Center, Lakewood, Colorado 80214

Shmuel A. Ben-Sasson (29), The Hubert H. Humphrey Center for Experimental Medicine and Cancer Research, The Hebrew University-Hadassah Medical School, Jerusalem 91120, Israel

Wolfgang Bielke (107), Department of Biology, University of Massachusetts, Amherst, Massachusetts 01003

Reid P. Bissonnette (153), Division of Cellular Immunology, La Jolla Institute for Allergy and Immunology, La Jolla, California 92037

Ralph Buttyan (369), Departments of Urology and Pathology, College of Physicians and Surgeons, Columbia University, New York, New York 10032

Peter G. H. Clarke (277), Institut d'Anatomie, Faculté de Medecine, Université de Lausanne, 1005 Lausanne, Switzerland

Marc C. Colombel (369), Departments of Urology and Pathology, College of Physicians and Surgeons, Columbia University, New York, New York 10032

Electra C. Coucouvanis (387), Department of Anatomy, University of California at San Francisco, San Francisco, California 94143

Monica Driscoll (323), Department of Molecular Biology and Biochemistry, Center for Advanced Biotechnology and Medicine, Rutgers University, Piscataway, New Jersey 08855

Alan Eastman (41), Department of Pharmacology and Toxicology, Dartmouth Medical School, Hanover, New Hampshire 03755

Maria Erecińska (217), Cell Biology Graduate Group and Department of Pharmacology, School of Medicine, University of Pennsylvania, Philadelphia, Pennsylvania 19104

Pamela J. Fraker (57), Department of Biochemistry and Department of Microbiology and Public Health, Michigan State University, East Lansing, Michigan 48824

Robert R. Friis (355), Laboratory for Clinical and Experimental Research, University of Bern, CH-3004 Bern, Switzerland

Yael Gavrieli (29), The Hubert H. Humphrey Center for Experimental Medicine and Cancer Research, The Hebrew University-Hadassah Medical School, Jerusalem 91120, Israel

F. Jon Geske (139, 355), Division of Laboratory Research, AMC Cancer Research Center, Lakewood, Colorado 80214

Glenda C. Gobé (1), Department of Pathology, University of Queensland Medical School, Herston, Queensland 4006, Australia

Douglas R. Green (153), Division of Cellular Immunology, La Jolla Institute for Allergy and Immunology, La Jolla, California 92037

Brian V. Harmon (1), School of Life Science, Queensland University of Technology, Brisbane, Queensland 4000, Australia

James E. Johnson (243), Bowman Gray School of Medicine, Department of Neurobiology and Anatomy, Program in Neuroscience, Winston-Salem, North Carolina 27157

John F. R. Kerr (1), Department of Pathology, University of Queensland Medical School, Herston, Queensland 4006, Australia

Louis E. King (57), Department of Biochemistry and Department of Microbiology and Public Health, Michigan State University, East Lansing, Michigan 48824

Deborah Lill-Elghanian (57), Department of Biochemistry and Department of Microbiology and Public Health, Michigan State University, East Lansing, Michigan 48824

Zheng-Gang Liu (99), Program in Molecular and Cellular Biology, University of Massachusetts, Amherst, Massachusetts 01003

Artin Mahboubi (153), Division of Cellular Immunology, La Jolla Institute for Allergy and Immunology, La Jolla, California 92037

Robert J. Mark (187), Department of Anatomy and Neurobiology, Sanders-Brown Research Center on Aging, University of Kentucky, Lexington, Kentucky 40536

Gail R. Martin (387), Department of Anatomy, University of California at San Francisco, San Francisco, California 94143

Seamus J. Martin (153), Division of Cellular Immunology, La Jolla Institute for Allergy and Immunology, La Jolla, California 92037

Mark P. Mattson (187), Department of Anatomy and Neurobiology, Sanders-Brown Research Center in Aging, University of Kentucky, Lexington, Kentucky 40536

Anne J. McGahon (153), Division of Cellular Immunology, La Jolla Institute for Allergy and Immunology, La Jolla, California 92037

Kelly A. McLaughlin (99), Program in Molecular and Cellular Biology, University of Massachusetts, Amherst, Massachusetts 01003

Carolanne E. Milligan (107), Department of Biology, University of Massachusetts, Amherst, Massachusetts 01003

Jason C. Mills (217), Cell Biology Graduate Group, School of Medicine, University of Pennsylvania, Philadelphia, Pennsylvania 19104

Rona J. Mogil (153), Division of Cellular Immunology, La Jolla Institute for Allergy and Immunology, La Jolla, California 92037

Joseph H. Nadeau (387), Department of Human Genetics, Montreal General Hospital, Montreal, Quebec, Canada H3G 1A4

Walter K. Nishioka (153), Division of Cellular Immunology, La Jolla Institute for Allergy and Immunology, La Jolla, California 92037

Ronald W. Oppenheim (277), Department of Neurobiology and Anatomy and Neuroscience Program, Bowman Gray School of Medicine, Wake Forest University, Winston-Salem, North Carolina 27157

Barbara A. Osborne (99), Department of Veterinary and Animal Sciences, Program in Molecular and Cellular Biology, University of Massachusetts, Amherst, Massachusetts 01003

Randall N. Pittman (217), Cell Biology Graduate Group and Department of Pharmacology, School of Medicine, University of Pennsylvania, Philadelphia, Pennsylvania 19104

Steven J. Robinson (107), Institute of Molecular Biology, University of Oregon, Eugene, Oregon 97403

Robert T. Schimke (77), Department of Biological Sciences, Stanford University, Stanford, California 94305

Lawrence M. Schwartz (99,107), Department of Biology, Program in Molecular and Cellular Biology, University of Massachusetts, Amherst, Massachusetts 01003

Yoav Sherman (29), Department of Pathology, The Hubert H. Humphrey Center for Experimental Medicine and Cancer Research, The Hebrew University-Hadassah Medical School, Jerusalem 91120, Israel

Steven W. Sherwood (77), Department of Biological Sciences, Stanford University, Stanford, California 94305

Yufang Shi (153), Division of Cellular Immunology, La Jolla Institute for Allergy and Immunology, La Jolla, California 92037

Sallie W. Smith (99), Department of Veterinary and Animal Sciences, University of Massachusetts, Amherst, Massachusetts 01003

Robert Strange (139,355), Division of Laboratory Research, AMC Cancer Research Center, Lakewood, Colorado 80214

William G. Telford (57), Department of Biochemistry and Department of Microbiology and Public Health, Michigan State University, East Lansing, Michigan 48824

Songli Wang (217), Department of Pharmacology, School of Medicine, University of Pennsylvania, Philadelphia, Pennsylvania 19104

Clay M. Winterford (1), Department of Pathology, University of Queensland Medical School, Herston, Queensland 4006, Australia

PREFACE

Any cell can be murdered by the application of some noxious treatment. These cells then die by necrosis, a passive process that involves disruption of membrane integrity, influx of calcium ions and water, and subsequently, cellular lysis (reviewed in Farber, 1990). In contrast, many cells die by a process that involves cellular condensation. In most cases, this occurs with the morphology of apoptosis, which is characterized by membrane blebbing, the deposition of electron-dense chromatin along the inner margin of the nucleus, and the pinching-off of membrane-bound bodies (Kerr *et al.,* 1972).

Historically, it was assumed that the death of cells within a developmental context represented pathological cellular loss. In fact, when pyknotic cells were observed during embryogenesis, it was assumed that these "granules" represented "mitotic metabolites" (Rabl, 1900; Jokl, 1920). Only much later was it accepted that cell death could be a normal developmental process (reviewed in Glucksmann, 1951). That large numbers of cells die during development was not appreciated until the landmark paper of Hamburger and Levi-Montalcini (1949), when careful quantitative cell counts were made within defined regions of the nervous system. Their studies of the dorsal root ganglia of the chick demonstrated that only about 60% of the neurons that were initially produced were maintained in the newly hatched chick. Studies by other investigators have shown that in various regions of the vertebrate nervous system, upward of 85% of the neurons are lost before or shortly after birth (Oppenheim, 1991). Such massive cell death is not restricted to the nervous system. In fact, programmed cell death can be found in every tissue. This has led to the hypothesis that all cells in animals are programmed to die unless they receive specific signals from neighboring cells that result in their reprieve (Raff, 1992).

There are many reasons why the extent of cell death has been so underestimated. At any given time, the number of cells identified as dying in histological material may be quite small. This is due to several observations, including: (1) there is an apparent lack of synchrony among dying cells within a population; (2) dying cells are usually rapidly phagocytosed, thereby removing them before they can be counted; (3) cell death is an efficient and rapid process that may be completed in a matter of minutes to hours; (4) given that other cells may be dividing or infiltrating the area (such as macrophages), the volume of the tissue being examined may actually increase during the period of cell death; and (5) the histological properties of dying cells may not be obvious upon casual inspection.

For many years, the study of cell death was undertaken by a relatively small number of investigators examining specific developmental systems. Investigators in one field largely overlooked the results from other disciplines. However, during the past few years there has been an explosion in the number of papers examining various aspects of cell death (Fig. 1). There are many reasons for this newfound interest in the field. First, the pioneering work of Horvitz and his colleagues demonstrated conclusively that programmed cell death requires the activity of specific genes (reviewed in Ellis *et al.*, 1991). This has given many laboratories the impetus to clone, characterize, and manipulate putative cell death genes. Second, the demonstration that the proto-oncogene *bcl-2* acts by blocking cell death rather than by promoting mitosis forced many investigators to appreciate that for tumor growth the loss of cells was as important as the addition of new ones (Tsujimoto *et al.*, 1985). This appreciation has recently been boosted by the demonstration that the tumor supressor gene p53 can act as a switch between cell proliferation and cell death (Lowe *et al.*, 1993; Clarke *et al.*, 1993). The fact that p53 was the 1994 Molecule of the Year in *Science* and the subject of over 1000 papers also has attracted attention to the field. In addition, the development of technical innovations for the study of cell death has allowed many investigators from a wide variety of disciplines to enter the field. A wealth of review articles cataloging the distribution and

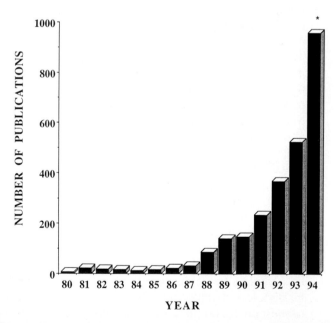

Fig. 1 While programmed cell death has been studied for a century (Beard, 1896), it was not a major focus of research in biology. As can be seen from this graph, this has changed dramatically during the past few years. The field is currently in a period of exponential growth that has yet to plateau. *, Estimate for 1994 based on publications from the first half of the year. (Data kindly provided by John J. Cohen.)

features of cell death throughout both phylogeny and development are available (Glucksmann, 1951; Saunders, 1966; Wyllie *et al.*, 1980; Oppenheim, 1991; Ellis *et al.*, 1991; Arends and Wyllie, 1991; Cohen, 1993; Schwartz and Osborne, 1993). What is not present in the literature, however, is a comprehensive collection of methods that can be applied to the study of cell death. With this volume we are attempting to fill this void. We have brought together chapters representing a broad range of technical approaches and model systems. The chapters presented cover topics from the cellular to the organismal and from molecular to anatomical. The protocols and insights presented are the product of years of study by experts in their respective fields. It is our hope that the methods and insights presented in this book will help convert us into students of cell death, rather than immunologists and neurobiologists. Since virtually all cells in an organism contain the genetic information required to commit suicide, the ability to regulate this process may ultimately offer the potential to treat a wide range of disorders. The ability to induce the rapid, noninflammatory death of deleterious cells in a lineage-specific manner cannot be overstated. Neither can the potential to rescue valuable cells that may inappropriately activate their endogenous cell death programs. The next few years offer great potential for this field.

In closing, we would like to take this opportunity to acknowledge some of the people who made invaluable contributions to the successful generation of this volume. We thank all of the investigators who wrote chapters for their considerable efforts in providing detailed comprehensive protocols and insights, Phyllis Moses and her staff at Academic Press for providing support and guidance to us during every phase of the project, Lisa Korpiewski for excellent technical assistance, and last, the members of our respective laboratories for all of their help and insight.

References

Arends, M. J., and Wyllie, A. H. (1991). Apoptosis: Mechanisms and roles in pathology. *Int. Rev. Exp. Pathol.* **32,** 223–254.

Clarke, A. R., Purdie, C. A., Harrison, D. J., Morris, R. G., Bird, C. C., Hooper, M. L., and Wyllie, A. H. (1993). Thymocyte apoptosis induced by p53-dependent and independent pathways. *Nature (London)* **362,** 849–852.

Cohen, J. J. (1993). Apoptosis. *Immunol. Today* **14,** 26–130.

Ellis, R. E., Yuan, J., and Horvitz, H. R. (1991). Mechanisms of cell death. *Annu. Rev. Cell Biol.* **7,** 663–698.

Farber, J. L. (1990). The role of calcium ions in toxic cell injury. *Environ. Health Persp.* **84,** 107–111.

Glucksmann, A. (1951). Cell deaths in normal vertebrate ontogeny. *Biol. Rev.* **26,** 59–86.

Hamburger, V., and Levi-Montalcini, R. (1949). Proliferation, differentiation and degeneration in the spinal ganglia of the chick embryo under normal and experimental conditions. *J. Exp. Zool.* **111,** 457–502.

Jokl, A. (1920). Zur Entwicklung des Anurenauges. *Anat. Hefte* **59,** 217.

Kerr, J. F. R., Wyllie, A. H., and Currie, A. R. (1972). Apoptosis: A basic biological phenomenon with wide ranging implications in tissue kinetics. *Br. J. Cancer* **26,** 239–257.

Lowe, S. W., Schmitt, E. M., Smith, S. W., Osborne, B. A., and Jacks, T. (1993). p53 is required for radiation-induced apoptosis in mouse thymocytes. *Nature (London)* **362,** 847–849.

Oppenheim, R. W. (1991). Cell death during the development of the nervous system. *Annu. Rev. Neurosci.* **14,** 453–501.

Rabl, C. (1900). *Ueber den Bau und die Entwicklung der Linse.* Leipzig.

Raff, M. C. (1992). Social controls on cell survival and cell death. *Nature (London)* **356,** 397–400.

Saunders, J. W. (1966). Death in embryonic systems. *Science* **154,** 604–612.

Schwartz, L. M., and Osborne, B. A. (1993). Programmed cell death, apoptosis and killer genes. *Immunol. Today* **14,** 582–590.

Tsujimoto, Y., Gorham, J., Cossman, J., Jaffe, E., and Croce, C. (1985). The T (14;18) chromosome translocations involved in B cell neoplasms result from mistakes in VDJ joining. *Science* **229,** 1390–1393.

Wyllie, A. H., Kerr, J. F. R., and Currie, A. R. (1980). Cell death: The significance of apoptosis. *Int. Rev. Cytol.* **68,** 251–306.

Lawrence M. Schwartz and Barbara A. Osborne

CHAPTER 1

Anatomical Methods in Cell Death

John F. R. Kerr,* Glenda C. Gobé,* Clay M. Winterford,* and Brian V. Harmon†

* Department of Pathology
University of Queensland Medical School
Herston, Queensland 4006, Australia

† School of Life Science
Queensland University of Technology
Brisbane, Queensland 4000, Australia

I. Introduction

In this chapter we will describe methods for studying the morphology of cell death and the criteria used in identifying apoptosis and necrosis. Apoptosis was originally distinguished from necrosis on the basis of its ultrastructure (Kerr, 1971; Kerr *et al.,* 1972). Electron microscopy still provides the most reliable method for recognizing the two processes; in many cases, however, they can be identified confidently using light microscopy alone.

We will restrict our account to mammalian cells. In lower animals, clear-cut correlations between the morphology of cell death and its circumstances of occurrence have not yet been defined. Programmed cell death that occurs

Copyright © 1995 by Academic Press, Inc. All rights of reproduction in any form reserved.

during the normal development of invertebrates has been shown to differ from apoptosis in a number of ways (Robertson and Thomson, 1982; Lockshin and Zakeri, 1991; Schwartz *et al.*, 1993; Zakeri *et al.*, 1993).

After the discovery was made that apoptosis of mammalian cells is accompanied by double-strand cleavage of nuclear DNA at the linker regions between nucleosomes (Wyllie, 1980), DNA electrophoresis was widely used for identification of this process (see Chapter 3); the development of a so-called "ladder" in agarose gels came to be regarded as a biochemical hallmark of the process. Several reports, however, indicate that such a ladder is not always demonstrable (Cohen *et al.*, 1992; Collins *et al.*, 1992; Falcieri *et al.*, 1993; Oberhammer *et al.*, 1993; Tomei *et al.*, 1993). When using DNA electrophoresis to look for apoptosis in a particular system, it is prudent to supplement this technique with morphological studies, at least initially. The same precaution applies to identification of apoptosis by the *in situ* end labeling or nick translation techniques that have recently been introduced (see Chapter 2). DNA electrophoresis is not a satisfactory technique for identifying necrosis. The random DNA degradation that accompanies this type of cell death is a late phenomenon (Duvall and Wyllie, 1986), often evident by electrophoresis only some hours after the onset of structural degeneration (Collins *et al.*, 1992). An additonal advantage of using morphological over biochemical methods for identifying cell death is that these techniques allow independent recognition of apoptosis and necrosis when the processes occur simultaneously.

Apoptosis and necrosis both involve a sequence of consecutive morphological events. The evolution of the processes with time must be considered when describing their diagnostic characteristics. We will not justify every statement in the morphological descriptions with lists of references. Comprehensive bibliographies and many additional illustrations can be found in reviews (Wyllie *et al.*, 1980; Kerr *et al.*, 1987; Wyllie, 1987; Walker *et al.*, 1988). Laboratory protocols for the morphological study of cell death are located at the end of the chapter.

II. Morphological Identification of Cell Death

The recognition of apoptosis and necrosis is based primarily on the distinctive changes that take place within the affected cells. However, when these two processes occur *in vivo,* they also differ in their distribution and in the tissue reactions that are associated with them. These latter features may be of subsidiary use in identification. Thus, apoptosis typically involves scattered individual cells in a tissue, whereas necrosis involves groups of adjoining cells. Further, necrosis is usually accompanied by an acute inflammatory response with exudation of neutrophil leukocytes and monocytes; this event is characteristically absent in apoptosis. Infiltrating mononuclear cells may, nevertheless, be present where apoptosis is being induced by cytotoxic lymphocytes, a situation that pertains, for example, in the cellular immune rejection of allografts, in graft-versus-host disease, in certain human skin diseases such as lichen planus, and in viral hepatitis (Kerr *et al.*, 1984, 1987). The differences just mentioned are, of course, not evident in cell cultures; under these circumstances identification must be based entirely on the morphology of the dying cells themselves. Since the diagnostic features of apoptosis and necrosis are most readily appreciated at the electron microscopic level, we will begin with an account of their ultrastructure.

A. Electron Microscopy

The sequential ultrastructural events in apoptosis and necrosis are shown in diagrammatic form in Fig. 1. The figure legend provides a synopsis of the processes.

1. Apoptosis

The earliest unequivocal morphological evidence of the onset of apoptosis is found in the nucleus. The chromatin condenses and becomes aggregated in sharply delineated, finely granular masses of uniform texture that abut the inner surface of the nuclear envelope (Fig. 2). The proportion of the nucleus occupied by condensed chromatin varies with cell type, being particularly high in lymphoid cells (Fig. 3) and much lower in cells such as HeLa cells that have little heterochromatin (Fig. 4). The early chromatin changes are often, but not always, accompanied by convolution of the nuclear outline (Figs. 2 and 4). In many cell types, the nuclear convolution then becomes extreme (Figs. 5 and 6) and is followed by budding to produce discrete nuclear fragments of varying size and chromatin content, which are still surrounded by double membranes (Figs. 7 and 8). Condensed chromatin occupies the whole of the cross-sectional area in some of the fragments and is confined to peripheral crescents in others (Figs. 7 and 8). However, nuclear convolution and budding tend to be restricted in cells with a high nucleus : cytoplasm ratio, for example, thymocytes (Fig. 3). Remnants of the nucleolus are evident in the nuclei and the nuclear fragments in some planes of section, and take two forms: clusters of dispersed granules (Figs. 3, 4, 7, and 8) and compact granular masses that are usually closely apposed to the inner surface of the condensed chromatin (Figs. 4, 5, and 7). The latter are probably derived from the nucleolar fibrillar center (Wyllie, 1987).

A series of equally dramatic events takes place in the cytoplasm concurrently with the nuclear changes just described. Overall condensation occurs; in cells with relatively abundant cytoplasm, this event is usually associated with extensive protrusion or blebbing of the cell surface (Figs. 4–6), a phenomenon that is well demonstrated by scanning electron microscopy (Fig. 9). However, as is the case with convolution of the nuclear outline, the cell surface protrusion tends to be limited in thymocytes and small lymphocytes (Fig. 3). This process

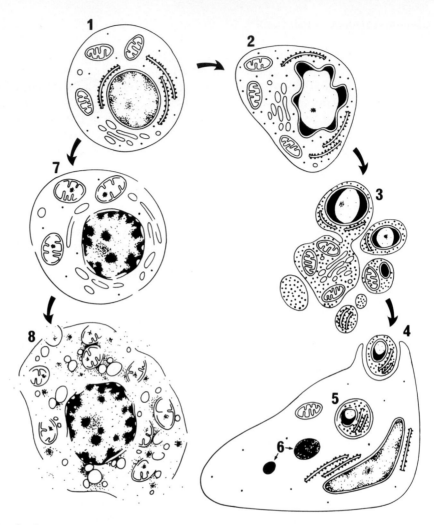

Fig. 1 Sequence of ultrastructural changes in apoptosis (2–6) and necrosis (7 and 8). A normal cell is shown in stylized form at 1. Early apoptosis (2) is characterized by compaction and segregation of chromatin in sharply circumscribed masses that abut on the inner surface of the nuclear envelope, convolution of the nuclear outline, condensation of the cytoplasm with preservation of the integrity of organelles, and the beginning of convolution of the cell surface. In the next phase (3), the nucleus fragments and further condensation of the cytoplasm is associated with extensive cell surface protrusion, followed by separation of the surface protuberances to produce membrane-bounded apoptotic bodies of varying size and composition. These bodies are phagocytosed (4) by nearby cells and are degraded by lysosomal enzymes (5), being rapidly reduced to nondescript residues within telolysosomes (6). In the irreversibly injured cell, the onset of necrosis (7) is manifest as irregular clumping of chromatin without radical change in its distribution, gross swelling of mitochondria with the appearance of flocculent densities in their matrices, dissolution of ribosomes, and focal rupture of membranes. At a more advanced stage of this process (8), all cellular components disintegrate. In tissues, the overall configuration of the cell is reasonably maintained until it is removed by mononuclear phagocytes, but in cell cultures dissolution eventually ensues.

4

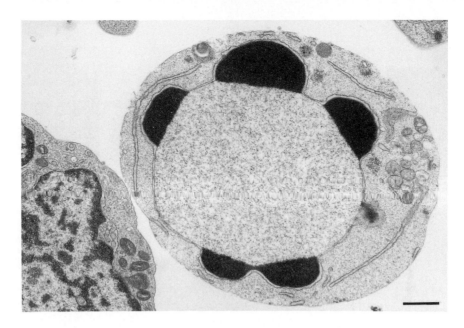

Fig. 2 Early stage of apoptosis of murine NS-1 cell occurring spontaneously in culture. Note convolution of the nuclear outline and segregation of condensed chromatin in sharply circumscribed, uniformly fine granular masses that lie against the inner surface of the nuclear envelope. Bar: 1 μm.

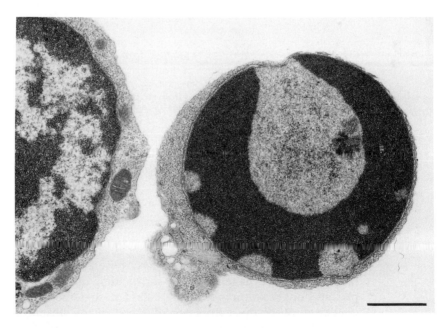

Fig. 3 Apoptosis of thymocyte from 21-day-old rat 4 hr after addition of hydrocortisone to culture at concentration of 10 μM. Bar: 1 μm.

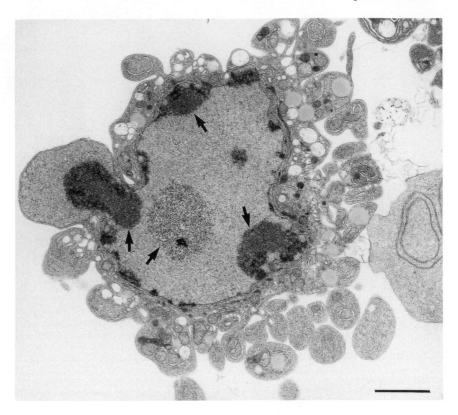

Fig. 4 Apoptosis of HeLa cell 24 hr after addition of actinomycin D to culture at concentration of 15 μg/ml. Note convolution of cell and nuclear outlines, small size of masses of condensed chromatin, and two types of nucleolar remnants (arrows). Despite the paucity of chromatin, agarose gel electrophoresis of DNA extracted from the culture showed a well-developed oligonucleosomal ladder. Bar: 2 μm.

also tends to be restricted in epidermal keratinocytes, probably because of splinting of much of the cytoplasm by abundant tonofilaments (Weedon *et al.*, 1979). Microvilli, if originally present, disappear by the time blunt protuberances develop on the cell surface. Clear vacuoles may be numerous in the condensing cytoplasm (Figs. 3 and 4) and can sometimes be found discharging their contents by exocytosis; this is manifest in scanning electron micrographs as the occurrence of surface craters (Wyllie, 1987). Cytoplasmic condensation is clearly apparent in sections of tissues, where these cells contrast with the cells with more dispersed organelles in the surrounding unaffected tissue (Fig. 6). Cytoplasmic condensation may be more difficult to detect in cell cultures. Note that condensation of the cytoplasm of a cell in the absence of compaction and segregation of its nuclear chromatin is not indicative of apoptosis; viable cells with relatively condensed cytoplasm have been found in many normal organs

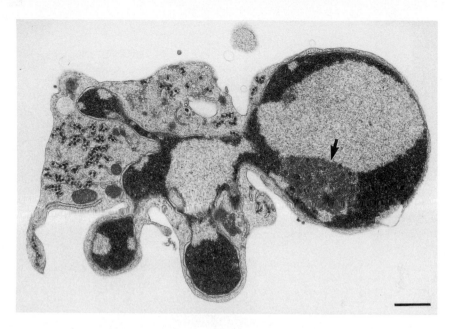

Fig. 5 Apoptosis of NS-1 cell; same culture illustrated in Fig. 2. Note convolution of cell and nuclear outlines and compact nucleolar remnant (arrow). The dark bodies in the cytoplasm are virus particles. Bar: 1 μm.

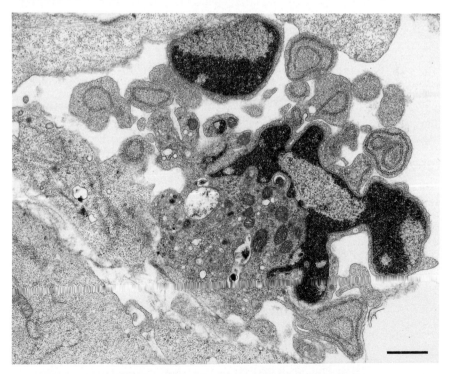

Fig. 6 Apoptosis occurring in murine EMT6 tumor nodule 2 hr after heating at 44°C for 30 min. Note condensation of cytoplasm with preservation of integrity of organelles and convolution of cell and nuclear outlines. Bar: 1 μm.

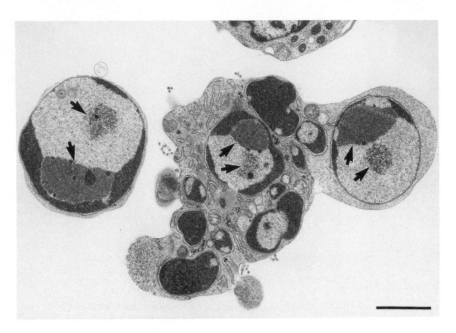

Fig. 7 Apoptosis of NS-1 cell; same culture illustrated in Figs. 2 and 5. Note multiple nuclear fragments, which are enclosed by double membranes and which have varying chromatin content. Nucleolar remnants are indicated by arrows. Bar: 2 μm.

and in tumors. The crowded organelles remain well preserved at this early stage of apoptosis (Figs. 6 and 7).

The protuberances that have formed on the cell surface soon separate by sealing of the plasmalemma to produce membrane-bounded apoptotic bodies of varying size and composition. In some cases, the cell is converted into many small bodies, whereas in others, one or two of the bodies are conspicuously larger than the rest. When cell surface protrusion has been restricted, one of the resulting apoptotic bodies is inevitably relatively large. Such large apoptotic bodies derived from keratinocytes have traditionally been termed Civatte bodies by dermatopathologists (Weedon *et al.*, 1979). The number of nuclear fragments in individual apoptotic bodies varies widely, and the presence of a nuclear component is not consistently related to the size of a body. The cytoplasmic organelles in newly formed apoptotic bodies are still intact.

Apoptotic bodies formed within tissues are mostly rapidly phagocytosed by resident macrophages (Figs. 10 and 11) or by other nearby cells, including epithelial cells (Fig. 12). However some of the bodies formed within single-layered lining epithelia are extruded into the adjacent lumen (Fig. 13). In tumors, many apoptotic bodies are taken up by the neoplastic cells. After their phagocytosis, apoptotic bodies are degraded by lysosomal enzymes derived from the cells in which they lie and are soon reduced to nondescript debris within telolyso-

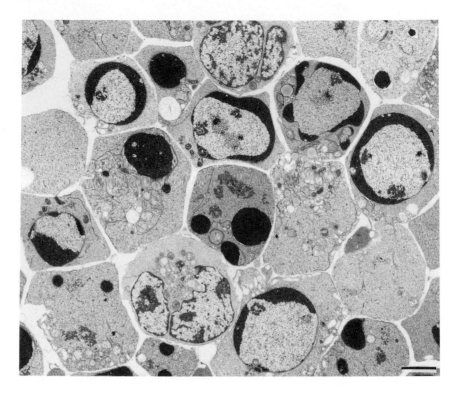

Fig. 8 Cells showing early apoptosis and multiple apoptotic bodies containing nuclear fragments of varying size in culture of BM13674 Burkitt's lymphoma cell line 4 hr after heating at 43°C for 30 min. Bar: 3 μm.

somes. During the process of degradation, their apoptotic origin can be recognized if they contain remnants of nuclear fragments in which masses of segregated, markedly condensed chromatin are still discernible (Figs. 14 and 15). The disposal of apoptotic bodies formed in tissues is characteristically effected without disruption of overall tissue architecture; the surrounding cells close ranks as the apoptotic bodies are phagocytosed.

Apoptotic bodies formed in cell cultures frequently escape phagocytosis and spontaneously degenerate within a few hours. Their bounding membranes rupture and swelling and dissolution of their organelles occurs; these changes resemble those observed in necrosis (Fig. 1). Such degenerate apoptotic bodies can, however, be distinguished from cells that have undergone necrosis *ab initio* if they contain apoptotic-type nuclear fragments (Fig. 16). Apoptotic bodies shed into lumina from epithelial surfaces suffer a similar fate. Note that newly formed apoptotic bodies with intact bounding membranes exclude vital dyes. Only when degeneration supervenes do these bodies lose the capacity to

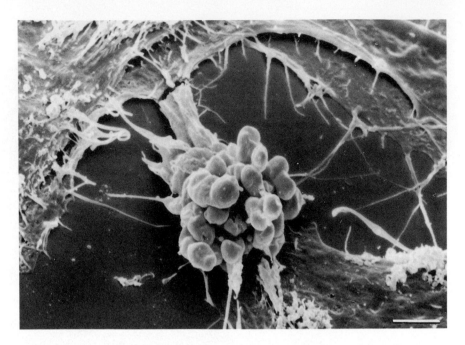

Fig. 9 Apoptosis in same HeLa cell culture illustrated in Fig. 4. The surface protuberances are graphically obvious using scanning electron microscopy. Bar: 2 μm.

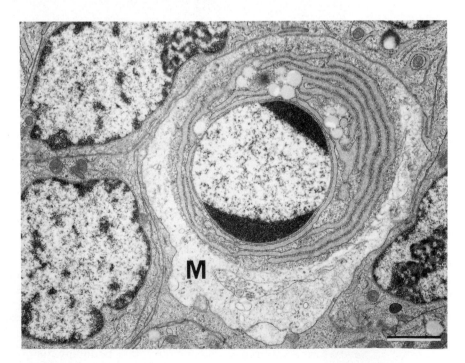

Fig. 10 Apoptotic body of epithelial cell origin, phagocytosed by intraepithelial macrophage (M) in rat ventral prostate 2 days after castration. The body shows early degenerative changes, but its apoptotic nature is still clearly evident. Bar: 2 μm.

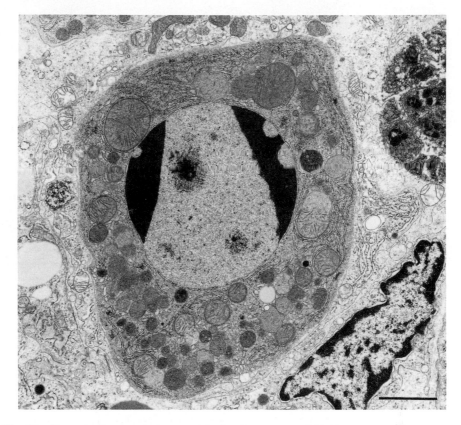

Fig. 11 Recently phagocytosed apoptotic body of hepatocyte origin within macrophage in median lobe of rat liver 3 days after ligation of the associated portal vein branch. In addition to a typical nuclear fragment, cytoplasmic condensation and preservation of the integrity of organelles are well shown. Bar: 2 μm.

exclude dyes (Sheridan *et al.*, 1981). The use of dye exclusion to assess the time of occurrence of cell death in culture systems may therefore give misleading results if apoptosis rather than necrosis is involved.

The frequency with which each of the phases of apoptosis just described is observed by electron microscopy under a particular set of circumstances is dependent on both the kinetics of the process and, in the case of a discrete wave of enhanced apoptosis, the point in the wave at which the sample is taken. Apoptosis characteristically affects the individual members of a cell population asynchronously. Phase contrast microscopy of cell cultures (Russell *et al.*, 1972; Sanderson, 1976; Matter, 1979) shows that the onset of this process is abrupt and is followed by violent convulsion of the cell surface; conversion of a cell into a cluster of apoptotic bodies is completed within several minutes. This observation correlates with the fact that convoluted budding cells such as

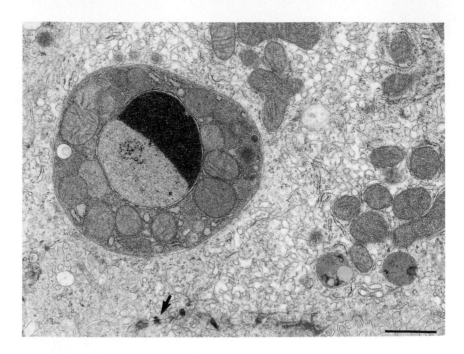

Fig. 12 Well-preserved apoptotic body within phagosome in cytoplasm of hepatocyte. Epithelial nature of cell is indicated by presence of desmosome (arrow). Same specimen illustrated in Fig. 11. Bar: 1 μm.

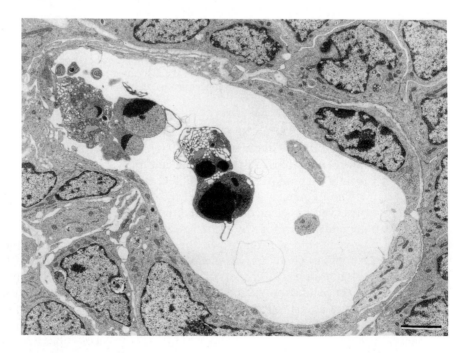

Fig. 13 Apoptotic bodies shed into tubule lumen in nephrogenic zone of kidney of 5-day old rat 3 hr after exposure to 5 Gy X-rays. Bar: 3 μm.

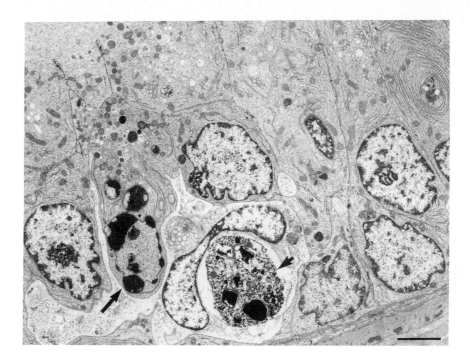

Fig. 14 Early apoptosis of epithelial cell (long arrow) and degraded apoptotic body within intraepithelial macrophage (short arrow) in rat ventral prostate 2 days after castration. Although at an advanced stage of degradation, the apoptotic body in the macrophage can still be recognized by the presence of typical nuclear fragments. Bar: 3 μm.

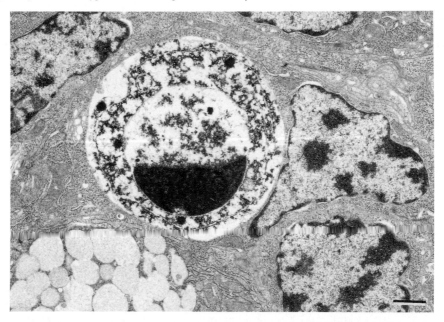

Fig. 15 Partly degraded apoptotic body containing recognizable nuclear fragment within mouse small intestinal crypt epithelial cell 2 hr after injection of cytosine arabinoside in dose of 250 mg/ kg body weight. The faintly stained secretory droplets lie in an adjacent goblet cell. Bar: 1 μm.

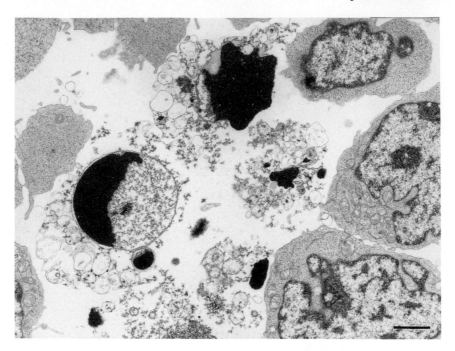

Fig. 16 Spontaneous degeneration of apoptotic bodies in culture of BM13674 Burkitt's lymphoma cell line 6 hr after exposure to 5 Gy X-rays. The cytoplasmic abnormalities resemble those observed at an advanced stage of necrosis, but apoptotic nuclear changes can still be recognized. Bar: 2 μm.

those illustrated in Figs. 4 and 6 are encountered relatively infrequently in electron microscopic studies. In tissues, phagocytosis and degradation of apoptotic bodies also proceed rapidly and are completed within a matter of hours (Wyllie *et al.*, 1980; Bursch *et al.*, 1990). Thus, unless samples for electron microscopy are taken early in a wave of apoptosis, the later degradative stages in the sequence will tend to dominate the sample observed.

2. Necrosis

Under certain circumstances, such as exposure of cells to very high temperatures or to concentrated solutions of corrosive chemicals, cell death is virtually instantaneous. In most cases of cellular injury, however, the damage is less catastrophic, so the development of necrosis is preceded by the appearance of morphological abnormalities that are indicative of grossly disturbed cellular homeostasis. These features include swelling of the cell sap, surface blebbing, dilatation of the cisternae of the endoplasmic reticulum, shrinkage followed by swelling of the inner compartment of mitochondria, dispersal of ribosomes, and mild clumping of nuclear chromatin (Trump *et al.*, 1981). The surface blebbing that sometimes follows injury must not be confused with that which occurs during apoptosis. The onset of necrosis is heralded by severe (high amplitude) swelling of mitochondria; this event, in turn, is followed by the appearance of flocculent densities in mitochondrial matrices; gross swelling of all cytoplasmic compartments; rupture and dissolution of plasma, organelle, and nuclear membranes; and disappearance of ribsomes (Figs. 17–19). The clumping of nuclear chromatin evident in the early stages of injury becomes more marked with the onset of cytoplasmic degeneration, but the chromatin is not significantly redistributed (Figs. 17 and 19); the aggregates differ from those seen in apoptosis because they are less uniform in texture, have less sharply defined edges, and are irregularly scattered through the nucleus rather than being sharply segregated at its periphery (compare Figs. 17 and 19 with Figs. 2–8). Moreover, the nuclei of cells undergoing necrosis never bud to form discrete, membrane-bounded fragments. At a late stage of necrosis, the chromatin is degraded (Fig. 18) and may disappear completely.

When necrosis affects cells in culture, the swollen degenerate cells eventually

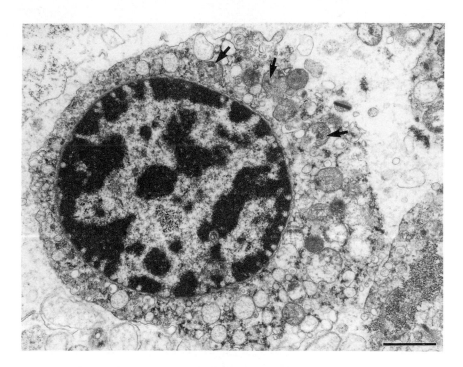

Fig. 17 Necrosis due to ischemia developing spontaneously in the center of murine P-815 tumor growing in muscle. Note that the masses of condensed chromatin have poorly defined edges and are irregularly scattered through the nucleus. Mitochondria are grossly swollen, and some contain flocculent densities (arrows). The overall configuration of the cell is maintained, despite the cytoplasmic degeneration and rupture of the plasma membrane. Bar: 1 μm

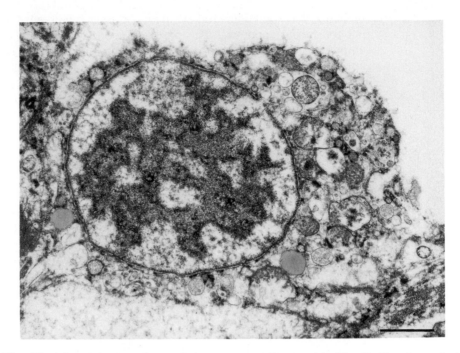

Fig. 18 Advanced stage of necrosis in same tumor illustrated in Fig. 17. The chromatin is degraded. Flocculent densities are prominent in swollen, degenerate mitochondria. The plasma membrane has disappeared. Bar: 1 μm

disintegrate; the debris is then dispersed in the medium. In tissues, however, necrotic cells tend to retain their overall configuration until they are taken up and digested by cells of the mononuclear phagocytic system. The latter comprise both resident macrophages and monocytes that have emigrated from blood vessels during the acute inflammatory response. Epithelial cells are never involved in mopping up the debris of necrotic cells. When the proportion of cells affected by necrosis is relatively small, when the vasculature and the connective tissue framework of the involved area are still intact, and when the adjacent surviving cells are capable of proliferation, the tissue may return to normal if the cause of the injury is removed rapidly. However, when confluent necrosis of all cells in a piece of tissue takes place, organization with replacement by granulation tissue occurs and the late outcome is the development of a collagenous scar.

B. Light Microscopy

1. Apoptosis

The light microscopic recognition of apoptosis mainly depends on the detection of discrete well-preserved apoptotic bodies. Although convoluted budding

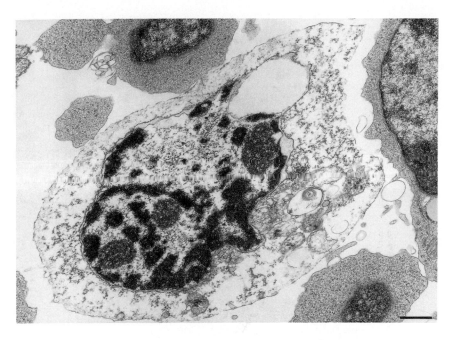

Fig. 19 Necrosis occurring spontaneously in culture of BM13674 Burkitt's lymphoma cell line. The abnormalities are very similar to those that occur *in vivo* (see Fig. 17). Bar: 1 μm.

cells are sometimes observed in smears, they are rarely seen in paraffin sections of immersion-fixed tissue. The process of separation of apoptotic bodies from a condensing cell is likely to progress to completion during the time required for fixatives to penetrate the relatively large blocks used for standard histology, even when fixation is begun soon after the tissue is obtained by biopsy. Apoptotic bodies that have undergone significant degradation following their phagocytosis are difficult to recognize, although those with a nuclear component tend to remain recognizable at a later stage of degradation than those without. In cell cultures, apoptotic bodies that have degenerated spontaneously eventually become indistinguishable from the debris of necrotic cells (Harmon *et al., 1990*).

The diversity of size and structure of apoptotic bodies that is evident by electron microscopy is mirrored in their light microscopic appearances (Fig. 20). Most are roughly round or oval in shape. Although the larger bodies are relatively conspicuous (Fig. 20a and b), the smallest are easily overlooked unless they contain a speck of condensed chromatin (Fig. 20a and b). Some of the apoptotic bodies derived from cells with copious cytoplasm may contain little or no chromatin (Fig. 20b), whereas those derived from cells with a high nucleus : cytoplasm ratio, such as thymocytes, often contain single, relatively large chromatin masses surrounded by narrow cytoplasmic rims (Fig. 20c). Still other bodies contain numerous nuclear fragments (Fig. 20e). The characteristic segregation of condensed chromatin that is seen in apoptotic nuclear fragments

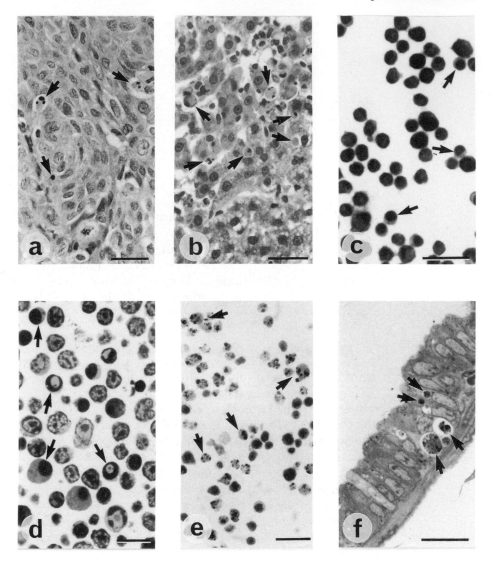

Fig. 20 Light microscopic appearances of apoptotic bodies (arrows). (a) Untreated, poorly differentiated, human carcinoma. Sample is a 5-μm paraffin section stained with hematoxylin and eosin (H and E). Bar: 40 μm. (b) Median lobe of rat liver 3 days after ligation of associated portal vein branch; same experiment illustrated in Figs. 11 and 12. Sample is a 5-μm paraffin section stained with H and E. Bar: 30 μm. (c) Thymocytes from 21-day-old rat 4 hr after addition of hydrocortisone to culture at concentration of 10 μM; same experiment illustrated in Fig. 3. Smear stained with H and E. Bar: 20 μm. (d) Same thymocyte culture illustrated in c. Sample is a 1-μm resin section stained with toluidine blue. Bar: 10 μm. (e) Sample from 7-day-old culture of murine NS-1 cell line. Smear stained with H and E. Bar: 40 μm. (f) Rat ventral prostate 3 days after castration. Some apoptotic bodies have been taken up by epithelial cells; others lie within intraepithelial macrophages with faintly stained cytoplasm. Sample is a 1-μm resin section stained with toluidine blue. Bar: 20 μm.

by electron microscopy is evident by standard light microscopy only in very thin sections (Fig. 20d); in relatively thick sections and smears, the nuclear fragments usually appear uniformly dense (Fig. 20a, b, c, and e). Fluorescence microscopy following staining with acridine orange (Dive *et al.,* 1992) or propidium iodide (Nicoletti *et al.,* 1991) is reported to be capable of demonstrating apoptotic chromatin segregation in smears; we have no personal experience with this technique.

In tissues, apoptotic bodies are often found in clusters (Fig. 20b), but scattered single bodies are also common (Fig. 20a and b). Determining whether a particular body is still extracellular or has been phagocytosed is sometimes difficult using a light microscope. Bodies that have been taken up by large, faintly stained macrophages may appear, at first sight, to lie within clear spaces (see Fig. 20f; compare with Fig. 10). Such large, apoptotic-body-laden phagocytic cells that occur in lymphoid germinal centers have long been referred to as tingible body macrophages. The shrinkage of cells that inevitably takes place during the preparation of paraffin blocks frequently results in the development of spaces around extracellular apoptotic bodies; this artifact tends to make these bodies stand out in paraffin sections (Fig. 20a and b).

2. Necrosis

The light microscopic recognition of necrosis usually poses little difficulty. As already indicated, groups of contiguous cells are typically affected when the process occurs in tissues (Fig. 21); infiltrating neutrophil leukocytes or mononuclear phagocytes are frequently present at the edges of the involved area (Fig. 21a). The irregular clumping of nuclear chromatin seen by electron microscopy may be apparent (Fig. 21b), but the nuclei often appear uniformly dense in relatively thick sections (Fig. 21b) and in smears. At a late stage of degeneration, when the chromatin has disappeared, the nucleus may be only barely discernible as a faintly stained ghost (Fig. 21a and b). The cytoplasm of necrotic cells loses its basophilia and organized structure, and becomes granular or finely vacuolated (Fig. 21a and b). Rupture of plasma membranes results in decreased definition of cellular outlines but, as mentioned earlier, necrotic cells in tissues tend to retain their shape (Fig. 21a and b) until removed by mononuclear phagocytes. In cultures, on the other hand, disintegration results in the production of amorphous debris.

III. Problems Encountered in the Morphological Identification of Cell Death

In this section, we will briefly discuss difficulties that may be encountered in the morphological identification of cell death, beginning with electron microscopy.

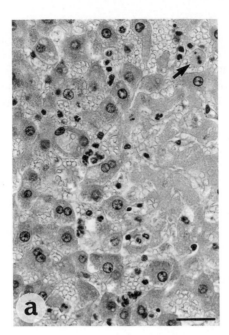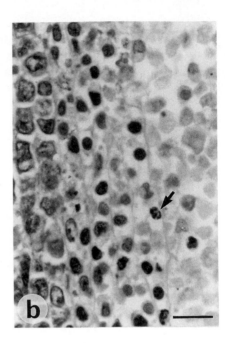

Fig. 21 Light microscopic appearances of necrosis. (a) Central part of a lobule of rat liver 24 hr after injection of the pyrrolizidine alkaloid heliotrine in a dose of 420 mg/kg body weight. The nuclei of the cells in the area of necrosis are only faintly stained. Infiltrating neutrophil leukocytes are present at the margins of the area. A single apoptotic body is arrowed. Sample is a 5-μm paraffin section stained with H and E. Bar: 40 μm. (b) Area of necrosis due to ischemia developing spontaneously in center of murine P-815 tumor growing in muscle. In cells at a late stage of degeneration, the nuclei no longer stain but those near the edge of the area of necrosis, where degeneration is less advanced, show chromatin condensation. Because of the thickness of the section, however, discrete clumps of chromatin as seen by electron microscopy are evident in only a few of these cells (arrow). Sample is a 5-μm paraffin section stained with H and E. Bar: 20 μm.

Well-preserved phagocytosed apoptotic bodies that lack a nuclear component may occasionally be difficult to distinguish from autophagic vacuoles. The latter contain portions of a cell's own cytoplasm that have become sequestered within membrane-bounded cytoplasmic vacuoles and are later digested by lysosomal enzymes. Most apoptotic bodies are, however, much larger than the portions of cytoplasm sequestered during autophagy. The presence of organelles that are foreign to the cell in which a vacuole lies would, of course, rule out the possibility of its being autophagic.

Apoptotic bodies that have undergone degradation after phagocytosis can be identified with certainty if they contain recognizable nuclear fragments (see, for example, Figs. 14 and 15). However, a large mass of partly degraded cytoplasm lying within an epithelial cell is virtually diagnostic of apoptosis, even when a nuclear fragment is absent. Similar masses within macrophages, on the other hand, might represent phagocytosed remnants of necrotic cells. Although unphagocytosed apoptotic bodies that have undergone spontaneous degeneration in cell cultures can often be differentiated from necrotic cells by the appearance of their nuclear components (Fig. 16), degeneration eventually reaches a stage at which the products of the two types of cell death become indistinguishable. When all the cell debris present in a culture is found to be at an advanced stage of dissolution, it may be necessary to repeat the experiment and take samples at earlier times to determine the type of cell death taking place. Moreover, remember that apoptosis and necrosis frequently occur together.

Finally, so-called type B dark cells (Fig. 22) have sometimes been misidentified as apoptotic cells. The similarities are, however, only superficial (Harmon, 1987). Although marked cytoplasmic condensation is a feature of type B dark cells, the mitochondria are typically grossly swollen (Fig. 22). Moreover, the distribution of compacted nuclear chromatin resembles that observed in necrosis rather than apoptosis (compare Fig. 22 with Figs. 6 and 17) and the cytoplasm becomes stretched out between adjacent cells (Fig. 22; Harmon 1987) instead of budding to produce rounded bodies. The significance of type B dark cell formation is uncertain, but this process may represent a variant of necrosis.

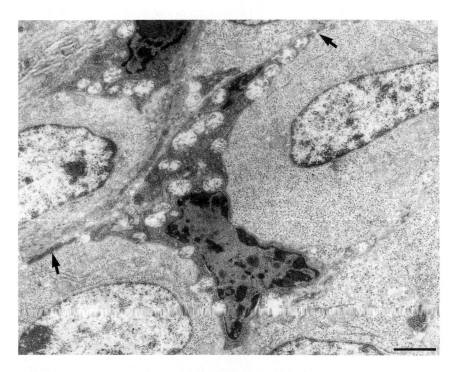

Fig. 22 Type B dark cells in murine P-815 tumor growing in muscle. Note that masses of condensed chromatin are irregularly scattered through the nuclei, that mitochondria are grossly swollen despite the overall cytoplasmic condensation, and that attenuated processes of cytoplasm (arrows) are stretched out between adjacent viable cells. Bar: 2 μm.

At the level of light microscopy, difficulty in distinguishing between apoptosis and necrosis mainly arises for small cells with a high nucleus : cytoplasm ratio, for example, thymocytes. Recall that these cells show restricted nuclear fragmentation or cellular budding during apoptosis. In smears and relatively thick sections, it may be impossible to determine whether the pattern of chromatin compaction present is of the apoptotic or the necrotic type. Where doubt remains about the type of cell death present under particular circumstances, electron microscopy should provide a definitive answer.

IV. Morphological Methods for Studying Cell Death

In this section we will provide laboratory protocols for the morphological study of cell death and will comment on practical points that require special attention.

A. Fixing and Processing Specimens for Transmission Electron Microscopy

Specimens for electron microscopy must be fixed rapidly to prevent autolysis. Immersion fixation is carried out at room temperature in a volume of fixative at least 25 times that of the specimen. After fixation, samples are transferred to buffer, in which they may be stored until it is convenient to process them further. Prolonged storage in fixative should be avoided.

1. Fixing Cultured Cells

Suspension cultures should be centrifuged (100 g for 5 min) and washed three times in sterile isotonic saline prior to fixation in buffered glutaraldehyde solution (details of the fixative are given in Table I). Cells growing as monolayer cultures tend to become detached and to float off into the medium when they undergo apoptosis. Unless the fixation method used for such cultures ensures that the detached cells are included in the sample, the number of cells undergoing apoptosis can be seriously underestimated. Thus, the supernatant is first poured off and retained. The remaining cells are detached by trypsin–versene treatment (0.2% trypsin, 1 mM versene in phosphate buffered saline) for several minutes at room temperature or 37°C. Prolonged treatment with this mixture will cause cytoplasmic vacuolation. The cells detached by the trypsin–versene treatment are then combined with the supernatant. The mixture is fixed in the same manner as a suspension culture.

2. Fixing Tissues

In experimental animal studies, the best results are obtained using vascular perfusion fixation. Animals are deeply anesthetized, the heart is exposed, and the needle of a butterfly infusion set (Venisystems Butterfly-25; Abbott Ireland,

Table I
Fixing and Processing Specimens for Transmission Electron Microscopy

Step	Solution	Time
1. Fixation		
Cell cultures	3% Glutaraldehyde in 0.1 M sodium cacodylate buffer, pH 7.2	2hr
Tissue samples	5% Glutaraldehyde + 4% paraformaldehyde in 0.067 M sodium cacodylate buffer, pH 7.2	2–4 hr
2. Buffer wash	0.1 M Sodium cacodylate buffer	1 × 15 min; overnight
3. Postfixation	1% Aqueous osmium tetroxide (carry out this step in a fume cupboard)	2 hr
4. Wash	Deionized water	3 × 5 min
5. *En bloc* staining	5% Aqueous uranyl acetate	30 min
6. Wash	Deionized water	2 × 5 min
7. Dehydration	50% Ethanol AR grade	2 × 5 min
	70% Ethanol AR grade	2 × 5 min
	90% Ethanol AR grade	2 × 5 min
	100% Ethanol AR grade	2 × 15 min
	100% Ethanol AR grade	1 × 60 min
8. Clearing	Propylene oxide	1 × 10 min; 2 × 20 min
9. Resin infiltration	50% Epoxy resin[a]: 50% propylene oxide	Overnight
	100% Epoxy resin[b]	2 hr
10. Embedding	In plastic capsules or rubber embedding molds with 1–2 days polymerization at 60°C	

[a] The resin should be made up just prior to use and should include the hardeners and accelerator. Most commercially available epoxy resins give satisfactory results.

[b] This step is omitted for cell pellets, the pellets being embedded directly in 100% resin after overnight infiltration with 50% resin : 50% propylene oxide. Moreover, cell pellets are best embedded in the microcentrifuge tubes used for earlier processing rather than in capsules or molds.

Ltd., Sligo, Republic of Ireland) is inserted into the left ventricle. The vascular system is then perfused with a solution of sterile heparinized isotonic saline; care is taken not to exceed mean arterial blood pressure. An incision made in the right atrium should, theoretically, allow the saline to flush the blood out of the whole body; in practice, the head and neck tend to be inadequately flushed by this method. For perfusion of this area, an incision in both jugular veins is preferable. Once the blood has been flushed out, a solution of 5% glutaraldehyde and 4% paraformaldehyde in 0.067 M sodium cacodylate buffer (pH 7.2) is infused to fix the tissues. A successful perfusion is accompanied by appreciable hardening of the animal's body. The required organs are removed from the animals as soon as possible after perfusion, sliced into pieces approximately 4 mm thick with a sharp, clean razor blade, and placed in fresh fixative for 2–4 hr to complete the fixation at room temperature. After this time, the pieces of tissue are placed on a sheet of dental wax and diced with a sharp, clean razor blade into blocks approximately 1 mm³ in size; then the tissues are stored in cacodylate buffer (Table I) prior to further processing. With large animals, perfusion via a catheter inserted into the thoracic aorta may give a better result. The latter approach, however, requires a greater degree of surgical skill.

Many specimens cannot be fixed by the perfusion method (for example, human biopsy material); for these, immersion fixation is used. In this method, freshly removed tissue is placed in a pool of fixative (Table I) on a sheet of dental wax and is diced into blocks approximately 1 mm³ in size. If a knowledge of specimen orientation is required, a strip of tissue approximately 4 × 1 × 1 mm may be cut. The freshly cut samples are then transferred to a specimen bottle containing fresh fixative and immersed for 2–4 hr.

3. Further Processing of Specimens

Cell pellets or tissue blocks are further processed for electron microscopic examination as detailed in Table I. Whereas blocks are conveniently processed in small glass vials, cell pellets are best processed in microcentrifuge tubes. Centrifugation to reform the pellet may be necessary before each decantation of fluid. Resin sections cut 0.5–1 μm in thickness on an ultramicrotome are picked up on a sharpened wooden histology applicator (orange stick), placed in a drop of distilled water on a standard microscope slide, and dried on a hot plate. The dried sections are stained (on the hot plate) with freshly filtered 1% toluidine blue in a 1% aqueous sodium borate solution. These sections are used for light microscopic selection of areas for study by electron microscopy, but are also useful for definitive light microscopy. Ultrathin sections (50–70 nm in thickness) are picked up on copper grids and stained with lead citrate (1.33 g lead nitrate, 1.76 g sodium citrate, 8 ml 1 M sodium hydroxide, made up to 50 ml with distilled water) prior to examination with the transmission electron microscope.

B. Fixing and Processing Specimens for Scanning Electron Microscopy

Cells to be used for scanning electron microscopy must be thoroughly washed in sterile isotonic saline prior to fixation in buffered glutaraldehyde solution (see Table II). Whereas cells growing as monolayers can be fully processed *in situ,* those growing as suspension cultures must be collected on a millipore filter after dehydration in ethanol (Table II). Cell monolayers or millipore filters containing cells are then critical point dried, coated with gold, and examined with the scanning electron microscope. Critical point drying can be carried out directly from absolute ethanol or after transferring the cells into amyl acetate prior to drying (Table II).

C. Fixing and Processing Specimens for Light Microscopy

As already indicated, light microscopy may be performed on 0.5- to 1-μm sections cut from the resin blocks prepared for transmission electron micros-

Table II
Fixing and Processing Specimens for Scanning Electron Microscopy

Step	Solution	Time
1. Fixation	3% Glutaraldehyde in 0.1 M sodium cacodylate buffer, pH 7.2	2 hr
2. Buffer wash	0.1 M Sodium cacodylate buffer	overnight
3. Dehydration	70% Ethanol AR grade	2 × 5 min
	90% Ethanol AR grade	2 × 5 min
	100% Ethanol AR grade	2 × 15 min
	100% Ethanol AR grade	1 × 60 min
	Amyl acetate (optional step)	1 × 60 min
4. Critical point drying		
5. Sputter coating with gold		

copy. On the other hand, smears of cell cultures or blocks of tissues may be processed exclusively for light microscopy.

1. Fixing and Staining Smears

Cell smears are made by placing a drop of cell suspension near one end of a glass slide, bringing a spreader slide up to the drop at an angle of 45°, and spreading the cells along the slide. The smear is allowed to dry in air and is fixed briefly in 70% methanol prior to staining.

Staining with hematoxylin and eosin (H and E) (Table III) is suitable for cells

Table III
Stains for Light Microscopy

Stain	Ingredients
Mayer's hematoxylin	
Solution 1	750 ml Distilled water
	1 g Hematoxylin
	50 g Potassium aluminium sulphate
Solution 2	200 ml Distilled water
	0.2 g Sodium iodate
Add Solution 2 to Solution 1 with constant stirring; then dissolve 50 g chloral hydrate in this mixture. Add 2 g citric acid. Dissolve, filter, and make up to 1 liter with distilled water.	
Eosin	0.3% Eosin Y in 70% ethanol
Modified Giemsa	
Giemsa stain	0.8 g Giemsa
	0.1 g Eosin Y
	100 ml 1 : 1 Glycerol and methanol
Dissolve and filter.	
Giemsa buffer	9 ml 0.1 M Citric acid in 25% methanol
	11 ml 0.2 M Na$_2$HPO$_4$ in 25% methanol
	380 ml Distilled water
Adjust pH to 6.4 with hydrochloric acid.	

Table IV
Fixing and Processing Specimens for Light Microscopy

Step	Solution	Time
1. Fixation	10% Phosphate-buffered formalin, pH 7.2	24–48 hr
2. Dehydration	70% Ethanol LR grade	1 × 1 hr
	95% Ethanol LR grade	2 × 1 hr
	100% Ethanol LR grade	3 × 1 hr
3. Clearing	Xylol	2 × 1 hr
4. Wax infiltration	Paraplast wax (MP 57–58°C)	2 × 1 hr at 60°C
5. Embedding	Embed tissue in molten paraplast wax	

with a low nucleus : cytoplasm ratio. Fixed smears are stained with Mayer's hematoxylin for 15 sec, blued in running tap water for 5 min, dipped briefly in eosin; then they are dehydrated in ethanol, cleared in xylol, and mounted in Depex (BDH Laboratory Supplies, Poole, England) medium. Cells with a high nucleus : cytoplasm ratio can also be stained with H and E, but identification of apoptotic cells is often difficult. A modified Giemsa stain (Table III) may give better resolution of nuclear detail. Fixed smears are stained for seconds to minutes, depending on the resultant staining intensity for a particular cell type, then washed in Giemsa buffer, followed by water, ethanol, and xylol prior to mounting.

2. Fixing and Processing Tissues

Slices of tissue that are 3 mm or less in thickness are cut and immersed in buffered formalin (Table IV). The volume of fixative used should be at least 20 times the volume of the tissue. After dehydration, clearing, wax infiltration, and embedding (Table IV), sections are cut at a thickness of 3 to 4μm. These sections are then dewaxed in xylol and rehydrated through graded ethanol solutions. H and E staining is suitable for identifying apoptosis and necrosis in such sections.

References

Bursch, W., Paffe, S., Putz, B., Barthel, G., and Schulte-Hermann, R. (1990). Determination of the length of the histological stages of apoptosis in normal liver and in altered hepatic foci of rats. *Carcinogenesis* (*London*) **11,** 847–853.

Cohen, G. M., Sun, S.-M., Snowden, R. T., Dinsdale, D., and Skilleter, D. N. (1992). Key morphological features of apoptosis may occur in the absence of internucleosomal DNA fragmentation. *Biochem. J.* **286,** 331–334.

Collins, R. J., Harmon, B. V., Gobé, G. C., and Kerr J. F. R. (1992). Internucleosomal DNA cleavage should not be the sole criterion for identifying apoptosis. *Int. J. Radiat. Biol.* **61,** 451–453.

Dive, C., Gregory, C. D., Phipps, D. J., Evans, D. L., Milner, A. E., and Wyllie, A. H. (1992). Analysis and discrimination of necrosis and apoptosis (programmed cell death) by multiparameter flow cytometry. *Biochim. Biophys. Acta* **1133,** 275–285.

Duvall, E., and Wyllie, A. H. (1986). Death and the cell. *Immunol. Today* **7,** 115–119.

Falcieri, E., Martelli, A. M., Bareggi, R., Cataldi, A., and Cocco, L. (1993). The protein kinase inhibitor staurosporine induces morphological changes typical of apoptosis in MOLT-4 cells without concomitant DNA fragmentation. *Biochem. Biophys. Res. Commun.* **193**, 19–25.

Harmon, B. V. (1987). An ultrastructural study of spontaneous cell death in a mouse mastocytoma with particular reference to dark cells. *J. Pathol.* **153**, 345–355.

Harmon, B. V., Corder, A. M., Collins, R. J., Gobé, G. C., Allen, J., Allan, D. J., and Kerr, J. F. R. (1990). Cell death induced in a murine mastocytoma by 42–47°C heating *in vitro:* Evidence that the form of death changes from apoptosis to necrosis above a critical heat load. *Int. J. Radiat. Biol.* **58**, 845–858.

Kerr, J. F. R. (1971). Shrinkage necrosis: a distinct mode of cellular death. *J. Pathol.* **105**, 13–20,

Kerr, J. F. R., Wyllie, A. H., and Currie, A. R. (1972). Apoptosis: A basic biological phenomenon with wide-ranging implications in tissue kinetics. *Br. J. Cancer* **26**, 239–257.

Kerr, J. F. R., Bishop, C. J., and Searle, J. (1984). Apoptosis. *Recent Adv. Histopathol.* **12**, 1–15.

Kerr, J. F. R., Searle, J., Harmon, B. V., and Bishop, C. J. (1987). Apoptosis. *In* "Perspectives on Mammalian Cell Death" (C. S. Potten, ed.), pp. 93–128. Oxford University Press, Oxford.

Lockskin, R. A., and Zakeri, Z. (1991). Programmed cell death and apoptosis. *In* "Apoptosis: The Molecular Basis of Cell Death" (L. D. Tomei and F. O. Cope, eds.), pp. 47–60. Cold Spring Harbor Laboratory Press, Cold Spring Harbor, New York.

Matter, A. (1979). Microcinematographic and electron microscopic analysis of target cell lysis induced by cytotoxic T lymphocytes. *Immunology* **36**, 179–190.

Nicoletti, I., Migliorati, G., Pagliacci, M. C., Grignani, F., and Riccardi, C. (1991). A rapid and simple method for measuring thymocyte apoptosis by propidium iodide staining and flow cytometry. *J. Immunol. Meth.* **139**, 271–279.

Oberhammer, F., Fritsch, G., Schmied, M., Pavelka, M., Printz, D., Purchio, T., Lassmann, H., and Schulte-Hermann, R. (1993). Condensation of the chromatin at the membrane of an apoptotic nucleus is not associated with activation of an endonuclease. *J. Cell Sci.* **104**, 317–326.

Robertson, A. M. G., and Thomson, J. N. (1982). Morphology of programmed cell death in the ventral nerve cord of *Caenorhabditis elegans* larvae. *J. Embryol. Exp. Morphol.* **67**, 89–100.

Russell, S. W., Rosenau, W., and Lee, J. C. (1972). Cytolysis induced by human lymphotoxin. Cinemicrographic and electron microscopic observations. *Am. J. Pathol.* **69**, 103–118.

Sanderson, C. J. (1976). The mechanism of T cell mediated cytotoxicity II. Morphological studies of cell death by time-lapse microcinematography. *Proc. R. Soc. London B* **192**, 241–255.

Schwartz, L. M., Smith, S. W., Jones, M. E. E., and Osborne, B. A. (1993). Do all programmed cell deaths occur via apoptosis? *Proc. Natl. Acad. Sci. USA* **90**, 980–984.

Sheridan, J. W., Bishop, C. J., and Simmons, R. J. (1981). Biophysical and morphological correlates of kinetic change and death in a starved human melanoma cell line. *J. Cell Sci.* **49**, 119–137.

Tomei, L. D., Shapiro, J. P., and Cope, F. O. (1993). Apoptosis in C3H/10T½ mouse embryonic cells: Evidence for internucleosomal DNA modification in the absence of double-strand cleavage. *Proc. Natl. Acad. Sci. USA* **90**, 853–857.

Trump, B. F., Berezesky, I. K., and Osornio-Vargas, A. R. (1981). Cell death and the disease process. The role of calcium. *In* "Cell Death in Biology and Pathology" (I. D. Bowen and R. A. Lockshin, eds.), pp. 209–242. Chapman and Hall, London.

Walker, N. I., Harmon, B. V., Gobé, G. C., and Kerr, J. F. R. (1988). Patterns of cell death. *Meth. Achiev. Exp. Pathol.* **13**, 18–54.

Weedon, D., Searle, J., and Kerr, J. F. R. (1979). Apoptosis. Its nature and implications for dermatopathology. *Am. J. Dermatopathol.* **1**, 133–144.

Wyllie, A. H. (1980). Glucocorticoid-induced thymocyte apoptosis is associated with endogenous endonuclease activation. *Nature (London)* **284**, 555–556.

Wyllie, A. H. (1987). Cell death. *Int. Rev. Cytol. Suppl.* **17**, 755–785.

Wyllie, A. H., Kerr, J. F. R., and Currie, A. R. (1980). Cell death: The significance of apoptosis. *Int. Rev. Cytol.* **68**, 251–306.

Zakeri, Z. F., Quaglino, D., Latham, T., and Lockshin, R. A. (1993). Delayed internucleosomal DNA fragmentation in programmed cell death. *FASEB J.* **7**, 470–478.

CHAPTER 2

Identification of Dying Cells—
In Situ Staining

Shmuel A. Ben-Sasson, Yoav Sherman,* and Yael Gavrieli

The Hubert H. Humphrey Center for Experimental Medicine and Cancer Research
and *Department of Pathology, The Hebrew University-Hadassah Medical School
Jerusalem 91120, Israel

I. Introduction

Programmed cell death (PCD) does not exist in prokaryotes or protozoa, which proliferate and expand perpetually under optimal conditions. This process is a unique feature of multicellular organisms that enables continuous renewal of tissues by cell division while maintaining the steady-state level of the various histological compartments under tight control (Raff, 1992). In other words, the biological principle of continuous turnover that universally applies to macromolecules is also manifested in metazoans with respect to whole cells. This turnover includes the replacement, on a regular basis, of worn-out cells in tissues such as the intestinal epithelium or the epidermis, both facing the hardships of the

Copyright © 1995 by Academic Press, Inc. All rights of reproduction in any form reserved.

external world. On the other hand, the mechanism of PCD enables a change in tissue profile, that is, a change in size or even a change in shape, as exercised during morphogenesis, for example, in the process of digit formation (Dvorak and Fallon, 1991).

Thus, the entire host of cells present in organized tissue should constitute the context of PCD research. This context includes a study of not only the interplay between cell proliferation and cell elimination within the same lineage, but also the interaction between various cell populations in the same tissue or even those in remote tissues that communicate through paracrine or endocrine loops, respectively. The basic requirements for any methodology of PCD detection are therefore: (1) resolution at the individual cell level and (2) *in situ* applicability while preserving the tissue architecture.

In an attempt to investigate PCD in its physiological context, we developed a method that satisfies both criteria that is referred to hereafter as TUNEL (Gavrieli *et al.,* 1992). In essence, this technique is based on the observation that PCD is associated with DNA degradation (Wyllie, 1980). Consequently, researchers have come to consider the appearance of the ladder of nucleosomal DNA on agarose gels as the hallmark of PCD. However, this biochemical approach involves a nondiscriminatory grinding of the entire tissue. In addition, the method is qualitative rather than quantitative in nature; for the individual cell, PCD is an all-or-none event and should be assessed accordingly. The TUNEL method relies on the *in situ* labeling of DNA breaks in individual nuclei in tissue sections processed through the routine procedures of histopathology. TUNEL stands for TdT-mediated dUTP-biotin nick end labeling and relies on the specific binding of terminal deoxynucleotidyl transferase (TdT) to exposed 3'-OH ends of DNA followed by the synthesis of a labeled polydeoxynucleotide molecule. Nuclear DNA on histological sections is first exposed by proteolytic treatment; then TdT is used to incorporate biotinylated deoxyuridine into the sites of DNA breaks. The signal is amplified by avidin–peroxidase, enabling conventional histochemical identification of PCD by light microscopy.

II. The TUNEL Procedure

This section presents a generalized protocol of the TUNEL method.

A. Materials and Methods

1. Preparing Slides

Fix tissues with 4% formaldehyde (buffered) or paraformaldehyde for 1–7 days prior to embedding in paraffin. Spread 5 to 10-μm thick sections on 3-aminopropyl triethoxysilane (TESPA; No. A3648, Sigma, St. Louis, MO)-coated slides.

Sub slides according to this procedure:

1. Dip slides in 10% HCl/70% ethanol, followed by distilled water, then 95% ethanol.
2. Dry in oven at 150°C for 5 min and allow to cool.
3. Dip slides in 2% TESPA in acetone for 10 sec.
4. Wash twice with acetone, then distilled water.
5. Dry at 42°C.

TESPA is superior to poly-L-lysine in preventing tissue detachment from the glass.

Heat paraffin sections for 10 min at 70°C or 30 min at 58–60°C. Perform hydration by transferring the slides through the following solutions: twice to 100% xylene for 5 min, then twice to 96% ethanol for 3 min, then 90% ethanol, 80% ethanol, and double-distilled water (DDW).

2. Nuclei Stripping

Immerse the tissue sections in 10 mM Tris-HCl, pH 8.0, for 5 mins. Then incubate with 20 μg/ml proteinase K in 10 mM Tris-HCl, pH 8.0, for 15 min at room temperature (RT). Finally wash in DDW four times for 2 min each.

Note: Tissue cultures or cell suspensions can be conveniently processed for staining by the TUNEL method, following a 30-min to 2-hr fixation. Standard 30-mm tissue culture dishes have useful dimensions and can be processed throughout.

For tissue culture cells (or cell suspensions fixed and attached to a slide), the proteinase K treatment can be omitted. (Otherwise nonspecific staining might be enhanced.)

3. H$_2$O$_2$ Treatment

Inactivate endogenous peroxidase by covering the sections with 3% H$_2$O$_2$ (100–200 μl) for 5 min at RT. Then rinse with DDW.

4. Elongation of DNA Fragments

Pre-incubate the sections in TdT buffer (1×: 30 mM Trisma base, 140 mM sodium cacodylate, pH 7.2, 1 mM cobalt chloride).

Reaction Mixture

Allow 75 μl or less (if possible) for each slide.

1. TdT buffer (10×: (0.3 M Trisma base, 1.4 M sodium cacodylate, pH 7.2, 10 mM cobalt chloride, 1 mM DTT); 10% of the final volume

2. Biotin-11-dUTP stock, prepared by dissolving 0.5 mg bio-11-dUTP in 1.0 ml 10 mM Tris-HCl, pH 7.0; use 0.07 μl stock solution per 1 μl final solution to 40 μM final concentration

3. TdT, 0.3 enzyme units/μl final solution

4. DDW to final volume

The reaction mixture is prepared in the following order: (1) DDW, (2) 10$\times$ TdT buffer, (3) bio-11-dUTP, and (4) TdT enzyme. Add 75 μl reaction mixture to each slide; add less if the section is small. Incubate the sections, covered with cover slips, in a humid atmosphere for 60 min at 37°C.

5. Termination

Terminate the reaction by transferring the slides to a bath of 300 mM NaCl and 30 mM sodium citrate (pH about 8) for 15 min at RT.

Wash the slides for 5 min at RT with phosphate buffered saline (PBS). Then block with 2% human serum albumin in PBS (100 μl for each slide) for 10 min at RT. Wash excess albumin with PBS by incubation for 5 min at RT.

6. Staining

Incubate the sections with avidin-conjugated peroxidase in PBS at 37°C for 30 min (100 μl for each slide). Then stain with 3-amino-9-ethylcarbazole (AEC) for 30 min at 37°C; wash with PBS. Light counterstaining with hematoxylin is recommended. After mounting with glycerol–gelatin (No. GG-1, Sigma), the covered slides can be stored for many months at RT. Alternatively, diaminobenzidine (DAB) and a nonaqueous mounting medium such as alkylacrylate in xylene can be used. Extra-avidin peroxidase (BioMakor, Rehovot, Israel) and strepavidin–peroxidase give lower background than native egg avidin. Also, conjugated alkaline phosphatase and the appropriate chromogen yield equally good results.

B. Comments

1. Other biotin, or digitoxin, conjugates of dNTP can replace bio-11-dUTP in the TdT reaction mixture.

2. Fluorescent-labeled (FITC) avidin can be used (instead of peroxidase–avidin) for fluorescence microscopy, which yields beautiful results (the ''light at the end of the TUNEL'').

3. *In general, fresh solvents are recommended since paraffin traces might interfere with the enzymatic reaction.*

4. This method can also be applied to cryostat sections. In this case, the proteolytic pretreatment might be milder or even omitted.

Note: DNase I pretreatment (0.1–1 μg/ml, 10 min, RT) is recommended as a positive control. However, make sure to wash all vials thoroughly; otherwise residual DNase I activity might introduce a high background noise to your system.

Comments Related to the TdT Reaction

The ideal substrate for TdT action is single-stranded DNA or a 3′ protruding end of double-stranded DNA; blunt ends are a poor substrate. However, when Cu^{2+} ions replace Mn^{2+} in the reaction mixture (as included in the preceding protocol), ~21% labeling yield can be achieved (for a detailed discussion, see Deng and Wu, 1983). Recessive ends, the ones expected following the introduction of single-strand nicks into nuclear DNA, are the least efficient substrates for TdT. Therefore, the ability to identify PCD reproducibly by the TUNEL method suggests the existence of double-strand nicks, as evidenced also by the end-point production of the nucleosomal ladder. In addition, although we were able to eliminate "background noise" by pretreatment with DNA ligase, even a large excess of this repair enzyme did not change the pattern of authentic PCD labeling (S. Ben-Sasson *et al.,* unpublished results). Another possibility is that nicks are formed initially at dA–dT tracts in the genome during PCD, since the 3′ ends of these "breathing" DNA sequences were shown to be readily accessible to the TdT reaction, even in the recessive form (Roychoudhury and Wu, 1980).

III. Related Methods

Shortly after the publication of the TUNEL method, a group from The Netherlands described a system that in principle is very similar to ours (Wijsman *et al.*, 1993). The only differences are: (1) pretreatment of the sections with 1 *M* NaSCN, a chaotropic agent, at 80°, in addition to proteolytic treatment with 0.5% pepsin in HCl (pH 2), and (2) the use of DNA polymerase I instead of TdT. The selection of this template-dependent enzyme enables the detection of recessed 3′-OH termini only, not protruding or blunt ends. On the other hand, TdT favors the latter, as already explained. Since single-strand nicks might exist normally in viable cells, the application of polymerase I can sometimes result in staining in non-apoptotic cells, as noted by these authors (Wijsman *et al.*, 1993). In general, the correlation of PCD with nucleosomal ladder of DNA fragments indicates that the process involves the formation of blunt or protruding ends.

Other work that deserves special mention is the pioneering demonstration of DNA breaks in the nucleus of terminally differentiated lens cells by Modak and colleagues (Modak and Bollum, 1972). The application of TdT in combination with radiolabeled deoxynucleotides and autoradiography marks the first illustration of lens PCD *in situ* (although we became aware of it only during the latest stage of the preparation of our manuscript).

IV. Tissue Distribution of PCD

Since the days when the term "apoptosis" was coined to account for the cytological changes associated with many examples of PCD (Kerr *et al.*, 1972), the morphological changes of nuclear chromatin condensation and subsequent DNA fragmentation were accepted as the standard criteria for PCD *in situ* (Wyllie *et al.*, 1984). For reasons of convenience, these morphological criteria will be referred to hereafter as apoptosis, which should be distinguished from the general concept of PCD.) However, an unbiased look at histological sections of rapidly renewing tissues by conventional techniques failed to reveal apoptotic cells (S. Ben-Sasson *et al.*, unpublished results). Our initial test was, therefore, an examination of the TUNEL method in "death certified" tissues, that is, a determination of whether TUNEL can label cells at sites where PCD is expected based on other findings. The following section summarizes our major findings. (For more details, see Gavrieli *et al.*, 1992.)

An outstanding model system for such an assessment is the small intestinal epithelium. This tissue has a high turnover rate; a complete replacement of the entire cell population is accomplished within 3–4 days (Wright and Alison, 1984). This tissue is a simple columnar epithelium with a typical kinetics related to its unique regular architecture. Cells proliferate at the lower part of the crypt and migrate up the villus toward the lumen. As the cells move, they differentiate, senesce, and are finally shed into the lumen. This scheme was established by careful pulse–chase experiments that followed the streaming pathway of newly formed cells (for a review, see Potten and Loeffler, 1990). Although no apoptotic cells can be recognized in general in small intestine histological sections, the application of the TUNEL method exclusively labels only the tips of the villi (Color Plate 1). A higher magnification reveals an internal gradient with respect to the reaction intensity in which the nuclear periphery is more intensely stained (Color Plate 2). The localization of TUNEL staining is not a result of an artifact related to some differential enzyme accessibility, since the deliberate introduction of DNA nicks into all nuclei by pretreatment of the sections with DNase I results in intensive labeling of all cells throughout the entire section (Color Plate 3). Further details related to the dynamics and pattern of labeling of the intestinal epithelium will be discussed subsequently.

Another tissue in which stratification of the various cellular compartments is unambiguously delineated is the epidermis. The proliferative zone is at the basal layer and cells progressively migrate toward the surface to become terminally differentiated. Only nuclei in the uppermost layer are stained by the TUNEL method.

Other rapidly renewing tissues in which PCD is implicated by independent means include lymphatic tissues and ovarian follicles. In both cases, the pattern of staining obtained by the TUNEL method is as anticipated. In the ovary, atretic follicles but not mature follicles are clearly demarcated via the labeling of nuclei of granulosa cells surrounding the oocyte. In Peyer's patches, which are also known for their massive PCD (Motyka and Reynolds, 1991), TUNEL-positive cells are dispersed in the tissue and, in many cases, appear in small clusters of 2–3 cells. PCD in the thymus, on the other hand, seems to be underrepresented as detected by the TUNEL method. A description of the sequence of events that apparently takes place in this lymphoid tissue is given subsequently.

In addition, PCD is occasionally observed by the TUNEL method in slowly renewing tissues such as the liver, the kidney, and the exocrine pancreas.

A separate group of tissues with interesting properties includes those that show rapid sloughing of cells, for example, the endometrium. In principle, the tissues should display PCD since apoptotic cells are observed. However, although classical apoptotic cells are intensely stained by the TUNEL method, such cells constitute only a tiny minority of the entire endometrial tissue. The rest of the endometrium is TUNEL negative during various stages of the human menstrual cycle. Thus, in this case the process of elimination does not seem to be executed at the single cell level, but encompasses the tissue as a whole without affecting individual cells. In other words, mechanistically this process might be closer to ischemia than to PCD.

In conclusion, we have been able to demonstrate, using the TUNEL method, that DNA degradation is taking place in dying cells, *in vivo*. The fraction of cells undergoing PCD that were revealed by this method far exceeds that of those presenting apoptotic morphology. The PCD patterns described were observed in human, rat, mouse, and chicken tissues.

V. Methodological Considerations

In most histochemical staining of specific cell populations, the presence of the target cells at the expected location is guaranteed under normal conditions. PCD-positive cells, on the other hand, have very short lives *in situ* and are therefore very difficult to follow. The reason for this is simple: in other cases we are looking at viable functioning cells that occupy their normal site of activity, whereas PCD not only involves cell suicide but also the elimination of the dead corpuscle. In most tissues, the initiation of PCD is coupled to cell removal. In the gastroinestinal tract, dying cells are immediately shed to the lumen and washed away. In the epidermis, they become part of the acellular layer of the stratum corneum. In lymphoid tissues, they are engulfed and disintegrated by macrophages. In fact, our ability to follow dying cells by the TUNEL method depends on the sequence of events, namely, that DNA degradation will precede cell elimination. This is not always the case: careful monitoring of the diurnal cycle of PCD in the small intestine revealed that the percentage of PCD-labeled villi at any given time cannot account collectively for the expected rate of PCD. Thus, only about 10% of the PCD population executes DNA fragmentation while still part of the tissue. In the rest, this process appears to take place after the cells are shed to the lumen (S. Ben-Sasson *et al.*, unpublished results). Consequently, if cells undergoing PCD are indeed removed quickly and efficiently, chances of identifying them in gradual long-term degenerative processes, for example, the aging brain, are very poor.

Our conclusion is, therefore, that DNA degradation is a relatively late event along the PCD pathway. Although it is still the best criterion available for PCD detection in many tissues, efforts should be made to identify earlier events. For example, in the nematode *Caenorhabditis elegans,* researchers showed that nuclease activation is a very late, independent event in PCD (Hedgecock *et al.*, 1983). Only in tissues in which PCD-positive cells are trapped for some time, for example, ovarian atretic follicles, can the entire spectrum of PCD manifestations be observed regularly. Moreover, in our view, morphological apoptosis might represent an *in vivo* microenvironmental pathology, that is, defective disposal of dead cells. Likewise, it might represent an *in vitro* epiphenomenon.

VI. Tissue Dynamics of PCD

A major strength of the TUNEL method is its ability to elucidate dynamic processes of tissue remodeling in which PCD plays a pivotal role. In principle, the most important information can be obtained by following changes in PCD patterns with respect to space and time, under steady-state conditions and in response to various challenges. To illustrate this aspect, we will describe our observations in the small intestine and the thymus.

Even a superficial look at longitudinal sections of the small intestine, stained by the TUNEL method, reveals that PCD appears in clusters. Several tens of villus tips are positively labeled on both sides of the lumen, as can be seen in Color Plate 1, whereas the remaining neighboring tissue is negative. This result suggests that waves of PCD occur in the form of circular bands, and was similarly noticed in the large intestine. Patches of TUNEL-positive tissue were also observed in the epidermis and include several tens of cells.

As mentioned earlier, a follow-up of the diurnal cycle of PCD by the TUNEL method failed to yield a full recovery of the expected rate of cell elimination, as deduced from the known rate of cell proliferation in the small intestine. Apparently in this tissue, two PCD pathways exist: (1) a more frequent one in which cells are being shed while still TUNEL negative, ensuing DNA degradation taking place in the lumen, and (2) a less frequent one in which cells become TUNEL positive *in situ,* facilitating their close investigation. The difference between these two pathways is not restricted to epithelium labeling but seems to reflect a much deeper difference in the dynamics of tissue turn-over. Whereas the first track involves the rapidly proliferating epithelium only, the appearance of TUNEL-positive epithelial cells in the other track is preceded by massive PCD in the lamina propria, as evidenced by the TUNEL method. In other words, the second less-frequent pathway represents a replacement of the entire intestinal mucosa, including both the epithelium and the underlying connective tissue. Moreover, in this case the process is guided by connective tissue elements, probably intraepithelial lymphoid cells (S. Ben-Sasson *et al.,* unpublished data). Thus, the broad cellular and histological context should be taken into account when a crucial well-coordinated process such as PCD is studied.

The thymus is another tissue in which fast turnover takes place that was subjected to an intensive investigation of PCD. Again, a significant discrepancy appears between the observed and expected rates of TUNEL-positive cells in the thymus. A closer look reveals aggregates of TUNEL-positive nuclei or their fragments around small vessels. Occasionally, a longitudinal section of such vessels highlights the intensity of PCD localized around it (S. Ben-Sasson *et al.,* unpublished data). Collectively, these findings suggest a well-organized microanatomy of the thymus that couples PCD to the disintegration and elimination of the involved cells. Whether the "kiss of death" occurs at a remote site, leaving the cells still TUNEL negative, and the affected cells then migrate toward the "sewage system" of the thymus, or whether the selection of cells is carried out near their point of entry into the tissue remains unknown. Cells that survive selection proceed inward while the rest are executed on site and are discarded through the same channels through which they enter.

Another intriguing issue related to PCD dynamics is its involvement in the response of tissues to changing conditions, which in turn cause a shift in their steady state. We chose the process of villus size adjustment during starvation as a model system. It is a well-established phenomenon that, following food deprivation, the villi are shortened and resume their height on refeeding (Altman, 1972). When this overall tissue adaptation was analyzed with the aid of the TUNEL method, it became apparent that the process is accompanied by a marked change in the PCD pattern, including an extension of PCD to lower parts of the villus, in an alternating fashion, enabling a gradual change in size without a traumatic exposure of the villus tip (S. Ben-Sasson *et al.,* unpublished data).

VII. PCD vs Necrosis

A question that arises from time to time is whether one can distinguish PCD or apoptosis from necrosis *in situ*. Biologically, the distinction is clear: PCD is a physiological process that takes place under normal conditions whereas necrosis represents a pathological death of cells as a result of ischemia. At the cellular level, however, the final stages might be the same: organellar disintegration that eventually leads to DNA degradation, accompanied by nuclear condensation and fragmentation.

Nevertheless, at the tissue level, PCD can be differentiatied by several criteria even when stained by the TUNEL method: PCD is an ordered process whereas necrosis represents a chaos. Therefore, in any given tissue, the pattern of TUNEL staining is distinct; for example, the epithelium undergoes PCD at the villus tips of the small intestine (Color Plates 1 and 2). On the other hand, an ischemic intestinal tissue typically includes all cellular elements of the mucosa at the affected site. Moreover, as stated earlier, PCD is coupled with dead cell elimination by convenient routes. In contrast, necrosis is typified by the *in situ* accumulation of the dead corpuscles; their evacuation takes much longer. Consequently, a necrotic site might occasionally display numerous nuclear fragments that are also TUNEL positive, yet the nature of the tissue damage cannot be mistaken for PCD.

VIII. Perspectives

PCD adds a new dimension to the biology of multicellular organisms. The entire cell becomes the elementary unit of tissue remodeling and cross-talk occurs between cells in a binary language of all or none. PCD is a "voluntary" erosion that occurs one step ahead of physical erosion and thereby protects the tissue from the continuous insults of the surrounding environment. In addition, PCD plays a key role in morphogenesis and selection of the cellular repertoire during development of the organism. Therefore, *in vivo*, the PCD process must be kept under tight control. In other words, the elucidation of the signals that regulate PCD bears consequences beyond academic curiosity since such discoveries help us manipulate physiological and pathological processes such as aging.

Further research aimed at the identification of earlier events along the PCD pathway should enhance our delineation of the involved cell populations. In the meantime, TUNEL should be combined with other methods, such as *in situ* hybridization, that address the tissue context as a whole. After all, the three-dimensional tissue is more complex than the sum of its individual constituent cells, or even the sum of the few macromolecules and their corresponding genes that have been identified to date.

References

Altman, G. (1972). Influence of starvation and refeeding on mucosal size and epithelial renewal in the rat small intestine. *Am. J. Anat.* **133,** 391–400.

Deng, G.-R., and Wu, R. (1983). Terminal transferase: use in the tailing of DNA and for in vitro mutagenesis. *Meth. Enzymol.* **100,** 96–116.

Dvorak, L., and Fallon, J. F. (1991). Talpid mutant chicken limb has anteroposterior polarity and altered patterns of programmed cell death. *Anat. Rec.* **231,** 251–260.

Gavrieli, Y., Sherman, Y., and Ben-Sasson, S. A. (1992). Identification of programmed cell death in-situ via specific labeling of nuclear DNA fragmentation. *J. Cell Biol.* **119,** 493–501.

Hedgecock, E. M., Sulston, J. E., and Thompson, J. N. (1983). Mutations affecting programmed cell death in the nematode *C. elegans. Science* **220,** 1277–1279.

Kerr, J. F. R., Wyllie, A. H., and Currie, A. R. (1972). Apoptosis: Basic biological phenomenon with wide-ranging implication in tissue kinetics. *Br. J. Cancer* **26,** 239–257.

Modak, S. P., and Bollum, F. J. (1972). Detection and measurement of single-strand breaks in nuclear DNA in fixed lens sections. *Exp. Cell Res.* **75,** 307–313.

Motyka, B., and Reynolds, J. D. (1991). Apoptosis is associated with the extensive B cell death in the sheep ileal Peyer's patch and the chicken bursa of Fabricius: A possible role in B cells selection. *Eur. J. Immunol.* **21,** 1951–1958.

Potten, C. S., and Loeffler, A. (1990). Stem cells: Attributes, cycles, spirals, pitfalls and uncertainties. Lessons for and from the crypt. *Development* **110,** 1001–1020.

Raff, M. C. (1992). Social controls on cell survival and cell death. *Nature* **356,** 397–400.

Roychoudhury, R., and Wu, R. (1980). Labeling of duplex DNA with terminal transferase. *Meth. Enzymol.* **65,** 42–62.

Wijsman, J. H., Jonker, R. R., Keijzer, R., Van de Velde, C. J. H., Cornelisse, C. J., and Van Dierendonck, J. H. (1993). A new method to detect apoptosis in paraffin sections: In situ end-labeling of fragmented DNA. *J. Histochem. Cytochem.* **42,** 7–12.

Wright, N., and Alison, M. (1984). "The Biology of Epithelial Cell Populations." Clarendon Press, Oxford.

Wyllie, A. H. (1980). Glucocorticoid induced thymocytes apoptosis is associated with endogenous endonuclease activation. *Nature* **284,** 555–556.

Wyllie, A. H., Morris, R. G., Smith, A. L., and Dunlop, D. (1984). Chromatin cleavage in apoptosis: Association with condensed chromatin morphology and dependence on macromolecular synthesis. *J. Pathol.* **142,** 67–77.

CHAPTER 3

Assays for DNA Fragmentation, Endonucleases, and Intracellular pH and Ca^{2+} Associated with Apoptosis

Alan Eastman

Department of Pharmacology and Toxicology
Dartmouth Medical School
Hanover, New Hampshire 03755

I. Introduction

Apoptosis was originally defined by specific morphological criteria, one of which was the condensation of chromatin within the nucleus (Kerr *et al.,* 1972; Wyllie *et al.,* 1980). This condensation is thought to be caused by an endonuclease activity that can cleave DNA between the nucleosomes to give fragments that resolve on electrophoresis as multiples of about 180 bp. This DNA digestion has been considered a hallmark of apoptosis, but a number of

Copyright © 1995 by Academic Press, Inc. All rights of reproduction in any form reserved.

investigators have observed apoptosis with no detectable internucleosomal DNA digestion. However, Arrends *et al.* (1990) demonstrated that injection of an endonuclease into nuclei produced apoptotic morphology, suggesting that the endonuclease activity is an essential event and not just the product of cell death. The discrepancy is resolved by the realization that DNA digestion to nucleosome-length fragments is probably "overkill" and that even a few breaks into high molecular weight fragments are adequate to produce the morphological changes associated with apoptosis. Using pulsed-field gel electrophoresis, DNA fragments of 50–300 kb have been observed that may represent cleavage of higher order chromatin domains (Brown *et al.*, 1993; Oberhammer *et al.*, 1993). Generally, hematopoietic cell lines produce more extensive DNA digestion, whereas epithelial and fibroblast cell lines are more likely to give faint nucleosome ladders or higher molecular weight smears.

If endonuclease digestion is an essential component of apoptosis, then establishing which endonuclease is involved will facilitate an understanding of the upstream signaling events. Earliest reports implicated a Ca^{2+}/Mg^{2+}-dependent endonuclease (Nikonova *et al.*, 1982), but since then other endonucleases have been implicated. Whether one single endonuclease is responsible for apoptosis in all circumstances or whether many endonucleases are responsible, depending on the particular cell system and inducing stimulus, remains to be determined. At least six different Ca^{2+}/Mg^{2+}-dependent endonucleases with molecular masses ranging from 18 to 120 kDa have been reported (Eastman and Barry, 1992; Peitsch *et al.*, 1993; Ribeiro and Carson, 1993); two of these have been identified as cyclophilin A (J. A. Cidlowski, unpublished observations) and deoxyribonuclease I (Peitsch *et al.*, 1993). Deoxyribonuclease II has also been implicated in apoptosis (Barry and Eastman, 1993; Barry *et al.*, 1993); this endonuclease is active at acidic pH and has no metal ion requirements. Certain cells such as interleukin 2 (IL-2)-dependent cytoxic T lymphocytes (CTLL-2) also show high levels of a Mg^{2+}-dependent endonuclease (Kawabata *et al.*, 1993).

A particular technical concern is that many studies have only used limited incubation conditions and therefore have observed only those endonucleases that are active under such selective conditions. For example, most studies have been performed around pH 7.4, a condition under which deoxyribonuclease II is inactive and therefore would not be detected. Note that other, as yet unidentified endonucleases might be involved in apoptosis; these also may be missed by current assays. A review of the many endonucleases has been published (Eastman and Barry, 1992).

The detection of an endonuclease does not establish its involvement in apoptosis. Manipulations of endonucleases in extracts may be misleading since any endonuclease can produce internucleosomal cleavage when active. For example, immunodepletion with an antibody against a particular endonuclease will certainly inhibit its *in vitro* activity (Peitsch *et al.*, 1993). However, this result does not confirm that the same endonuclease is responsible for this activity in apoptotic cells.

Another characteristic of endonucleases is the structure of the ends of DNA that are produced during digestion. Most endonucleases produce ends with 3' hydroxyl and 5' phosphate groups. In contrast, deoxyribonuclease II produces 5' hydroxyl and 3' phosphate ends. Investigators have generally observed the former end structures. However, other nuclear enzymes such as phosphatases can modify the end structures and thereby confound the determination of which endonuclease was responsible.

An alternative means by which to implicate any endonuclease is to determine whether appropriate intracellular conditions exist during apoptosis. Hence Ca^{2+}/Mg^{2+} dependent endonucleases are thought to be activated by increases in intracellular Ca^{2+}, and deoxyribonuclease II is activated by intracellular acidification. The Mg^{2+}-dependent endonuclease appears to be an anomaly, since it should be constitutively active in cells at physiological pH and Mg^{2+} concentration; presumably some intracellular inhibitor must exist.

The work from this laboratory has focused on deoxyribonuclease II because it was the predominant endonuclease detected in Chinese hamster ovary (CHO) cells (Barry and Eastman, 1993). No Ca^{2+}/Mg^{2+}-dependent endonuclease was detected in these cells. Subsequent experiments have concentrated on human promyelocytic HL-60 cells because these were excellent candidates for measuring intracellular ion concentrations. Intracellular pH and Ca^{2+} are conveniently measured using fluorescent dyes. Initially, we used a spectrofluorimeter to measure the average intracellular pH and Ca^{2+} in a population of cells (Barry and Eastman, 1992). Subsequently we have used a flow cytometer to measure these ions in individual cells (Barry et al., 1993). This latter method has been invaluable for identifying subpopulations of cells that have undergone intracellular acidification. We have identified such acidification in HL-60 cells following incubation with cytotoxins and in IL-2-dependent CTLL-2 cells following withdrawal of IL-2. The results have demonstrated a selective loss of pH regulation with an intracellular acidification of close to 1 pH unit, which is sufficient to activate deoxyribonuclease II. However, acidification and DNA digestion occur concurrently, so no cause and effect relationship can be established. The final evidence for which endonucleases are associated with apoptosis will have to be acquired when genetically manipulated cells are produced that express only specified endonucleases.

II. Detection of Internucleosomal DNA Fragmentation

Internucleosomal DNA fragmentation is measured by electrophoresis in an agarose gel. Most methods involve some sort of DNA purification, possibly involving separation of low from high molecular weight DNA, with phenol extractions, protease and ribonuclease digestions, and ethanol precipitations. These manipulations are not only time consuming but also limit any quantitative assessment of the amount of DNA fragmentation. The method we established obviates these concerns: a defined number of cells is pipetted directly into the wells of an agarose gel, in which the cells are lysed, digested with protease and ribonuclease, and then electrophoresed (Barry and Eastman, 1993). All the high molecular weight DNA is retained in or near the well, while the nucleosome fragments are resolved in the gel. DNA is usually detected by staining with ethidium bromide, but can also be detected by autoradiography if radioactivity was introduced at some earlier step in cell culture. This method was originally adapted from Eckhardt (1978).

A. Solutions

$10\times$ TBE: 0.89 M Tris, 0.89 M boric acid, 25 mM EDTA, pH 8

TE: 10 mM Tris, 0.1 mM EDTA, pH 7.4

20% SDS (w/v): 100 g sodium lauryl sulfate (SDS) in 400 ml water; stir until dissolved and dilute to 500 ml

10 mg/ml RNase A: 100 mg ribonuclease A in 10 ml TE; place in boiling water for 15 min to inactivate any deoxyribonuclease; aliquot to 200 μl; store at $-20°$C

16 mg/ml proteinase K: 100 mg proteinase K in 6.25 ml TE; aliquot to 200 μl; store at $-20°$C

Sample buffer: 10% glycerol, 10 mM Tris, pH 8, 0.1% (w/v) bromophenol blue

B. Method

1. Pour 350 ml 2% agarose in TBE buffer in a large (20 × 34 cm) horizontal gel support. Insert the comb 2–3 cm from the top of the gel.

2. Once the gel solidifies, remove the section of gel immediately above the comb by cutting along the top side of the comb with a scalpel. Fill this top section with 1% agarose, 2% SDS, 64 μg/ml proteinase K (0.5 g agarose, 5 ml $10\times$ TBE, 5 ml 20% SDS, 40 ml water, 200 μl of 16 mg/ml proteinase K). Add proteinase K only after the gel cools below 50°C to avoid inactivating the enzyme. If some of this agarose solution runs out onto the prepoured 2% agarose, let it solidify and then remove it from below the wells while wearing gloves. This excess usually comes off easily by peeling the layer off the gel.

3. We routinely use 10^6 cells per lane. The use of more cells (up to 10^7) may increase the sensitivity for detecting DNA fragments, but would only be necessary if very few cells were undergoing DNA digestion. If the required number of cells is in a volume of medium greater than 1 ml, reduce the volume by centrifuging the cells at 1000 rpm for 5 min and remove the excess medium. Transfer the cells to a 1.5-ml Eppendorf microcentrifuge tube and store them on ice. Once the cells are harvested, it is important to keep them on ice suspended in either medium or phosphate buffered saline. Incubating the cells at room

temperature after harvest can allow continued endonuclease activity beyond the desired time point. It is best to run the cells within 6 hr of harvest.

4. When ready to run the gel, pellet the cells (5 sec at 14,000 rpm) and aspirate the medium without disturbing the pellet. Resuspend the pellet with 15 μl sample buffer/RNase A (mixed at 1:1 volume) and load directly into the well of the submerged gel. Add sample buffer/RNase to each sample just prior to loading, because the cells will lyse with continued incubation and become very difficult to pipette.

5. Once all the samples are loaded, electrophorese for 14 hr at 60 V at room temperature. When voltage is applied, the SDS and proteinase K are electrophoresed through the wells, facilitating hydrolysis of cellular components.

6. Stain the gel with 2 μg/ml ethidium bromide in water for 1 hr. Excess ethidium bromide, digested RNA, and SDS are removed from the gel by washing overnight twice with 3 L distilled water (this step can be accelerated by more frequent changes of water). Using a Fotodyne FCR-10 instant camera with accompanying Wratten UV filter and Polaroid 665 film, good photographs can be obtained by exposure at f = 4.5 for 30 sec. Resolution on photographs is enhanced if they are printed as the negative image (dark bands on light background; Fig. 1).

C. Comments

This method can be modified for any size gel. We initially used a 100-ml gel successfully (Sorenson *et al.,* 1990), but found somewhat better resolution on the larger apparatus described here.

Although quantification is not very accurate, a relative assessment of the amount of DNA digestion can be obtained by densitometry of the photographic negatives. These values were compared with a standard curve produced by electrophoresis of a known amount of sheared DNA under identical conditions. Minimally detectable degradation represented about 50 ng, whereas amounts up to 2 μg have been observed. The majority of DNA always remains in the well and stains far too intensely to be quantified by densitometry. However, knowing the number of cells added per lane permits an assessment of the total amount of DNA added. More accurate quantification can also be obtained if radioactivity is incorporated into the DNA prior to experimentation; gel sections are then analyzed by scintillation counting.

An alternative method of obtaining a quantitative assessment of DNA digestion is lysing cells in a hypotonic buffer plus detergent (e.g., 5 mM Tris, pH 7.4, 5 mM EDTA, 0.5% Triton X-100, 1 mM phenylmethysulfonyl fluoride) and then centrifuging at 27,000 g (Arrends *et al.,* 1990). The modern modification uses a microcentrifuge at full speed and quantification is provided either by initial incorporation of radioactivity into the DNA or by colorimetric assay (Duke and Cohen, 1992). Note that none of these methods quantifies the number

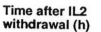

Fig. 1   DNA digestion induced in CTLL-2 cells following withdrawal of IL-2. Cells were incubated without IL-2 for 0, 8, 12, or 16 hr. Then cellular DNA digestion was analyzed by electrophoresis. The results show an early increase in high molecular fragmentation near the well, followed by increasing internucleosome fragmentation.

of cells that may be digesting their DNA; many cells could be partially digesting their DNA or a few cells could be undergoing extensive digestion (for analysis on a cell by cell basis, see Chapter 2). Furthermore, none of these techniques detects rare breaks such as the production of 50–300 kb fragments. This type of DNA digestion requires analysis by pulsed-field gel electrophoresis, as published by Walker *et al.* (1993).

III. Extraction and *in Vitro* Assays for Endonucleases

In addition to measuring DNA digestion in cells undergoing apoptosis, researchers can also mimic this effect by incubating isolated nuclei under conditions that lead to autodigestion of their genomic DNA. Nuclei can also be extracted and assayed for endonuclease activity, but in this case plasmid DNA is added as a substrate. These assays have been used to identify the endonucleases that might be involved in the apoptotic process (Barry and Eastman, 1993).

Many methods have been reported for making nuclei and nuclear lysates. The following method was modified from Shapiro *et al.* (1988) and was used in our studies on deoxyribonuclease II in CHO cells (Barry and Eastman, 1993). Extracts made with this method have also exhibited Ca^{2+}/Mg^{2+}-dependent endonuclease in rat thymocytes and Mg^{2+}-dependent endonuclease in CTLL-2 cells.

A. Solutions

HSSE buffer: 20 mM HEPES, pH 7, 0.75 mM spermidine, 0.15 mM spermine, 0.1 mM EDTA, 1 mM DTT, 1 mM phenylmethylsulfonyl fluoride (PMSF)

APB buffer: 10 mM sodium acetate, 10 mM sodium phosphate, 10 mM bistrispropane, 1 mM DTT, 1 mM PMSF; adjust pH to required value

B. Methods

1. Nuclei Preparation

All steps are performed at 4°C or on ice. Pellet approximately 10^8 cells at 150 g for 5 min. Resuspend the cells in HSSE buffer at approximately 10^7 cells/ml and swell for 20 min. Homogenize the cells with 3 strokes in a dounce homogenizer, followed by immediate addition of 1/3 volume of 1.5 M sucrose with 2 strokes of the homogenizer. Centrifuge the homogenate for 10 min at 150 g and discard the supernatant. Resuspend the pellet in 0.75 ml 0.5 M sucrose in HSSE and layer on a 0.75-ml cushion of 1.5 M sucrose/HSSE. Centrifuge the nuclei through the cushion for 20 min at 13,000 g.

2. Nuclear Autodigestion

Resuspend the nuclei in 1 ml APB buffer, pH 7, and count the nuclei on a Coulter counter. Recentrifuge to remove any residual spermine and spermidine, resuspend at 2×10^7 nuclei/ml in APB buffer, pH 7. Add 10^6 (50 μl) nuclei to 200 μl APB buffer at either pH 5 for acidic endonuclease or pH 0 for ion-dependent endonucleases. For Mg^{2+}-dependent endonuclease, add $MgCl_2$ to 10 mM and EGTA to 5 mM. For Ca^{2+}/Mg^{2+}-dependent endonuclease, add $CaCl_2$ to 2 mM and $MgCl_2$ to 2 mM. Alternatively, to inhibit both ion-dependent endonucleases, add EDTA to 5 mM. Incubate the nuclei at 37°C for 1 hr; then add 15 μl sample buffer/RNase and load the entire sample into a large well in an agarose gel. Assess DNA digestion as described for whole cells. Examples of this assay were shown by Barry and Eastman (1993).

3. Nuclear Extracts

Resuspend 10^8 nuclei/ml in APB buffer, pH 7, containing 0.5 M NaCl. Incubate on ice for 1 hr. Add an equal volume of 20% polyethylene glycol 8000, 0.5 M NaCl. Incubate on ice for 1 hr. Pellet the precipitated DNA by centrifugation at 14,000 rpm for 10 min. Recover the supernatant.

4. Endonuclease Assay

We routinely use the BCA Protein Assay Reagent (Pierce Chemical Company, Rockford, IL) to quantify protein in extracts. The nuclear extracts should be assayed at a dilution greater than 1/10 to reduce the NaCl concentration such that it does not inhibit the endonuclease assays. To quantify the level of an endonuclease, further dilutions are made until no digestion is observed. Incubate appropriate dilutions of the extract with 100 ng supercoiled or linearized plasmid DNA in 20 μl APB buffer at varied pH and varied Ca^{2+}/Mg^{2+}, as described earlier. Incubate the samples at 37°C for 60 min and then electrophorese the DNA in a 1% agarose gel containing ethidium bromide. The best photographs are obtained if the gels are destained overnight in distilled water and photographed on a UV transilluminator. An example of the use of this assay is shown in Fig. 2.

We have defined 1 unit of endonuclease activity as the amount of endonuclease that converts supercoiled or circular plasmid DNA to a smear of fragments lacking any discrete substrate bands in 15 min at 37°C under optimum conditions for each endonuclease (Barry and Eastman, 1993). Over digested DNA is generally not detected or consists of very small fragments. In this procedure, 10-fold dilutions of a sample are assayed for activity; the dilution that just digests all the plasmid DNA is designated as having 1 unit of activity. Total activity is calculated from the dilution required to yield 1 unit of activity.

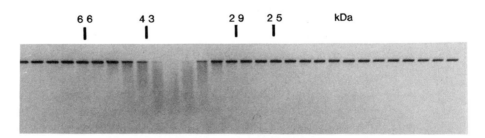

Fig. 2 An example of the endonuclease assay performed in cell extracts. In this experiment, cell extracts were fractioned over a gel filtration column; elution was compared to a series of molecular weight standards. Aliquots of the fractions were then assayed for DNase II activity by incubation with linearized plasmid DNA; the products were separated by electrophoresis. The peak of DNA digestion occurred around 40 kDa; the most active fraction contains 1 unit of activity, as defined in the text.

IV. Measurement of Intracellular pH and Ca^{2+}

Intracellular ion concentrations are frequently measured using ion-sensitive fluorescent dyes. A number of different dyes can be used for each ion; many of these dyes are described in the Molecular Probes catalog (Eugene, OR). The best dyes for this purpose are those that have a shift in their fluorescence spectrum on binding the appropriate ion. The ion concentration is then measured as a ratio at two wavelengths and avoids complications from variations in the amount of dye loading into individual cells. These dyes contain an acetoxymethyl ester that facilitates uptake, but that is cleaved by intracellular nonspecific esterases so the dye is trapped within the cell. To measure pH or Ca^{2+}, BCECF and fura-2, respectively, have frequently been used. For BCECF, emission is measured at 535 nm with excitation of the dye at 439 nm and 505 nm. The ratio of these two values is used to obtain the intracellular pH. Accordingly, BCECF is considered an "excitation ratio" dye. Fura-2 is also an excitation ratio dye. These dyes can readily be used to measure pH and Ca^{2+} in a spectrofluorimeter. The result is an average ion concentration for all the cells in the population. Unfortunately, cells undergo apoptosis heterogeneously, so any change in ion concentration measured as an average of the population is likely to be a significant underestimation of the change in individual cells.

The pH and Ca^{2+} in individual cells can be measured by flow cytometry. However, flow cytometry requires the use of "emission ratio" dyes. The selection of emission ratio dyes is more limited; we chose carboxy-SNARF-1 for pH measurements and indo-1 for Ca^{2+} measurements. The use of these dyes has clearly resolved subpopulations of cells undergoing ion changes. The flow cytometer has also been used to sort cells based on intracellular ion concentrations; this distinction has correlated intracellular acidification with apoptosis in several systems. Note that some cells (e.g., CHO cells and rat thymocytes) inefficiently load carboxy-SNARF-1 and other fluorescent probes, making these types of assay cell line dependent.

A. Intracellular pH Measurement

1. Method

In a typical experiment, 1×10^6 cells are incubated with 1 μM acetoxymethylester derivative of carboxy-SNARF-1 in 2 ml regular culture medium at 37°C in a CO_2 incubator. This incubation is for 30–60 min prior to analysis. The cells are then pelleted and resuspended in fresh culture medium, and are assayed from a vial maintained at 37°C on a Becton Dickinson Facstar Plus flow cytometer (Braintree, MA). Intracellular carboxy-SNARF-1 is excited at 488 nm and the emission is measured at both 585 nm and 640 nm with 5-nm band-pass filters using linear amplifiers. The results are displayed on a two-dimensional dot plot with 585-nm fluorescence on the ordinate and 640-nm fluorescence on the abscissa (Fig. 3). Since the distance of each cell from the origin is directly proportional to the amount of carboxy-SNARF-1 loaded in the cell, cells on the same line out from the origin possess the same intracellular pH. The distribution shows that cells have significantly different loading characteristics but a fairly homogeneous intracellular pH. A shift to the right represents cells with lower intracellular pH whereas a shift to the left represents cells with higher intracellular pH.

Intracellular pH of a sample population is estimated by comparison of the mean ratio values of the sample with a calibration curve of intracellular pH generated by incubation of carboxy-SNARF-1 loaded cells with 10 μg/ml of the proton ionophore nigericin in K buffer (17.3 mM HEPES, 17.3 mM MES, 30 mM NaCl, 115 mM KCl, 1 mM MgCl$_2$, 0.1 mM CaCl$_2$) with the pH adjusted to a series of values between pH 6 and 8 (Fig. 3A). The 585/640 ratio can also

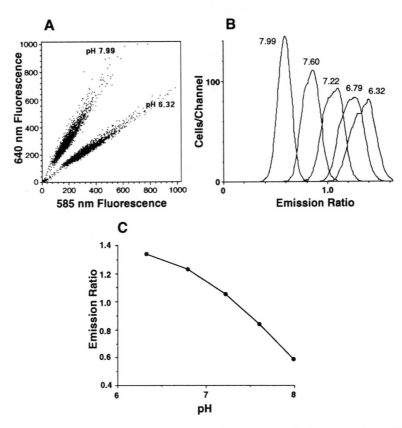

Fig. 3 Calibration of intracellular pH by flow cytometry. Cells were loaded with carboxy-SNARF-1 for 1 hr and then incubated with a proton ionophore at the indicated extracellular pH to obtain (A) a dot plot distribution, (B) the histogram representation of the results, and (C) a calibration curve.

be calculated electronically as an instrument parameter and the results plotted as a histogram (Fig. 3B). Cells with carboxy-SNARF-1 fluorescence of less than 50 units are excluded to prevent the derivation of artifactual ratio values. The pH standard curve can then be expressed as shown in Fig. 3C. Although calibration is usually performed on control cells, it can also be performed on cells that have undergone apoptosis. In the latter case, a mixed population of normal and acidic cells (see subsequent discussion) becomes a single population once the cells are incubated with the proton ionophore.

Examples of experimental results are presented in Fig. 4. To quantify the

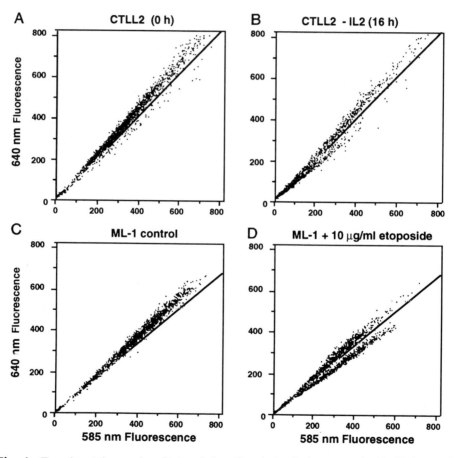

Fig. 4 Experimental examples of intracellular pH variation during apoptosis. (A, B) Apoptosis was induced in CTLL-2 cells by removal of IL-2 and intracellular pH was measured after 16 hr (A, control; B, after removal of IL-2). (C, D) Apoptosis was induced in myeloid leukemia ML-1 cells with the topoisomerase II inhibitor etoposide and were analyzed for intracellular pH changes 4 hr later (C, control; D, treated). A short diagonal line that extrapolates through the origin has been included to emphasize the shift in fluorescence that occurs in acidic cells.

number of acidic cells produced during apoptosis, regions can be selected on the dot plot. In other words, a box can be drawn to contain 90 or 99% of the control population, and the number of acidic cells assessed as the number of cells in an adjacent box containing all acidic cells. In Fig. 4, a partial diagonal line has been drawn that extrapolates through the zero point; the complete line is not drawn since it would obscure the data points. In both examples of control cells (Fig. 4A, C), most of the cells lie to the left of this diagonal whereas, following induction of apoptosis (Fig. 4B, D), a significant population of cells has shifted to the right of the diagonal. Note that at low levels of dye loading, considerable overlap occurs between the normal and apoptotic cells.

2. Comments

The difference in intracellular pH between normal and apoptotic cells is up to 1 pH unit, but this distinction is not reflected in a very large change in the fluorescence ratio. However, the change is clearly detectable because of the bimodal character of the distribution. Rather than a gradual change in fluorescence ratio as the cells become acidic, a characteristic shift of the whole apoptotic population to a lower pH occurs. This change is explained by the selective loss of pH regulation while maintaining an electrochemical gradient across the cell membrane (Barry *et al.*, 1993). This gradient sets up an equilibrium in which the intracellular pH drops by about 0.7 pH units.

The fluorescence intensity (i.e., distance from the origin) of the acidic cells compared with normal cells appears to vary with cell line. For example, myeloid ML-1 cells, after damage with etoposide, show an acidic population with fluorescence intensity similar to that of normal cells (Fig. 4D). However, in CTLL-2 cells after withdrawal of IL-2, a marked reduction is seen in the fluorescence intensity of the acidic cells (Fig. 4B). This change may be due to several parameters: (1) the acidic cells may not load as much dye as the normal cells or (2) the dye may leak more rapidly as the cells lose membrane integrity. However, the fluorescence values still lie on a line out from the origin and the fluorescence ratio still yields an accurate assessment of the intracellular pH.

B. Intracellular Ca^{2+} Measurement

Intracellular Ca^{2+} is measured using 1 μM acetoxymethyl ester of indo-1, as described for carboxy-SNARF-1, but with flow cytometer excitation set at 355 nm and emission measured at 405 nm and 485 nm. Results can also be plotted in a two-dimensional dot plot, demonstrating both the difference in loading of individual cells and the spectrum of Ca^{2+} concentrations in the population (Fig. 5). Intracellular Ca^{2+} is then calculated using the Ca^{2+} dissociation equation (Grynkiewicz *et al.*, 1985):

$$[Ca^{2+}]in = K_d \cdot \frac{(R - R_{min})\, S_{f2}}{(R_{max} - R)\, S_{b2}}$$

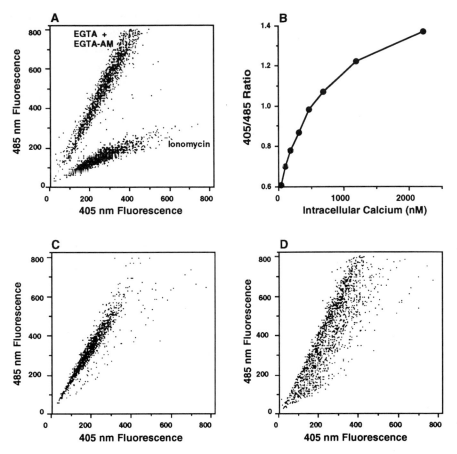

Fig. 5 Analysis of intracellular Ca^{2+} by flow cytometry. (A) Cells were loaded with indo-1; then Ca^{2+} was either increased to a saturating value with ionomycin or decreased to zero with EGTA plus ionomycin. (B) The calibration curve for intracellular Ca^{2+} concentration derived from the values in A and the Ca^{2+} dissociation equation. (C) Ca^{2+} in CTLL-2 cells in the presence of IL-2. (D) Ca^{2+} in CTLL-2 cells 16 hr after withdrawal of IL-2.

Six measurements are needed to resolve this equation, all of which are obtained from the flow cytometry results (Fig. 5): the fluorescence for the sample at 405 and 485 nm to give the ratio R; the fluorescence for the Ca^{2+}-free form of the dye at 405 (S_{f1}) and 485 (S_{f2}); and fluorescence for the Ca^{2+}-bound form of the dye at 405 (S_{b1}) and 485 (S_{b2}). R_{max} is the fluorescence ratio for the dye bound to Ca^{2+} (S_{b1}/S_{b2}), which is obtained at saturating Ca^{2+} concentration by incubating cells in 20 μM of the Ca^{2+} ionophore ionomycin in excess extracellular Ca^{2+}, that is, at least 1 mM Ca^{2+}; this condition is satisfied in normal medium. R_{min} is the ratio of fluorescence of free dye ($S_{f1}S_{f2}$) that is obtained by incubating cells with > 2 mM EGTA (depending on the concentration in the medium) and 30 μM EGTA-AM to chelate both extracellular and intracellular Ca^{2+}, respectively. The value S_{f2}/S_{b2} is therefore the ratio of the 485 nm fluorescence for free and Ca^{2+}-bound dye. The Ca^{2+} dissociation constant K_d for indo-1 is 250 nM at 37°C at an ionic strength of 0.1 at pH 7.08; this value will change significantly at different temperatures, pH values, and ionic strengths (June and Rabinovitch, 1990).

Examples of the dot plot distributions of cells treated with ionomycin plus excess Ca^{2+} or ionomycin with EGTA are shown in Fig. 5A; a calculated calibration curve is shown in Fig. 5B. Cells with increasing intracellular Ca^{2+} shift to the right in the indo-1 dot plot, whereas cells with less intracellular Ca^{2+} shift to the left. An example of intracellular Ca^{2+} distribution in cells during apoptosis is displayed in Fig. 5 C, D. In this example, apoptosis was induced in CTLL-2 cells after removal of IL-2. A heterogeneous increase in the Ca^{2+} level is seen. These cells are the same cells that showed an increase in pH (Fig. 4B). However, chelation of Ca^{2+} prevented the Ca^{2+} increase, but did not prevent either acidification or DNA digestion (J. Li and A. Eastman, unpublished data).

V. Summary

The methods described here facilitate assessment of the DNA digestion that was once considered a hallmark of apoptosis, but is now recognized as a common overdigestion of genomic DNA; less frequent breaks also lead to the morphological appearance of an apoptotic nucleus. Whatever level of DNA digestion occurs, it may be brought about by activity of one of a number of endonucleases. Whether one endonuclease is responsible in all cases of apoptosis or whether different endonucleases are responsible in various systems and circumstances remains unknown. In attempting to identify changes in ion concentrations that might activate these endonucleases, we have observed intracellular acidification to correlate with apoptosis consistently. This observation may have other implications since the acidification could be responsible for activating other proteins, such as proteases, which are also associated with apoptosis. The importance of such events to the onset of apoptosis remains to be fully established.

References

Arrends, M. J., Morris, R. G., and Wyllie, A. H. (1990). Apoptosis: The role of the endonuclease. *Am. J. Pathol.* **136,** 593–608.

Barry, M. A., and Eastman, A. (1992). Endonuclease activation during apoptosis: The role of cytosolic Ca^{2+} and pH. *Biochem. Biophys. Res. Commun.* **186,** 782–789.

Barry, M. A., and Eastman, A. (1993). Identification of deoxyribonuclease II as an endonuclease involved in apoptosis. *Arch. Biochem. Biophys.* **300,** 400–450.

Barry, M. A., Reynolds, J. E., and Eastman, A. (1993). Etoposide-induced apoptosis in human HL-60 cells is associated with intracellular acidification. *Cancer Res.* **53,** 2349–2357.

Brown, D. G., Sun, X.-M., and Cohen, G. M. (1993). Dexamethasone-induced apoptosis involves cleavage of DNA to large fragments prior to internucleosomal fragmentation. *J. Biol. Chem.* **268,** 3037–3039.

Duke, R. C., and Cohen, J. J. (1992). Morphological and biochemical assays of apoptosis. *In* "Current Protocols in Immunology" (J. E. Coligan, A. M. Kruisbeek, D. H. Margulies, E. M. Shevach, and W. Strober, eds.), Vol. 1, pp. 3.17.1–3.17.16. John Wiley and Sons, New York.

Eastman, A., and Barry, M. A. (1992). The origins of DNA breaks: A consequence of DNA damage, DNA repair or apoptosis. *Cancer Invest.* **10,** 229–240.

Eckhardt, T. (1978). A rapid method for the identification of plasmid desoxyribonucleic acid in bacteria. *Plasmid* **1,** 584–588.

Grynkiewicz, G., Poenie, M., and Tsien, R. Y. (1985). A new generation of Ca^{2+} indicators with greatly improved fluorescence properties. *J. Biol. Chem.* **260,** 3440–3450.

Juno, C. H., and Rabinovitch, P. S. (1990). Flow cytometric measurement of intracellular ionized calcium in single cells with Indo-1 and Fluo-3. *Meth. Cell Biol.* **33,** 37–58.

Kawabata, H., Anzai, N., Masutani, H., Hirama, T., Yoshida, Y., and Okuma, M. (1993). Detection of Mg^{2+}-dependent endonuclease activity in myeloid leukemia cell nuclei capable of producing internucleosomal DNA cleavage. *Biochem. Biophys. Res. Commun.* **191,** 247–254.

Kerr, J. F. R., Wyllie, A. H., and Currie, A. R. (1972). Apoptosis: A basic biological phenomenon with wide-ranging implications in tissue kinetics. *Br. J. Cancer* **26,** 239–257.

Nikonova, L. V., Nelipovich, P. A., and Umansky, S. R. (1982). The involvement of nuclear nucleases in rat thymocyte DNA degradation after γ-irradiation. *Biochim. Biophys. Acta* **699,** 281–289.

Oberhammer, F., Wilson, J. W., Dive, C., Morris, I. D., Hickman, J. A., Wakeling, A. E., Walker, P. R., and Sikorska, M. (1993). Apoptotic death in epithelial cells: Cleavage of DNA to 300 and/or 50 kb fragments prior to or in the absence of internucleosomal fragmentation. *EMBO J.* **12,** 3679–3684.

Peitsch, M. C., Polzar, B., Stephan, H., Crompton T., MacDonald, H. R., Mannerherz, H. G., and Tschopp, J. (1993). Characterization of the endogenous deoxyribonuclease involved in nuclear DNA degradation during apoptosis (programmed cell death). *EMBO J.* **12,** 371–377.

Ribeiro, J. M., and Carson, D. A. (1993). Ca^{2+}/Mg^{2+}-dependent endonuclease from human spleen: Purification, properties, and role in apoptosis. *Biochemistry* **32,** 9129–9136.

Shapiro, D. H., Sharp, P. A., Wahli, W. W., and Keller, M. J. (1988). A high-efficiency HeLa cell nuclear transcription extract. *DNA* **7,** 47–55.

Sorenson, C. M., Barry, M. A., and Eastman, A. (1990). Analysis of events associated with cell cycle arrest at G_2 and cell death induced by cisplatin. *J. Natl. Cancer Inst.* **82,** 749–754.

Walker, P. R., Kokileva, L., LeBlanc, J., and Sikorsak, M. (1993). Detection of the initial stages of DNA fragmentation in apoptosis. *BioTechniques* **15,** 1032–1040.

Wyllie, A. H., Kerr, J. F. R., and Currie, A. R. (1980). Cell death: The significance of apoptosis. *Int. Rev. Cytol.* **68,** 251–306.

CHAPTER 4

Quantification of Apoptotic Events in Pure and Heterogeneous Populations of Cells Using the Flow Cytometer

Pamela J. Fraker, Louis E. King, Deborah Lill-Elghanian, and William G. Telford

Department of Biochemistry and
Department of Microbiology and Public Health
Michigan State University
East Lansing, Michigan 48824

I. Introduction

Apoptosis is a form of cell suicide that is receiving considerable attention because of the many roles it plays in regulation of immune function, embryogen-

Copyright © 1995 by Academic Press, Inc. All rights of reproduction in any form reserved.

esis, and neuronal development (Goldstein *et al.*, 1991; Cohen *et al.*, 1992). In the immune system, anti-self and nonsense clones of cells evidently undergo apoptosis as they arise, thereby ensuring the destruction of these potentially dangerous cells. Moreover, natural killer and cytolytic T cells can induce apoptosis in their target cells (Cohen *et al.*, 1992). This fact, along with the finding that some efficacious chemotherapeutic agents work by inducing apoptosis in tumor cells, indicates that this form of cell death plays a major role in cancer defense (Cohen *et al.*, 1992). Researchers have determined that apoptosis also plays a role, albeit a negative one, in the etiology of AIDS. Healthy helper T cells incubated with HIV^+ cells or a coat protein from the HIV virus, gp120, along with cross-linking antibodies against gp120, undergo apoptosis (Banda *et al.*, 1992). Cell suicide may, therefore, play a significant role in the loss of $CD4^+$ T cells in the AIDS patient. The important role that apoptosis plays in AIDS, cancer defense, and regulation of the immune system indicates that methods that rapidly and quantitatively assess apoptosis are clearly needed.

In the laboratory, murine thymocytes or cell lines have become prevalent models in which to study apoptosis, primarily because of limitations in previous methodology that required that pure populations of cells be used. Pharmacological doses of dexamethasone and prednisolone—used in the treatment of a variety of cancers of the immune system, arthritis, inflammation, autoimmunity, and chronic allergic responses—cause substantial apoptosis among the immature thymocytes of young rodents (Telford *et al.*, 1991; Cohen *et al.*, 1992). However, the endogenous production of natural glucocorticoids also plays a role in regulating the immune system during stress, trauma, and malnutrition, which results in thymic atrophy and lymphopenia (Fraker *et al.*, 1993). Thus, we also must learn more about how these natural steroids affect thymocyte development and lymphopoiesis in the bone marrow. However, current methodology has limited such studies because of the heterogeneity of bone marrow and other tissues of interest.

The nuclei of cells of the immune system condense in size as they enter apoptosis. Unlike necrotic cells, apoptotic cells have mitochondria and endoplasmic reticulum that appear normal. Indeed, mitochondria of apoptotic cells stain brightly with rhodamine 123, a dye taken up by mitochondria that are maintaining normal transmembrane potential (Darzynkiewicz *et al.*, 1992). This form of cell suicide is also thought to depend on the induction of genes that have yet to be characterized (Cohen *et al.*, 1992; Schwartz *et al.*, 1993). Thus, cycloheximide and actinomycin D, as well as other inhibitors of protein and mRNA synthesis, can be used to block inudction of apoptosis in a variety of systems including murine thymocytes and bone marrow (Cohen *et al.*, 1992; Garvy *et al.*, 1993). In the cells of the immune system, an endogenous nuclease is activated that causes rapid cleavage of the DNA at the internucleosomal linker regions. As a result, 200-bp multimers of DNA are generated; few fragments larger than 4000 bp are observed on DNA gels (Goldstein *et al.*, 1991;

Telford *et al.*, 1991; Cohen *et al.*, 1992; see Chapter 3). The complete destruction of genomic material indicates that this form of cell death excludes any possibility of significant repair or survival on the part of the cell. Although apoptotic cells exclude trypan blue, they nevertheless develop membrane-bounded vesicles known as apoptotic bodies that can break away from the parent cell late in apoptosis (Darzynkiewicz *et al.*, 1992).

In the past, pathologists have relied heavily on phase contrast or electron microscopy to verify the presence of apoptotic cells using morphological criteria such as chromatin condensation and membrane blebbing. Such techniques, however, do not lend themselves to rapid quantitative work since they are very labor intensive and expensive. For this reason, many immunologists and biochemists have preferred to analyze populations of cells for apoptosis using DNA gels to demonstrate the presence of a ladder of DNA fragments. Unfortunately, this methodology also has its limitations. After the cells are exposed to an apoptotic cue, the DNA is extracted from the whole population and electrophoresed on agarose gels; this process is followed by ethidium bromide staining to identify the DNA fragments. The DNA ladder formed from the 200-bp multimers, generated by the endonuclease in apoptotic cells, is the usual criterion for the occurrence of apoptosis. This procedure requires a 3- to 4-day effort and is semiquantitative at best, giving no information about the actual proportion of cells in the population that have become apoptotic.

Perhaps the greatest constraint of DNA gels and, to some extent, of microscopy is that these techniques tend to confine studies to relatively pure populations of cells, which explains the current focus on thymocytes and cell lines. For individuals interested in pursuing the role of apoptosis in AIDS, cancer, regulation of the immune system, lymphopoiesis, glucocorticoid therapy, and other biological processes, these methods are wholly unsatisfactory and are an impediment to the advancement of these fields of study.

However, a solution to these methodological problems has come into use. In the past, the flow cytometer often has been used to analyze the cell cycle status of populations of cells. This machine can readily quantify the proportion of cells in G_0/G_1, S, and G_2/M phase when fluorescent DNA dyes are used. For example, the cytometer can easily identify cells in G_2/M phase since they have twice the amount of DNA and, therefore, twice the fluorescent intensity of cells in the G_0 or G_1 compartment. In addition to emitted light, the flow cytometer also analyzes scattered light from the cells passing through the laser beam. Forward scatter and side scatter indicate cell size and granularity (or density), respectively. Since apoptotic cells typically decrease in size, in part because of nuclear condensation, the cytometer ought to be able to detect apoptotic cells as smaller, less fluorescent entities. Indeed, in an earlier study, Compton and colleagues (1988), using rat thymocytes treated with dexamethasone, found a peak of DNA fluorescence in the hypodiploid region of the cell cycle, which in a flow cytometric histogram is seen as a region to the left of G_0/G_1. In this case, unfixed thymocytes were incubated in acridine orange for less than 5 min and were analyzed immediately for cell cycle status. Although a peak of DNA fluorescence was observed in the hypodiploid region of the cell cycle for dexamethasone-treated thymocytes, this fluorescent region was often variable in width and occasionally overlapped with the G_1 peak. Thus, the degree of separation of apoptotic cells and the precision of quantification was not as significant as desired. Nevertheless, this work encouraged us to try to develop a fixation–staining method that would provide a sharp, well-separated peak containing apoptotic cells. In the following sections, we discuss the use of the flow cytometer to identify the proportion of cells that are undergoing apoptosis in pure and heterogeneous populations.

II. Staining and Fixative Methods for Quantification of Apoptosis in Pure Populations of Cells: One-Color Analysis

To determine the proportion of cells that are apoptotic in a relatively homogeneous population, such as thymocytes or cell lines, collection of simple scatter profiles coupled with staining the DNA with a suitable fluorescent dye is sufficient for analyzing these cells using the flow cytometer. This method requires only a few hours of effort, utilizes intact cells, and results in highly quantitative information regarding the degree of apoptosis in a population. In addition, a complete cell cycle profile is obtained simultaneously (Telford *et al.*, 1991).

A. Precautions

When beginning a new protocol, it is important to determine the viability of populations that are presumed to include apoptotic cells. Unlike necrotic cells, apoptotic cells should exclude trypan blue dye if suitable induction conditions have been used (Telford *et al.*, 1991; Darzynkiewicz *et al.*, 1992). Overall losses in cell numbers, excessive levels of debris, uptake of trypan blue, and other characteristics may indicate that the conditions used to induce apoptosis were too harsh or too long in duration. In such cases, a wide distribution of cell sizes is typically seen in the scatter profile. With a little practice, investigators can detect many apoptotic murine immune cells by simple phase contract microscopy, which also allows inspection for debris, aggregation, and other unwanted features prior to fluorescence-activated cell sorting (FACS) analysis. Although a few hours to several days may be required to induce apoptosis, one must be aware that, once generated, the apoptotic cells have only a finite lifetime before they disintegrate.

B. Solutions

Phosphate buffered saline (PBS): 1 part 0.1 *M* phosphate buffer, pH 7.4; 9 parts isotonic saline

Hanks balanced salt solution (HBSS) (gm/l): 8 g NaCl, 0.4 g KCl, 0.047 g Na_2HPO_4, 0.06 g KH_2PO_4, 1 g glucose per 1 liter distilled water

DNA staining reagent: PBS, pH 7.4, containing 0.1% Triton X-100 (optional), 0.1 mM EDTA (pH 7.4), 0.05 mg/ml RNase A (50 units/mg), 50 μg/ml propidium iodide (PI) 70% Ethanol

Chicken Red Blood Cells (CRBCs): Stored at $-70°C$ in 5% dimethylsulfoxide, 250 mM sucrose, 50 mM sodium citrate buffer, pH 7.6

C. Methods

1. Fixation

If cell retrieval and viability are acceptable after induction of apoptosis, then $\sim 2 \times 10^6$ cells should be aliquoted per sample, pelleted by slow centrifugation (350 g, and washed in PBS or HBSS (without phenol). After a second centrifugation, the pellet should be resuspended in ice cold 70% ethanol, with rapid but gentle mixing, to a final cell density of 1×10^6 cells/ml. A minimum fixation period of 30 min at 4°C is required. Fixation times that exceed 5 hr should be avoided to limit aggregation of cells (Telford *et al.*, 1991). In many cases, the cells can be stored in ethanol at $-20°C$ for several days before staining and analysis, if desired (Darzynkiewicz *et al.*, 1992). Although many other fixative methods were tried, ethanol fixation gave the sharpest DNA fluorescent apoptotic peaks, which were well separated from the G_0/G_1 region for reasons to be discussed subsequently.

2. Staining

Prior to staining, the ethanol-fixed cells are centrifuged at 400 g, decanted, and blotted to remove as much ethanol as possible; then the cells are washed at least once in PBS or HBSS. The pelleted cells are rapidly but gently resuspended in 1 ml DNA staining reagent, described earlier. If rapid analysis is desired, 1 hr of staining at room temperature is sufficient (Telford *et al.*, 1991). Triton X-100 is optional, but in some cases the detergent may make the cells more permeable to the stain, as well as helping prevent cell aggregation. RNase A should be protease and DNase free (Boehringer-Mannheim, Indianapolis, IN). Since PI and other DNA dyes do not bind covalently to the DNA and can diffuse out of the cells, the cells must remain in the staining solution prior to analysis. Cells can usually be stored overnight in the dark at 4°C in the staining solution prior to analysis. PI is excited with the 488-nm line of an argon laser; emission is detected at 620-700 nm. As will be discussed later, many other DNA dyes can be used in place of PI. The effective concentration of dye will vary as indicated in another publication (Telford *et al.*, 1992).

D. Standards, Controls, and Gating

The flow cytometric analysis of a cell population for apoptotic cells, based on DNA staining alone, is relatively straightforward. A rapidly growing isolate of culture cells or a fresh preparation of thymic T cells should be fixed and stained in parallel with experimentally treated samples. The fresh isolates are used to establish the position of the G_1 peak for the normal cell population. In the case of mouse and human cells, an aliquot of 2×10^6 CRBCs (Colorado Serum Co., Denver, CO) stained in parallel with experimental cells can serve as an approximate indicator of apoptotic cell DNA fluorescence (see Fig. 1). CRBCs are frequently used as internal standards for cell analysis, having a DNA content about 35% that of normal human diploid cells (Vindelov *et al.*, 1983). These cells help define the hypodiploid region in which apoptotic cells are found. Materials with less fluorescence than CRBCs may represent debris or artifacts. DNA fluorescence gating is done in a cytogram of integrated fluorescence area versus fluorescence width signal. The DNA fluorescence of single cells will increase linearly in fluorescence area without significant increase in fluorescence width. Cell doublets and large aggregates will cause increased signal width at each cell cycle phase. In the DNA cytogram, apoptotic cells and CRBCs will give a fluorescence peak that is completely separated from the normal G_1 peak at approximately 50% of G_1 peak fluorescence (see Fig. 1). A DNA fluorescence gate including CRBCs and apoptotic cells through the normal G_2/M peak should be drawn. Cell aggregates and nuclear fluorescence greater than G_2 should be excluded. This gate should contain greater than 85% of all DNA fluorescence events and is used to initiate FACS analysis. The histogram of the integrated fluorescence signal area derived from this gate can than be partitioned into apoptotic hypofluorescent (A_0), G_0/G_1, S, and G_2/M regions (see Fig. 1). The coefficient of variation (CV) for the G_0/G_1 peak for properly prepared normal cells can run from 1.8–2.5% for FACS units using a flow cell to 2.5–3.2% for jet in air FACS units. Not unlike the peaks from column chromatography and other related techniques, the peaks on DNA histograms have an expected standard deviation in their width (CV) that must also be taken into account. Analysis should be done on 10,000 cells (events) within the fluorescence gate to give optimum accuracy. A typical DNA histogram that includes apoptotic as well as CRBCs is shown in Fig. 1. In this case, murine thymocytes that have undergone apoptosis after treatment with dexamethasone show up as a sharp peak in the hypodiploid region and are well separated from intact cells in G_0/G_1. The CRBCs which have about half the DNA fluorescence of murine cells in G_0/G_1, reside further to the left in the histogram. As indicated, materials with less fluorescence than the CRBCs should be excluded since they might represent blebs, debris, or necrotic cells.

If a forward versus side scatter cytogram has been established to include debris, CRBCs, and normal cells, then apoptotic cell scatter will be found below the scatter profile of normal cells at about the location noted for CRBCs but

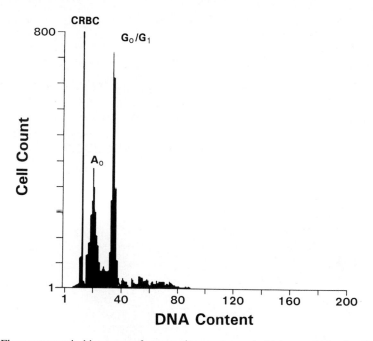

Fig. 1 Flow cytometric histogram of mouse thymocytes and chicken red blood cells (CRBC) analyzed for DNA content. Dexamethasone-treated thymocytes and CRBCs were ethanol fixed and stained with propidium iodide reagent, as described in the text. Samples were analyzed for 10,000 single events on an Ortho Cytofluorograph using an Intel 30386 computer with Acqcyte and Multiflow cytometric analysis software (Phoenix Flow Systems, San Diego, CA). Excitation was with the 488-nm line of an argon laser; emission detection was at 620–700 nm. Linear DNA fluorescence is given.

above that for cell debris. The percentage of cells in the scatter profile of the apoptotic population (which is separate from the scatter of normal cells) will be very similar to the percentage of cells found in the apoptotic DNA fluorescence peak. FACS analysis based solely on forward scatter may also be done with pure populations of cells that are fairly uniform in size. Figure 2 shows the scatter profile of thymocytes incubated for 8 hr in the presence or absence of 1 μM corticosterone, a glucocorticoid that induces substantial apoptosis in immature thymic T cells. Note the shift in the proportion of cells in the lower portion of the diagram, which defines the small apoptotic cells as 10% of all untreated thymocytes and 63% of cells exposed to steroid.

Both DNA histograms and scatter profiles of dexamethasone-treated rat thymocytes are shown in Fig. 3. Note that the small amount of apoptosis (1.3% of all gated cells) in freshly prepared thymocytes (time 0) rises to 14% after 6 hr of incubation without exposure to dexamethasone. This change is commonly

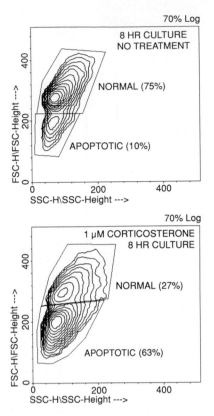

Fig. 2   Flow cytometric cytograms of mouse thymocytes analyzed for forward scatter (cell size) versus side scatter (cell granularity of density) after an 8-hr incubation in the presence or absence of 1 μM corticosterone. Normal and apoptotic regions are identified; percentage values are for the entire population.

observed in thymocytes and is probably due to the ongoing accumulation of apoptotic cells that would be expected to occur among thymocytes that contain nonsense or anti-self clones. The difference also may reflect the fact that cultures do not contain all the appropriate cytokines and growth factors. However, 6 hr of exposure to dexamethasone escalates apoptosis to 43% of all events, as seen in the third panel of Fig. 3. Note that the apoptotic peak for rat thymocytes, although well separated from G_0/G_1, is not as sharp as that for the murine thymocytes, representing a species variation that becomes more problematic in the case of human cells. The far right panel of Fig. 3 readily demonstrates the problems encountered when assessment of apoptotic events is delayed too long. Both the scatter profile and the DNA histogram indicate degeneration of cells in all phases of the cell cycle.

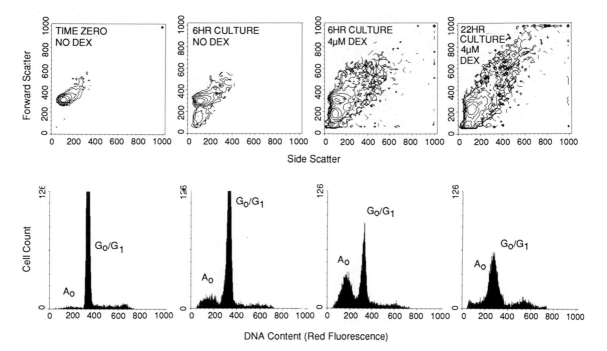

Fig. 3 DNA histogram and scatter profiles of rat thymocytes stained with propidium iodide before and after exposure to dexamethasone. Forward scatter (size) versus side scatter (granularity) is shown for freshly prepared rat thymocytes and those incubated with or without 4 μM dexamethasone for 6 hr or 22 hr. DNA fluoresence (red) of propidium iodide is shown for all treatment groups (excitation at 488 nm with detection at 620–700 nm for 10,000 events per histogram). The channel numbers at which cells in the A_0 and G_0/G_1 stages of the cell cycle reside are indicated. Samples were analyzed on a FACS Vantage fluorescence-activated flow cytometer (Becton Dickinson, San Jose, CA). The scatter profile and DNA histogram of thymocytes exposed to dexamethasone for 22 hr indicate degeneration of cells in all phases of the cell cycle, whereas distinct regions are still apparent after 6 hr of exposure to steroid.

III. Verification of the Fidelity of the Apoptotic Region

Since flow cytometry represents a valuable but nevertheless new approach to assessing apoptosis, it is necessary to validate the method further. Using murine thymocytes, the apoptotic region (A_0) has been shown to increase in proportion as the dose or time of exposure to glucocorticoids or irradiation was increased (Telford et al., 1991). Further, the proportion of cells in the apoptotic region of the FACS correlates with the amount of fragmented DNA observed on DNA gels prepared from the same populations of cells. Standard inhibitors of apoptosis such as cycloheximide, actinomycin D, aurintricarboxylic acid, and high zinc salts reduced the percentage of cells in the A_0 region to levels observed in untreated thymocytes (Telford *et al.*, 1991). Thus, the appropriate correlates exist. In another investigation, cells from both the A_0 and G_0/G_1 region were sorted, and DNA was extracted from the cells and analyzed by gel electrophoresis (Telford *et al.*, 1994). Ladder-like DNA fragments were noted for cells found within the A_0 whereas intact genomic DNA was found in cells from the G_0/G_1 region. Since its 1991 publication by Telford *et al.*, this method has been successfully used by a number of other laboratories to study not only thymocytes but also HL-60 cells, MOLT-4, B-lineage bone marrow cells, and BAF-3 hematopoietic cells (Darzynkiewicz *et al.*, 1992; Ormerod *et al.*, 1992; Sun *et al.*, 1992; Garvy *et al.*, 1993).

IV. Other DNA Dyes

A variety of intercalating and external binding DNA dyes have been tested, all of which are as successful as PI for analysis of apoptosis (Telford *et al.*, 1992). These DNA dyes, the laser line usually used for excitation, and the emission ranges are provided in Table I. In all cases, fixed apoptotic cells exhibit reduced background fluorescence, giving a clear and sharp peak in the same region of the cell cycle as PI. The ability of each dye listed in Table I to detect apoptotic cells is remarkably similar; the percentage of events noted below G_0/G_1 is quite similar regardless of the DNA dye used (Telford *et al.*, 1992). These features give the method great flexibility, since the investigator can select a DNA dye that has little spectral overlap with other fluorochromes used to study other cell parameters.

Because all the dyes work well, researchers wanted to determine why fixed and stained apoptotic cells have less DNA fluorescence than control cells. Since apoptotic cells condense in size and exclude trypan blue, they are assumed to

Table I
Common DNA Dyes Used in Flow Cytometry

DNA binding dye	Binding mechanism	Laser excitation line (nm)	Emission range (nm)
Propidium iodide	Intercalative	488	550–640
Ethidium bromide	Intercalative	488	550–640
Acridine orange	Intercalative	488	510–530
7-Amino-actinomycin D	Intercalative	488, 541	640–680
Daunomycin	Intercalative	488	560–600
Mithramycin A	External	457	510–530
Chromomycin A_3	External	457	550–580
Olivomycin	External	457, 480–500	530–560
Hoechst 33258	External	351–364	380–420
Hoechst 33342	External	351–364	380–420
DAPI	External	351–364	380–420

maintain membrane integrity and not to leak their contents. Thus, they may take up less DNA dye than normal cells. However, these cells should be fairly permeable to the dyes after fixation in ethanol. Furthermore, it seems unlikely that such diverse DNA dyes, which vary significantly in size and structure, would all be excluded to the same extent. Other investigators argue that the genomic DNA of apoptotic cells is digested into fragments small enough to leak out of the cells when the membranes are permeabilized with detergents or ethanol (Afanasyev et al., 1993; Gong et al., 1993). In the latter case, only small amounts of DNA are thought to be left behind in the ethanol-fixed apoptotic cells, which would presumably account for the lower fluorescence on DNA staining. In another study, Afanasyev et al., (1993) demonstrated that oligonucleotide multimers of DNA can be found in the supernatant surrounding apoptotic cells. Unfortunately, no quantitative information was given regarding the proportion of total cellular DNA represented by the extracellular material. Nevertheless, the ethanol fixation method may permeabilize cells in such a way that these small DNA fragments can leak out readily. However, the spectral pattern for apoptotic cells, especially in the case of glucocorticoid-treated or irradiated murine thymocytes, is surprisingly uniform, with peaks of modest CVs (Telford et al., 1991). The variance or standard deviation in the width of the histogram peaks is less than 3% for the murine system, which is quite good. That all apoptotic cells should lose such consistent amounts of DNA seems unlikely. Alternatively, ethanol may alter the chromatin structure of apoptotic cells so less DNA is available for staining by the dyes.

V. Ethanol Fixation

In murine lymphoid cells, ethanol fixation of apoptotic cells appears to be critical for providing a clearly separated apoptotic (A_o) peak. Examination of populations of fixed and unfixed apoptotic cells by phase contrast or electron microscopy reveals a diverse group of phenotypes (Darzynkiewicz et al., 1992). Although acid denaturation was often used to facilitate staining of DNA in the past, these preparative methods fail to provide distinct A_o peaks. Ethanol is thought to cause cell shrinkage, which may make the condensed apoptotic cells even smaller, thus making them more identifiable by forward scatter (Fig. 4). Ethanol is also presumed to cause coagulation of proteins and nucleic acids, although the precise effects of ethanol on chromatin are unknown. Ethanol also dissolves some of the lipids of the membrane of cell, which is thought to permeabilize the cells, so the dyes may have more access to the DNA. This increase in permeability might also facilitate the efflux of small DNA fragments from the apoptotic cells, as already discussed. Regardless, electron micrographs of ethanol-fixed apoptotic cells reveal interesting changes in their morphology. In Fig. 4, the dark condensed nucleus of the apoptotic cells is readily evident. Ethanol fixation clearly reduces the size of the apoptotic cell; the cytoplasm seems devoid of secondary structure. However, there is little discernible change

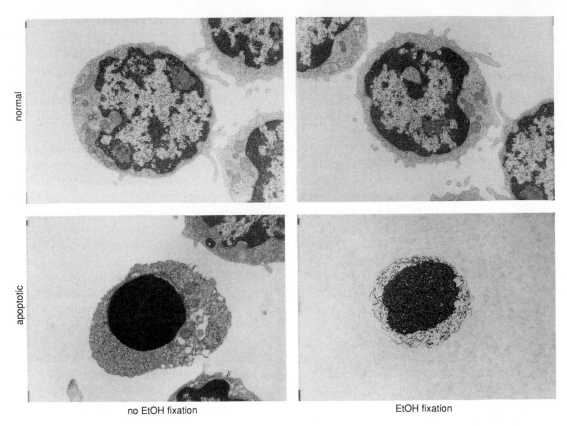

Fig. 4 Electron micrographs of normal and apoptotic murine thymocytes. Ethanol fixation of apoptotic cells further reduces their size, leaving a cytoplasm devoid of organelles.

in either the size or the physical appearance of normal thymocytes following ethanol fixation. Thus, ethanol seems to make a number of changes in the physical size and appearance of the apoptotic cell that may account for the ability of the flow cytometer to detect them as a clear and distinct subset of the cell cycle. However, as will be discussed, the staining and fixation protocols may have to be adjusted for different cell types as well as for cells from different species.

VI. Studying Apoptosis in Heterogeneous Populations: Two-Color Analysis

Without a doubt, the most important advantage of FACS analysis is its ability to quantify apoptotic events in specific subsets of cells in a heterogeneous population. Using FACS, we have shown that T-lineage cells developing in the murine thymus are very prone to glucocorticoid-induced apoptosis. Using two-color analysis in conjuction with scatter assessments, the degree of glucocorticoid-induced apoptosis in CD4$^+$ or CD8a$^+$ cells is readily quantifiable. Two-color analysis is done using PI to stain DNA and fluorescein isothiocyanate (FITC)-conjugated (CD4) or biotinylated anti-CD8a made fluorescent with avidin–FITC to detect T-cell subpopulations (See Fig. 5). Using the methodology demonstrated here, we have also shown that B-lineage cells developing in murine bone marrow are likewise prone to glucocorticoid- or radiation-induced apoptosis (Garvy et al., 1993, 1994). Broad population subsets (e.g., CD4$^+$ cells) can be further defined developmentally using additional phenotypic markers sinultaneously (e.g., CD4$^+$CD8$^+$, CD4$^+$CD8$^-$, CD4$^+$CD3$^-$). The technique we are demonstrating can be easily expanded to three-color analysis using two fluorochromes for antibody detection [FITC and phycoerythrin (PE)] in combination with DNA cell cycle analysis using 7-amino-actinomycin D (7-AAD) and a single argon laser. Application of DNA standards and checks for cell viability are very similar to the processes in one-color analysis and will not be repeated here. However, additional controls are needed to ensure that gating and color compensation are done properly.

A. Solutions

Label buffer: PBS, pH 7.4, 0.15% sodium azide, 10% heat-inactivated, fetal bovine serum (FBS) previously absorbed against the appropriate cell suspension; filter buffer through a 0.22-μm filter to remove any particulates

Fixation buffer: PBS, pH 7.4, containing 50% heat-inactivated FBS; filter buffer through a 0.22-μm filter to remove any particulates

DNA staining reagent: PBS, pH 7.4, containing 0.1% Triton X-100 (optional), 0.1 mM EDTA (pH 7.4), 0.05 mg/ml RNase A (50 units/mg), 50 μg/ml PI

B. Methods

1. Phenotypic Labeling

In the case of murine lymphoid cells, aliquots of 2×10^6 cells were incubated for ~30 min at 4°C with appropriately titered phenotypic antibodies in label buffer. Fluorescein-conjugated antibodies or biotinylated antibodies made detectable with avidin–FITC work well in combination with PI staining of DNA. The DNA dye 7-AAD, which excites with the 488-nm line of an argon laser and emits at ≥650 nm, may be used in a dual antibody phenotyping system in which FITC is used for the weaker fluorescent signal and PE for the strongest phenotypic marker. Some loss of PE fluorescence will be experienced with simultaneous staining with 7-AAD. Alternatively, a UV-excited blue-fluorescing dye such as 4′-6-diaminido-2-phenylindole (DAPI) may be used in combination with PE and FITC to reduce spectral overlap and/or fluorescence quenching.

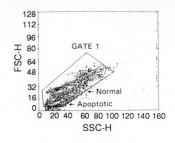

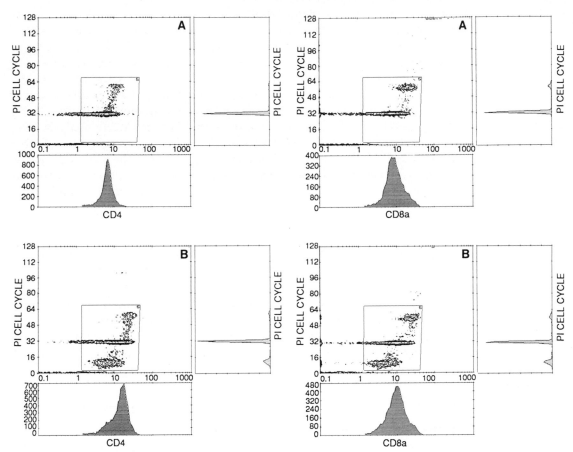

Fig. 5 Anti-CD4 or Anti-CD8a phenotypically labeled thymocytes were analyzed for apoptosis 8 hr after exposure to 0.1 μM dexamethasone (Dex). An unincubated and an 8-hr 0.1 μM Dex-treated culture were used to define the scatter gate (gate 1, *top*), which is drawn to include apoptotic and normal thymocytes. Debris and large cell aggregates are excluded. Cells within gate 1 are examined for the presence of FITC-antiCD4 or biotinyl-antiCD8a–avidin FITC. Unincubated normal thymocyte samples that were phenotyped, fixed, and DNA stained in parallel with treated samples were used to define the phenotypic positive box (gate 2) in cytogram panels A–C. Cells to the left of the box are phenotypically negative. Data below the box that are FITC positive but propidium iodide (PI) negative are considered debris or cell fragments. Histograms present the green (antibody) and red (PI) fluorescence of cells selected through gates 1 and 2. The red fluorescence histogram was analyzed for apoptotic cells using Multiplus software. Phoenix Flow, San

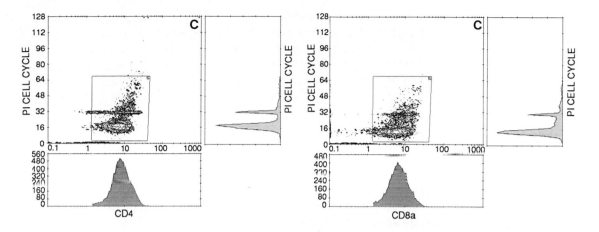

		Percent Apoptotic	
Treatment	Panel	CD4$^+$	CD8a$^+$
No Incubation	A	0%	0%
8 Hr Culture No Treatment	B	13%	17%
8 Hr Culture 0.1 μM Dexamethosone	C	69%	77%

Fig. 5 *(continued).*

Diego CA Greater than 90% of the analyzed events meet gate 1 specification, whereas 80–90% of gate 1 meet gate 2 specification. The percentage coefficient of variation (CV) for G_0/G_1 is 2.1% for unincubated to 3% for dexamethasone-treated samples. Samples were analyzed on a FACScan without DNA signal processing. Ten thousand events were analyzed for each sample.

2. Fixation

The cells are washed twice after phenotypic labeling and are resuspended in fixing buffer to which ice cold 70% ethanol is added dropwise with gentle mixing to a final concentration of 50% ethanol. This process constitutes a more gentle fixation method that ensures excellent retention of fluorescent surface antibody label in murine thymocytes, B-lineage marrow cells, and splenocytes. However, currently it may be necessary to fix human cells first with low concentrations of paraformaldehyde prior to ethanol fixation to ensure adequate retention of the fluorescent antibodies used to label surface markers.

3. DNA Staining

After 1 hr of fixation, the cells can be DNA stained or retained in fixatives for up to 2 days at 4°C. Prior to FACS analysis, samples must be washed to remove ethanol and any precipitated protein. DNA staining with PI is done as described earlier. If 7-ADD is used, the staining solution contains 5 μg/ml 7-AAD, 0.1% Triton-X 100, 0.1 mM EDTA(Na)$_2$ in PBS, pH 7.4. DAPI staining solution contains 1 μg/ml DAPI, 0.1 mM EDTA (Na)$_2$. If loss of phenotype fluorescence occurs, the detergent may be omitted. RNase is not required when using DAPI or 7-AAD, since these reagents do not stain RNA as do PI and acridine orange.

C. Standards, Controls, and Gating

Phenotypic identification of apoptotic populations is more complex because of spectral overlap between phenotyping fluorochromes and DNA dyes. Thus, a series of phenotypic and DNA stained controls is required. In two-color experiments using FITC-phenotype-gated PI-stained preparations, a sample of untreated cells is required that has been singly labeled with FITC or PI. These cells are used to establish detection of FITC and to set electronic color compensation for spectral overlap of PI in the FITC detector. A phenotype-negative sample is also required to determine the boundary between phenotype-positive and -negative cells. Color compensation is accomplished according to these steps:

1. Run a phenotype-positive sample. Set the FITC detector to log amplification and detector voltage to position the phenotype-positive population roughly in mid-log scale (Fig. 5; phenotype histogram).

2. Run the phenotype-negative sample (typically an isotype-matched immunoglobulin) and adjust the detector voltage to set the negative sample in the first decade of the log scale. Rerun the phenotype-positive population and confirm clean separation of phenotype-positive and -negative populations and detection of the appropriate percentage of positive phenotype.

3. Set up the PI fluorescence detector as described for one color in Section II,C,3. The PI-only stained samples should be separated cleanly into the hypodiploid region (where apoptotic cells or CRBCs are found) and normal stages of cell cycle (G_0/G_1, S, G_2/M; Fig. 5, PI cell cycle histograms).

4. Examine a two-color cytogram of PI fluorescence vs FITC fluorescence and remove the PI fluorescence sensed in the FITC detector by electronic color compensation. Less than 5% compensation correction of PI in FITC will be required to remove PI fluorescence from the FITC fluorescence detector when phenotype-negative cells are within the first decade of FITC log fluorescence.

5. Four cell populations should now be detected in the two-color cytogram: (1) phenotype-negative apoptotic cells, if any exist, will be found in the lower left quadrant; (2) the cell cycle of the phenotype-negative population will be in the upper left quadrant; (3) the cell cycle of the phenotype-positive population will be in the upper right quadrant; and (4) the apoptotic phenotype-positive population will be in the lower right quadrant.

When multiple fluorochromes (FITC, PE, and others) are used in phenotyping, a single-color-labeled population with its isotype control will be required to determine positive/negative separation for each fluorochrome. Additionally, all combinations of fluorochrome-pair-stained samples will be required to set color compensation for each fluorochrome (e.g., FITC/PE pair, FITC/7-AAD pair, PE/7-AAD pair). A DNA-stained sample without other fluorochromes will be required to monitor for DNA fluorescence in each fluorochrome detector being used in the analysis. Finally, a sample of untreated cells, positive for all phenotypic markers and stained for DNA, must be used to confirm that the appropriate phenotypic values are obtained for each marker and agree with the literature values for that marker.

Once color compensation is completed, gating decisions may be made and data can be collected. Figure 5 demonstrates gating based on scatter and phenotypic fluorescence. Gate 1 is set in the scatter cytogram to include all single cell events for further analysis. Only debris and cell aggregates are left outside the gate (see Fig. 5, gate 1). The selected population is then examined for phenotypic fluorescence and a gate (gate 2) is drawn around the phenotype-positive population (box in two-color cytograms; Fig. 5 A–C). The cell cycle of the scatter and phenotype-selected population is then analyzed (y axis histogram). This approach is useful when using FACS units of limited analytical capacity.

Additional refinement of the data may be obtained using FACS units that allow pulse processing of the DNA signal into area and width signals. In this case, the first gate selects for single cell events as described in Section II,C,3. Single cells are then analyzed for phenotypic fluorescence, as just described and a similar gate can be defined (gate 2). A scatter cytogram of the phenotype-positive population may be obtained and a third gate defining the light scatter of the appropriate population may be drawn. This gate should only be used to remove a low percentage of cell types that clearly have inappropriately bound an antibody as indicated by light-scatter signature (e.g., granulocyte scatter present in a lymphoid cell population). This gate must not remove apoptotic cells of low scatter or G_2/M cells of higher scatter. The cell cycle of cells meeting all three gate definitions may be obtained and analyzed.

VII. Problems Encountered with Human Lymphoid Cells

Several laboratories have reported that apoptotic human lymphocytes do not give a sharp, well-separated peak when the fixation–staining procedures described earlier are used. Indeed, on fixing and staining irradiated human peripheral lymphocytes, a very broad peak that extended as a shoulder from G_1 was obtained in the region that potentially contained apoptotic cells. The urgent need to be able to assess rates of apoptosis in humans who have AIDS or are undergoing cancer therapy has, unfortunately, led some investigators to rely on such data. Indeed, some individuals have included for consideration materials one or more log decades below G_1 of the cell cycle histogram. This is unfortunate because debris, nuclei, and apoptotic blebs may also be found in these regions, thus giving an erroneous impression of the degree of apoptosis in a population. DNA gels and electron micrographs, which are needed to ascertain exactly what materials are found in these regions, have yet to be performed.

This situation has led some investigators to propose that the DNA of apoptotic human lymphoid cells may not be fragmented in the same manner as that of murine cells, or that the chromatin structure of human cells is significantly different than that of murine cells. We have found that varying the fixation conditions, the amount of DNA dye, the length of staining time, and other parameters had little effect on the quality of data for apoptotic human cells. However, incubating fixed human cells for 0.5 hr at 30°C prior to staining did provide a better separated peak, although it remained quite broad and its actual composition remains unknown. However, some progress has been made with the human promyelocytic cell line HL-60. Investigators have incubated apoptotic ethanol-fixed HL-60 cells at room temperature in a phosphate–citric acid buffer containing 0.1% Triton X-100 for 30 min prior to staining (Gong *et al.*, 1993). These investigators feel that this additional step facilitates the extraction of low molecular weight DNA from apoptotic cells. Using this method, very sharp, well-separated peaks were obtained that were well separated from G_1 (almost adjacent to the *y* axis in the histograms). Whether this approach works well with normal human lymphocytes has not been determined. Furthermore, the investigators themselves acknowledge that the method is not always reliable because of lysis of apoptotic cells in some cases (Darzynkiewicz *et al.*, 1994). Thus, a reliable quantitative means for staining and fixing human apoptotic cells still has not been developed. Hopefully, this unfortunate situation will be resolved quickly.

VIII. Molecular and Biochemical Assessments of Apoptotic Cells

Yet another advantage of flow cytometry for the field of apoptosis is that it is possible to sort or retrieve cells in the apoptotic region for further study. This is no small advantage, since we know surprisingly little about the status of various vital functions in cells undergoing apoptosis. For such studies, however, one would probably not want to use cells that have been altered by ethanol fixation and are no longer viable. Furthermore, some of the DNA dyes are themselves toxic to cells and, in fact, do not readily enter intact or nonpermeabilized cells (e.g., PI). However, the Hoechst dyes provide good alternatives since they are taken up at a moderate rate by normal cells and have relatively low toxicity, at least to cells of the immune system. Normal cells actively efflux the dye, whereas most apoptotic cells seem to retain more of the dye and, thus, are more brightly labeled than actively functioning normal cells (Ormerod *et al.*, 1992; Sun *et al.*, 1992). The methodology is simple, requiring only that cells be incubated at 37°C with Hoechst dye at a low concentration (usually Hoechst 33342 at 1 μg/ml) for 15 min. The cells are immediately placed on ice to reduce efflux of the dye and are immediately sorted. To use sorting as a routine preparative method by which to obtain large numbers of apoptotic cells would be too expensive. Nevertheless, sorting can be a valuable approach when biochemical and functional evaluations can readily be done on a small number of apoptotic cells.

IX. Summary

The rapid and highly quantitative nature of flow cytometric cell cycle analysis for determining the proportion of apoptotic cells in a population makes it the method of choice for a variety of studies requiring quantitative information about cell death. Furthermore, by employing multiparameter analysis including phenotypic labeling, FACS makes it possible to study apoptosis in specific subsets of cells within a heterogeneous population. Live sorting of cells in the apoptotic region offers the possibility of studying the effects of this form of cell death on key biochemical functions of the cell. Nonetheless, further modification of the fixing–staining methods presented here will be needed to make FACS useful for analysis of apoptosis in human cells.

Acknowledgments

The authors thank Teresa Vollmer for preparation of the manuscript, and the MSU Biotechnology Program as well as the National Institutes of Health (HD10586-15) for their financial support.

References

Afanasyev, V. N., Korol, B. A., Matylevich, N., Pechatnikou, V. A., and Umansky, S. R. (1993). The use of flow cytometry for the investigation of cell death. *Cytometry* **14**, 603–609.

Banda, N. K., Bernier, J., Kwahara, D. K., Kunle, R., Hargwood, N., Sekaly, R. P., and Finkel, T. H. (1992). Crosslinking CD4 by human immunodeficiency virus gp120 primes T cells for activation induced apoptosis. *J. Exp. Med.* **176**, 1099–1106.

Cohen, J. J., Duke, R. C., Fadok, V. A., and Sellins, K. S. (1992). Apoptosis and programmed cell death in immunity. *Annu. Rev. Immunol.* **10**, 267–293.

Compton, M. M., Haskill, J. S., and Cidlowski, J. A. (1988). Analysis of glucocorticoid actions on rat thymocyte deoxyribonucleic acid by fluorescence-activated flow cytometry. *Endocrinology* **122**, 2158–2164.

Darzynkiewicz, Z., Bruno, S., Del Bino, G., Gorczyca, W., Hotz, M. A., Lassota, P., and Traganos, F. (1992). Features of apoptotic cells measured by flow cytometry. *Cytometry* **13**, 795–808.

Darzynkiewicz, Z., Li, X., and Gong, J. (1994). Assays of cell viability. Discrimination of cells dying by apoptosis. *In* "Methods in Cell Biology" (Darzynkiewicz, Z., Robinson, J. P., Chrissman, H. A., eds.) Vol. 41, p 15–38, Academic Press, New York.

Fraker, P. J., King, L. E., Garvy, B. A., and Medina, C. (1993). The immunopathology of zinc deficiency in humans and rodents: A possible role for programmed cell death. *In* "Nutrition and Immunology" (D. Klurfield, ed.), Vol. 8, p. 267–283. Plenum Press, New York.

Garvy, B. A., Telford, W. G., King, L. E., and Fraker, P. J. (1993). Glucocorticoids and irradiation induced apoptosis in normal murine bone marrow B-lineage lymphocytes as determined by flow cytometry. *Immunology* **79,** 270–277.

Garvy, B. A., King, L. E, Telford, W. G., Morford, L. A., and Fraker, P. J. (1994). Chronic levels of corticosterone reduces the number of cycling cells of the B-lineage in murine bone marrow and induces apoptosis. *Immunology* **80,** 587–592.

Goldstein, P., Ojcius, D. M., and Young, D. E. (1991). Cell death mechanisms and the immune system. *Immunol. Rev.* **12,** 29–57.

Gong, J., Li, X., and Darzynkiewicz, Z. (1993). Different patterns of apoptosis of HL-60 cells induced by cycloheximide and comptothean. *J. Cell Physiol.* **157,** 263–268.

Ormerod, M. G., Collins, M. K. L., Rodriguez-Tarduchy, G., and Robertson, D. (1992). Apoptosis in interleukin-3-dependent haemopoietic cells: Quantitation by two flow cytometric methods. *J. Immunol. Meth.* **153,** 57–65.

Sun, X. M., Snowden, R. T., Skiletter, D. N., Dinsdale, D., Ormerod, M. G., and Cohen, G. M. (1992). A flow-cytometric method for the separation and quantitation of normal and apoptotic thymocytes. *Anal. Biochem.* **204,** 351–356.

Telford, W. G., King, L. E., and Fraker, P. J. (1991). Evaluation of glucocorticoid-induced DNA fragmentation in mouse thymocytes by flow cytometry. *Cell Prolif.* **24,** 447–459.

Telford, W. G., King, L. E., and Fraker, P. J. (1992). Comparative evaluation of several DNA binding dyes in the detection of apoptosis-associated chromatin degradation by flow cytometry. *Cytometry* **13,** 137–143.

Telford, W., King, L., and Fraker, P. (1994). Rapid quantitation of apoptosis in pure and heterogenous populations using flow cytometry. *J. Immunol. Meth.* **172,** 1–16.

Vindelov, L., Christensen, I., and Nissen, N. (1983). Standardization of high resolution flow cytometric DNA analysis by the simultaneous use of chicken and trout red blood cells as internal references standards. *Cytometry* **3,** 328–331.

CHAPTER 5

Cell Cycle Analysis of Apoptosis Using Flow Cytometry

Steven W. Sherwood and Robert T. Schimke

Department of Biological Sciences
Stanford University
Stanford, California 94305

I. Introduction

Although the involvement of oncogene and tumor suppressor products (e.g., c-myc, p53, bcl-2) in apoptosis points to the importance of cell growth/proliferation control in the induction of apoptosis, the relationship of cell cycle control mechanisms to physiological cell death remains unclear (e.g., Sentman *et al.*, 1991; Williams, 1991; Evans *et al.*, 1993; Hockenberry *et al.*, 1993; Lowe *et al.*, 1993; Vaux, 1993; Yonish-Rouach *et al.*, 1993). This relationship is clearly evident in the context of cell death induced by cytotoxic agents (Barry *et al.*, 1990; Eastman, 1990), which frequently disrupt cell cycle progression in specific ways prior to apoptosis.

For many years, we have been examining the cell cycle effects of various cytotoxic drugs from the perspective of drug resistance. Because drug resistance represents the converse of cell killing, we have extended these studies to an examination of the relationship between cell cycle perturbation and cell killing (Kung *et al.*, 1990b; Schimke *et al.*,1991). More recently, the focus of these studies has been on the cytotoxic end point of drug action—apoptosis (Sherwood and Schimke, 1994; Sherwood *et al.*, 1994b). These studies have exploited the unique capabilities of flow cytometry to distinguish and isolate specific subsets of cells in heterogeneous cell populations. When combined with experimental protocols permitting analysis of cell cycle kinetics and the temporal aspects of cell response to treatment, these techniques provide a detailed picture of the relationship of cell cycle events to apoptosis within a well-defined cell cycle context. An important caveat to the latter point, however, is the difficulty in defining cell cycle states in cells undergoing perturbed cell cycles (discussed subsequently).

Using flow cytometry to analyze asynchronous and synchronized cell populations, we have observed a number of features of drug-induced apoptosis in cultured cells: (1) apoptosis can occur in any cell cycle phase; (2) a given agent can induce apoptosis in more than one cell cycle phase; (3) there is a clear temporal relationship between cell cycle "arrest" induced by different agents and the onset of apoptosis; (4) proteolysis of specific intracellular proteins represents an early event of apoptosis; (5) aberrant mitosis precedes apoptosis under a variety of conditions, including exposure to inhibitors of DNA synthesis; (6) cells with sub-G_1 DNA content are not always apoptotic cells; (7) DNA content can be an unreliable index of cell cycle position; and (8) under all conditions we have examined, cell cycle "stasis" precedes drug-induced apoptosis (Sherwood and Schimke, 1994; Sherwood, *et al.*, 1994b). In this chapter, we describe specific examples of applications of standard flow cytometric methods to the analysis of drug-induced apoptosis in HeLa S3 cells.

II. Methods

A. General Considerations for Flow Cytometry

A wide variety of physiological and structural changes associated with apoptosis are amenable to flow cytometric analysis; several of these are listed in Table I.

All the parameters listed can be analyzed using straightforward flow cytometric methods and basic technologies (ultraviolet laser is required for some applications). Because flow cytometry permits simultaneous measurement of multiple parameters on a cell-by-cell basis, the combinatorial possibilities are many. Researchers are aided in sorting through these possibilities by a plethora of general references detailing theoretical and applied aspects of flow cytometry, as well as applications of the technology to cell biological problems relevant to apoptosis (Gray and Darzynkiewicz, 1987; Shapiro, 1988; Yen, 1989; Mel-

Table I
Phenotypic Elements of Apoptosis
that Can Be Readily Analyzed by
Flow Cytometry

Physiochemical state of chromatin
DNA nicking/breaking and degradation
Proliferative/cell cycle state
Expression of specific gene products
Degradative changes in specific cellular proteins
Changes in intracellular ions (pH, Ca,$^{2+}$ others)
Plasma and mitochondrial membrane polarization
Light scatter properties/cell size
Plasma membrane permeability/cell viability

amed *et al.*, 1993; Robinson, 1993, Haugland, 1994). Additionally, several papers specifically addressing the application of flow cytometry to the study of apoptosis have been published (Darzynkiewicz *et al.*, 1992; Telford *et al.*, 1992; Afanasyev *et al.*, 1993). In this chapter, we focus our discussion on methods that permit determination of the cell cycle context of apoptosis, and we present specific examples of results obtained with these methods.

B. Flow Cytometric Analysis of the Cell Cycle Phase of Apoptosis Using Correlated DNA Content and Light Scatter Measurements

Flow cytometry has long been an important tool for the analysis of cell cycle kinetics. Establishment of general methods for staining intracellular proteins and the increasing commercial availability of antibody probes that permit detection of specific proteins simultaneously with cellular DNA content generate additional possibilities for cell cycle analysis, and allow the examination of the relationship of cell proliferation to apoptosis using flow cytometry. We have made extensive use of correlated DNA content and light scatter measurements to analyze the kinetics and cell cycle phase of apoptosis in cultured mammalian cells. We have also examined changes in intracellular Ca^{2+} associated with apoptosis, as well as proteolysis of cyclin B and other intracellular proteins in apoptotic cells. These methods are described in a later section.

1. Harvesting and Fixing Cells for DNA Content Analysis

1. Wash cells with 37°C phosphate-buffered saline (PBS) and trypsinize with 0.05% trypsin–EDTA (Gibco Gaithersburg, MD). Typically, we wash the cells with trypsin, remove trypsin, and incubate at 37°C for 1–5 min (depending on specific cell type). It is *critical* to retain the floating cells present in the population to ensure obtaining a complete picture of the cell population; therefore, *pool floating cells and trypsinized attached cells*. Centrifuge cells 5 min to pellet

(1200 rpm in a clinical desktop centrifuge). Note that under some conditions, separate analysis of floating cells and adherent cells may be useful.

2. Wash cell pellet in 1 : 1 mixture of PBS and McIlvaine's buffer 0.2 M Na_2HPO_4, 0.1 M citric acid, pH 7.5). Triturate cells 3–4 times with a pipetter and add 2 volumes cold (4°C) ethanol, with gentle mixing. Cells fixed in this manner can be stored for prolonged periods (>1 month) at 4°C. Cell fixation for simultaneous analysis of DNA content and immunochemical staining for analysis of cyclin B protein are described next.

2. Cell Staining

1. Pellet fixed cells, decant ethanol, and wash cells in distilled water (approximately 1 ml/10^6 cells).

2. Pellet cells again and resuspend in staining solution. Incubate a minimum of 1 hr at 37°C.

3. After staining, pellet cells, and resuspend in fresh propidium iodide staining solution. [PBS containing 100 μg/ml RNase A (DNase-free) and 10 μg/ml propidium iodide (PI)].

Adequate RNase treatment is essential for good results with PI, which binds to RNA as well as DNA. RNase must be free of DNases, which can be achieved by boiling stock solution of RNase (10 mg/ml in 10 mM Tris, pH 7.5, 10 mM NaCl) for 10 min (Sambrook et al., 1989). Chromomycin A3 can be used to stain DNA in fixed cells and does not require RNase treatment, although buffer composition is important for optimal staining (Crissman et al., 1979). Viable cells (i.e., plasma membrane intact) can be stained for DNA content with Hoechst 33242. Fluorochrome binding to DNA can reflect DNA base composition and also can be affected by the physical state of chromatin (Crissman et al., 1979; Darzynkiewicz, 1990; Latt and Langlois, 1990; Telford et al., 1992). Hence different DNA dyes may give somewhat different pictures of chromatin changes during apoptosis. All reagents are available through common sources (e.g., Sigma Chemical Co., St. Louis, MO; Molecular Probes, Eugene, OR).

3. Analysis

Cells are analyzed suspended in staining solution. DNA content and light scatter are measured simultaneously as correlated parameters. Data are displayed as univariate plots of individual parameters as well as bivariate plots of DNA content against scatter.

a. Fluorescence
PI-stained cells are analyzed using 488-nm excitation. The PI signal is filtered through a 635-nm band-pass filter to ensure removal of scattered light Linear amplification of fluorescence signals is utilized. Pulse processing electronics

(peak integral vs peak height) are used to process the DNA signals and remove cell doublets from analysis by gating.

b. Light Scatter

Both forward angle (1–19°) and orthogonal (side) light scatter signals (FALS and 90LS, respectively) are collected as correlated parameters of DNA content. Although light scatter theory is complex, change in cell "size" is reflected in light scatter changes (Salzman *et al.*, 1990) and we have observed very good linear correlation of light scatter (the arithmetic mean of scatter distributions for both FALS and 90LS, jet-in-air flow cell configuration) with total cellular protein content and Coulter volume over a wide range of cell sizes (Kung *et al.*, 1993; Sherwood *et al.*, 1994a). We would stress that our results are derived from measurements of cells of a single type (e.g., HeLa S3, CHO, or 3T3) undergoing unbalanced cell growth and not from comparisons of different cell types. In the latter situation, the correlation of scatter to cell volume can become highly variable (see Freyer and Habbersett, 1994). Thus, although scatter signals contain more than size information, light scatter provides a useful estimate of cell size. Under the conditions we describe, this estimation is helpful in identifying apoptotic cells, which undergo "shrinkage" or "compaction" as part of the apoptotic process (Kerr *et al.*, 1972; Wyllie *et al.*, 1980). Additionally, this estimation provides a measure of cell volume or total cellular protein content to which immunofluorescence measurements of specific proteins can be normalized (see subsequent discussion; see also Jaccobberger, 1989; Sherwood *et al.*, 1994a). Note that differences among cytometers with respect to scatter detectors and optical geometry may result in patterns different from those shown.

4. Results

a. Apoptosis Induced by Cell Cycle "Arrest"

Apoptosis induced by cell cycle "arresting" concentrations of colcemid and aphidicolin is shown in Figs. 1 and 2, respectively. Under these conditions, cells arrest at a cell cycle checkpoint identified by cell response to antitubulin drugs (Kung *et al.*, 1991; Hoyt *et al.*, 1992; Li and Murray, 1992) or at the G_1/S transition, a cell cycle transition also associated with checkpoint function. Apoptotic cells are clearly distinguishable as distinct subpopulations using DNA content and light scatter measurements, whatever the cell cycle phase of cells at the time of apoptosis. Coulter volume distributions are shown for mitotically arrested cells undergoing apoptosis. The data show that this measure of cell size is similar to 90LS (and to FALS, to a lesser degree). Single-parameter DNA histograms for aphidicolin-induced apoptosis at the G_1/S boundary demonstrate the sub-G_1 DNA content of cells undergoing apoptosis from the G_1/S boundary (Fig. 2, but see subsequent discussion). The frequency of apoptotic cells, determined by integrating the area "occupied" by apoptotic cells in DNA vs scatter histograms, is well correlated with the frequency of apoptotic cells

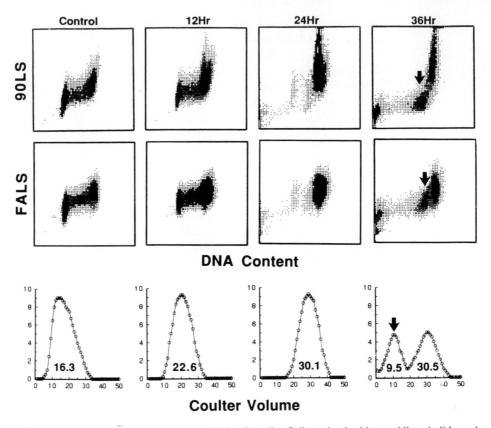

Fig. 1 Colcemid-induced apoptosis in HeLa S3 cells. Cells stained with propidium iodide and correlated DNA are plotted against light scatter data collected. (*Top*) DNA vs 90LS (*Bottom*) DNA vs FALS histograms. Cell frequency/channel (z axis) is indicated by intensity of stippling. Asynchronous cultures were treated with 70 ng/ml colcemid and harvested at the times indicated. Shown below the histograms are Coulter volume distributions for the cell populations shown, with the mean channel of the distributions (y axis = cells/channel $\times$ 50). Apoptotic cells are indicated by arrows.

determined by scoring apoptotic cells (on the basis of nuclear morphology) using fluorescence microscopy (Fig. 3).

b. Apoptosis Associated with Aberrant Mitosis

Under certain conditions of drug exposure, the appearance of cells with sub-G_1 DNA content (as in Fig. 2) is not directly associated with apoptosis, but with an aberrant multipolar mitosis that precedes cell death and apoptosis (Fig. 4). We have observed this phenomenon in HeLa (and other) cells after prolonged

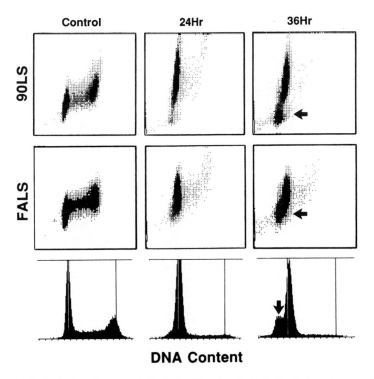

Fig. 2 Aphidicolin-induced apoptosis in HeLa S3 cells. Aphidicolin-induced apoptosis occurs in cells at the G_1/S boundary approximately 18 hr after cells become arrested with a G_1/S DNA content. Asynchronous cultures were treated with 2 $\mu g/ml$ aphidicolin and harvested at the times shown. Prior to apoptosis, cells have a G_1/S DNA content but have undergone unbalanced cellular growth and therefore begin to undergo apoptosis at a cell size similar to that of mitotic cells. Single parameter DNA histograms are shown at the bottom. Apoptotic cells are indicated by arrows.

exposure to very low concentrations of aphidicolin, colcemid, and cisplatin; pulse exposure to high concentrations of colcemid (Sherwood *et al.*, 1994b); and continuous exposure to some protease inhibitors (Sherwood *et al.*, 1993, 1994). The cytotoxicity of other agents has also been reported to involve aberrant mitosis (Vidair *et al.*, 1992); we suspect this phenomenon to represent a fairly widespread mechanism of drug toxicity. The importance of the phenomenon in the context of utilizing flow cytometry to analyze apoptosis is that the initially viable hypodiploid daughter cells of such mitoses closely resemble apoptotic cells in both size and DNA content. Although these microcells subsequently undergo apoptosis (approximately 18 hr postmitosis in HeLa S3 cells), at the time of their formation the cells are intact, viable, and nonapoptotic. Confusing so-called microcells with apoptotic cells will, therefore, lead to misin-

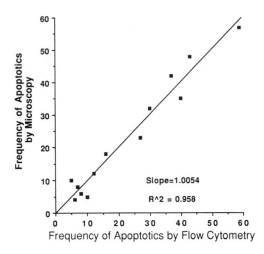

Fig. 3 Correlation of the frequency of apoptotic cells determined by flow cytometry and by microscopic examination using propidium iodide staining and fluorescence microscopy. Apoptosis was induced in HeLa S3 cells using several different drugs. The frequency of apoptotic cells was determined by analysis of DNA vs 90LS histograms. These same samples were subsequently scored for the frequency of apoptotics by fluorescence microscopy, using nuclear morphology to identify apoptotic cells.

terpretation of the mechanism of cell killing and the cell cycle context of apoptosis.

C. Simultaneous Analysis of Specific Intracellular Proteins and DNA Content during Apoptosis: Cyclin B

Various studies have shown that proteolysis of specific intracellular proteins is a significant process early in apoptosis (Bruno *et al.,* 1992; Kaufmann *et al.,* 1993; Miller *et al.,* 1993a,b). Investigators have suggested that proteolysis of nuclear matrix proteins may represent a key step in initiating DNA degradation (Bruno *et al.,* 1992; Tomei *et al.,* 1993). Therefore to be able to measure changes in specific protein levels under conditions inducing apoptosis would be useful. When antibody probes suitable for use in whole cells are available, immunofluorescence measurement of protein levels can be combined with measurements of DNA content and light scatter, providing detailed information of the kinetics of proteolysis, cell size/morphological changes (light scatter), and DNA content. We have used flow cytometry to examine cyclin B protein levels in mitotically arrested cells undergoing apoptosis as well as changes in the nuclear matrix protein NuMA, and to observe proteolysis of these proteins during apoptosis. We describe our method for analyzing cyclin B levels during apoptosis.

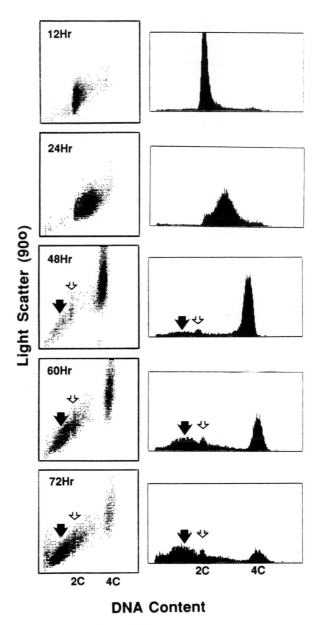

DNA Content

Fig. 4. Microcell mitosis in synchronized HeLa S3 cells treated with very low concentrations of aphidicolin. Synchronized HeLA cells were grown in the presence of 0.2 μg/ml aphidicolin. Cell cycle progression is strongly delayed (normal cell cycle time 18 hr), resulting in multipolar mitoses. Daughter microcells have G_1 DNA content and size less than that of normal G_1 cells. Microcells undergo apoptosis beginning 18 hr after their formation. Solid arrow, microcells; open arrow, G_1 cells.

1. Harvesting and Fixing Cells for Immunochemical Detection of Cyclin B Protein

The method for immunochemical measurement of cyclin B protein in flow cytometry has been published (Gong *et al.*, 1993a,b; Sherwood *et al.*, 1994a); the reader is directed to these references for details.

1. Trypsinize and wash cells with PBS (4°C).
2. Resuspend cells in PBS containing 1% or 3% paraformaldehyde (4°C). When fixed with 1% paraformaldehyde, cells can be stored at 4°C overnight before staining. Cells fixed with 3% paraformaldehyde must be tested the same day, since prolonged fixation reduces antibody binding and, hence, staining. The virtue of the former method is that it allows a number of samples to be accumulated for staining at one time, minimizing measurement variation.

2. Cell Staining

1. Wash fixed cells free of paraformaldehyde with 4°C PBS and resuspend in permeabilization buffer (150 mM NaCl, 1 mM HEPES, pH 7.0, 4% fetal bovine serum (FBS), 0.1% Triton X-100) on ice for 10 min.
2. Gently wash cells with PBS and pellet. Resuspend in primary antibody solution [PBS containing 200 μg/ml RNase A, 10 μg/ml PI, 1% bovine serum albumin (BSA), anti-cyclin B antibody (Pharmingen, San Diego, CA; 1 : 250 dilution, approx. 0.5 μg/ml final concentration].
3. Incubate cells for 1–8 hr (can go overnight) at room temperature.
4. Wash in large volume (>10× staining volume) PBS containing BSA (1%).
5. Resuspend in staining buffer containing approximately 1 μg/ml final concentration FITC-labeled goat anti-mouse IgG and incubate for 0.5–1 hr at room temperature.
6. Wash cells in large volume of PBS.
7. Resuspend in PBS containing RNase (200 μg/ml), PI (10μg/ml), and BSA (1%). Cells retain good staining for at least 24 hr if kept at 4°C.

3. Analysis

Cells are analyzed using standard filters for simultaneous "red and green" analysis. Filter FITC signal through a band-pass filter (we utilize a 525-nm/20-nm band-pass filter) and PI through >610-nm filter (we use a 635-nm/20-nm band-pass filter). Data are collected as DNA vs cyclin B and DNA vs scatter (90LS and FALS) correlated histograms, as well as univariate histograms of all signals. Data can also be collected in listmode, which is particularly useful for measuring ratios of changes in specific protein content relative to changes in light scatter (cell size). Pulse processing electronics are used for gating (DNA signal) to remove cell doublets from analysis.

4. Results

An example of apoptosis-associated cyclin B degradation occurring in cells arrested in mitosis (induced by colcemid) is shown in Fig. 5. Cyclin B protein can be seen to accumulate and subsequently undergo a rapid decline as apoptosis begins (Fig. 5). The large "gap" between 4C cells with high levels of cyclin B and those with essentially background levels indicates that cyclin proteolysis occurs rapidly, relative to cellular changes associated with loss of fluorochrome binding to DNA. The relative change in cyclin B content in apoptotic cells is approximately 6-fold, whereas the change in light scatter is 2-fold and in Coulter volume is 3-fold. Hence, the decline in cyclin B content is not simply a reflection of apoptotic cell shrinkage. This pattern is also seen in cells immunochemically stained for the nuclear matrix protein NuMA (S. W. Sherwood, unpublished data).

D. Measurement of Intracellular Ca^{2+} Changes Associated with Drug-Induced Cell Cycle "Arrest"

The onset and/progression of apoptosis is known to be associated with elevation of intracellular free calcium levels in a wide variety of experimental systems. Although the exact role of Ca^{2+} in apoptosis remains somewhat unclear, we were interested in determining whether or not this physiological change represents an element of apoptosis associated with drug-induced cell cycle perturbation. The availability of several Ca^{2+} chelating molecules that undergo changes in their fluorescence excitation and emmission properties makes Ca^{2+} change measurements by flow cytometry quite simple. We have utilized fluo-3, which has spectral properties very similar to those of fluorescein (FITC), and indo-1 to examine Ca^{2+} changes occurring in cells arrested in cell cycle progression by antitubulin agents as well as by inhibitors of DNA synthesis. Because fluo-3 staining does not require an ultraviolet excitation source, we describe our

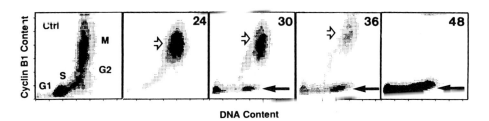

Fig. 5 Degradation of cyclin B protein in HeLa S3 cells undergoing apoptosis in mitosis, resulting from mitotic arrest induced by colcemid. Asynchronous cultures were treated with 70 ng/ml colcemid for times shown; then cells were harvested and fixed for simultaneous DNA and cyclin B staining as described. The first panel shows untreated cells with cell cycle phases indicated. Open arrows mitotically arrested cells; closed arrows, apoptotic cells.

method for combined analysis of cell size (light scatter) and staining with propidium iodide (a charged molecule that will only enter cells with permeabilized
plasma membranes) to determine the relationship between changes in cell size,
Ca^{2+}, and plasma membrane permeability. The method described is similar to
that of Deckers *et al.,* (1993).

1. Dye Preparation and Cell Loading

Fluo-3 can be obtained from Molecular Probes, and a 1 mM stock made
with 100% dimethylsulfoxide (DMSO). PI is dissolved in water at 10 mg/ml
concentration. Cells are incubated (37°C) with Fluo-3 for 1 hr immediately prior
to analysis using a final concentration of 1 μM in culture medium. After dye
loading, cells are washed, trypsinized, and resuspended in PBS containing 4%
FBS plus 2–5 μg/ml PI, and kept on ice until analysis. Immediately following
analysis, cells are fixed with ethanol (70% final concentration) for subsequent
cell cycle analysis using PI to stain for DNA content, as described earlier.

2. Analysis

The machine set-up used for this analysis is identical to that described earlier
for immunofluorescence analysis of protein content, with the exception that
the fluo-3 and PI signals are logarithmically amplified. Data are collected as
correlated histograms (PI vs FALS, Fluo-3 vs FALS). Listmode data are also
collected to examine the level of intracellular free Ca^{2+} relative to the functional
state of the plasma membrane (permeable/"intact").

3. Results

Figure 6 shows the changes induced by aphidicolin in HeLa S3 cells undergoing apoptosis at the G_1/S boundary (see Fig. 2). Shown are histograms for
cell populations 24–48 hr after initiation of continuous exposure to 2 μg/ml
aphidicolin. The top three panels (PI content vs cell size) represent changes in
plasma membrane permeability. The figure shows that with increasing time of
drug exposure, progressively greater numbers of cells take up PI. By 48 hr of
treatment, essentially all cells have become permeable to PI. The middle three
panels (fluo-3 vs cell size) show that a similar increase in the frequency of cells
with elevated free intracellular Ca^{2+} occurs during apoptosis. The bottom three
panels show cell cycle analysis using PI staining of cellular DNA content (as
described already) for the same samples. These panels can be compared with
those shown in Fig. 2. A similar change in Ca^{2+} levels accompanies apoptosis
induced in mitotically arrested cells.

The use of listmode analysis to examine the relationship between changes
in intracellular Ca^{2+} and plasma membrane permeability is shown in Fig. 7. In
this figure, the same sample as the 36-hr time point in Fig. 6 was collected in

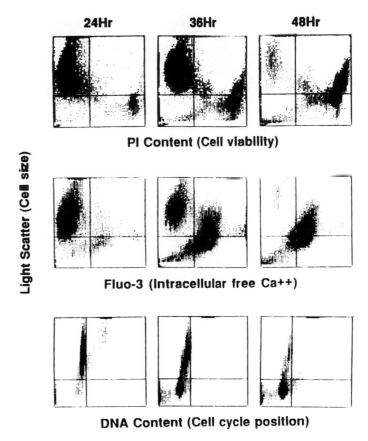

Fig. 6 Correlated changes in plasma membrane permeability and intracellular free calcium during aphidicolin-induced cell cycle arrest and apoptosis. Asynchronous cultures were continuously exposed to drug and were harvested at the times indicated. Cells were loaded with 1 μM fluo-3 1 hr prior to harvest. Immediately prior to analysis, 2 $\mu g/ml$ propidium iodide (PI) was added to the cells. Following analysis, cells were fixed by addition of ethanol to 70% final concentration and were re-analyzed for cell cycle position as described earlier.

listmode format. The intracellular calcium levels for different subsets of cells were then determined by gating on the PI vs FALS histogram (i.e., on cell viability measured by vital dye exclusion). The Ca^{2+} levels in the total (ungated) population and three subsets of cells (PI-excluding viable cells; unidentified cells intermediate in PI content; and PI-permeable "dead" cells) are shown as fluo-3 vs FALS (Ca^{2+} content vs cell size) and univariate fluo-3 (Ca^{2+}) distributions. This analysis demonstrates the significant heterogeneity in Ca^{2+} level in the total cell population of asynchronous cells undergoing apoptosis. It is also possible to establish that Ca^{2+} and cell size changes occur with similar kinetics (within the time frame shown) and that both changes precede loss of

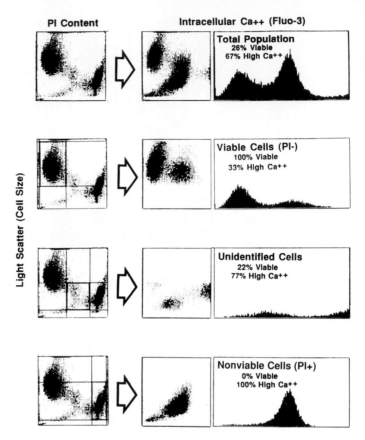

Fig. 7 Analysis of intracellular free Ca^{2+} as a function of cell viability. Data shown in Fig. 6 were collected in list-mode format and re-analyzed by gating on propidium iodide (PI) content. (*Top*) Total cell population. (*Bottom*) Three subpopulations based on PI content (plasma membrane permeability to PI) were isolated by gating; Ca^{2+} levels and cell size were measured.

membrane permeability. By sorting viable cells with elevated Ca^{2+}, we have found that changes in nuclear morphology precede or are coincident with increased Ca^{2+} levels (M. T. Boitrell, unpublished observations). Thus, using this protocol, the temporal relationships between several aspects of cellular changes associated with drug-induced apoptosis are rapidly obtained.

III. Comments

Using a variety of simple flow cytometric methods, we have "mapped" drug-induced apoptosis within the context of cell cycle progression in HeLa S3 cells treated with a variety of therapeutically relevant drugs. The results of experiments specifically addressing the question of when in the cell cycle apoptosis begins are summarized in the diagrams in Fig. 8. Shown are the positions of apoptotic cell populations (open ellipses) beginning at the point of cell cycle "arrest" and at time points thereafter, as cells proceed to eventual disintegration via apoptotic body formation. The data from which this figure was constructed were obtained using synchronized cell populations (cells synchronized by mitotic shake-off, which does not perturb cell cycle progression; see Sherwood *et al.,* 1994 a,b) but they are shown superimposed on the distribution of asynchronous cells (G_1 and G_2/M cells, black areas; S phase, between dotted lines). The numbers shown in the apoptotic subpopulations indicate the

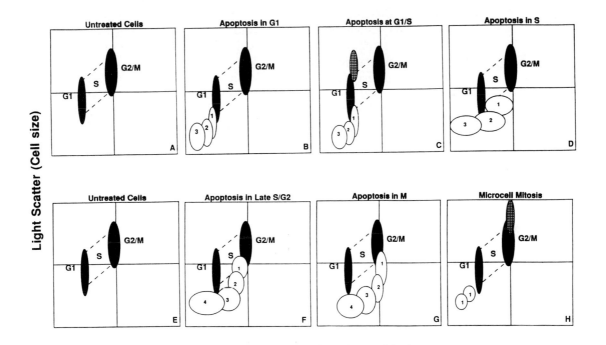

DNA Content (Cell cycle position)

Fig. 8 Changes in the position of apoptotic cells in DNA histograms during continued progression of apoptosis. Apoptosis was induced by different drugs in synchronized HeLa S3 cells and analyzed using DNA staining and light scatter analysis. The positions of synchronous populations of apoptotic cells in the phenotypic space defined by propidium iodide (PI) fluorescence and 90LS are indicated by the numbered stippled areas (1 being earliest). Solid black areas represent normal G_1 (2C) and G_2/M (4C) cells for reference. (C, H) Populations shown in cross-hatching are abnormally large, reflecting unbalanced cellular growth prior to apoptosis. Treatments shown are (A) control; (B) nutrient starvation; (C) high dose aphidicolin; (D) moderate-dose aphidicolin, low- and high-dose cisplatin; (E) control; (F) low-dose aphidicolin; (G) high-dose colcemid; (H) low-dose continuous colcemid, high-dose pulse exposure colcemid, very low dose aphidicolin, protease inhibitor (acetyl-leucyl-leucyl-norleucinal). Compare these diagrams with data shown in Figs. 1, 2, and 4.

temporal order in which apoptotic cells are seen at different times after the onset of apoptosis (1 being the earliest time point). Unbalanced growth may precede onset of apoptosis so that, prior to the onset of apoptosis, the treated cells are larger than normal (indicated in the figure by crosshatched areas in C and H). The conditions of treatment under which we have observed each example of apoptosis are described in the legend.

In general, we have used DNA content and cell size to describe the cell cycle stage at the time of cell cycle "arrest" and the onset of "stasis" (Sherwood and Schimke, 1994; Sherwood et al., 1994b). We also have used a variety of morphological and biochemical features in addition to flow cytometric analysis of cell size and DNA content. We would caution, however, that under such perturbed conditions, definition of cell cycle state can become problematic. For example, CHO cells treated with aphidicolin will accumulate cyclin B to high levels and become mitotically "competent," even though the cells have the DNA content of G_1/early S phase cells. HeLa S3 cells behave differently; although they grow to mitotic size while "arrested" at the G_1/S boundary, they do not accumulate cyclin B under identical circumstances of drug exposure (Schimke et al., 1992; Kung et al., 1993). Other cyclins show similar differences as a function of cell line (L. Urbani, unpublished observations). Consequently, although we utilize DNA content to define where in the cell cycle apoptosis occurs, we do so with an awareness of the indeterminate nature of such descriptions. We would also anticipate significant cell line differences in these patterns and although we have not examined this question in detail, results with CEM (lymphoblastic leukemia) and Tera-2 (embryonic carcinoma) cells tend to show less extreme changes in light scatter, presumably reflecting the larger nucleocytoplasmic ratio of these cells. Having established these patterns with synchronized cell populations, we now find that it is possible to identify quickly significant features of cell cycle perturbation and the induction of apoptosis by cytotoxic agents in cell populations treated during asynchronous growth.

The single most obvious aspect of the temporal changes in these histograms is that apoptotic cells move "down and to the left," reflecting cell shrinkage and loss of the DNA fluorochrome binding that accompanies chromatin changes (chromatin condensation and degradation). Both changes are prominent features of the apoptotic process. All cells eventually end up as debris near the origin of the axes. Note that because of these temporal changes in the histograms, the picture of apoptosis obtained will depend on the degree of specificity and synchrony in the induction of apoptosis and on the time frame in which it is analyzed. Thus, we emphasize that studying apoptosis in flow cytometry is most productive when it includes kinetic analysis of the process. The importance of adequate temporal analysis is indicated in experiments with cisplatin in which we observed apoptosis in (synchronous) S phase cells in both the first S phase after initiation of continuous exposure to an LD_{99} concentration and in the second S phase, when cells were exposed to an LD_{50} concentration of the drug (Sherwood et al., 1994b).

The movement of apoptotic cells through the phenotypic space defined by DNA content and light scatter may be important when using single parameter DNA histograms in tumor diagnosis/prognosis. Becasue apoptotic cells originating in late S, G_2, or M phase come to underlie cycling S phase cells, failure to account for this source of "contamination" of the S phase fraction can lead to misleading estimates of the fraction of cells in S phase using single parameter DNA histogram analysis alone. For example, cell cycle analysis based on single parameter DNA histograms for the histograms shown in Fig. 1 give S phase estimates of 32, 17, 19 and 36% for the control, 12-, 24-, and 36-hr time points, respectively. By 24–36 hr of exposure to this concentration of colcemid, essentially all of the cycling cells present will have accumulated in mitosis; hence the calculated 36% "S phase" fraction is composed entirely of apoptotic cells. Apoptosis may also represent a soruce of cells that fail to label with bromodeoxyuridine but have S phase DNA content.

Aberrant (multipolar) mitosis presents a flow cytometric picture similar or identical to that of apoptosis in G_1 or G_1/S cells (Sherwood et al., 1994a,b). because in both situations, the cells of interest (microcells or apoptotic cells) are seen as distinct populations of cells with a DNA content and size less than those of G_1 cells (see Figs. 1, 2, and 5). Although the microcells do eventually undergo apoptosis, at the time of their formation they are intact, viable cells (morphologically and as measured by vital dye exclusion and cytoplasmic esterase activity). Distinguishing microcells from apoptotic cells is readily done by fluorescence microscopy utilizing DNA fluorochromes to examine nuclear structure and/or by antitubulin immunofluorescence staining of mitotic spindles. Obviously, as microcells undergo apoptosis, this distinction becomes more problematic. Thus, differentiating between microcells and apoptotic cells must be done relatively soon after multipolar mitosis occurs. This distinction is very important in defining mechanisms of cell killing and induction of apoptosis, the cell cycle context of apoptosis, and the timing of apoptosis relative to cellular events, particularly since this event may represent a fairly widespread aspect of antiproliferative drug action. The mechanism of this process involves disturbances in centrosome/centriole replication associated with altered cell cycle progression (S. W. Sherwood, unpublished observations). Although some potential exists for confusing microcells with aneuploid subpopulations (we frequently obtain two very distinct subpopulations of microcells; see Sherwood and Schimke, 1994), in general this possibility can be eliminated based on the relatively high variance in DNA content of microcell populations (measured as the coefficient of variation, CV) and the absence of a corresponding hypotetraploid peak.

Although quantification of intracellular molecules by flow cytometry is less sensitive than methods such as immunoblotting, quantitative data can be very reliable (for discussion, see Jaccobberger, 1989; Sherwood et al., 1994a). Because such data can be collected in a correlated multiparameter manner, the temporal relationships between the events of apoptosis can (sometimes) be established relatively easily. For example, we can deduce from the data shown in Fig. 4 (1) that cyclin B disappears rapidly from apoptotic cells and (2) that it does so prior to cells undergoing chromatin changes large enough to be visualized as a ''loss'' of DNA staining. Such conclusions require collecting correlated DNA content, cell size (light scatter), and cyclin B immunofluorescence (including isotype controls) measurements to ensure that observed changes in protein content are specific changes and do not simply reflect cell size changes.

In similar fashion, it is easily demonstrated that elevations in intracellular free Ca^{2+} and cell shrinkage are closely correlated in time, and that both precede loss of membrane permeability (Figs. 6 and 7). Measurement of apoptosis-associated changes in intracellular free Ca^{2+} levels are also enhanced by the multiparameter capabilities of flow cytometry. Qualitative Ca^{2+} measurements are quite simple, and quantitative measurements can be made with rigorous calibration and appropriate controls. Analyses such as those shown provide simple means by which to establish rapidly the temporal order of cellular events associated with apoptosis. In principle, the analysis shown in Fig. 6 could be performed at one time using the DNA stain Hoechst 33242, fluo-3, and PI and a dual laser (UV + 488 nm) cytometer. Perhaps the most significant point regarding the analysis shown in Fig. 7 is that total population measurements of parameters such as Ca^{2+} level (e.g., as would be obtained with a fluorimeter) will ''average across'' heterogeneity in the populations, precluding accurate assessment of the cellular changes occurring during apoptosis. As shown, the Ca^{2+} distributions of viable and inviable cells are very different and a whole-population measurement of Ca^{2+} levels is significantly less informative.

In conclusion, the multiparameter capabilities of flow cytometry are well suited to analyzing the events of apoptosis. Because phenotypic heterogeneity in cell populations can be made readily apparent by flow cytometry, the multiparameter possibilities frequently can compensate for the lower quantitative sensitivity inherent in some flow cytometric measurements. Often, relatively simple and well-established methods quickly provide a substantial amount of information. Clearly, cultured cells represent ''ideal'' material for such analysis. An important issue is the extent to which these kinds of analyses can be applied to studying apoptosis in other kinds of material. We have utilized the methods described here to examine developmentally regulated apoptosis in embryonic mouse germ cells (Coucouvanis *et al.*, 1993) and have observed similar patterns of change during insect gametogenesis (Sherwood *et al.*, 1989). Early reports point to the feasibility of using such methodologies to examine the induction of apoptosis by therapeutic agents *in vivo* (Gorczya *et al.*, 1993). We would expect that further analyses will show that flow cytometry, particularly in combination with imaging techniques, will prove useful in understanding apoptosis in the contexts of developmental biology as well as in tumor biology and in prognosis and analysis of tumor response to therapy.

Acknowledgments

We thank Daphne Rush, Jamie Sheridan, Andrew Kung, Catherine Carswell-Crumpton, Lal Yilmaz, and Louie Kim for their significant contributions to this work. A special thanks to Barbara Osborne for her patience and comments during revision of the chapter. This work was supported by National Institutes of Health Grant CA16318 to R. T. Schimke.

References

Afanasyev, V. N., Korol, B. A., Matylevich, N. P., Pechatnikou, V. A., and Umansky, S. R. (1993). The use of flow cytometry for the investigation of cell death. *Cytometry* **14(6)**, 603–609.

Barry, M. A., Behnke, C. A., and Eastman, A. (1990). Activation of programmed cell death (apoptosis) by cisplatin, other anticancer drugs, toxins and hyperthermia. *Biochem. Parmacol.* **40(10)**, 2353–2362.

Bruno, S., Del Bino, G., Lassota, P., Giarretti, W., and Drazynkiewicz, Z. (1992). Inhibitors of proteases prevent endonucleolysis accompanying apoptotic death of HL-60 leukemic cells and normal thymocytes. *Leukemia* **6(11)**, 1113–1120.

Coucouvanis, E. C., Sherwood, S. W., Carswell-Crumpton, C., Spaack, E. G., and Jones, P. P. (1993). Evidence that the mechanism of prenatal germ cell death in the mouse is apoptosis. *Exp. Cell Res.* **109**, 238–247.

Crissman, H. A., Stevenson, A. P., Kissane, R. J., and Tobey, R. A. (1979). Techniques for quantitative staining of cellular DNA for flow cytometric analysis. *In* "Flow Cytometry and Sorting" (M. R. Melamed, P. F. Mullaney, M. L. Mendelsohn), pp. 243–261. John Wiley and Sons, New York.

Darzynkiewicz, Z. (1990). Probing nuclear chromatin by flow cytometry. *In* "Flow Cytometry and Sorting" (M. R. Melamed *et al.*, eds.), pp. 315–340. Wiley-Liss, New York.

Darzynkiewicz, Z., Bruno, S., Del Bino, G., Gorczyca, W., Hotz, M. A., Lassota, P., and Traganos, F. (1992). Features of apoptotic cells measured by flow cytometry. *Cytometry* **13(8)**, 795–808.

Deckers, C. L. P., Lyons, A. B., Samuel, K., Sanderson, A., and Maddy, A. H. (1993). Alternative pathways of apoptosis induced by methylprednisolone and valinomycin analysed by flow cytometry. *Exp. Cell Res.* **208**, 362–370.

Eastman, A. (1990). Activation of programmed cell death by anticancer agents: Cisplatin as a model system. *Cancer Cells* **2**, 275–280.

Evans, G., Wyllie, A. H., Gilbert, C. S., Littlewood, T. D., Land, H., Brooks, M., Waters, C. M., Penn, L. Z., and Hancock, D. C. (1992). Induction of apoptosis in fibroblasts by c-*myc* protein. *Cell* **69**, 119–128.

Freyer, J. P., and Habbersett, R. C. (1994). Utility of electronic cell volume compared to light scatter. *National Flow Cytometry Resource Newsletter,* February.

Gong, J., Traganos, F., and Darzynkiewicz, Z. (1993a). Expression of cyclins B and E in individual MOLT-4 cells and in stimulated human lymphocytes during their progression through the cell cycle. *Int. J. Oncol* **3**.

Gong, J., Traganos, F., and Darzynkiewicz, Z. (1993b). Simultaneous analysis of cell cycle kinetics at two different ploidy levels based on DNA content and cyclin B measurements. *Cancer Res.* **53**, 5096–5099.

Gorcyza, W., Bigman, K., Mittleman, A., Ahmed, T., Gong, J., Melamed, M. R., and Darzynkiewicz, Z. (1993). Induction of DNA strand breaks associated with apoptosis during treatment of leukemias. *Leukemia* **7(5)**, 659–670.

Gray, J. W., and Darzynkiewicz, Z. (eds.) (1987). "Techniques in Cell Cycle Analysis." Humana Press, Clifton, New Jersey.

Haugland, R.P. (1994). "Molecular Probes Handbook of Fluorescent Probes and Research Chemicals." Molecular Probes, Eugene, Oregon.

Hockenberry, D., Nunez, G., Milliman, C., Schreiber, R. D., and Korsmeyer, S.J. (1990). Bcl-2 is an inner mitochondrial membrane protein that blocks programmed cell death. *Nature* **348,** 334–336.

Hoyt, M. A., Totis, L., and Roberts, B. T. (1991). *S. cerevisiae* genes required for cell cycle arrest in response to loss of microtubule function. *Cell* **66,** 507–517.

Jaccobberger, J. W. (1989). Cell cycle expression of nuclear proteins. *In* "Flow Cytometry: Advanced Research and Clinical Applications." (A. Yen, (ed.), Vol. 1, pp. 305–326. CRC Press, Boca Raton, Florida.

Kaufmann, S. H., Desnoyers, S., Ottaviuano, T., Davidson, N. E., and Poirer, G. G. (1993). Specific proteolytic cleavage of poly(ADP-ribose) polymerase: An early marker of chemotherapy induced apoptosis. *Cancer Res.* **53,** 3972–3985.

Kerr, J. F., Wyllie, A. H., and Currie, A. R. (1972). Apoptosis: A basic biological phenomenon with wide-ranging implications in tissue kinetics. *Br. J. Cancer* **26,** 239–257.

Kung, A. K., Sherwood, S. W., and Schimke, R. T. (1990a). Cell-line differences in the control of cell cycle progression in the absence of mitosis. *Proc. Natl. Acad. Sci. USA* **87,** 9553–9557.

Kung, A. K., Zetterberg, A., Sherwood, S. W., and Schimke, R. T. (1990b). Cytotoxic effects of cell cycle phase specific agents: Results of cell cycle perturbation. *Cancer Res.* **50,** 7307–7317.

Kung, A. K., Sherwood, S. W., and Schimke, R. T. (1993). Differences in regulation of protein synthesis, cyclin B acccumulation and cellular growth in response to inhibition of DNA synthesis in Chinese hamster ovary and HeLa S3 cells. *J. Biol. Chem.* **268(31),** 23072–23080.

Latt, S. A., and Langlois, R. G. (1990). Fluorescent probes of DNA microstructure and DNA synthesis. *In* "Flow Cytometry and Sorting," 2d Ed. (M. R. Melamed *et al.*), pp. 249–290. Wiley-Liss, New York.

Li, R., and Murray, A. W. (1991). Feedback control of mitosis in budding yeast. *Cell* **66,** 519–531.

Lowe, S. W., Schmitt, E. M., Smith, S. W., Osborne, B. A., and Jacks, T. (1993). p53 is required for radiation-induced apoptosis in mouse thymocytes. *Nature* **362,** 847–852.

Melamed, M. R., Mullaney, P. F., and Mendelsohn, M. L. (1979). "Flow Cytometry and Sorting," 1st Ed. John Wiley and Sons, New York.

Melamed, M. R., Lindmo, T., and Mendelsohn, M. L. (1990). "Flow Cytometry and Sorting," 2d Ed. Wiley-Liss, New York.

Miller, T. E., Beausang, L. A., Winchell, L. F., and Lidgard, G. P. (1993a). Detection of nuclear matrix proteins in serum from cancer patients. *Cancer Res.* **52,** 422–427.

Miller, T. E., Beausang, L. A., Meneghini, M., and Lidgrad, G. P. (1993b). Death induced changes to the nuclear matrix: The use of anti-nuclear matrix antibodies to study agents of apoptosis. *BioTechniques* **15(6),** 1042–1047.

Robinson, J. P. (ed.) (1993). "Handbook of Flow Cytometric Methods." John Wiley-Liss, New York.

Salzman, G. C., Singham, S. B., Johnston, R. C., and Bohren, C. F. (1990). Light scattering and cytometry. *In* "Flow Cytometry and Sorting." (M. R. Melamed, T. Lindmo, M. L. Mendelsohn eds.) pp. 81–108. Wiley-Liss, New York.

Sambrook, J., Fritsch, E. F., and Maniatis, T. (1989). "Molecular Cloning." Cold Spring Harbor Laboratory Press, Cold Spring Harbor, New York.

Schimke, R. T., Kung, A. K., Rush, D. R., and Sherwood, S. W. (1991). Differences in mitotic control among mammalian cells. *Cold Spring Harbor Symp. Quant. Biol.* **56,** 417–425.

Sentman, C. L., Shutter, J. R., Hockenberry, D., Kanagawa, O., and Korsmeyer, S. J. (1991). Bcl-2 inhibits multiple forms of apoptosis but not negative selection in thymocytes. *Cell* **67,** 879–888.

Shapiro, H.M. (1988). "Practical Flow Cytometry." Liss, New York.

Sherwood, S. W., and Schimke, R. T. (1994). The induction of apoptosis by cell cycle phase specific drugs. *In* "Apoptosis" (R. T. Schimke and E. Mihich, (eds.) Proc. 6th Pezcoller Foundation Symposium Plenum Press, N. Y. and London pp. 223–236.

Sherwood, S. W., Atkinson, R. C., Schimke, R. T., and Loher, W. (1989). Flow cytometric analysis of spermatogenesis in *Teleogyrillus commodus. J. Insect Physiol.* **35(12),** 975–980.

Sherwood, S. W., Kung, A.K., Rush, D. F., and Schimke, R. T. (1994a). Cyclin B expression in HeLA S3 cells studied by flow cytometry. *Exp. Cell Res.* (*in press*).

Sherwood, S. W., Sheridan, J. P., and Schimke, R. T. (1994b). Induction of apoptosis by the anti-tubulin drug colcemid: Relationship of mitotic checkpoint control to the induction of apoptosis in HeLa S3 cells. *Exp. Cell. Res.* **215,** (*in press*).

Telford, W. G., King, L. E., and Fraker, P. J. (1992). Comparative evaluation of several DNA binding dyes in the detection of apoptosis-associated chromatin degradation by flow cytometry. *Cytometry* **13(2),** 137–142.

Tomei, L. D., and Cope, F. O. (1991). "Apoptosis: The Molecular Basis of Cell Death," Current Communications in Cell and Molecular Biology. Cold Spring Harbor Laboratory Press, Cold Spring Harbor, New York.

Tomei, L. D., Shapiro, J. P., and Cope, F. O. (1993). Apoptosis in C3H/T10$^{1/2}$ mouse embryonic cells: Evidence for internucleosomal DNA modification in the absence of strand-cleavage. *Proc. Natl. Acad. Sci. USA* **90,** 853–857.

Vaux, D. (1993). Towards an understanding of the molecular mechanisms of physiological cell death. *Proc. Natl. Acad. Sci. USA* **90,** 275–280.

Williams, G. T. (1991). Programmed cell death: Apoptosis and oncogenesis. *Cell* **65,** 1097–1098.

Wyllie, A.H., Kerr, J. F., and Currie, A. R. (1980). Cell death: The significance of apoptosis. *Int. Rev. Cytol.* **68,** 251–300.

Yen, A. (ed.) (1989). "Flow Cytometry: Advanced Research and Clinical Applications," 2 vols. CRC Press, Boca Raton, Florida.

Yonish-Rouach, E., Grunwald, D., Wilder, S., Kimchi, E. M., Lawrence, J-J., May, P., and Orens, M. (1993). p53-mediated cell death: Relationship to cell cycle control. *Mol. Cell. Biol.* **13(3),** 1415–1423.

CHAPTER 6

Transient Transfection Assays to Examine the Requirement of Putative Cell Death Genes

Barbara A. Osborne★,†, Sallie W. Smith★, Zheng-Gang Liu†, Kelly A. McLaughlin†, and Lawrence M. Schwartz†,‡

★Department of Veterinary and Animal Sciences
†Program in Molecular and Cellular Biology
‡Department of Biology
University of Massachusetts
Amherst, Massachusetts 01003

I. Introduction

In many systems examined to date, the process of apoptosis has been shown to require new gene synthesis. This recognition has led many investigators to attempt to identify genes that regulate apoptosis. Some of the strategies used to identify new genes are described in Chapters 7 and 8. However, cloning the genes that are differentially expressed in dying cells is only the first step necessary for determining the role(s) such genes might play in the induction of

Copyright © 1995 by Academic Press, Inc. All rights of reproduction in any form reserved.

apoptosis. Clearly, some of these genes may play critical roles in the process whereas others may be involved more peripherally. The identification of essential cell death genes will provide a better understanding of the events that lead to cell death.

To demonstrate a requirement for a particular gene, it is necessary to inactivate the gene and determine if this loss affects apoptosis. One obvious mechanism for gene inactivation is targeted deletion by means of homologous recombination. Although targeted gene deletion will definitively determine whether the gene is essential for apoptosis, these experiments are time consuming, technically demanding, and expensive. Another approach to gene inactivation is the use of antisense oligonucleotides. This technique requires the use of phosphorothioate oligonucleotides capable of entering cells as well as of being able to anneal to and inactivate the targeted mRNA (Takayama and Inoye, 1990). Although this technique has worked quite well in several instances, it is unlikely to be effective if the targeted gene is expressed at high levels. Additionally, the use of antisense oligonucleotides is expensive and the cost of derivatized oligonucleotides restricts the general use of these potentially useful compounds.

We have utilized an alternative approach to determine the requirement for particular genes in the induction of apoptosis. Our strategy is based on the elegant experiments of the Harlow laboratory. These researchers demonstrated that co-transfection of a gene encoding a cell surface molecule, CD20, with constructs encoding mutations of cyclin-dependent kinases (van den Heuvel and Harlow, 1994) or p107 (Zhu *et al.*, 1993) allowed the identification of transfected cells by expression of CD20 cell surface protein. These experiments suggested that co-transfection of CD20 serves as an excellent marker for transfection. We have used this technique to transfect various antisense constructs transiently, in conjunction with CD20, into the T-cell hybridoma DO11.10 (Liu *et al.*, 1994). Following transfection, the cells are treated with agents known to induce apoptosis and are analyzed by FACScan. Since only a small percentage (i.e., ~ 4%) of the cells is transfected routinely, these experiments normally would be difficult, if not impossible, to interpret. However, by analyzing only those cells that express CD20, we have been able to assess the effects of the antisense constructs on the induction of apoptosis in transfected cells. This strategy is diagrammed in Fig. 1.

Briefly, our approach is to co-transfect CD20 and an antisense construct of the gene under investigation. The concentrations of the two plasmids are adjusted so that it is likely that cells receiving the antisense construct will also receive CD20. Following transfection, Percoll gradients are used to remove cells that die during the transfection procedure. After culturing the cells in the presence of agents capable of inducing apoptosis for varying amounts of time, cells are examined by FACScan for the percentage of CD20$^+$ cells that survive induction of cell death.

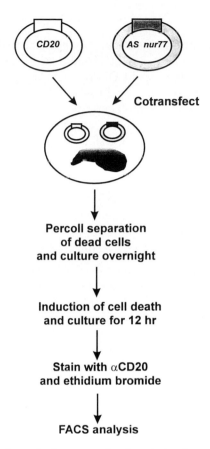

Fig. 1   Flow diagram of transfection protocol and analysis of putative cell death genes.

II. Methods

A. Cell Culture

   Cells are maintained in appropriate medium and protein supplements (in the case of DO11.10, a mixture of 50% RPMI 1640 + Dulbecco's Modified Eagle's Medium supplemented with 10% horse serum is used). We find it useful to clone DO11.10 periodically and select clones that are particularly sensitive to the induction of apoptosis. Many cell lines become resistant to cell death over long periods of cell culture; therefore, selection of a sensitive clone is critical for the maintenance of cell lines that are easily induced to die.

B. Transfection

A culture of logarithmically growing cells is harvested and washed twice in medium without serum. Then 1×10^7 cells are mixed with purified circular plasmid in 0.5 mg/ml DEAE–dextran (Ausubel *et al.*, 1987) and incubated at room temperature for 30 min. The concentration of plasmid used will vary depending on the molecular weight of the DNA to be transfected. We recommend using a 2- to 4-fold excess of the test plasmid (antisense *nurr77* in this illustration) over the CD20 reporter plasmid. A typical experiment will use 3–5 μg test plasmid, which will ensure that CD20$^+$ cells also are likely to carry the test plasmid. Cells are then washed once in RDG medium + 50 mM Tris, pH 7.5, and incubated in 10 ml RPMI 1640/DME + 10% horse serum overnight at 37°C.

C. Percoll Separation of Dead Cells Generated by Transfection

Cells are passed over Percoll to remove dead cells 24 hr after transfection. Cells are washed once and resuspended in 1 ml serum-free medium and layered on top of 2.5 ml 40% Percoll (Pharmacia). The Percoll gradient is spun at 300–340 g with no brake for 30 min at room temperature. Live cells are harvested from the bottom of the tube. Note that different cell types may require different concentrations of Percoll to achieve satisfactory separation of live from dead cells. Live cell concentration is determined by trypan blue counting. Our experience is that 90–95% viable cells are obtained by Percoll separation.

D. Induction of Apoptosis in Transfected Cells

Following Percoll separation, live cells are placed in fresh medium + serum and incubated with appropriate stimuli to induce apoptosis. In the examples provided here, cells are placed in 1-ml wells of a 24-well Costar tissue culture plate that has been precoated with antibody against the T-cell receptor found on the surface of DO11.10 cells (see Chapter 9 for more details). Other agents that induce cell death, such as dexamethasone, may be substituted.

E. Detection of Cell Surface Expression of CD20

Following 18 hr of incubation with apoptosis-inducing agent, cells are harvested and washed twice with phosphate-buffered saline (PBS) + 2% horse serum + 0.1% NaN$_3$ (0.1% BSA may be substituted for horse serum) and are counted in a hemacytometer. Aliquots of 1×10^5 cells are placed into 10 × 75 mm tubes suitable for use on the FACScan, centrifuged at 1500 rpm for 10 min, and resuspended in 100 μl anti-CD20 antibody diluted according to the manufacturer's suggestion in PBS + 50 ng/ml ethidium bromide to stain chromosomal DNA. (Note: Ethidium bromide is a suspected carcinogen and a known mutagen.) It is essential to titer any antibody that is purchased to ensure that the results obtained represent maximal antibody binding. We have used fluorescein-conjugated anti-CD20 purchased from both Pharmingen (La Jolla, CA) and Becton Dickinson (Mountain View, CA) with satisfactory results. Cells are incubated with the antibody at 4°C for 30 min in 0.1% NaN_3. Cells are then washed twice with PBS + 2% serum and resuspended in 0.5 ml PBS for analysis on the FACScan. As a negative control, samples should include a fluorescently labeled antibody of the same immunoglobulin class at the same concentration as the anti-CD20.

F. FACScan Analysis

Cells are analyzed on a FACScan or comparable fluorescence analyzer. The important factor is that the device used must have a laser that excites at 488 nm and photomultiplier tubes that collect emitted light at 525–540 nm and 550–640 nm for fluorescein and ethidium bromide, respectively. Although, in principle, this analysis could be performed by fluorescence microscopy, our experience has been that the eye is not sensitive enough to detect the low levels of fluorescence emitted by the anti-CD20 antibody.

For analysis, live cells are gated by forward scatter versus FL3 (ethidium bromide), as shown in Fig. 2. As cells begin to die, they become smaller and thus lose the ability to scatter light (see Chapter 4). This feature is reflected by a loss of forward or 180° light scatter on the FACScan. These cells also take up molecules such as the DNA intercalating dye ethidium bromide. Thus, a dying or dead cell becomes smaller and ethidium bromide positive. The live gated cells are then examined for the appearance of the CD20 surface antigen. In the experiment shown in Fig. 2A, cells were co-transfected with CD20, in addition to either sense or antisense constructs of the immediate early gene *nur77*, a gene identified to be induced during apoptosis in mouse T cells (Liu *et al.*, 1994). Following transfection, the cells were subjected to the techniques just described and examined by FACScan. When the *nur77* sense construct is used and live cell gating is performed, few if any CD20[+] cells are detected (Fig. 2B). In contrast, when antisense *nur77* is used for transfection, approximately 13% CD20[+] cells are found (Fig. 2C). Considering that the efficiency of transfection is low, the appearance of 13% of live cells bearing the CD20 antigen is highly significant. In fact, many of the live cells are not transfected, so one would not expect to find a high number of CD20[+] cells.

The use of a reporter plasmid such as CD20 allows the investigator to examine the fate of only those cells that have been transfected with the DNA of choice. This is a powerful strategy, since most transient transfections result in the introduction of DNA into a very minor proportion of cells. In our case, using the T-cell hybridoma DO11.10 we are able to transfect only 2–4% cells. Clearly, this proportion of cells is not tractable to standard biochemical or anatomical analysis under normal conditions. However, the use of a reporter plasmid allows these rare cells to be monitored. Another distinct advantage of this system is

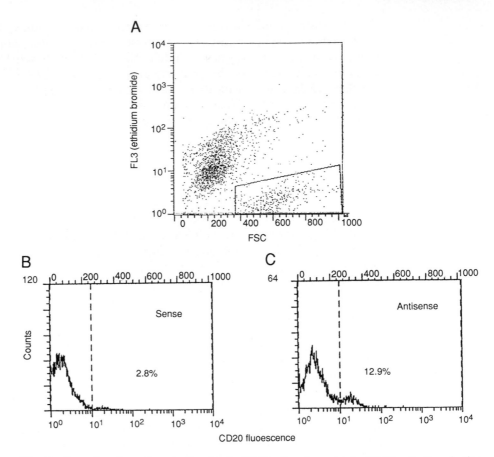

Fig. 2 Detection of transfected cells using the CD20 cell surface protein. (A) After the transfection strategy outlined in Fig. 1, cells are analyzed on the FACScan for viability using forward scatter and FL3 (ethidium bromide fluorescence). Those cells that are determined to be viable by scatter and lack of FL3 fluorescence are gated electronically by the FACScan (this population is found within box) and analyzed for CD20 expression. (B) D011.10 cells co-transfected with CD20 and *nur77* sense, gated on viable cells, and analyzed for CD20 fluorescence. (C) D011.10 cells co-transfected with CD20 and *nur77* antisense and analyzed as in B.

the ability to test a large number of DNA constructs for their ability to affect various cellular processes in a rapid fashion. The time frame for these experiments is quite short, consuming no more than 48–60 hr, in contrast with the normal 2–3 wk required to establish stably transfected cell lines. This technique also obviates problems that frequently arise in the establishment of stable transfectants. For example, if one transfects a construct that inhibits cell death, a transfectant resistant to induction of apoptosis is the expected phenotype. However, frequently a cell becomes resistant to induction of apoptosis independent of the transfection. The frequency of isolation of cell death mutants is quite high, even without the use of chemical mutagens. Therefore it becomes essential to be able to demonstrate that the defect in apoptosis is the consequence of the transfected DNA rather than of isolating a naturally occurring mutant. This can be accomplished, but such experiments are tedious and, given the transient transfection strategy described in this chapter, such protocols may be unnecessary.

Note that CD20 is one of a number of reporters that might be used for such experiments. For example, Nolan *et al.,* (1988) described the use of lacZ reporter constructs that are detected by the viable fluorescent dye FDG (fluorescein di β-D-galactopyranoside). Since this dye is a substrate for β galactosidase, live cells transfected with lacZ-expressing constructs can be analyzed for enzyme activity. The cells typically are loaded with FDG in a hypotonic solution and are analyzed by FACS for cleavage of the substrate. Additionally, other surface markers might be used as reporters for transfection. Note, however, that the use of CD20 was carefully considered and that this marker was chosen for its high level of expression. The reporter chosen must be able to express the gene product of choice at sufficiently high levels that detection of the protein product is easy and reliable.

Yet another simple detection system that could be adapted for the experimental protocols described in this chapter is suggested by the recent data of Chalfie *et al.* (1994). This group used a naturally occurring bioluminescent compound called green fluorescent protein (GFP), isolated from the jellyfish *Aequorea victoria,* to label specific cells in *Caenorhabditis elegans*. This reporter is attractive since it is naturally fluorescent, is non toxic, and is readily diffusible throughout cells, even those with complex structures such as neurons. To adapt this marker for the experiments described in this chapter, a plasmid would be constructed containing an appropriate mammalian promoter upstream from GFP. This construct could then be used in co-transfection assays with antisense constructs.

III. Summary

In conclusion, this chapter provides a convenient and efficient method for the detection and analysis of transiently transfected cells. Such strategies allow a fast and simple analysis of the requirement for particular genes that have been identified as being induced during apoptosis. Our experience has been that, when screening for "cell death genes," it is easy to isolate genes induced during apoptosis but far more difficult to determine the requirement for any given gene. These protocols have rendered such determinations much simpler to perform.

References

Ausubel, F. M., Brent, R., Kingston, R. E., Moore, D. D., Smith, J. A., Siedman, J. G., and Struhl, K. (1987). "Current Protocols in Molecular Biology" Wiley Interscience, New York

Chalfie, M., Tu, Y., Euskirchen, G., Ward, W. W., and Prasher, D. C. (1994). Green fluorescent protein as a marker for gene expression. *Science* **263,** 802–805.

Liu, Z.-G., Smith, S. W., McLaughlin, K. M., Schwartz, L. M., and Osborne, B. A. (1994). Apoptotic signals generated through the T cell receptor require the immediate-early gene *nur77*. *Nature* **367,** 281–284.

Nolan, G. P., Fiering, S., Nicolas, J.-F., and Herzenberg, L. A. (1988). Fluorescence-activated cell analysis and sorting of viable mammalian cells based on β-D-galactosidase activity after transduction of *Escherichia coli lacZ*. *Proc. Natl. Acad. Sci. USA* **85,** 2603–2607.

Takayama, K. M., and Inoye, M. (1990). Antisense RNA. *Crit. Rev. Biochem. Mol. Biol.* **25,** 155–184.

van den Heuvel, S., and Harlow, E. (1994). Distinct roles for cyclin-dependent kinases in cell cycle control. *Science* **262,** 2050–2054.

Zhu, L., van den Heuvel, S., Helin, K., Fattaey, A., Ewen, M., Livingston, D., Dyson, N., and Harlow, E. (1993). Inhibition of cell proliferation by p107, a relative of the retinoblastoma protein. *Genes Dev.* **7,** 1111–1125.

CHAPTER 7

Cloning Cell Death Genes

Lawrence M. Schwartz,★ Carolanne E. Milligan,★ Wolfgang Bielke,★ and Steven J. Robinson†

★ Department of Biology
University of Massachusetts
Amherst, Massachusetts 01003

† Institute of Molecular Biology
University of Oregon
Eugene, Oregon 97403

Copyright © 1995 by Academic Press, Inc. All rights of reproduction in any form reserved.

I. Introduction

In the 1960s, several laboratories demonstrated that the programmed death of certain tissues could be delayed or prevented by treatment with inhibitors of RNA or protein synthesis (Tata, 1966; Lockshin, 1969). These observations that metabolic poisons were protective rather than toxic were surprising, and suggested that programmed cell death may require *de novo* gene expression (reviewed by Schwartz, 1991). Irrefutable evidence that genes were required for programmed cell death came from experiments done by the Horvitz laboratory examining the development of the tiny free living nematode *Caenorhabditis elegans* (reviewed by Ellis *et al.,* 1991; Driscoll, 1992). These researchers demonstrated that mutations in *ced* genes (*cell death* abnormal) abolished programmed cell death. Several of these genes have been cloned from *C. elegans* and two have identified human homologs (Vaux *et al.,* 1992; Yuan and Horvitz, 1992; Yuan *et al.,* 1993; Hengartner and Horvitz, 1994). These experiments have prompted many laboratories to attempt to identify and characterize the genes that may be responsible for cellular suicide. This chapter outlines a series of protocols that can be used to clone differentially expressed genes.

Several issues should be considered before attempting to clone putative cell death genes. First, do data suggest that the pattern of cell death observed is dependent on *de novo* gene expression? If cell death cannot be inhibited by the transcriptional inhibitor actinomycin D, the protein products required for death may already be resident in the cell and post-transcriptional mechanisms such as phosphorylation may activate the program. Second, can relatively pure populations of synchronously dying cells be obtained for study? Since cell death is often not synchronous within a given tissue, only a small number of cells expressing the genes of interest may be present at any given time. Complicating the issue is the fact that the bulk of the cells present in a tissue may not be on a pathway of differentiation that leads to death. For example, although 12,000 of the 22,000 lumbar spinal motor neurons die during embryogenesis in the chick (Hamburger and Oppenheim, 1982), the majority of cells in the spinal cord are not motor neurons. Consequently, if spinal cords are used as the starting material, less than a fraction of 1% of the cells will be dying motor neurons. Therefore, although isolating differentially expressed genes from this tissue is possible, in all likelihood these genes will not be "cell death" genes. Third, is sufficient mRNA available from the cells? If large numbers of cells die in a relatively synchronous manner, such as muscle cells in metamorphosing tadpole tails, mRNA availability is not an issue. In contrast, in the interdigital cells of the embryonic mammalian hand, mRNA abundance may be a major limitation. Fourth, are the protocols of interest sufficiently straightforward that reasonable success can be anticipated?

All the protocols provided are based on the assumption that the regulation of cell death will result in changes in the abundance of a small set of specific mRNAs. The difficulty is in isolating these specifically altered transcripts from the thousands of constitutive transcripts within every eukaryotic cell. For regulated but abundantly expressed genes, a relatively straightforward approach such as plus/minus screening is sufficient. However, for rare transcripts that may be involved in the initial steps of a cell's commitment to die, more subtle techniques are necessary, even if they are technically more difficult. Each of the protocols introduces bias in some way. For example, plus/minus screening (Sargent, 1987) is particularly biased toward abundant transcripts whereas differential display (Liang and Pardee, 1992) is biased toward genes that hybridize to specific arbitrary sequences. Since it is impossible to avoid these biases completely, it is necessary to be mindful of how each of these approaches may affect the types of clones isolated. Rather than being a comprehensive survey of available methods, these protocols represent different strategies that have been used successfully to clone putative cell death genes in our laboratory (Schwartz et al., 1990; Liu et al., 1994; Bielke et al., 1995; C. E. Milligan, K. A. McLaughlin, B. A. Osborne, and L. M. Schwartz, unpublished data). The use of these protocols is obviously not limited to the study of cell death. Any developmental process that depends on changes in transcript abundance can be examined with these methods.

The general strategy for each of the various cloning techniques will be described here, followed by detailed methodological protocols at the end of the chapter. For plus/minus screening, a cDNA library is generated from the poly (A)$^+$ RNA of "condemned" tissues. To isolated the up- or down-regulated sequences within the library, radiolabeled cDNA probes are generated from mRNA isolated from the tissue of interest at each of two different developmental stages. These mRNA probe populations are then used to screen the library. The strategy is to evaluate, for each recombinant cDNA clone in the library, the relative abundance of its corresponding mRNA at two different developmental stages. This technique is similar to performing a "reverse Northern," in which the RNA rather than the recombinant DNA serves as the probe. In this manner, large numbers of recombinants can be tested for their apparent levels of expression at the two stages under investigation. The advantages of this protocol are (1) that it works (Schwartz et al., 1990; Liu et al., 1994), (2) that it is mechanically straightforward, and (3) that it can yield full-length recombinants for subsequent study. The limitations are (1) that sufficient amounts of mRNA are required for both library construction and probe generation; (2) that it is not a sensitive screen, so only the most abundantly expressed genes will be detected; and (3) that although it is easy, it is tedious and relatively expensive to perform.

The Differential Display protocol relies on using arbitrary 3' and 5' synthetic oligonucleotide primers that will anneal to only a subset of mRNA molecules isolated from the cells (Liang and Pardee, 1992). These mRNAs are then amplified by reverse transcriptase/polymerase chain reaction (RT/PCR) in the presence of ^{35}S-labeled nucleotide, are fractionated in DNA sequencing gels, and are visualized by film autoradiography. Bands that are detected at only one stage of development (i.e., mRNA from dying cells) are then isolated from the gel, reamplified, and subcloned for further study. The virtues of this protocol are (1) that it is mechanically easy; (2) that it is inexpensive (when using the kit; RNAmap Kit, GenHunter Corp., Brookline, MA); and (3) that it can detect rare sequences. The disadvantages are (1) that it is prone to artifacts that produce many false positives; (2) that as applied, it will only allow the isolation of a subset of the relevant genes; and (3) that only small portions of the gene are isolated, necessitating subsequent library screens to obtain full-length recombinants.

A third cloning strategy, the PCR-based subtractive hybridization Gene Expression Screen developed by Wang and Brown (1991), requires that mRNA be isolated from each of two developmental stages to be examined (i.e., before and after the commitment to die). The mRNA is converted into short stretches of cDNA, to which adapters are ligated. Since both ends of the cDNAs are then composed of known sequences, they can be amplified by PCR. One pool of cDNA is designated the "driver" and the other the "tracer." By alternating which stage is the driver, either up- or down-regulated sequences can be obtained. The driver is biotinylated and used to subtract out sequences that are common to both developmental stages. By subtracting these cDNAs, reamplifying the remaining recombinants and then subtracting again, a dramatic enrichment of differentially expressed sequences, as well as of rare transcripts, is achieved. By subsequently determining how many times the same genes are re-isolated, a Poisson distribution can be generated that allows one to determine the relative number of genes that are differentially expressed during the developmental process. To obtain new sequences, additional rounds of subtraction can be undertaken. As advantages, this protocol (1) allows one to calculate the number of up- and down-regulated genes involved in a developmental process, (2) generates fragments or "tags" of these genes to be cloned (and sequenced) for study, (3) avoids the problems inherent in PCR generation of lengthy regions by consciously selecting short fragments for PCR and subtraction, (4) permits rare sequences to be isolated, and (5) allows the starting pool of material to be regenerated by PCR, so an unlimited number of rounds of subtraction can be performed to isolate all the genes involved in a developmental process. However, disadvantages are that this protocol (1) requires sufficient starting material to allow for dual library construction, (2) yields cDNA fragments rather than full-length recombinants, and (3) is complicated and technically demanding.

A final protocol provided in this chapter is based on work from the Eberwine laboratory (Van Gelder et al., 1990). This method amplifies populations of mRNAs that can then be used as hybridization probes or for library construction. The basic method involves annealing a single-stranded primer to the poly (A)$^+$ tail of mRNA. Using standard protocols, double-stranded cDNA is generated. This protocol is useful because a T7 bacterial polymerase promotor sequence is generated 3′ to the poly (A)$^+$ tail of the cDNA. When the double-stranded cDNA is used in an *in vitro* transcription assay with T7 polymerase, about 2000 copies of RNA are generated per DNA template. This complementary RNA (cRNA) is the antisense strand of the gene and can be used as a probe for mRNA. Alternatively, the cRNA can be converted to double-stranded cDNA and used in any of the preceding protocols as the starting material for library construction.

Independent of the screening method used, once differentially expressed genes have been isolated, a range of experiments should be performed to confirm that they are in fact "cell death" genes. Obviously it is beyond the scope of this chapter to provide detailed protocols for these analyses. In addition, many of the studies that one would want to perform would be specific to the system under analysis. For example, in the nematode *Caenorhabditis elegans,* a range of powerful genetic analyses can be applied that are not available for studies of tadpole tail resorption. Nevertheless, standard assays can be employed. First, the researcher should demonstrate that the cloned gene is either up- or down-regulated in cells at the time they become committed to die. This can be done with Northern blots for populations of pure synchronously dying cells, such as cultures of PC12 cells. However, since cell populations in most tissues *in vivo* are heterogeneous, Northern blots can provide misleading results. In these situations, *in situ* hybridization should be used to obtain cellular resolution. If coupled with other histological techniques such as TUNEL (Gavrieli *et al.,* 1992; Chapter 2), a stronger correlation with the commitment to die can be obtained.

Once evidence is available that the gene is differentially expressed with death, other tissues should be examined to determine whether expression is confined to dying cells. Again, this can be done by *in situ* hybridization, Northern blots, or immunocytochemistry, when an antibody is available against the gene product. A "coincidence" approach using differential display has been used to increase the probability of detecting genes involved in cell death, rather than in some other developmental processes taking place in the tissue under investigation (Bielke *et al.,* 1995). These authors simultaneously examined two regressing tissues: postlactation mammary gland and postcastration ventral prostate. By selecting those differentially expressed sequences that were common to both tissues, they enhanced the likelihood of detecting common apoptosis-associated genes. If the gene is expressed in nondying cells, this does not mean that the gene is not an essential component of the cell death pathway. Instead, it may mean that other gene products are also required to regulate the cell death phenotype. In fact, this appears to be true for many genes involved in mammalian cell death, including *p53* (Clark *et al.*, 1993; Lowe *et al.*, 1993), the transcription factor *nur77* (Liu *et al.*, 1994; Woronicz *et al.*, 1994), and c-*myc* (Evan *et al.*, 1992; Shi *et al.*, 1992).

These differentially expressed genes should be sequenced. Since some of the screening protocols such as the gene expression screen (Wang and Brown, 1991) generate large numbers of putative cell death gene fragments, it may be best to sequence first and examine tissue expression subsequently. This approach will ensure that the same gene is not inadvertently analyzed several times. Once the gene has been sequenced, database analysis may reveal its identity in a manner similar to the "Expression Sequence Tag" (EST) approach (e.g., Adams *et al.*, 1993). Obviously knowing the identity of a number of cloned genes will help the investigator prioritize them for analysis. Presumably the identification of a transcription factor will be of greater interest for defining regulatory pathways than the cloning of a cytoskeletal element. In many cases, the gene will not be readily identified from database analysis. In these cases, several options for subsequent analysis remain. First, it is important to check as many databases as possible. No one database contains all known DNA sequences. For example, several of our novel (or "pioneer" genes) could be found in EST databases, thereby giving us some information about tissue expression and phylogenetic conservation. Second, the presumptive protein product can be examined for functional motifs such as zinc-finger domains or calcium-binding motifs. This analysis can facilitate the prioritization of genes as well as suggest experiments designed to define function (i.e., $^{45}Ca^{2+}$-binding studies). Third, although the gene product may be novel, the proteins with which it interacts may not be. These proteins can be identified using the yeast two-hybrid method to identify protein–protein interaction (see Chapter 8).

If the gene in question is identifiable, a wealth of valuable reagents may be available for experimentation, including antibodies and mutant animals. Defining the role(s) of the cloned genes in the cell death process is a major undertaking. At present, mutational studies have only been performed for a small number of genes. Given the current interest in the regulation of cell death, this list will grow rapidly.

II. Differential Display

This protocol is based on the published procedures of Liang and Pardee (1992) and Liang *et al.* (1992). Differential Display allows one to clone the 3'-most region of genes that are differentially expressed in a given tissue. The virtues of this protocol are that it requires only small amounts of total RNA and does not require the construction of cDNA libraries. The disadvantages are that the sequences isolated are not full-length clones, they must be subcloned into useful vectors, and the rate of false positives is relatively high.

A commercial kit (RNAmap Kit; GenHunter Corp. (Brookline, MA)) contains many of the reagents required to begin the Differential Display procedure. As described next, many additional reagents and primers can be used. The procedures for using these novel reagents are described by Liang et al., (1993).

A. DNaseI Treatment (Message Clean Kit; GenHunter Corp.)

1. To 50 μg total RNA (in 50μl DEPC-water), add 5.7 μl 10× reaction buffer and 1 μl RNAse-free DNaseI (10 units/μl).
2. Incubate 30 min at 37°C.
3. Phenol/chloroform extract.
4. Precipitate supernatant with sodium acetate (pH 4.0; final concentration 0.3 M) with 3× total volume of EtOH at −80°C for 20 min or overnight at −20°C. Spin and wash pellet with 70% EtOH.
5. Redissolve the dried RNA in 50 μl DEPC-treated water.
6. Measure an aliquot of the RNA by OD_{260} and also run an aliquot (5–10 μg) on a gel to check integrity.
7. Dilute remaining RNAs to 0.1 μg/μl with DEPC-treated water and store in aliquots at −80°C.

B. Reverse Transcription

In a microfuge tube, combine the following reagents:

2.0 μl DNase-treated total RNA (0.1 μg/μl)
1.6 μl dNTP mix (250 μM)
2.0 μl 1st strand primer (e.g., T_{12}MN; 10 μM)
2.0 μl 10X reverse transcriptase buffer (provided with enzyme)
11.4 μl DEPC-treated dH_2O

Notes. M indicates a degenerate position, consisting of G, C, or A. N indicates one defined base; (Liang et al., 1993). Concentrations are provided for stock solutions except as noted.

C. Primer Annealing

1. Heat mixture for 5 min at 65°C, then for 10 min at 37°C.
2. Add 1 μl MMLV reverse transcriptase (RT) (100 U/μl).

3. Continue incubation at 37°C for 50 min.

4. Incubate at 75°C for 10 min to heat-inactivate the reverse transcriptase.

5. Store on ice or − 20°C.

D. Polymerase Chain Reaction

1. Set up 20 μl PCR reactions in $1\times$ PCR buffer (use master mixes as often as possible, to keep the conditions constant).

   1.6 μl dNTP mix (25 μM)

   2.0 μl arbitrary 10-mer 5′ primer (2 μM; preferably 50% GC content)

   2.0 μl 1st strand primer (10 μM)

   2.0 μl RT reaction

   1.0 μl [^{35}S] dATP (1200 Ci/mmol)

   0.2 μl AmpliTaq (5 U/μl, Perkin Elmer, Norwalk, CA)

2. Add 25 μl paraffin on top.

3. Run 40 cycles of PCR, conditions: 94°C for 30 sec, 40°C for 2 min, 72°C for 30 sec.

4. Add 1 cycle at 72°C for 5 min.

5. Mix 3.5 μl PCR reaction with 2 μl acrylamide gel loading buffer.

6. Heat for 5 min at 80°C.

7. Load directly onto a 6% denaturing DNA sequencing gel.

8. Run gel at 60 W for 3–4 hr.

9. Without fixation, dry gel to filter paper (wear gloves).

10. Mark filter paper with India ink mixed with a small amount of ^{32}P or ^{35}S (old probes work fine). This is an important step to ensure that the bands can be located again later for isolation.

11. Expose to X-ray film (i.e., Kodak XAR film) overnight at − 70°C with intensifying screens (see Fig. 1).

12. Align film and gel precisely.

Fig. 1 Differential Display using primers (T_{12}MC/5′-AGCCAGCGAA-3′) and total RNA from the ventral prostates of rats. Unmanipulated controls (lanes 1–4); 2 days postcastration (lanes 5–8); 4 days postcastration (lanes 9–12), and 6 days postcastration (lanes 13–16). Lanes 1, 2, 5, 6, 9, 10, 13, and 14 were duplicates derived from the same cDNA source and were handled in parallel. cDNAs for lanes 4, 8, 12, and 16 were prepared at separate times, but were amplified using the same PCR conditions used for the other samples. Note the differences in the products formed from the samples in lanes 3, 7, 11, and 15 relative to the other reactions derived from the same RNA sources but amplified on a different PCR machine using the same cycling parameters. Consequently, even subtle changes such as the use of different PCR machines can introduce significant variability into the products obtained.

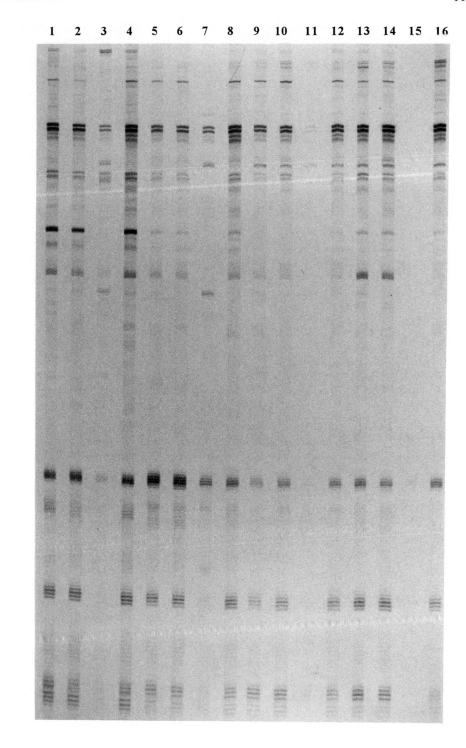

13. Use a scalpel to cut out bands that appear to be differentially expressed. Re-expose the gel to ensure that the correct fragment has been isolated.
14. Transfer the gel slice into 100 μl dH$_2$O and incubate for at least 15 min.
15. Boil the tube for 15 min and spin for 2 min to remove debris.
16. Transfer supernatant into a tube containing 10 μl 3 M sodium acetate, pH 5.3. Add 5 μl glycogen (10 mg/ml) and 450 μl cold 100% ethanol.
17. Precipitate for 30 min to 1 hr at $-70°C$.
18. Centrifuge. Wash the pellet in cold 85% ethanol. Air dry. Resuspend in 10 μl dH$_2$O.

E. Reamplification

Note. Use the same primers as in the first PCR reaction.

1. Combine the following reagents in a 500-μl microfuge tube:
   4 μl purified band from preceding isolation step
   20.4 μl dH$_2$O
   4 μl 10× PCR buffer
   3.2 μl dNTP mix (250 μM)
   4 μl 10-mer sense primer (2 μM)
   4 μl T$_{12}$MN primer (10 μM)
   0.4 μl AmpliTaq polymerase
2. Use the PCR conditions described in Section II,D (40 cycles).
3. Analyze about 50% of each reamplification mix on a preparative agarose gel. It is useful to run molecular weight markers on the gel to ensure that a fragment of the correct size is generated. If no band is visible, you may use an aliquot from the reamplification mix for a second round of PCR.
4. Fragments can be isolated in 1% low melt agarose for either probe preparation or subcloning. We use the blunt-end cloning procedure of Liu and Schwartz (1992). A kit that utilizes the same strategy has been produced by Stratagene, La Jolla, CA (Cat. no. 211190).

F. Tips and Hints

We initially followed the description for the GenHunter Differential Display kit. The method works well using self-designed primers and reagents from other distributors. We always start with a few control reactions to check the quality of the RNAs and the purity of the reagents. Always run duplicates of each Differential Display PCR on the sequencing gel to confirm that a change in expression really occurs. Using RNA from *in vivo* tissues, you can also run multiple reactions with RNAs from successive developmental stages to follow a time course of expression. To avoid variability, only compare reactions prepared in parallel (i.e., using the same reaction mixes, PCR apparatus, etc., see also Fig. 1).

In addition to the standard controls that would be used with any PCR, it is important to ensure that there is no DNA contamination from the original RNA sample. To do this, perform the PCR on a sample of the DNase I-treated RNAs that have not been subjected to first-strand cDNA synthesis and run it on a Differential Display gel.

Another possible source of contamination is the presence of additional PCR fragments in the Differential Display gel that run at the same size as the PCR product of interest. These fragments can derive from different RNAs that contain similar priming sites. Their signal may be too weak to be detected by overnight exposures, but they can nevertheless be reamplified and inadvertently subcloned. For ease of analysis, PCR fragments used for reamplification should be larger than 200 bp, but not so large that they cannot be resolved from other bands at the top of the gel. The presence of additional smaller fragments following reamplification may reflect priming from internal sites within the desired PCR product.

For subsequent analysis of the PCR product on Northern blots, the probe can be random-prime labeled without subcloning by adding 1 μl T$_{12}$MN primer (10 μM) to the reaction. For example, Northern blots of rat mammary gland were generated using 5 μg poly (A)$^+$ RNA per lane (see Figs. 2 and 3). For about 80% of the tested clones, transcripts could be detected on Northern blots and were visualized after 1–5 days exposure using intensifier screens at −80°C.

If problems arise with the reamplification PCR, the following reaction conditions have given good results when the Perkin–Elmer protocol has not worked well.

$10 \times$ PCR Buffer

100 mM Tris, pH 8.3
500 mM KCl
0.01% gelatin
15 mM MgCl$_2$
0.1% NP-40
0.1% Tween 20

Assuming approximately 15,000 different mRNAs per cell type and the display of 50–100 fragments per reaction under the previously described conditions, researchers suggested that 100–150 reactions with different primer combinations should be sufficient to detect almost all differentially expressed mRNAs using the Differential Display protocol (Liang and Pardee, 1992). However, since different 5′ primers may hybridize to multiple sites within the same fragment, different sized bands can be derived from the same RNA. cDNAs with complementary A/T-rich internal tracts may be amplified by the T$_{12}$MN primers alone.

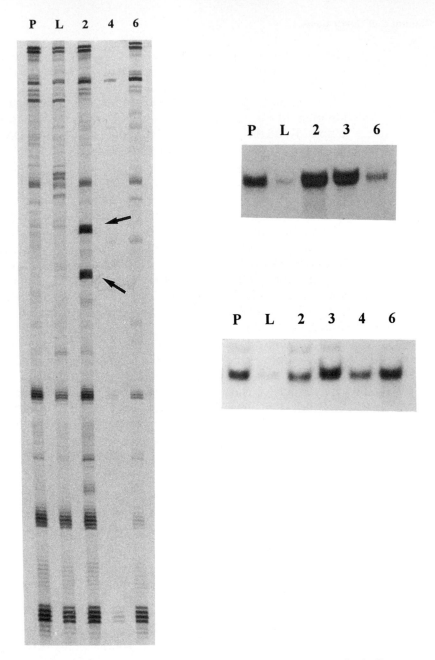

Fig. 2 (*Left*) Differential Display using total RNA from rat mammary glands: P, pregnant; L, lactating; 2, 4, 6, days after forced weaning. Small arrows mark genes that may be expressed differentially during mammary development. Primer combination used was $T_{12}MC/5'$-AGCCAGC-GAA-3'. (*Right*) Northern blots generated using poly A^+ mRNA (5 μg) from mammary glands of pregnant, lactating, and postweaning animals. Blots were hybridized with the cDNAs obtained from the Differential Display procedure. Expression of these genes is clearly reduced during the lactation period.

P L 2 4 6

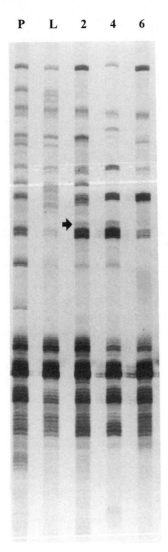

P L 2 3 4

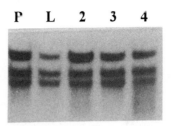

Fig. 3 (*Left*) Differential Display screen using the same mammary gland developmental stages used in Fig. 2 but with a different primer set ($T_{12}MG/5'$-GGTACTCCAC-3'). A pair of genes with up-regulated expression apparent following weaning (arrow) was isolated. (*Right*) Northern blot probed with the marked cDNAs from the screen shown at left. However, despite their apparent developmental regulation when analyzed by Differential Display, by Northern analysis these genes appear to be expressed constitutively.

Furthermore, since many tissue-derived samples may include multiple cell types, this too can increase the complexity of the problem. Consequently, the stringency conditions (e.g., annealing temperature, primer length) may have to be adjusted to the complexity of the RNA source.

III. Plus/Minus Screening

A. Basic Protocol

As for many cloning strategies designed to identify differentially expressed genes, we assume that a cDNA library is already in hand from the stage in development when cells are committed to die. The term "committed" implies that cells can no longer be rescued by treatment with the transcriptional inhibitor actinomycin D suggesting that the genes whose products are required for death have already been expressed and therefore will be represented in the library. Since libraries can be constructed with any of a number of commercially available kits, these methods will not be addressed here.

1. Infect the appropriate bacterial host with phage containing cDNAs from the library, mix with top agarose, and plate out on agar plates. The density should be sufficiently low that individual plaques can be clearly identified.

2. Using sterile toothpicks, transfer phage from individual plaques to the top agarose of two separate agar plates. The plates should be marked to identify top and bottom and then placed on numbered graph paper. That way, the same square can be found on the two plates. Touch the toothpick onto the two plates and then discard. In this manner, you will generate two plates with the same set of ordered recombinants.

3. Repeat the process until the desired number of recombinants has been plated. Plating 1000–2000 recombinants is not unreasonable during the course of a screening.

Note. If you have recombinants that you know are up- or down-regulated, place them in the same location at the bottom of each plate. They will serve as internal reference points to monitor the types of signals you can reasonably anticipate from the screen.

4. Incubate the plates at 37°C overnight.

5. Move plates to 4°C for 1 hr or more. Using a pencil, label the edge of either nylon or nitrocellulose filters with the plate number and then perform filter lifts on each plate for 1 min. Mark the filters and plates with a needle dipped in India ink to properly align the filter on the plate once a hybridization signal has been obtained.

6. Treat the filters as recommended by the manufacturer using the protocols designed for Southern blots. This will require denaturation and neutralization steps. The DNA will need to be immobilized on the filters by baking at 80°C for 2 hr or by UV cross-linking.

7. Filters should then be prehybridized and hybridized in the solutions appropriate for the filter type used. Add high specific activity single-stranded cDNA probes that are made from each of the two developmental stages to be compared (see Section III,B).

8. Hybridize the membranes overnight with the labeled cDNA probes. Follow the manufacturer's recommendations regarding hybridization temperature and solutions. It is probably better to err on the side of reduced stringency rather than high stringency, since a developmentally regulated gene of interest may be in low abundance. (If this is the case, it may be valuable to allow the hybridization to continue for 48–72 hr rather than overnight). As well, the single-stranded cDNA probes may be short and may not hybridize well under stringent conditions. Elevated backgrounds can always be corrected by subsequent washes.

9. While the filters are still wet, wrap them in plastic wrap, tape them to a large sheet of paper, and expose to X-ray film. (Remember that once the filter is dried, the label cannot be washed off if it becomes necessary to re-wash the filters to reduce background).

10. After developing the autoradiograms, compare the pair of "spots" for each recombinant (see Fig. 4). Background may preclude seeing dramatic differences between filters. As well, since there may be subtle (or dramatic) differences in plaque transfer efficiency or in the specific activity of the two probes (a real problem), it is important to compare a number of spots on the two films to be able to identify a recombinant that is potentially differentially expressed. An example of a typical (i.e., unattractive) screening is shown in Fig. 4. Note that all the plaques give a background that may be below the level of detection

Pre-Commitment   Post-Commitment

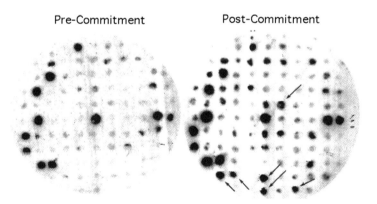

Fig. 4 Plus/minus screening. A cDNA library generated from the intersegmental muscles of the tobacco hawkmoth was screened for sequences that were up-regulated with the muscle's commitment to die. Random recombinants from the library were transferred to the same coordinates on two identical agar plates. After the plaques grew, nylon filter lifts were performed and the membranes were hybridized with single-stranded ^{32}P-labeled cDNA probes generated from the poly A$^+$ RNA from muscles either precommitted to die or postcommitted. After film autoradiography, many of the recombinants were detectable. Note that many of the recombinants were not labeled above background levels, suggesting that they were at low abundance in the RNA pool used for probe generation. Also note that the intensity of hybridization on the two filters is different. Nevertheless, several putative up-regulated sequences could be detected (arrows).

for specific transcripts. Also note that some intensely labeled recombinants, representing constitutively expressed abundant genes, can serve as landmarks. Those that appear to be dramatically different between the two stages (marked with arrows) can be picked for rescreening. In a recent screen, we obtained four different up-regulated recombinants from 1200 plaques examined (Sun *et al.*, 1995).

11. Those recombinants that appear to be differentially expressed should be transferred to a new pair of plates and rescreened. This will allow some of the false positives to be selected against, making subsequent analyses easier.

B. Generating ^{32}P-Labeled Single-Stranded cDNA Probes

To screen the filters for recombinants encoding differentially regulated genes, mRNA from the tissue at the two stages of interest must be converted into radio-labeled cDNA. A variety of methods are available for generating single-stranded [^{32}P] cDNA probes. The method outlined here is based on a protocol from Bethesda Research Laboratories (Technical Bulletin 8020-1).

1. Use care when handling mRNA. Into a sterile tube, pipet 1 μl 5 mM dCTP, 1 μl 5 mM dTTP, 1 μl 5 mM dGTP, 1 μl **0.1** mM dATP, 6 μl [^{32}P] dATP, and 2 μl 500 mM KCl.

2. Speed-Vac the sample to dryness. Add the following reagents to the tube:

   poly A$^+$ mRNA (0.5 μg)

   2 μl oligo dT (12–18-mer; 1 μg)

   2 μl actinomycin D (250 μg/ml)

   2 μl AMV reverse transcriptase (5 U/μl)

   1 μl RNase inhibitor (i.e., RNasin; Promega, Madison, WI)

   2 μl 5× AMV RT buffer (provided with the enzyme)

3. Incubate the tube at 37°C for 60 min. Terminate the reaction and hydrolyze the mRNA by adding 40 μl dH$_2$O, 20 μl 500 mM EDTA (pH 8.0), and 25 μl 150 mM NaOH.

4. Incubate the sample 1 hr at 65°C. Then neutralize by adding 25μl 1 M Tris-Cl, pH 8.0, and 25 μl 1 N HCl.

5. Remove the unincorporated label using a spin column or other means. Boil for 5 min, ice cool, and use as a probe for one set of filters. Repeat the process with the other probe and set of filters.

IV. Gene Expression Screen

The protocol described by Wang and Brown (1991) is a cloning strategy that allows one to estimate the total number of genes that are differentially expressed between developmental stages, as well as to clone a portion of them. The method allows detection of genes which are increased or decreased 6-fold during the particular cellular process and have at least 15 copies of the individual message per cell. In this technique, double-stranded cDNA is cut by restriction endonucleases to produce small fragments or "tags" (50–800 bp fragments). Multiple rounds of hybridizations of varying duration (long or short) are followed by PCR amplification with each cycle of screening. For each round of subtraction, the long hybridization is essential for the removal of low abundance, complex, common (constitutive) cDNAs, whereas the short hybridization is effective for the depletion of high abundance, common cDNAs. The key to this protocol is to run simultaneous reactions, one to isolate up-regulated messages and the other to isolate down-regulated messages. For example, to isolate up-regulated messages, after the first round of subtraction the subtracted "down"-regulated material is biotinylated and then hybridized with the subtracted "up"-regulated material (from the first round) to remove less abundant, complex, common messages. Following this step, the resulting "subtracted" cDNA is hybridized with biotinylated starting material (for up-regulated genes, use "precommitment" material) to remove abundant common messages that may be still present. The remaining material is enriched for fragments or "tags" of messages that should be up-regulated. The reverse reaction is done simultaneously to obtain enriched down-regulated messages. This complex procedure is outlined in Fig. 5. The important point in this procedure is to remove common cDNAs that are equally abundant in the two symmetrical reactions. PCR amplification allows uncovering of rare messages; common ones are subtracted out, leaving those that may be differentially expressed. Multiple gene fragments are considered analogous to multiple alleles in a genetic screen. The frequency with which different portions or fragments of differentially expressed genes are identified independently is calculated, and by applying Poisson analysis, the total number of genes that are up- or down-regulated during the developmental program can be estimated.

After the first round of subtraction, the material is enriched for either up- or down-regulated messages. If one is searching for a tag for a particular biological process, such as cell death, cloning a portion of this enriched material may be useful. However, if one wishes to determine the number of differentially expressed genes as well as clone to them, multiple rounds of subtraction are required. Using this method to isolate genes that are differentially expressed during tadpole tail regression in response to thyroid hormone, Wang and Brown (1991) found that after their first round of subtraction for up-regulated genes, they isolated two inserts that hybridized with greater than 90% of their clones. Further rounds of subtraction removed these and enriched for more up-regulated messages. After three rounds of subtraction, 25 cDNA fragments were isolated from reactions designed to obtain up-regulated sequences and 5 from those to isolate down-regulated messages. These cDNA fragments were probed against full-length cDNA library; fragments that hybridized against the same clone

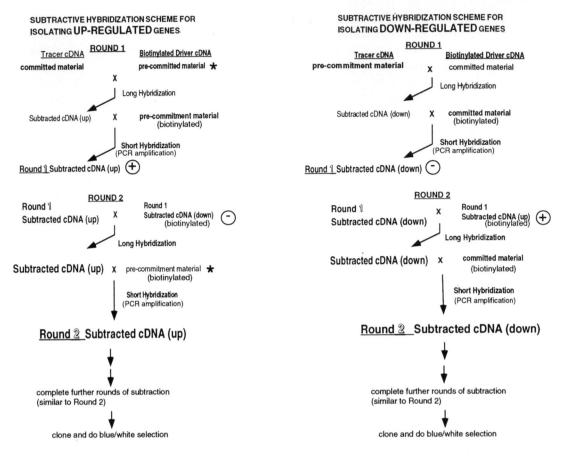

Fig. 5 Subtractive hybridization protocol described by Wang and Brown (1991). Each reaction has two components: the tracer cDNA (contains both common and putatively unique messages) and the driver cDNA (contains messages common to each stage). The driver cDNA pool is biotinylated and then hybridized with the tracer cDNA, so common messages in each cDNA pool will anneal. Each cycle of subtraction includes rounds of both long and short hybridizations. The common messages are subsequently removed with streptavidin (not illustrated), and the remaining "subtracted" cDNA is amplified by PCR.

The key to this protocol is to run simultaneous reactions; one to isolate up-regulated genes and the other to isolate down-regulated ones. For example, after round 1 of subtraction for each arm of the protocol, the subtracted down-regulated material (−) is biotinylated (now considered driver cDNA) and then hybridized with subtracted up-regulated material from the other reaction (tracer cDNA), as indicated, to remove less abundant common messages. Following this step, the subtracted putatively up-regulated cDNA is hybridized with the initially biotinylated driver cDNA [for up-regulated genes, use precommitment material (*)] to remove abundant common messages that may be still present. With each round of subtraction, the material becomes further enriched for unique messages. Two rounds are illustrated; however, several more should be performed to estimate the number of genes involved in the process under study.

were considered to be derived from the same cDNA. All in all, 16 up-regulated and 4 down-regulated genes were identified; Possion analysis estimates that there should be about 30 genes that are up-regulated during that particular developmental process (Wang and Brown, 1991).

The protocol involves:

1. Conversion of mRNA to ds cDNA.
2. digestion of the ds cDNA with blunt-end 4-base-cutting restriction enzymes. The best restriction enzymes for each species must be determined. For our purposes, we use *Alu*I and *Hae*III. The object is to have cDNA in a size range of 50–800 bp.
3. ligation of the linkers to the cut cDNA to allow PCR amplification. The original protocol called for linkers with an *Eco*RI site. We replaced this site with a *Hind*III site. *Alu*I cuts in the middle of the *Hind*III site; therefore, the cDNA will not be internally cut by the restriction enzyme used later to remove the linker-primers.
4. Size selection to isolate fragments 50–800 bp in length.
5. PCR amplification of the cDNA.
6. Digestion of the linker ends on the driver cDNA, followed by biotinylation.
7. A series of long and short hybridizations.
8. Removal of common sequences.
9. Cloning of subtracted cDNA.
10. Analysis of cloned sequences to verify that they are differentially expressed.

Before beginning this method, refer to the original papers (Wang and Brown, 1991; Buckbinder and Brown, 1992). A protocol that details the procedures from generating ds cDNA through the first round of subtraction is presented here.

A. Generating Double-Stranded cDNA

Any method for the generation of blunt-ended double-stranded cDNA can be used, including a number of commercially available kits. The one presented here is based on the single tube reaction protocol of McClelland and co-workers (1988) and Don and co-workers (1993).

1. For first strand synthesis, add the following to an RNase-free microfuge tube:

5 μl mRNA in dH$_2$O (approximately 1 μg poly A$^+$ mRNA)

1 μl RNase inhibitor (20 units; Perkin-Elmer)

5 μl × KGB buffer (see Section IV, H)

    0.4 μl oligo dT (20 pM final concentration; Perkin-Elmer)

    2 μl 5× dNTP mix (2.5 mM stock; final, 0.5 mM each)

    0.5 μl MMLV reverse transcriptase (50–100 units)

Incubate 30 min at 42°C and then cool on ice.

2. For second strand synthesis, add the following to the same tube:

    10 μl 0.5× KGB buffer

    1 μl *E. coli* DNA polymerase I (10 units)

    0.5 μl RNase H (0.4 units)

3. Incubate 12°C for 30 min, then at room temperature for 30 min.

4. Incubate at 70°C for 10 min.

5. Cool on ice.

6. To fill in the ends, add 1μl T4 DNA polymerase (1 unit). Incubate at 37°C for 15 min, then at 70°C for 15 min.

The ds cDNA should now be restriction digested to generate 50–800 bp fragments. (Note: The restriction enzymes used to generate these fragments should be determined empirically for each species to be studied. Digest genomic DNA with various restriction enzymes and use agarose gel electrophoresis to determine which combination of enzymes generates the desired range of fragments).

7. Split the sample volume in half. Cut one sample with *Alu*I and the other with *Alu*I and *Hae*III, by adding

    13 μl ds cDNA

    37 μl 1× KGB buffer

    1 μl *Alu*I or 1 μl *Alu*I + 1 μl *Hae*III

Incubate at 37°C overnight. The restriction digested ds cDNA is now ready for linker ligation.

B. Kinasing and Ligating Linkers

1. Linker Primers

      21-mer (*Hin*dIII): 5'-TGACGACTTAAGCTTGGACTA-3'

      25-mer (*Hin*dIII): 5'-TAGTCCAAGCTTAAGTCGTCATACA-3'

2. Linker Kinase Reaction

    1. Combine

        25 μl 10× ligase buffer (Promega)

        15 μl 25-mer (3 μg/μl stock)

        205 μl dH$_2$O

        5 μl T4 kinase (10 U/μl; Promega)

    2. Incubate for 1 hr at 37°C.

    3. Add 15 μl 21-mer (2.56 μg/μl stock).

4. Mix and incubate at 45°C for 10 min. This will form a duplex oligonucleotide with one blunt end and one 4-base protruding end at the 3′ end.

$$5′ \qquad \text{TGACGACTT}\textbf{AAGCTT}\text{GGACTA } 3′$$
$$3′ \text{ ACATACTGCTGAATT}\underline{\textbf{TCGAA}}\text{CCTGAT } 5′$$

Note. The *Hin*dIII site is in boldface letters and the *Alu*I site is underlined.

3. Linker–Ligation Reaction

1. Combine

   10 μl digested cDNA (see preceding step)

   55 μl freshly prepared duplex phosphorylated linker

   30 μl dH$_2$O

   2.5 μl 10× ligase buffer (Promega)

   5 units T4 DNA ligase

2. Mix and incubate overnight at 15°C.
3. Incubate 2 hr at room temperature.
4. Add 5 μl 0.5 M EDTA (pH 8.0) to stop the reaction.
5. Phenol/chloroform extract and ethanol precipitate.
6. Wash pellet with 70% ethanol, air dry, and resuspend the pellet in 12 μl TE.

C. Purifying Ligated cDNA from Unligated Linker

1. Make a 1.4% low melt agarose minigel in 1× TBE.
2. Load half of the ligated cDNA onto the gel. (Do not stain the ligated cDNA with ethidium bromide since this will interfere with subsequent PCR amplification and ligation. To determine molecular weight, use pre-stained molecular weight markers.)
3. Fractionate the cDNA so that material runs about 1 cm from the loading well. Locate the region between 50 bp and 800 kb and cut out this portion of the cDNA. (The gel slice can be stored at 4°C for up to 1 wk.)
4. Melt the cut out gel at 65°C.
5. Use 1–3 μl for each PCR (see reaction subsequent step). The remaining unused cDNA in the gel can be stored at -20°C.

D. PCR Protocol

1. In a 500 μl microfuge tube, mix (do at least 10–20 tubes per cDNA sample):

   3 μl primer-ligated cDNA from the gel (0.5–1.0 μg/sample)

   4 μl MgCl$_2$ solution (25 mM stock; Perkin-Elmer)

     5 μl 10× PCR buffer (Perkin-Elmer)

     5 μl 10× dNTPs (200 μM each)

     1 μl 21-mer primer (1 μg)

     27.5 μl dH$_2$O (final volume of 50 μl)

2. Heat tubes at 95–100°C for 5 min.

3. Cool on ice.

4. Add 0.5 μl Taq polymerase (2.5 units; Perkin-Elmer) and, 50 μl mineral oil.

5. Perform 30 PCR cycles: 1 min at 94°C, 1 min at 50°C, and 2.25 min at 72°C.

6. If the PCR product is going to be cloned (after the final round of subtractions), perform 30 cycles: 1 min at 94°C, 1 min at 50°C, and 2.25 min at 72°C, followed by 1 cycle: 5 min at 94°C, 1 min at 50°C, and 2 hr at 72°C.

7. After PCR, combine reaction products and roll across a piece of parafilm to remove mineral oil.

8. Phenol/chloroform extract and ethanol precipitate.

9. Wash pellet with 70% ethanol and air dry.

10. Resuspend in 30 μl dH$_2$O and determine the amount of cDNA by the ethidium bromide dotting procedure outlined in Section V,B.

E. Removing Linkers on Driver cDNA

Before biotinylation, the driver cDNA must be cut with *Hin*dIII. This procedure cuts the linkers, which will prevent the driver cDNA from being PCR amplified after subtraction.

1. Combine

     115 μg PCR amplified cDNA

     15 μl 10× *Hin*dIII buffer (Promega)

     dH$_2$O to 150 μl

     5 μl *Hin*dIII (Promega)

2. Incubate overnight at 37°C.

3. Add 150 μl dH$_2$O to the tube.

4. Phenol/chloroform extract and ethanol precipitate.

5. Resuspend in 100 μl 1 mM EDTA (pH 8.0).

Note. Start with 115 μg cDNA to get a final yield of 100 μg after biotinylation. In our laboratory we have successfully used much less starting material (10–15 μg).

F. Biotinylating Driver cDNA

1. Mix 100 μl driver cDNA (1 μg/ μl in 1 mM EDTA) and 100 μl photobiotin (1 μg/μl; see Section IV,H).
2. Irradiate the tube with its cap open for 15 min, 10 cm from the 275 W sunlamp.
3. Add 30 μl 1 M Tris, pH 9.1.
4. Extract with an equal volume of butanol. (The organic phase after this extraction is the upper phase).
5. Repeat the butanol extractions until the upper organic phase is colorless. This usually requires four extractions.
6. Extract the cDNA with chloroform to remove residual butanol.
7. Ethanol precipitate the cDNA, wash pellet in 70% ethanol, and air dry.
8. Resuspend DNA in 100 μl 1 mM EDTA (pH 8.0).
9. Repeat the biotinylation, butanol extractions, and ethanol precipitation (as described).
10. Resuspend the twice biotinylated cDNA in 100 μl TE.

G. Subtractive Hybridization

Note. The subtractive hybridizations involve a long hybridization of tracer cDNA with biotinylated driver cDNA (to remove less abundant common messages), isolation of the subtracted cDNA, and a short hybridization of the subtracted cDNA from the long hybridization with biotinylated driver cDNA (to remove abundant common messages), followed by isolation of the final subtracted cDNA. The protocol for hybridization of driver and tracer cDNA and isolation of subtracted cDNA follows. This protocol should be performed twice for each round of subtraction, once for long hybridization and once for short hybridization.

For short hybridizations, start with all the subtracted cDNA from the previous long hybridization and mix it with 50 μg biotinylated driver. The amounts described in the Wang and Brown protocol are provided; however, we have successfully used much less cDNA while keeping the proportions consistent.

1. Mix 2.5 μg tracer cDNA with 50 μg biotinylated driver cDNA.
2. Ethanol precipitate the cDNAs, wash pellet in 70% ethanol, and air dry.
3. Resuspend the mixed cDNA pellet in 10 μl HE (see Section IV,H).
4. Place the cDNA solution in a boiling water bath for 3 min.
5. Cool on ice.
6. Centrifuge briefly.
7. Add 10 μl prewarmed (to 55°C) 2$\times$ hybridization buffer (see Section IV,H).

8. Vortex and centrifuge briefly.

9. Add 40–50 μl mineral oil to top.

10. Heat mixture at 100°C for 3 min.

11. Transfer mixture to 68°C and incubate 20 hr for *Long* hybridization or 2 hr for *short* hybridization.

12. Add 103 μl prewarmed HE (55°C).

13. Incubate at 55°C for 5 min.

14. Transfer the mixture to a piece of parafilm and roll it around to remove mineral oil and then transfer to a clean tube.

15. Cool to room temperature.

16. Add 15 μl streptavidin (see Section IV,H).

17. Incubate 20 min at room temperature.

18. Phenol/chloroform extract (biotinylated DNA–streptavidin is visible as orange insoluble material at interface between organic and aqueous phases).

19. Add 10 μl streptavidin to collected aqueous phase.

20. Incubate 20 min at room temperature.

21. Phenol/chloroform extract.

22. Add 10 μl streptavidin to collected aqueous phase.

23. Incubate 20 min at room temperature.

24. Phenol/chloroform extract.

25. Chloroform extract.

26. Ethanol precipitate the subtracted cDNA, wash pellet in 70% ethanol, and air dry.

27. After the long and short hybridizations are complete, resuspend pellets in 30–70 μl TE (depending on starting amount).

28. Remove an aliquot of the subtracted cDNA and store it as a reserve for short term at 4°C or for long term at -20°C. The remaining subtracted cDNA should then be subjected to PCR as described earlier (use 3 μl sample/PCR).

29. Combine the PCR products. A portion of this sample can be *Hin*dIII digested and subcloned into a vector (we use pBluescript; Stratagene). The remaining portion is used for subsequent rounds of subtraction.

H. Reagents

2× KGB Buffer

200 mM potassium glutamate
50 mM Tris-acetate (50 mM Tris adjusted to pH 7.6 with ammonium acetate)
20 mM magnesium acetate

100 mg/ml BSA

1 mM 2-mercaptoethanol

1. Filter sterilize and store at 4°C.
2. Dilute to appropriate concentration with dH$_2$O.

Photobiotin (1 mg/ml final)

0.5 mg photobiotin (Vector Labs, Burlingame, CA, Cat. no. SP-1000)
500 μl 0.1 mM EDTA, pH 8.0.

Store at -20°C in the dark.

HE

0.003 g HEPES (10 mM; pH 7.3)
0.037 g EDTA (1 mM)
100 ml dH$_2$O

HE + NaCl

10 ml HE
0.88 g NaCl (0.15 M)

2× Hybridization Buffer

4.38 g NaCl (1.5 M)
0.007 g HEPES (50 mM; pH 7.3)
0.186 g EDTA (10 mM)
0.1 g SDS (0.2%)
50 ml dH$_2$O

Streptavidin (2 mg/ml)

1 mg streptavidin (Vector Labs, Cat. no. SA-5000)
500 μl HE + NaCl

V. Amplification of RNA

The RNA amplification protocol provided here is based on the one described by Eberwine and co-workers (Van Gelder *et al*, 1990). This method produces amplified RNA from a limited quantity of cDNA. Briefly, RNA is primed with a poly-dT oligonucleotide that has the T7 RNA polymerase promoter sequence 5′ to the oligo-dT region. Second-strand cDNA synthesis is performed to generate a double-stranded cDNA cassette that contains a functional T7 polymerase promotor downstream from the gene. The addition of this polymerase then results in the *in vitro* synthesis of approximately 2000 copies of the gene in the *antisense* orientation. This molecule can then be used as a hybridization probe on tissues or blots or as the starting material for the resynthesis of double-stranded cDNA.

A. Testing that Reagents Are RNase-Free

1. Mix 1–5 μl each reagent with 2 μl RNA standard (7.5 kb poly (A)-tailed RNA; Gibco-BRL, Cat. no. 15621-014), plus 1 μl ethidium bromide (1 mg/ml) in a total volume of 8 μl.
2. Incubate the sample for 30 min at 37°C.
3. Run the samples on a formaldehyde gel.
4. Examine under UV light. Samples in which the RNA standard cannot be detected presumably are contaminated with RNase and should be discarded.

Note.   This protocol may be valuable for many of the preceding procedures as well.

B. Quantifying mRNA or cRNA

This is from Maniatis *et al.,* (1982).

1. Add 0.5 μ/ml ethidium bromide to 1% agarose in diethylpyrocarbonate (DEPC)-treated dH_2O.
2. Pour the cooled solution into a 150-mm plastic petri dish. Allow the agarose to solidify.
3. Using a pipetting device, dot a series of DNA and/or RNA standards (e.g., 0.05, 0.1, 0.5, 1.0, 2.5, and 5.0 μg in a total volume of 5 μl/standard) onto the agarose slab.
4. Dot the sample (in a total volume of 5 μl) to be quantified onto the plate.
5. Let the samples sit on the plate at room temperature for several hours (this allows small contaminating particles to diffuse away).
6. Examine and photograph the plate under UV light. Compare the UV intensity of the sample to that of the standards to make an estimate of the concentration. This method has proven to be accurate and reproducible in our hands.

C. Generating Single-Stranded cDNA

1. Mix 40 μg total RNA and 100 ng poly-dT/T7 primer (sequence from Van Gelder *et al.,* 1990).

5'-TTTTTTTTTTTTTTTCGCGGATATCACTCAGCATAATGTT-
AAGTGACCGGCAGCAAA-3'

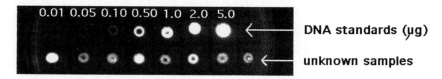

Fig. 6 Ethidium-bromide agarose-plate method of quantitative analysis of nucleic acids. Pour 1% agarose containing 0.5 μg/ml ethidium bromide into a petri dish and place known concentrations of nucleic acids (DNA is illustrated) on the plate in a total volume of 5 μl sample. Then place samples to be quantified on the plate (in a total volume of 5μl) and photograph under UV light. By comparing the intensity of fluorescence of the unknown nucleic acid sample(s) with that of the standards, the quantity of nucleic acid can be estimated.

    2. Incubate 80°C for 5 min.

    3. Cool on ice.

    4. Repeat the incubation and cooling twice.

    5. Add

            1 μl RNAsin RNase inhibitor (40 U) (Promega)

            16 μl 10× KGB buffer (see Section V,J)

            2 μl each dNTP (25 mM stock concentration)

            DEPC-dH$_2$O to 160 μl total volume

            0.5 μl reverse transcriptase (50–100 U)

    6. Incubate at 42°C for 30 min.

    7. Cool on ice.

    8. Ethanol precipitate.

    9. Resuspend pellet in 20 μl DEPC-dH$_2$O.

D. Second Strand cDNA Synthesis

    1. Heat at 95°C for 3 min.

    2. Cool on ice.

    3. Add:

            2.5 μl 10× second strand buffer (see Section V,J)

            2.5 μl 10× dNTP mix (see Section V,J)

            1.0 μl T4 DNA polymerase (5–10 units)

            1.0 μl Klenow polymerase (5 units)

            DEPC-dH$_2$O to a final volume of 25 μl

    4. Incubate at 14°C for at least 5 hr.

E. Removing Single-Stranded Molecules

1. Dilute sample to 94 μl with DEPC-dH$_2$O.
2. Add

   5 μl 10× S1 buffer (see Section V,J)

   2 μg yeast tRNA (or any other carrier, such as glycogen)

   1 unit S1 nuclease
3. Incubate at 37°C for 5 min.
4. Phenol/chloroform extract.
5. Ethanol precipitate.
6. Resuspend pellet in 21.5 μl DEPC-dH$_2$O.

F. Filling in the Ends

1. Add

   2.5 μl 10× KFI buffer (see Section V,J)

   0.5 μl 10× dNTP mix

   0.5 μl Klenow polymerase (2.5 units)

   0.5 μl T4 DNA polymerase (2.5–5.0 units)
2. Incubate at 37°C for 15 min.
3. Phenol/chloroform extract.
4. Ethanol precipitate.
5. Dissolve pellets in 20 μl DEPC-dH$_2$O.
6. Dialyze in 10 μl batches against 500 ml DEPC-dH$_2$O for 4 hr at 4°C to remove salts.
7. Recover the sample and place into an RNase-free microfuge tube.

Note. To dialyze such small quantities, take a 500 μl microfuge tube and bore a hole through the cap with a pasteur pipet that has been heated in a flame. Place the sample in the tube. Place a piece of dialysis tubing across the top of the tube and close the cap. Invert the tube so that the sample sits on the dialysis tubing and secure the microfuge tube in beaker with the dH$_2$O so the cap and dialysis tubing are submerged in the dH$_2$O.

G. Amplifying cRNA

We have been using the Ambion MEGAscript, Austin, TX *In Vitro* Transcription Kit (Cat. no. 1334).

1. Divide the dialyzed sample in half.
2. To each half, add

   2 μl 10× transcription buffer (supplied in kit)

   2 μl each dATP, dCTP, dGTP, dUTP (75 mM stock supplied in kit)

DEPC-dH$_2$O to a final volume of 20 μl

2 μl enzyme mix (supplied in kit)

3. Incubate at 37°C for 4–6 hr.
4. Phenol/chloroform extract.
5. Ethanol precipitate.
6. Dissolve pellet in 20 μl DEPC-dH$_2$O.
7. Check yield by dotting a sample onto an ethidium bromide/agarose plate, as described earlier.

H. Converting to Single-Stranded cDNA

1. Heat denature sample at 75–90°C for 3 min.
2. Cool on ice.
3. Add

   10–100 ng random hexamers (amount depends on yield of RNA)

   3 μl 10× KGB (see Section V,J)

   3 μl 100 mM DTT

   3 μl 10×dNTP stock (see Section V,J)

   DEPC-dH$_2$O to 30 μl

   2 μl reverse transcriptase (50–100 units)

4. Incubate at 37°C for 1 hr.
5. Phenol/chloroform extract.
6. Ethanol precipitate.
7. Resuspend pellet in 10 μl DEPC-dH$_2$O.

I. Resynthesizing Double-Stranded cDNA

1. Heat denature sample at 95°C for 2 min.
2. Cool on ice
3. Add

   100 ng oligo-dT/T7 promoter oligonucleotide (see preceding section for sequence)

   2 μl 10× KFI buffer (see Section V,J)

   2 μl 10× dNTP mix (see Section V,J)

   DEPC-dH$_2$O to 20 μl

   1 μl T4 DNA polymerase (5–10 units)

   1 μl Klenow polymerase (5–10 units)

4. Incubate for 2 hr at 14°C.
5. Phenol/chloroform extract.

6. Ethanol precipitate (with 5 μg yeast tRNA or other carrier).
7. Resuspend pellet in 19.5 μl DEPC-dH$_2$O.
8. Repeat the fill-in protocol described in Section V,F.
9. Phenol/ chloroform extract.
10. Ethanol precipitate.
11. Dissolve pellet in 20 μl DEPC-dH$_2$O.
12. Dialyze and recover sample as described earlier (check yield by placing a sample onto an ethidium bromide/agarose plate). Sample can be reamplified as described in Section V,D.

J. Reagents

10× KGB Buffer

1 M potassium glutamate
250 mM Tris-acetate (50 mM Tris adjusted to pH 7.6 with ammonium acetate)
100 mM magnesium acetate
500 mg/ml RNase-free BSA
5 mM 2-mercaptoethanol

1. Filter sterilize and store at 4°C.
2. Dilute to appropriate concentration with DEPC-dH$_2$O.

10× Second Strand Buffer

1 M Tris-HCl, pH 7.4
200 mM KCl
10 mM MgCl$_2$
500 mM DTT
500 mM (NH$_4$)$_2$SO$_4$

10× S1 Buffer

1 M NaCl
500 mM NaAc, pH 4.5
10 mM ZnSO$_4$

10× KFI Buffer

200 mM Tris-HCl, pH 7.4
10 mM MgCl$_2$
50 mM DTT
50 mM NaCl

10× dNTP Stock Mix

25 mM stock solution of all four dNTPs

Acknowledgments

We are grateful to many investigators who have shared their detailed laboratory protocols. In particular, we thank Drs. P. Liang, A. Pardee, E. Wang, D. Brown, T. Jacks, and J. Eberwine. We thank Kelly McLaughlin for both sharing her results and critically reading the manuscript. We thank Danhui Sun for the autoradiographs used for the plus/minus figure.

References

Adams, M. D., Kerlavoge, A. R., Fields, C., and Ventnar, J. C. (1993). 3,400 new expressed sequence tags identify diversity of transcripts in human brain. *Nature Genetics* **4**, 256–267.

Bielke, W., Ke, G., Saurer, S., and Friis, R. R. (1994). Apoptosis in the rat mammary gland and ventral prostate: Detection of "death"-associated genes using a coincident-expression cloning approach (*submitted*).

Buckbinder, L., and Brown, D. D. (1992). Thyroid hormone-induced gene expression changes in the developing frog limb. *J. Biol. Chem.* **267**, 25786–25791.

Clarke, R. R., Purdie, C. A., Harrison, D. J., Morris, R. G., Bird, C. C., Hooper, M. L., and Wyllie, A. H. (1993). Thymocyte apoptosis induced by p53-dependent and independent pathways. *Nature* **362**, 849–852.

Don, R. H., Cox, P. T., and Mattick, J. S. (1993). A 'one tube reaction' for synthesis and amplification of total cDNA from small numbers of cells. *Nucleic Acids Res.* **21**, 783.

Driscoll, M. (1992). Molecular genetics of cell death in the nematode *Caenorhabditis elegans*. *J. Neurobiol.* **23**, 1327–1351.

Ellis, R., Yuan, J., and Horvitz, H. R. (1991). Mechanisms and functions of cell death. *Annu. Rev. Cell Biol.* **7**, 663–698.

Evan, G. I., Wyllie, A. H., Gilbert, C. S., Littlewood, T. D., Land, H., Brooks, M., Waters, C. M., Penn, L. Z., and Hancock, D. C. (1992). Induction of apoptosis in fibroblasts by c-*myc* protein. *Cell* **69**, 119–128.

Gavrieli, Y., Sherman, Y., and Ben-Sasson, S. (1992). Identification of programmed cell death in situ via specific labeling of nuclear DNA fragmentation. *J. Cell Biol.* **119**, 493–501.

Hamburger, V., and Oppenheim, R. W. (1982). Naturally occuring neuronal death in vertebrates. *Neurosci. Commun.* **1**, 39–55.

Hengartner, M. O., and Horvitz, H. R. (1994). *C. elegans* survival gene *ced-9* encodes a functional homolog of the mammalian proto-oncogene *bcl-2*. *Cell* **76**, 665–676.

Liang, P., and Pardee, A. B. (1992). Differential display of cukaryotic messenger RNA by means of the polymerase chain reaction. *Science* **257**, 967–971.

Liang, P., Averboukh, L., Keyomarsi, K., Sager, R., and Pardee, A. B. (1992). Differential display and cloning of messenger RNAs from human breast cancer versus mammary epithelial cells. *Cancer Res.* **52**, 6966–6968.

Liang, P., Averboukh, L., and Pardee, A. B. (1993). Distribution and cloning of eukaryotic mRNAs by means of differential display: Refinements and optimization. *Nucleic Acids Res.* **21**, 3269–3275.

Liu, Z-G., Smith, S., McLaughlin, K. A., Schwartz, L. M., and Osborne, B. A. (1994). Apoptotic signals delivered through the T cell receptor require the immediate early gene *nur77*. *Nature* **367**, 281–284.

Lockshin, R. A. (1969). Programmed cell death: Activation of lysis by a mechanism involving the synthesis of protein. *J. Insect Physiol.* **15**, 1505–1516.

Lowe, S. W., Schmitt, E. M., Smith, S. W., Osborne, B. A., and Jacks, T. (1993). p53 is required for radiation-induced apoptosis in mouse thymocytes. *Nature* **362**, 847–849.

Maniatis, T., Fritsch, E. F., and Sambrook, J. (1982). "Molecular Cloning, A Laboratory Manual." Cold Spring Harbor Laboratory Press, Cold Spring Harbor, New York.

McClelland, M., Hanish, J., Nelson, M., and Patel, Y. (1988). KGB: A single buffer for all restriction enzymes. *Nucleic Acids Res.* **16**, 364.

Sargent, T. D. (1987). Isolation of differentially expressed genes. *Meth. Enzymol.* **152**, 423–432.

Schwartz, L. M., Kosz, L., and Kay, B. K. (1990). Gene activation is required for developmentally programmed cell death. *Proc. Natl. Acad. Sci. USA* **87,** 6594–6598.

Schwartz, L. M. (1991). The role of cell death genes during development. *BioEssays* **13,** 389–395.

Shi, Y. F., Glynn, J. M., Guilbert, L. J., Cotter, T. G., Bissonnette, R. P., and Green, D. R. (1992). Role for c-*myc* in activation-induced apoptotic cell death in T cell hybridomas. *Science* **257,** 212–214.

Sun, D., Ziegler, R., Milligan, C. E., Fahrbach, S., and Schwartz, L. M. (1995). Apolipophorin III is dramatically up-regulated during the programmed death of insect skeletal muscle and neurons. *J. Neurobiology* **26,** 119–129.

Tata, J. R. (1996). Requirement for RNA and protein synthesis for induced regression of the tadpole tail in organ culture. *Dev. Biol.* **13,** 77–94.

Van Gelder, R. N., von Zastrow, M. E., Yool, A., Dement, W. C., Barchas, J. D., and Eberwine, J. H. (1990). Amplified RNA synthesized from limited quantities of heterogeneous cDNA. *Proc. Natl. Acad. Sci. USA* **87,** 1663–1667.

Vaux, D. L., Weissman, I. L., and Kim, S. K. (1992). Prevention of programmed cell death in *Caenorhabditis elegans* by human bcl-2. *Science* **258,** 1955–1957.

Wang, Z., and Brown, D. D. (1991). A gene expression screen. *Proc. Natl. Acad. Sci. USA* **88,** 11505–11509.

Woronicz, J. D., Calnan, B., Ngo, V., and Winoto, A. (1994). Requirement for the orphan steroid receptor Nur77 in apoptosis of T-cell hybridomas. *Nature* **367,** 277–281.

Yuan, J. Y., and Horvitz, H. R. (1992). The *Caenorhabditis elegans* cell death gene *ced-4* encodes a novel protein and is expressed during the period of extensive programmed cell death. *Development* **116,** 309–320.

Yuan, J., Shaham, S., Ledoux, S., Ellis, H. M., and Horvitz, H. R. (1993). The *C. elegans* cell death gene *ced-3* encodes a protein similar to mammalian interleukin-1 beta-converting enzyme. *Cell* **75,** 641–652.

CHAPTER 8

Use of the Yeast Two-Hybrid System for Identifying the Cascade of Protein Interactions Resulting in Apoptotic Cell Death

Lynne T. Bemis, F. Jon Geske, and Robert Strange
Division of Laboratory Research
AMC Cancer Research Center
Lakewood, Colorado 80214

I. Introduction

The involuting mammary gland provides a useful model in which to study protein interactions required for the apoptotic process. Involution is the developmental stage following lactation when the mammary gland remodels from a milk-producing gland to a quiescent gland. Of the secretory epithelium, which is no longer needed to produce milk, 70% is removed by the process of apoptosis. Studies in our laboratory have shown that many changes in gene expression and protein expression occur during involution of the mammary gland (Strange *et al.,* 1992; L. T. Bemis, F. J. Geske, and R. Strange, unpublished data).

We have used the yeast two-hybrid system as a tool to identify protein interactors for proteins expressed during mammary gland involution. The goal of this project is to define the cascade of protein interactions that leads to apoptotic cell death in the mammary gland. The yeast two-hybrid technique takes advantage of a wealth of research on the separable domains of transcriptional regulatory proteins. A DNA binding domain and a transcriptional activation domain are essential for a functional transcriptional activator; however, these domains can be exchanged between proteins to create hybrid transcriptional regulatory proteins (Fig. 1). Of the several variations of the yeast two-hybrid system, methods described by Chien *et al.* (1991) or by Zervos *et al.* (1993) are generally used. The yeast two-hybrid system has been used successfully to identify protein interactors with protein kinases (Yang *et al.,* 1992; Harper *et al.,* 1993; Jackson *et al.,* 1993), p53 (Iwabuchi *et al.,* 1993), RB (Durfee *et al.,* 1993), and others.

Two areas of interest are readily addressed using the yeast two-hybrid system. The first involves protein interactions between known protein interactors and the second involves the identification of novel protein partners for specific target proteins. Novel protein interactions are identified by screening cDNA libraries with a target protein of interest. The target protein is fused in the same reading frame with a DNA binding domain (Fig. 1). The DNA binding domain is usually from GAL4 or LexA. The reporter gene construct has an upstream activation site required for binding by the specific DNA binding domain. The target protein fusion is expressed in yeast and should bind the upstream activation site, but should not be able to activate transcription in the absence of a transcriptional activation domain. The transcriptional activation domain is provided by the interacting protein and each cDNA is fused to the GAL4 activation domain or some equivalently well-characterized activation domain (also called acid patch; Zervos *et al.,* 1993). The cDNA is generated from mRNA isolated during the window of interest, that is, during mammary gland involution. The library is constructed by inserting the candidate cDNAs in the same reading frame as the activation domain either: (1) directly, reasoning that reverse transcription terminated at a variety of bases, generating a population of cDNAs, or (2) using defined linkers to generate different reading frames. Yeast target proteins have been screened successfully using a mouse cDNA library, suggesting that some proteins may be conserved evolutionarily so they can be screened with a library from another species (C. Denis and M. Drapper, personal communication). We have screened a mouse embryo library successfully (Chevray and Nathans, 1992) using a mouse target protein.

Several advantages of using the yeast two-hybrid system are that protein purification is not required to identify the novel interactor, that antibody production against the unknown protein is not required, and that the cDNA of the novel interactor is isolated at the time of interaction identification. The choice of two-hybrid system vectors may depend on the availability of an appropriate cDNA library or the choice of vector to be used in the generation of a new cDNA library. Whichever yeast two-hybrid reagents are chosen, the basic methods are the same.

II. Target Protein Fusions

The target protein is the protein of interest for which interactors are identified using the yeast two-hybrid system (TAR; Fig. 1.) The target protein is fused in the same reading frame with the DNA binding domain, resulting in a target protein that has a DNA binding domain at the amino terminus.

1. A cDNA of the target protein must be fused in frame with the DNA binding domain, usually GAL4 (Fields and Song, 1989; Chevray and Nathans, 1992) or LexA (Gyuris *et al.*, 1993; Zervos *et al.*, 1993). The target protein should be characterized for transcriptional activation domains, secretion signals, or

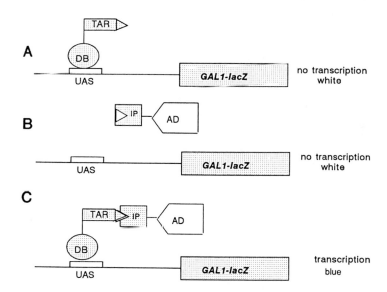

Fig. 1 The yeast two-hybrid system. (A) The DNA binding domain (DB) is fused to the target protein (TAR). There is no transcriptional activation domain (which is required for transcription); thus, there is no transcription from the DNA binding site in the promoter of the reporter gene. The yeast colonies remain white. (B) The protein interactor (IP) is fused to a transcriptional activation domain (AD). When this fusion protein is expressed in yeast, transcription of the reporter gene cannot take place without the DNA binding domain. (C) Transcription can occur because the interaction between the protein "partners" brings the DNA binding domain and the transcriptional activation domain together to act as one transcription activating protein. The yeast colonies are blue because activation of the reporter gene allows metabolism of the indicator X-GAL.

other targeting signals. Commercially available sequencing software can be used to identify protein domains. If possible any targeting domain should be deleted before fusion to the DNA binding domain.

2. The junction of the DNA binding domain and the target protein should be sequenced to confirm that the fusion protein will be translated correctly (Fig. 2).

3. If an antibody is available, protein extracts from yeast strains expressing the DNA-binding fusion should be subjected to Western analysis. Western analysis is used to verify protein expression and fusion size. Protein extraction for Western analysis can be done using the protocol described in the Appendix, although other equally applicable protocols are available.

4. The LexA repression assay can identify possible pitfalls before extensive use of the fusion protein (Brent and Ptashne, 1984). The repression assay allows the detection of the target protein in the nucleus and determines whether the LexA target fusion protein can bind DNA. This assay can only be used with LexA fusions.

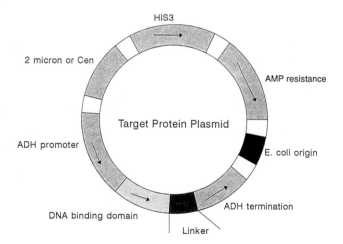

Fig. 2 Target protein plasmid. The required components for a target protein plasmid include those represented on this generalized plasmid. A yeast promoter is required and is often the constitutive ADH promoter, which is upstream from the start site for transcription of the DNA binding domain. A polylinker region with several unique restriction sites follows the DNA binding domain, allowing directional fusion of the target cDNA into the plasmid. The target cDNA will then be followed by stop codons and the correct termination sequences for ADH (ADH termination). Other yeast sequences required for plasmid maintenance by the yeast include either the 2 micron region or a centromere region (CEN) and a selectable marker, often HIS3, LEU2, or TRP1. The selectable marker HIS3 is shown here and is used to select for those yeast carrying the target protein plasmid. The *Escherichia coli* origin and AMP resistance (ampicillin resistance) gene are required for amplification of the plasmid in *E. coli*.

5. If a protein interacting partner has already been identified for the target protein, it can be helpful to fuse a known interactor in frame with an activation domain to use as a positive control.

6. Positive and negative control interaction plasmids are often available from the laboratories distributing the yeast two-hybrid reagents. We recommend using a positive control to check that plating conditions are correct.

7. Test the target protein in a yeast strain to be used for the interaction with the reporter construct present. This assay controls for the possibility that the target protein can activate transcription of the reporter by itself in the absence of a protein partner. If the target protein fusion can activate transcription of the reporter construct, it cannot be used in the yeast two-hybrid system. Deletion of the transcriptional activating region of the target protein may be possible, thus allowing use of the chosen target protein (e.g., the use of p53; Iwabuchi *et al.*, 1993).

A. Methods

All routine molecular biology techniques can be found in "Molecular Cloning: A Laboratory Manual" (Sambrook *et al.*, 1989). These techniques include plasmid amplification, isolation, and manipulation. Polymerase chain reaction (PCR) techniques have been clearly outlined in "PCR Protocols" (Innis *et al.*, 1990). Depending on the yeast two-hybrid system in use, it could be necessary to transform with two or three plasmids. Some yeast strains contain an integrated reporter construct, whereas others must be transformed with a reporter-containing plasmid. We find that individual transformation for each plasmid is the most reliable method. Also, having both the target plasmid and the reporter plasmid present in the yeast when transforming with a cDNA library is convenient. Transformation is carried out according to the methods of Schiestl and Gietz (1989) with the following exception: the yeast are grown in the selective media appropriate for maintaining plasmids that are required for testing the interaction. The yeast can be maintained on selective media when carrying the plasmids, and can be frozen at −80°C in 25% glycerol. Media must be appropriate for the selection of each plasmid, and are listed subsequently. The media must select for all plasmids. If β-galactosidase activity is to be tested, the plating media must be buffered.

III. cDNA Library Preparation and Screening

1. Screening for protein interaction with a known protein is accomplished by fusing the cDNA of the known interactor in frame with a transcriptional activation domain (Fig. 3) such as the GAL4 activation domain (amino acids 768–881).

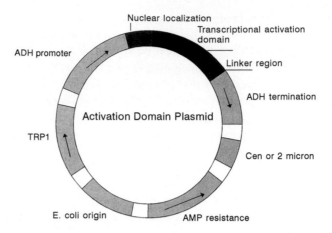

Fig. 3 Activation domain plasmid. The required components of the library plasmid are similar to those described for the target protein plasmid. There are several differences in the region required for the yeast two-hybrid system. The ADH promoter region is the same on both plasmids; however, a nuclear localization signal and transcriptional activation domain have been included to increase transport to the nucleus and to allow the interaction. A polylinker region follows the transcriptional activation domain and allows directional insertion of the cDNAs of the interactors. The ADH termination sequences are included to allow correct processing in yeast. A selectable marker different from that on the target protein plasmid is required and is shown here as TRP1.

2. The target protein plasmid (Fig. 2) and the known interactor on the transcriptional activation plasmid (Figure 3) are transformed into the yeast, which also contain a reporter gene. Interaction is measured by blue color and growth on selective media. β-gal activity can be quantified by the β-gal activity assay (Guarente, 1983).

3. Screening a cDNA library requires production of the library by reverse transcribing mRNA into cDNA and then cloning into the linker region of a plasmid similar to that represented in Fig. 3. Every cDNA fused in this manner will not result in a protein product, but one-third of the library should contain correctly fused proteins because of the triplet code for amino acids. Alternatively, linkers can be added that alter the number of nucleotides at the 5′ end of the cDNA to adjust for variable reading frames.

4. The number of transformants required should be estimated by the size of the cDNA library. The most effective initial screen is the use of a selectable marker such as growth in the absence of leucine (Zervos *et al.*, 1993) or growth in the absence of histidine (Iwabuchi *et al.*, 1993). β-Galactosidase activity or blue color should be used as a second screen. Only the colonies that activate transcription of both the selectable amino acid and the β-gal reporter genes should be screened further.

IV. Plating Protocol

The plating protocol is used to exclude certain types of false positives. Some of the putative interactors will be general factors that would give a positive result in the absence of either the target protein or the interacting protein. These false positives can be safely excluded by the plating protocol.

1. Freeze the putative interactor-containing yeast strains for future use ($-80°C$ in 25% glycerol).

2. Streak to single colonies on nonselective media (described in the Appendix). When these single colonies have grown, replica plate them to selective media. A separate replica plate to the selective media for each plasmid is required. Single colonies that have lost either the target-protein-containing plasmid or the library plasmid but still express β-galactosidase activity can be eliminated from the next round of screening. These are false positives because if either plasmid is lost and the reporters are still activated, then one of the interactors is not required for reporter activity.

3. Plasmid isolation is easiest from one of the colonies isolated by the plating protocol. If positive, the colony should not be able to activate transcription of a reporter gene. A colony that has lost its target protein plasmid and thus only contains the activation domain plasmid can be subjected to the "Smash and Grab" protocol. (An adaptation of the protocol by Hoffman and Winston, 1987, is included here.) The plasmid DNA is amplified *Escherichia coli* and digested with the appropriate restriction enzymes, to determine whether the isolated plasmid is a derivative of the cDNA library and to determine the insert size of the cDNA. One's first instinct may be to sequence all the remaining putative interactors, but it is better to continue screening for false positives.

V. Rescreening Controls and Sequencing

A. Rescreening

This procedure has been described previously by Bartel *et al.* (1993). The isolated plasmids that have been confirmed by restriction digest to contain an insert in the activation domain plasmid are transformed into a panel of yeast. One yeast strain should express the DNA binding protein with no target protein fusion. This allows screening of proteins that may interact with the DNA binding portion of the fusion protein. The second yeast strain should harbor some other target fusion to show that the interaction is not simply one of a protein that can interact with any target protein. Finally, the target-protein-containing strain is retransformed to verify that the library plasmid that has been isolated is indeed the one encoding a protein interactor for the target protein.

B. Sequencing Putative Interactors

Sequencing primers for the activation domain plasmids and the DNA binding domain plasmids are usually defined by the providers of the two-hybrids system in use. GAL4-based vectors (Bartel *et al.*, 1993) have been offered commercially by CLONTECH Laboratories (Palo Alto, CA). Sequencing primers are provided with the reagents acquired with the Matchmaker® kit from CLONTECH. We have used the Sequenase II® kit from USB (United States Biochemical, Cleveland, OH) to sequence cDNA inserts successfully.

VI. Summary

Use of the yeast two-hybrid system allows rapid identification of interacting protein or proteins for a specific target protein. The technique is readily applied and allows immediate isolation of a cDNA encoding the interacting protein. One consideration might be to outline criteria for continued study of the interactors once they are identified. Our criterion for further study of an interactor is its presence in the mammary gland at a developmental time when the target protein is also present. Further characterization of interactors may involve immunoprecipitation, enzyme assays, or other techniques applicable to the specific protein.

Appendix

1. Yeast Media

a. Nonselective Media

Plasmids will not be retained on this media. Do not use with transformed strains unless purposely trying to lose plasmids. Nonselective media is also called YEP and was described by Denis and Young (1983).

b. YEP or YD Plates

For 1 liter, weigh

> 10 g yeast extract (Difco, Detroit, MI)
> 20 g bactopeptone

Bring to 960 ml with deionized distilled water and stir until dissolved.
To a 2-liter flask, add

> 20 mg adenine
> 20 mg uracil
> 20 g agar (Difco)

Add 960 ml of the preceding solution and autoclave 20 min. Cool to 50°C and add 40 ml sterile 50% glucose. Mix well and pour plates.

For liquid media, do not add agar. After autoclaving, store in air-tight bottles. Add glucose prior to use to the correct concentration. Liquid media does not store well if glucose is added and left to sit for long periods of time.

c. Selective Media

Selective media (also called Drop-Out Plates) is adapted from that described for tryptophan selection by Williamson *et al.* (1981). For 1 liter, weigh

6.7 g yeast nitrogenous base without amino acids (Difco)

Add 950 ml deionized distilled water and mix till dissolved.
To this flask, add

20 mg adenine

20 mg uracil

20 mg tyrosine

20 g agar

Autoclave 20 min and cool to 50°C. Add appropriate drop-out solution (described next) to 10 ml/liter and 40 ml sterile 50% glucose. Pour plates.

For selection of adenine or uracil plasmid markers, leave out adenine or uracil as appropriate. Add 10 ml complete drop-out solution per liter.

d. Complete Drop-Out Solution

200 mg arginine

200 mg histidine

600 mg isoleucine

600 mg leucine

400 mg lysine

200 mg methionine

600 mg phenylalanine

500 mg threonine

400 mg tryptophan

500 mg valine

Add to 100 ml sterile water, dissolve by stirring, then filter sterilize. Amino acids are available from Sigma (St. Louis, MO). This recipe is for complete drop out. For specific drop out, add all amino acids except the one or ones for which your plasmid marker is selective.

e. X-Gal Plates

For plates containing X-gal (5-bromo-4-chloro-3-indolyl β-D-galactoside), spread 100 μl/10-cm plate of 2% X-gal in dimethylformamide or, after cooling to 50°C, add 4 ml 2% X-gal/liter. Note that pH 7.0 buffered plates are required to use X-gal.

f. pH 7.0 Buffered Plates

Prior to adding yeast nitrogenous base without amino acids, dissolve the following salts in water:

5.31 g/liter KH_2PO_4 monobasic
10.63 g/liter K_2HPO_4 dibasic

Then, complete as described for drop-out plates.

2. "Smash and Grab"

This protocol is for the isolation of yeast plasmids to be used in transformation of *E. coli* for plasmid amplification.

1. Grow an overnight 1.5-ml culture of the plasmid-bearing yeast in an appropriate selective media (one that will maintain the plasmid).
2. Spin culture in microcentrifuge for 1 min and remove supernatant. (Yeast pellet can be washed with cold sterile water at this point.)
3. Resuspend cells in
   0.2 ml "Smash and Grab" buffer
   0.1 ml buffer-saturated phenol
   0.1 ml chloroform/isoamyl alcohol (24:1)
   0.3 g acid-washed glass beads
4. Vortex vigorously for 2 min.
5. Spin in microcentrifuge for 5 min (12,000 rpm). Take clear top layer.
6. Apply the supernatant to a commercial clean-up kit column and follow directions for clean-up, or continue as directed in step 7. The Wizard® clean-up kits from Promega (Madison, WI) have worked well for us.
7. If you do not use a clean-up kit, add an equal volume of cold isopropanol to the supernatant from Step 5 and precipitate at −20°C for 30 min.
8. Spin in microcentrifuge for 20 min (12,000 rpm or greater) and dry pellet.
9. Resuspend pellet in 50 μl water.
10. Transform competent *E. coli* immediately.

"Smash and Grab" Buffer

1% SDS
2% Triton X-100
100 mM NaCl
10 mM Tris HCl, pH 8.0
1 mM EDTA

3. Yeast Extracts for Western Blots

1. Grow a 4-ml culture of cells overnight using appropriate selective media to maintain plasmids encoding fusion proteins. If cells are growing slowly, it may be necessary to grow them longer.

2. Pellet cells at 4000–5000 rpm for 5–10 min (until pellet forms). Keep cells on ice for all subsequent steps.

3. Wash pellet once with 2 ml cold sterile water; spin as in Step 2.

4. To pellet, add 400 μl SDS Harvest Buffer (SDSHB).

5. Add 1 g glass beads. Do not use a microcentrifuge tube for this step because the glass beads do not have enough room to mix. Use 13-ml plastic tubes with a round (not pointed) bottom.

6. Vortex in cold room 1.5 min in 30-sec bursts at the highest rpm of the vortex.

7. Add 300 μl SDSHB, depending on the size of the pellet. (This step is optional.)

8. Remove all liquid to microcentrifuge tube and put on ice.

9. Rinse glass beads 4× with one 300-μl aliquot SDSHB (work up and down with a pipetter).

10. Remove to the sample microcentrifuge tube from Step 8.

11. Boil sample for 5 min.

12. Spin at room temperature for 10 min in a microcentrifuge at 12,000 rpm.

13. Transfer supernatant to a microcentrifuge tube on ice that contains
    100 μl 10% Triton X-100 or 10% NP-40
    10 μl phenylmethylsulfonyl fluoride (PMSF) (100 mM)

Other protease inhibitors may also be included. Freeze at −20°C or add sample buffer, boil, and load. To do a Bradford protein calculation, use 10 μl extract per lane on a 1-mm thick minigel.

a. SDS Harvest Buffer

Make this fresh and keep on ice. For 5 ml stock, combine

100 μl 50X PMSF stock (100 mM in isopropanol or absolute ethanol)

50 μl 1 mg/ml leupeptin stock

50 μl 1 mg/ml aprotinin stock (optional)

50 μl 1 mg/ml pepstatin stock

100 μl 100 mM DTT stock

4.65 ml A.L. lysis buffer

Note: PMSF is **toxic.** Use carefully and read product information before using. PMSF has a short half-life in aqueous solution, so we suggest using the SDSHB within 45 min. All stock solutions are kept frozen at −20°C.

b. A. L. Lysis Buffer (Roussel et al., *1991*)

0.5% SDS

10 mM Tris C, pH 7.4

1 mM EDTA

Acknowledgments

After submission of this chapter, another chapter of a similar nature was published and may be a useful reference (Bartel *et al.,* 1993: "Cellular Interactions in Development: A Practical Approach." Oxford University Press, Oxford).

References

Bartel, P., Chien, C.-T, Sternglanz, R., and Fields, S. (1993). Elimination of false positives that arise in using the two-hybrid system. *Biotechniques* **14**, 920–924.

Brent, R., and Ptashne, M. (1984). A bacterial repressor protein or a yeast transcriptional terminator can block upstream activation of a yeast gene. *Nature* **312**, 612–615.

Chevray, P., and Nathans D. (1992). Protein interaction cloning in yeast: Identification of mammalian proteins that react with the leucine zipper of Jun. *Proc. Natl. Acad. Sci.* **89**, 5789–5793.

Chien, C.-T., Bartel, P. L., Sternglanz, R., and Fields, S. (1991). The two-hybrid system: A method to identify and clone genes for proteins that interact with a protein of interest. *Proc. Natl. Acad. Sci.* **88**, 9578–9582.

Denis, C. L., and Young, E. T. (1983). Isolation and characterization of the positive regulatory gene ADR1 from *Saccharomyces cerevisiae. Mol. Cell. Biol.* **3**, 360–370.

Durfee, T., Becherer, K., Chen, P.-L., Yeh, S.-H., Yang, Y., Kilburn, A., Lee, W-H., and Elledge, S. J. (1993). The retinoblastoma protein associates with the protein phosphatase type 1 catalytic subunit. *Genes Dev.* **7**, 555–569.

Fields, S., and Song, O. (1989). A novel genetic system to detect protein–protein interactions. *Nature* **340**, 245–246.

Guarente, L. (1983). Yeast promoters and *lacZ* fusions designed to study expression of cloned genes in yeast. *Meth. Enzymol.* **101**, 181–191.

Gyuris, J., Golemis, E., Chertkov, H., and Brent, R. (1993). Cdi1, a human G1 and S phase protein phosphatase that associates with Cdk2. *Cell* **75**, 791–803.

Harper, J. W., Adami, G R., Wei, N., Keyomarsi, K., and Elledge, S. J. (1993). The p21 Cdk-interacting protein Cip1 is a potent inhibitor of G1 cyclin-dependent kinases. *Cell* **75**, 805–816.

Hoffman, C. S., and Winston, F. (1987). A ten-minute DNA preparation from yeast efficiently releases autonomous plasmids for transformation of *E. coli. Gene* **57**, 267–272.

Innis, M. A., Gelfand, D. H., Sninsky, J. J., and White, T. J. (1990). "PCR Protocols: A Guide to Methods and Applications." Academic Press, San Diego.

Iwabuchi, K., Li, B., Bartel, P., and Fields, S. (1993). Use of the two-hybrid system to identify the domain of p53 involved in oligomerization. *Oncogene* **8**, 1693–1696.

Jackson, A. L., Pahl, P. M., Harrison, K., Rosamond, J., and Sclafani, R. A. (1993). Cell cycle regulation of the yeast Cdc7 protein kinase by association with the Dbf4 protein. *Mol. Cell. Biol.* **13**, 2899–2908.

Roussel, R. R., Brodeur, S. R., Shalloway, D., and Laudano, A. P. (1991). Selective binding of activated pp60[c-src] by an immobilized synthetic phosphopeptide modeled on the carboxyl terminus of pp60[c-src]. *Proc. Natl. Acad. Sci.* **88**, 10696–10700.

Sambrook, J., Fritsch, E. F., and Maniatis, T. (1989). "Molecular Cloning: A Laboratory Manual," 2d ed. Cold Spring Harbor Laboratory Press, Cold Spring Harbor, New York.

Schiestl, R. H., and Gietz, D. (1989). High efficiency transformation of intact yeast cells using single stranded nucleic acid as a carrier. *Curr. Genet.* **16,** 339–346.

Strange, R., Li, F., Saurer, S., Burkhardt, A., and Friis, R. R. (1992). Apoptotic cell death and tissue remodelling during mouse mammary gland involution. *Development* **115,** 49–58.

Williamson, V. M., Young, E. T., and Ciriacy, M. (1981). Tranposable elements associated with constitutive expression of yeast alcohol dehydrogenase II. *Cell* **23,** 605–614.

Yang, X., Hubbard, E. J. A., and Carlson, M. (1992). A protein kinase substrate identified by the two-hybrid system. *Science* **257,** 680–682.

Zervos, A. S., Gyuris, J., and Brent, R. (1993). Mxi1, a Protein that specifically interacts with Max to bind Myc–Max recognition sites. *Cell* **72,** 223–232.

CHAPTER 9

The End of the (Cell) Line: Methods for the Study of Apoptosis *in Vitro*

Anne J. McGahon, Seamus J. Martin, Reid P. Bissonnette, Artin Mahboubi, Yufang Shi, Rona J. Mogil, Walter K. Nishioka, and Douglas R. Green

Division of Cellular Immunology
La Jolla Institute for Allergy and Immunology
La Jolla, California 92037

I. Introduction

The concept of a type of cell death in which the dying cell is an active participant in its own demise is now well established. The most common form of this type of cell death has been termed apoptosis and is defined by certain morphological features (see Chapter 1, this volume; Kerr *et al.*, 1972). Early reports in the literature of the incidence of apoptosis ranged from its role in embryonic development of higher vertebrates (Harmon *et al.*, 1984) to its role in maintaining normal tissue homeostasis (Lynch *et al.*, 1986). As our understanding of the relevance of the discovery that normal tissue turnover required this type of programmed cell death grew, so did the number of circumstances under which apoptosis was reported to occur. From these studies grew the need for *in vitro* studies using cell lines derived from the tissues studied in order to dissect the complexities of the apoptotic process. It is beyond the scope of this chapter to describe the multiple advantages of using cell lines as tools for biological research. However, their importance as fundamental tools in the understanding of apoptosis is unquestionable.

One of the main constraints on studying apoptosis *in vivo* is the rapid rate of elimination of apoptotic cells by phagocytes such as macrophages. In tissues, apopotic cells and their smaller fragments (apoptotic bodies) are rapidly taken up by neighboring cells and are degraded within lysosomes. The precise mechanics of this process are still not clear; however, a variety of cell types including epithelial cells may be involved in this tissue maintenance process. Recognition and engulfment of apoptotic cells by phagocytes occurs before membrane integrity is undermined and protects the surrounding tissues or cells from the damaging effects of released intracellular contents. The mechanisms controlling the process in leukocytes are only partially understood (Cohen *et al.*, 1992; Savill *et al.*, 1993). Although researchers know that over 90% of thymocytes never reach maturity and that these cells almost certainly die via an apoptotic mechanism, such levels are not reflected in the number of apoptotic cells seen in the thymus at any given time. The paucity of apoptotic cells in a tissue in which almost all the cells die in a relatively short period of time is a good example of the problem of studying apoptosis *in vivo*, where a dead cell is removed almost as soon as it is formed.

Several immediate and obvious advantages accompany using cell lines for the study of apoptosis, including cost effectiveness and ease of manipulation. An example in this case is the use of cell lines to study the role of apoptosis in carcinogenesis. Features such as accelerated growth rates and a reduction in growth requirements make cell culture studies highly cost effective relative to animal models. Increasing evidence in the literature indicates that derangements in the cell cycle machinery may contribute to the uncontrolled cell growth characteristic of a tumor. Investigators have reported that the cyclin D1 gene can participate in cellular transformation (Hinds *et al.*, 1994). Many oncogenes are thought to encode components of the pathways through which growth factor signals feed into the cell cycle to stimulate cell division. Tumor suppressor genes such as *p53*, which normally regulate cell growth, may also operate through the cell cycle (Prives and Mandreidi, 1993). Our understanding of apoptosis in relation to the cell cycle has been augmeted by the use of such temperature-sensitive mutant cell lines as the FT210 line which possesses a defect in the cdc2 kinase. This defect causes the p34^{cdc2} protein to become inactivated and degraded in these cells at the restrictive temperature (Th'ng *et al.*, 1990). Using this cell line, researchers found that under certain conditions, premature cdc2 kinase activation is required for apoptosis (Shi *et al.*, 1994). In addition, the role of oncogenes such as v-*abl* in transformation have also been studied using temperature-sensitive mutants (Kipreos *et al.*, 1987). The primary advantage of such systems is the relative ease with which expression of the genes that are potentially involved in the process under study can be controlled.

Another extremely useful tool in apoptosis research has been the generation of transfected cell lines that express various candidate oncogenes. Such methods have led to the discovery of a role for c-*myc* in activation-induced cell death of T-cell hybridomas (Shi *et al.*, 1992) and for *bcl*-2 as an apoptosis repressor Sentman *et al.*, 1991). The interactions between such oncogenes in the cell death pathway has also been examined using such techniques (Bissonette *et al.*, 1992; Fanidi *et al.*, 1992). Discoveries using transfection techniques include the induction of apoptosis in fibroblasts by interleukin (IL)-1β converting enzyme (ICE) (Miura *et al.*, 1993) and the prevention of vertebrate neuronal death by the *crmA* gene, which is a specific inhibitor of ICE activity (Gagliardini *et al.*, 1994). Microinjection techniques used in the latter study also provide a sophisticated means for examining the role of certain genes in the induction or suppression of apoptosis in cell culture. In addition, tranfection of dominant negative inhibitors of gene function into cells or the transfection of genes with inducible promoters, such as the β-estradiol responsive GAL 4 promoter, allows the researcher to control the "switching" on or off of genes that are potentially involved in apoptosis and to examine their impact on the cell death process.

In this chapter, we describe some of the techniques commonly used to study cell death using *in vitro* cell culture systems. These techniques include a range of simple dye exclusion assays that give crude information on cell viability but do not necessarily discriminate between apoptotic or necrotic modes of cell death. Additionally, several cell death assays are described that are based on the observation that apoptosis is often accompanied by DNA fragmentation, either into the classical 'ladder' pattern of 200 bp integer multiples (Wyllie, 1980), 50kb fragments (Walker *et al.*, 1993), or single-stranded DNA cleavage (McGahon *et al.*, 1994). Assays used to measure this fragmentation include gel electrophoresis of total DNA, quantification of the release of radioactively labeled DNA, examination of cell cycle profile, and *in situ* nick translation (see Chapters 2–5, this volume). However, since apoptotic cells are morphologically highly distinctive and are easily distinguishable from both viable and necrotic cells, ideally most cell death assays should be coupled with direct morphological evaluation of the cell population under study to define with certainty the mode of cell death that is occurring.

II. Induction of Apoptosis

Apoptosis is observed primarily under physiological conditions or as a result of mildly pathological stimuli, in contrast to necrosis, a pathological form of cell death that is characterized by cell swelling, chromatin flocculation, rapid loss of membrane integrity, and rapid cell lysis, which occurs when cells are subjected to high doses of pathological stimuli. Whether a cell dies by apoptosis or necrosis on injury has been shown to be dictated largely by the dosage of the insult the cell recieves (Lennon *et al.*, 1991; Martin and Cotter, 1991). For example, using promyelocytic human leukemia HL-60 cells, researchers have observed that for many drugs, the type of cell death induced depends on the level of stress the cell endures. At high levels of insult the death is necrotic, whereas at lower levels of insult the cell engages its apoptotic machinery. This observation also applies to a number of other hematopoietic cell lines including U937, a monoblastoid line, Daudi, a B-lymphoblastoid line, and Molt-4, a T-lymphoblastoid line.

Note that each cell line varies in its sensitivity to cytotoxic stimuli. For instance, HL-60 cells exhibit extreme sensitivity to UV irradiation, the majority of the cells becoming apoptotic 4 hr after a 5-min exposure to the UV B source. In contrast, K562 cells, a chronic myelogenous leukemia cell line, exhibit a high degree of resistance to UV irradiation (Martin and Cotter, 1991). The relative sensitivities of different cell lines to the induction of apoptosis is probably dependent on a number of factors including the presence or absence of intracellular inducers or repressors of apoptosis and the presence or absence of extracellular signaling molecules or survival signals.

On a molecular level, the apoptotic process is thought to be directed by one or more "death genes." One such candidate gene is the c-*myc* proto-oncogene which can act as an inducer of both apoptosis (Evan *et al.*, 1992; Shi *et al.*, 1992) and proliferation. What dictates the cellular decision between these two responses is still unclear, but may be determined by cosignals such as the presence of growth factors or other survival signals in the cellular environment. This type of dual outcome on activation of an oncogene is also seen for the *p53* tumor suppressor gene. In thymocytes, *p53* appears to be an essential component of the pathway leading from DNA damage to apoptosis, although in other cell types *p53* activity can lead to growth arrest of the cell, DNA repair, and survival (Clarke *et al.*, 1993).

An equally important factor governing the sensitivity of a given cell type to the induction of apoptosis is the presence of intracellular repressors of apoptosis. *bcl*-2 has been shown to suppress apoptosis in a range of cell types. The expression of *bcl*-2 in tumor cells has been shown to increase the resistance of these cells to cytotoxic agents (Miyashita and Reed, 1993). Several genes with *bcl*-2 related sequences have been identified, including a human *bcl*-2 related gene *MCL*-1 isolated from a myeloid cell line (Kozopas *et al.*, 1993), although it is not yet clear whether this gene can also modulate apoptosis. Although the *bcl*-2 gene family is an important element in the control of apoptosis, other unrelated genes have also been identified as apoptosis repressors. These genes include the adenovirus *E1b* gene, activated T24-*ras*, v-*abl*, and the baculovirus gene *p35*. Finally, other factors that may determine the sensitivity of a particular cell to undergo apoptosis include its developmental stage, its point in the cell cycle, and its metabolic state.

The literature documenting apoptosis has increased almost exponentially in recent years. The modes of induction of apoptosis are both numerous and diverse, ranging from treatment with cytotoxic drugs (Lennon *et al.*, 1991) to growth factor withdrawal (Askew *et al.*, 1991). One of the first methods published for induction of apoptosis was the finding that immature T lymphocytes undergo apoptosis on exposure to glucocorticoids (Wyllie, 1980; Cohen and Duke, 1984). Other modes of induction include heat shock (Sellins and Cohen, 1991), certain toxins including dioxin (Bell and Jones, 1982), and antibodies that cross-link antigen receptors (Shi *et al.*, 1989). For the purposes of this chapter, we divide modes of induction of apoptosis into two basic categories: "physiological" induction by such means as antibody stimulation (Shi *et al.*, 1991) and growth factor withdrawal and "nonphysiological" induction by agents such as macromolecular synthesis inhibitors (Martin *et al.*, 1990b) and DNA damaging agents (Sellins and Cohen, 1987).

A. Physiological Stimuli

This category of agents is thought to induce apoptosis by mimicking physiological processes that occur *in vivo*. Unlike cytotoxic stimuli that can influence the type of cell death that occurs, depending on the dose given to the cell, physiological inducers of apoptosis such as glucocorticoids preferentially induce the apoptotic mode of cell death. These agents may be further subdivided into those that are extracellular or those that are intracellular inducers of this mode of cell death, both of which are dealt with in this chapter.

1. Extracellular Induction of Apoptosis

External signals that lead to the induction of apoptosis include the withdrawal of extracellular signals as well as their appearance. Dependence on essential survival factors for the suppression of apoptosis appears to be very widespread (reviewed by Raff, 1992). In hematopoietic stem cells, an important function of colony-stimulating factors appears to be the suppression of apoptosis (Williams *et al.*, 1990). In contrast, the stimulation of some cell surface molecules, such as the tumor necrosis factor receptor and the Fas/Apo-1 molecule, can often induce cell death by apoptosis (reviewed by Nagata, 1994). Other external information controlling the induction of apoptosis includes the presence or absence of interactions with other cells (Sellins and Cohen, 1991) and the presence or absence of soluble signaling molecules (reviewed by Collins and Lopez-Rivas, 1993).

a. Glucocorticoid-induced Cell Death

The mechanism of the induction of cell death by glucocorticoids has received much study. Dexamethasone treatment of immature rat thymocytes was first shown by Wyllie in 1980 to induce apoptosis. Glucocorticoids also induce apoptosis in human lymphoblastic cell lines, in which they induce a block in the G_1 phase of the cell cycle. The biological effects of glucocorticoids are exerted after interaction with their specific and high-affinity receptors. The mechanism of cell killing by glucocorticoids remains unclear but transcriptional regulation is implicated since an intact transcriptional transactivation domain of the glucocorticoid receptor is known to be required. This observation implies roles for either the induction or the repression of gene expression during this type of cell death. In A1.1 T-cell hybridoma cells, a dose of 500 nM dexamethasone is sufficient to induce apoptosis after a 6-hr incubation at 37°C. Many other T-cell hybridoma and lymphoma lines are normally susceptible at this dose as well.

b. Activation-Induced Cell Death

Cell cycle arrest and activation-induced apoptosis following stimulation of surface antigen receptors has been observed both in B-cell lymphomas (Scott *et al.,* 1986) and T-cell hybridomas (Shi *et al.,* 1991; Green *et al.,* 1992). In contrast with cell cycle arrest, which has been shown to be independent of calcium influx and is not inhibited by cyclosporin A (Mercep *et al.,* 1988), the induction of apoptosis in thymocytes treated with either anti-CD3 or calcium ionophores correlates with an increase in intracellular calcium (McConkey *et al.,* 1989). Although experiments with thymocytes suggest that neither protein kinase C nor tyrosine kinases are required for activation-induced death via the T cell receptor (TCR), the process in T-cell hybridomas appears to require both (Iseki *et al.,* 1991). Other activation signals such as cross-linking of Thy-1 or Ly6 have been shown to induce lymphokine production and apoptosis in T-cell hybridomas (Ucker *et al.,* 1989). Cell lines that have lost expression of the TCR do not produce lymphokines in response to anti-Thy-1 or anti-Ly6 antibodies, yet still die by apoptosis (Nickas *et al.,* 1992). The mode of signal transduction by these surface proteins is unknown, as is their relationship to the signals produced by the TCR, yet if related, these observations suggest that activation-induced death and activation-induced lymphokine production can also be dissociated.

i. Method   Apoptosis may be induced in the A1.1 T-cell hybridoma after activation of the cells via the CD3 domain of the TCR (Shi *et al.,* 1991). In this assay, T-cell hybridoma cells are exposed to antibody immobilized on plastic, the extent of apoptosis being dependent on the amount of antibody absorbed.

ii. Materials

> *Coating buffer: 50 mM* Tris-HCl, pH 9
>
> Anti-CD3 mAb: stock 1 mg/ml (other activating antibodies such as anti-TCR can be substituted)

iii. Procedure

1. Coat the plates with sufficient coating buffer to allow for complete coverage of the wells. Final antibody concentration in the coating buffer should be 1 μg/ml. For a 96-well plate, coat with 100 μl per well; for a 24-well plate, 500 μl

coating buffer is sufficient. The plates may be coated in advance and stored at 4°C for up to 1 wk.

2. Allow the antibody to adsorb to the plate for at least 4 hr at 37°C or preferably overnight at 4°C.

3. Wash the plates 5 times with culture medium to remove unbound antibody before seeding the cells at a density of 2–3 × 10^5/ml. In A1.1 cells, activation-induced cell death via CD3 stimulation can be detected morphologically in cell cultures after 4–6 hr.

c. The Fas/Apo-1 System

Incubating cells that are positive for the Fas molecule with a monoclonal antibody against Fas is sufficient to trigger apoptotic death (reviewed by Nagata *et al.*, 1994). Killing requires that the immunoglobulin molecule cross-links beyond bivalency; thus Fab_2 fragments are not effective. Death mediated by the Fas molecule is extremely rapid in the Jurkat T-cell lymphoma line, which undergoes extensive apoptosis 3–4 hr after stimulation with anti-Fas antibody (Kamiga Biomedical Co., Thousand Oaks, CA) at a concentration of 10^{-7} g/ml.

2. Apoptosis Induced by Growth Factor Withdrawal

Many cell lines in culture undergo extensive cell death in response to the withdrawal of serum and/or other growth factors from the culture medium; in most cases, the death observed is apoptotic. Signaling molecules such as epidermal growth factor (EGF), granulocyte–macrophage colony stimulating factor (GMCSF), and cytokines are some of the many signaling molecules present in serum. The dependence of a particular cell line on one or more of these moleules for the supression or induction of apoptosis may vary. Cells that undergo apoptosis following growth factor withdrawal include T lymphocytes on withdrawal of interleukin (IL)-2; eosinophils on IL-5 withdrawal; and neurons on nerve growth factor withdrawal (reviewed by Collins and Lopez-Rivas, 1993). In many cases these growth factors are not specific in their inhibition of apoptosis in their target cells and can be replaced by a variety of other growth factors.

One mechanism by which growth factor removal leads to the induction of apoptosis has been identified in fibroblasts engineered to express c-*myc* protein constitutively. When serum is removed from these cells, they fail to arrest (unlike their parental counterparts) but undergo apoptosis instead. This result suggests that other activated cells that cannot down-regulate c-*myc* or other cell cycle controlling molecules enter apoptosis when they cycle in the absence of appropriate growth factors. Overexpression of c-*myc* in some cell lines leads to death by apoptosis; however, this death was shown to be blocked by coexpression of *bcl*-2 in the same cell line (Bissonette *et al.*, 1992; Fanidi *et al.*, 1992), suggesting a cooperative role for these two oncogenes in cellular transformation.

a. Mouse IL-2-Dependent T Cells (CTLL-2)

Antigen specific helper (T_H) and cytotoxic T cells (CTLs) are dependent on IL-2 for their growth *in vitro*. On withdrawal of this cytokine, these cells synthesize proteins that result in their death by apoptosis. This behavior was first shown by Duke and Cohen (1986) in the CTLL-2 and HT-2 cell lines, which grow continuously in the presence of exogenously added IL-2. These cell lines are extremely sensitive to IL-2 withdrawl and undergo apoptosis as early as 6 hr postwithdrawal. Antigen-specific T cell lines are slightly more resistant to IL-2 deprivation, and apoptosis may be detected after a 24-hr withdrawal period.

b. IL-3-Dependent Cell Lines

In hematopoietic progenitor cells deprived of IL-3, the onset of apoptosis occurs some 12 hr after IL-3 removal. This form of apoptosis is not inhibited by protein synthesis inhibitors. Some cell lines that grow in IL-3 or serum have been shown to arrest on withdrawal of growth factors, but constitutive expression of c-*myc* in the same cells causes them to undergo apoptosis rapidly following growth factor withdrawal (Askew *et al.*, 1991).

3. CTL–Induced Death

(CTLs) kill primarily by inducing apoptosis in their targets (Masson *et al.*, 1985). Although DNA damage can be detected in most target cells, fragmentation into oligonucleosomes is not always observed (Howell and Vartz, 1987). For example Raji, a human B-cell lymphoma line, undergoes rare double-stranded breaks when exposed to CTLs, whereas single-stranded nicks are generated in the DNA of most murine targets of nonhematopoietic origin, including fibroblasts and epithelial cells (Sellins and Cohen, 1991). The CTLs may activate a target endonuclease that differs among cell types in the extent of damage that it can cause (Cohen and Sellins, 1992). A second possibility is that the CTLs inject the same enzyme into all cells, but the intracellular milieu of the target cell determines the activity of the enzyme and results in different patterns of DNA cleavage. Whether CTLs can induce apoptosis in all target cells or whether all targets are capable of undergoing apoptosis still remains unclear at the present time.

a. The Role of Granules, Protease, and Other Enzymes in CTL Killing

Cytotoxic effector cells induce apoptosis in their targets by means of cytoplasmic granules. These granules include pore-forming proteins called perforins (Podack *et al.*, 1991), a variety of proteases, in particular serine proteases known as granzymes or fragmentins, and the calcium-binding protein calreticulum. Much debate has taken place over the relative significance of perforin and granzymes in the induction of cell death. Contrary to initial reports, highly purified perforin has been found to induce necrosis rather than apoptosis (Duke *et al.*, 1989); thus CTL killing probably results from the combined involvement of both perforin and granzymes.

As early as 1980 studies were reported demonstrating that CTL- and natural killer (NK) cell-mediated killing involves the rapid breakdown of the DNA in target cells into oligonucleosomal fragments (Russell *et al.*, 1980). The purification of a 32-kDa protein called fragmentin from the granules of a rat NK leukemia (Shi *et al.*, 1992b) was one of the first observations that actually implicated an NK cell granule protein in the induction of such DNA fragmentation. Fragmentin-mediated DNA damage is Ca^{2+} dependent, which is consistent with the Ca^{2+}-dependent DNA fragmentation mediated by NK cells and lymphokine-activated killer cells (Henkart *et al.*, 1985). Two additional lymphocyte proteases implicated in the induction of DNA fragmentation have been purified by this group. These tryptases differ from the Asp-ase fragmentin in their substrate specificity, kinetics of action, requirement for protein synthesis, and target cell preference. Death of the target cell is the ultimate goal of the killer lymphocyte. Therefore it seems likely that these cells use multiple cytolytic mechanisms to achieve rapid killing (Ostergaard and Clark, 1989; Ju *et al.*, 1990).

4. The Role of Calcium Ions in the Induction of Apoptosis

Ca^{2+} influx plays an important role in the triggering of apoptosis in some but not all cell types (Kizaki *et al.*, 1988). Anti-CD3-stimulated DNA fragmentation in T-cell hybridomas (Odaka *et al.*, 1990) and in human thymocytes (McConkey *et al.*, 1989) is dependent on the presence of extracellular calcium. The induction of apoptosis in the HL-60 cell line by a range of apoptosis inducing agents including UV irradiation and ethanol was shown to be dependent on the presence of extracellular calcium (Lennon *et al.*, 1991). Ca^{2+} ionophores such as A23187, are known to be potent inducers of apoptosis in many cell types, indicating that Ca^{2+} ions can act as extracellular inducers of apoptosis. The role of intracellular calcium in the induction of the endonuclease(s) associated with apoptosis is addressed in Chapter 3. Martin and colleagues (1994) have reviewed the role of Ca^{2+} ions in the induction of apoptosis and the potential sites of action of these ions in the cell.

B. Nonphysiological Stimuli

Induction of cell death under nonphysiological conditions may result in necrosis if the cell has been subjected to a highly toxic insult, so it is important to evaluate the mode of cell death that is occurring by morphological assessment under light microscopy. Many nonphysiological stimuli induce apoptosis, including protein or RNA synthesis inhibitors, ethanol, azide, and numerous drugs, as well as viral and bacterial pathogens. Some of these agents are discussed in this section.

1. Apoptosis Induced by Macromolecular Synthesis Inhibitors

Considerable debate has occurred over the requirement for *de novo* synthesis of new gene products during apoptosis. Duke and colleagues (1983) were among the first groups to report that apoptosis can proceed in the presence of protein synthesis inhibitors. In the study in question, these investigators observed that apoptosis of CTL targets was not blocked in the presence of such drugs. Moreover, protein or RNA synthesis inhibitors such as actinomycin D and cycloheximide have been reported as inducers of apoptosis in many cell lines (Martin *et al.*, 1990b). For example, cycloheximide has been reported to trigger apoptosis in T-cell hybridomas (Cotter *et al.*, 1992), Burkitt's lymphoma cells (Takano *et al.*, 1991), and a variety of other cell types including rodent macrophages (Waring, 1990). The spontaneous apoptosis of B chronic lymphocytic human leukemia cells and peripheral blood neutrophils is also enhanced by cycloheximide (Collins *et al.*, 1991). Note that the protein synthesis requirement for apoptosis induced by the same stimulus varies among different cell types, suggesting that different cell types may have distinct apoptotic machinery or may regulate it differently (Martin, 1993). This topic is discussed in depth elsewhere (Martin, 1993), but the ultimate conclusion seems to be that whether a cell needs to synthesize proteins to undergo apoptosis is dicated by a combination of the stimulus in question and the particular cell type under study. Note that not all forms of apoptosis require protein synthesis, contrary to what investigators initially thought.

2. Induction of Apoptosis via Direct DNA Damage

a. Radiation-Induced DNA Damage

Radiation damage caused by ionizing or thermal radiation is well documented as a potent inducer of apoptosis in numerous cell types, most notably thymocytes (Sellins and Cohen, 1987). Apoptosis in mouse intestinal cell populations after irradiation has also been well documented (Hendry and Potten, 1982).

b. UV Irradiation

Some aspects of UV radiation-induced apoptosis in the human leukemia HL-60 cell line have been detailed (Martin and Cotter, 1991). Note from this study that the period of UV B irradiation is a determining factor in the type of cell death that occurs. Short periods of exposure lead to a massive induction of apoptosis in HL-60 cells and various other cell lines including U937, a monoblastoid line, and Molt-4, a T lymphoblastoid line. However, prolonged exposure of these cells to the same level of UV exposure results in necrosis. This observation reinforces the notion that the level and longevity of the insult, rather than the nature of the insult itself, determine the mode of cell death triggered.

i. Method.

1. Wash cells 3 times in growth medium and seed at 5×10^5 cells/ml in either 24-well plates or 25-cm^2 culture flasks.

2. Allow the cells to settle to a monolayer before exposing to a 302 nm UV transilluminator (as commonly used for viewing ethidium bromide-stained DNA gels) for varying time periods.

3. On removal from the UV light source, culture cells at 37°C in a humidified 5% CO_2 incubator. Apoptosis typically occurs within 3–4 hr.

c. Thermal Radiation

Hyperthermia has classically been regarded as inducing a form of necrotic cell death. However, researchers have demonstrated in several publications that mild hyperthermia induces apoptosis in many cell types. For example, hyperthermia-induced apoptosis has been reported in the human lymphoid cell line CCRF-CEM-C7 (Dyson *et al.*, 1986) and in several human and murine tumor cell lines (Harmon et al., 1991). Following heating at 43°C, murine thymocytes require further incubation at 37°C for DNA fragmentation to occur (Sellins and Cohen, 1991). This lag period suggests that a metabolic process similar to that observed in irradiated thymocytes is involved.

3. Hydrogen Peroxide-Induced Cell Death

Hydrogen peroxide has been reported to be inducer of cell death in various cell systems including murine blastocysts (Pierce and Parchment, 1991) and myeloid leukemia cell lines (Lennon *et al.*, 1991). This type of cell death has been attributed to the direct cytotoxicity of H_2O_2 and other oxidant species generated from H_2O_2. Cell injury by low level insult of this agent (typically 100 μM in HL-60 cells) allows the cell to activate an apoptotic cell death mechanism, whereas higher level insult has been shown to induce necrosis in the same cell type.

Although the mechanism of hydrogen peroxide-induced death has not been fully elucidated, investigators have suggested that oxidants activate endonucleases that lead to the DNA damage associated with apoptosis (Ueda and Shah, 1992). The requirement for protein or RNA synthesis in this mode of induction of cell death is dependent on cell type: renal tubular epithelial cells exposed to oxidant stress do not require macromolecular synthesis, whereas T-cell hybridomas do (Warter, 1992).

4. Apoptosis Induced by Cytotoxic Drugs

Most of the cytotoxic drugs used as chemotherapeutic agents have been shown to induce apoptosis in susceptible cells. The fact that disparate agents that interact with different targets induce cell death with some common features (DNA fragmentation and chromatin condensation) suggests that cytotoxicity is determined by the ability of the cell to engage this so-called programmed cell death pathway. It is important to titrate the drug being used against each cell line in a particular study since cell lines vary in their susceptibility to particular cytotoxic agents.

a. Topoisomerase Inhibitors

Topoisomerases are essential nuclear enzymes that function to resolve topological problems in DNA such as overwinding, underwinding, and concatenation, which normally arise during replication. Topoisomerase-targeting drugs appear to interfere with the breakage–reunion reaction of DNA topoisomerases. In the presence of these drugs, an aborted reaction intermediate termed the "cleavable complex" accumulates. DNA metabolic machinery apparently transforms reversible single-strand cleavable complexes to overt strands breaks, which may be an initial event in the cytotoxic pathway.

i. Topoisomerase-I-Targeting Drugs: Camptothecin. In mammalian as well as in yeast cells, toposiomerase I appears to be the sole target of camptotecin action. Topoisomerase I poisons produce breaks at replication forks that appear to be the equivalents of breaks in duplex DNA. A model has been proposed in which camptothecin inhibits the resealing step of the strand-passing reaction by binding simultaneously to DNA and topoisomerase I in a ternary complex. A dramatic S-phase cytotoxicity for camptothecin has been observed.

In experiments using camptothecin, drug concentrations titrate over the 100 mM to 100 nM range for mammalian cells.

ii. Topoisomerase-II-Targeting Drugs. Topoisomerase II has been identified as the site of action of many of the most widely used chemotherapeutic drugs including adriamycin, actinomycin D, and the epipodophylotoxins VP-16 (etoposide) and VM-26 (tenoposide). The topoisomerase II strand-passing reaction is interrupted by the intercalative antitumor drugs (adriamycin, m-AMSA, actinomycin D) and certain other nonintercalative drugs (VP-16 and VM-26) so that in the presence of a strong protein denaturant, the yield of topoisomerase-linked strand breaks increases. For a detailed review of topoisomerase inhibitors, see D'Arpa and Liu (1989).

In experiments using mammalian cells, titrate VP-16 or VM-26 over a range of 100 nM to 100 μM for the induction of apoptosis. Other cytotoxic drugs used as chemotherapeutic agents include cisplatin, araC, and adriamycin. For a detailed review on cytotoxic drugs, see Dive (1992).

III. Assessment of Apoptosis

During apoptosis a series of striking degenerative changes occurs within the cell including cell shrinkage, membrane blebbing, chromatin condensation, and nuclear fragmentation, as well as biochemical changes that are not obvious on morphological examination. The precise series of changes that occurs within a cell undergoing apoptosis may vary from cell line to cell line. For example, the type and extent of DNA fragmentation can vary. Walker and colleagues (1993) have reported on a number of model systems providing the morphological evidence of apoptosis occurring in the absence of detectable internucleosomal DNA cleavage. Other forms of DNA degradation not involving double-strand breaks have been shown to occur in C3H10T1/2 cells. Differences in apoptotic cell morphology can be readily visualized under the light microscope. Although some apoptotic cells do not appear to fragment into apoptotic bodies or do not exhibit extensive nuclear fragmentation, most do exhibit a marked chromatin condensation. The methods with which to assess apoptosis are numerous and varied, and may be based on one or more of the criteria mentioned earlier. We recommend that classification of cell death in a given system should be based on morphological assessment of the cells by light or electron microscopy, coupled with one or more of the following assays.

A. Cell Viability Assays

Loss of cell viability is most often measured as loss of membrane integrity. (Some of these assays also are addressed in Chapter 12 for neuronal cultures.) This event may be due to primary necrosis or secondary apoptosis. Loss of membrane integrity is measured by uptake of certain dyes such as trypan blue or propidium iodide or by release of radioactive chromium. Note that assays that rely on uptake of vital dyes will underestimate the extent of apoptotic cell death since apoptotic cells maintain membrane integrity for several hours. Other cell viability assays using dyes such as Alamar blue or MTT are based on biochemical parameters and measure the reduction in metabolic activity of the dying cell. These assays do not discriminate between necrosis and late apoptosis. To distinguish between these types of cell death, an assay based on DNA fragmentation and/or direct morphological evaluation of the dying cells under light microscopy is required.

1. Exclusion of Vital Dyes

a. Trypan Blue Exclusion

Trypan blue exclusion is a cell viablity assay based on the ability of live cells to exclude the vital dye trypan blue. Since one of the features of apoptotic cells is that they retain their membrane integrity for some time after the apoptotic program has started, loss of membrane integrity usually occurs only in the final stages of the process. Therefore cells in the early stages of apoptosis retain their ability to exclude the vital dye and may score as viable cells. As noted earlier, this type of assay should be combined with a morphological assay or a DNA fragmentation assay.

i. Method. To assess cell viability, an equal volume of trypan blue dye solution (0.1% w/v) is added to an aliquot of cells and is allowed to sit briefly at room temperature to allow for dye uptake. Cells are then loaded into a hemocytometer and scored for dye uptake (blue cells = dead cells). Score a minimum of 100 cells in three separate fields. Calculate cell viability as a percentage of the three separate determinations.

$$\% \text{ viable cells} = \frac{\text{number of cells excluding dye}}{\text{total number of cells}} \times 100$$

b. Flow Cytometric Analysis Using Vital Dyes

Cell shrinkage and an increase in granularity as a consequence of apoptosis may be detected and quantified by flow cytometry (see also Chapters 4 and 5). Cell shrinkage, possibly due to the extrusion of water, may be correlated to a decrease in forward light scatter. Depending on the cell line studied, this change may be paralleled by no change (HL-60 cells) or an increase (thymocytes) in right angle scatter, correlating to an increase in granularity of the cell as the chromatin condenses. These features alone may be used to quantify apoptosis, but more sophisticated methods such as propidium iodide staining, alone or in combination with the *in situ* nick translation assay (discussed later), allows quantification of DNA content and position in the cell cycle.

Nonviable cells may be recognized by their uptake of DNA-binding dyes such as propidium iodide or ethidium bromide. Nonviable cells lose their membrane integrity and are therefore permeable to these dyes and are easily detected and quantified on a flow cytometer (Fig. 1). For a more detailed discussion of the features of apoptotic cells as measured by flow cytometry, see Darzynkiewicz *et al.* (1992).

c. Propidium Iodide

Cells are incubated with the DNA-intercalating dye propidium iodide followed by flow cytometry analysis. Samples should be analyzed immediately since this dye binds reversibly to DNA and can leak out of dead cells and be taken up by viable cells which have been permeabilized during the fixation process. Ideally, cells should not be fixed; however, safety considerations often dictate that this is not possible.

i. Method

1. Incubate the samples with 5 μg/ml propidium iodide in phosphate-buffered saline (PBS), pH 7.2, for 5 min at room temperature.
2. Wash cells twice in PBS and fix in PBS, 1% paraformaldehyde.
3. Analyze samples immediately after staining. Nonviable cells appear as red fluorescing cells (see Fig. 1).

d. Ethidium Monoazide Bromide

Ethidium monoazide bromide (EMA) staining is advantageous over propidium iodide staining for cell viability, since this dye can be covalently coupled to DNA in dead cells during the staining procedure, thus preventing dye leakage on fixation. To achieve covalent coupling of EMA to DNA in dead cells, the cell population is subjected to a UV or fluorescent white light source (40 W) during the staining period (Martin *et al.*, 1994b).

i. Method

1. Cells are stained for 10 min at room temperature in 5–10 μg/ml EMA in PBS under simultaneous exposure to the UV or fluorescent light source.

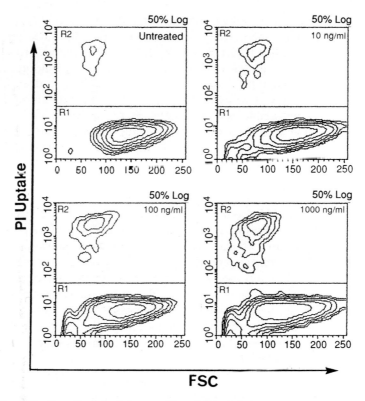

Fig. 1 Quantification of cell death by propidium iodide (PI) dye uptake. CEM T lymphoblastoid cells were cultured for 18 hr with the indicated concentrations of anti-Fas (IgM) antibody. Dead cells (*top*) are clearly separated from the live cell population with respect to their uptake of the fluorescent PI dye. Note that the dead cell population also exhibits decreased forward angle light scattering (FSC) properties, roughly equating to decreased cell size. This method lends itself to objective and highly quantitative estimates of cell death since large numbers of cells (>10,000) can easily be analyzed. In addition, the raw data can be stored on disk and retrieved for re-analysis at a later date.

2. Cells are then washed twice in PBS/azide.
3. Cells are fixed in PBS/azide containing 1% paraformaldehyde and analyzed on a flow cytometer to determine the percentage of nonviable (EMA-binding, red fluorescing) cells.

e. Hoechst 33342

The Hoechst dyes are benzimidazole derivatives that emit blue fluorescence when excited by UV light at about 350 nm. They have a high specificity for DNA and bind preferentially to A–T base regions but do not intercalate. Viable cell staining for DNA is usually performed by direct addition of Hoechst 33342 (final concentration 2–5 μg/ml) to cells in culture medium (37°C) for incubation periods of 30–90 min, depending on the cell type. Samples stained at 4°C produce poor DNA distributions, possibly indicating that conditions favoring active dye transport and retention are required. However, even under optimal conditions dye uptake, cytotoxicity, DNA binding, and analytical resolution are cell type dependent. A useful technique described by Sun *et al* (1992) combines two dyes, Hoechst 33342 and propidium iodide, to differentiate between viable and dead cells. Early apoptotic cells exclude propidium iodide and exhibit high blue fluorescence whereas dead cells (i.e., cells with damaged membranes) fluoresce both red and blue. Viable cells exhibit low blue fluorescence and also exclude propidium iodide.

2. MTT Assay

The MTT assay is a quantitative colorimetric assay for mammalian cell survival and proliferation, based on the ability of live cells to utilize a pale yellow substrate (a tetrazolium salt, MTT) and its subsequent modification into a dark blue formazan product by these cells. Tetrazolium salts are attractive candidates for this purpose, since they measure the activity of various dehydrogenase enzymes. The tetrazolium ring is cleaved in active mitochondria, so the reaction occurs only in living cells. The assay detects living but not dead cells, and the signal generated is dependent on the degree of activation of the cells. The main advantages of this assay are its rapidity and the lack of any radioisotope.

i. Materials

Stock solution: Thiazolyl Blue (MTT), 2.5 mg/ml, in deionized H_2O; store at 4°C until needed

0.04 *M* HCl in isopropanol, stored at room temperature

ii. Method

1. Grow cells in 96-well microtiter plates (typically 200 μl containing 10^5 cells).
2. After appropriate incubation with putative death inducers, determine viability by the addition of 10 μl stock MTT to each well.
3. Incubate the plate at 37°C for up to 4 hr. The reaction may be stopped earlier if the color develops rapidly (this depends on the cells in question). Appropriate positive and negative controls must be run each time in parallel.
4. At the end of the incubation period, carefully remove the media from the wells. To do so, centrifuge the plate at 1200 rpm for 5 min, then pipet off the supernatant, leaving 25–50 μl in each well to ensure that no cells are aspirated.
5. Add 100 μl 0.04 *M* HCl in isopropanol to each well. This solution solubilizes the formazan product and produces a homogeneous solution suitable for measurement of optical density.
6. After a 5-min incubation at room temperature to ensure that all the crystals

are dissolved, read the plates on a microELISA reader using a test wavelength of 595 nm.

3. Alamar Blue

The Alamar blue assay measures proliferation by means of a fluorimetric/colorimetric growth indicator that detects metabolic activity. Specifically, the system incorporates an oxidation–reduction (redox) indicator that both fluoresces and changes color in response to reduction of the growth medium that results from cell growth. Reduction related to growth causes the redox indicator to change from its oxidized form (nonfluorescent blue) to the reduced form (fluorescent red). A major advantage of this system is that it allows for continuous and kinetic monitoring of cell viability for extended periods of time, in contrast to MTT, XTT reduction, vital dye uptake, or tritiated thymidine uptake, all of which are single end point, sample-destructive methods. A disadvantage of this type of assay is that it does not measure cell death directly but measures the inhibition of cellular proliferation that accompanies death. Thus, agents that are merely cytostatic may give misleading results in this assay.

i. Materials

Alamar blue dye concentrate (Biosource International, CA)

ii. Procedure

1. Use 10^5 cells/ml seeded in a 96-well plate. Cells should be in log phase growth.
2. Aseptically add Alamar blue concentrate in an amount equal to 10% of the culture volume.
3. Return the cultures to the incubator for at least 4 hr. Incubation times vary depending on the metabolic rates of the cell lines being tested. For optimum results using HL-60, A1.1, and Jurkat cells, leave for 6 hr.
4. Measure absorbance on a microELISA plate reader at a wavelength of 570 nm. Subtract background absorbance measured at 600 nm. Alternatively, measure fluorescence with excitation wavelength at 560 nm and emission wavelength at 590 nm.

Note. It is important to work aseptically throughout this assay. Microbial contaminants will also reduce Alamar blue and will yield erroneous results if contaminated cultures are tested by this method. Samples with protein concentrations equivalent to 10% fetal calf serum do not interfere with the assay. In addition, there is no interference from the presence of phenol red in the growth medium.

Calculation. To calculate the percentage inhibition of growth by the cytoxic agent:

$$\frac{(OD_{570} - OD_{600} \text{ of test agent dilution}}{(OD_{570} - OD_{600}) \text{ of untreated control}} \times 100$$

B. Apoptosis Assays

No one discriminating feature allows the researcher to distinguish between the apoptotic and necrotic modes of cell death. The determination of whether a cell dies by apoptosis rather than necrosis may best be made on the basis of the structural changes in cellular chromatin that occur prior to membrane lysis. The majority of apoptosis assays are based on the quantification and visualization of DNA fragmentation in the cell undergoing apoptosis. Until recently, cleavage of the cellular DNA in the linker region between nucleosomes, producing fragments of nucleosomal dimensions, (or multiples thereof), was accepted as the biochemical hallmark of apoptosis. Recent reports in the literature have questioned the absolute requirement for DNA fragmentation during apoptosis, however, and reports of incidences in which cells have undergone the morphological features of apoptosis without the characteristic DNA fragmentation are increasing (Walker *et al.*, 1993). In these studies, a number of techniques designed to measure single-stranded nicks (TUNEL assay; Darzynkiewicz, 1993; Chapter 2) and the larger DNA fragments of 300 kb and 50 kb that are identifiable during the early stages of apoptosis (Walker *et al.*, 1993) have been used. Walker and colleagues (1993) present a method for detection of the initial stages of DNA fragmentation in apoptosis based on pulsed gel electrophoresis. This group has observed that DNA fragmentation during apoptosis proceeds through an ordered series of stages, commencing with the production of 300-kb fragments, that are then degraded to fragments of 50 kb. The 50-kb fragments are further degraded in some but not all cells to smaller fragments (10–40 kb) and then into the small oligonucleosome fragments that are recognized as the characteristic DNA ladder on conventional agarose gels. This technique should prove highly useful in cells that do not show the typical DNA fragmentation pattern observed during apoptosis. The following DNA fragmentation assays, when run in parallel with a cell viability assay and microscopic examination of the cells, should enable one to distinguish between the apoptotic and necrotic modes of cell death.

1. Morphological Assessment of Apoptosis

a. Cytospin Preparations of Cells

The morphological features of apoptotic cells can be identified most easily by light microscopy on cytospun cell preparations. Typical apoptotic features that may be identified include membrane blebbing, chromatin condensation, DNA fragmentation, and the formation of apoptotic bodies (see Chapter 1). The importance of microscopic examination of the cell line being studied cannot be overemphasized in light of the fact that DNA fragmentation in cell lines undergoing apoptosis is not a universal occurrence.

i. Materials
Staining

Leukostat Fixative Solution (CS#30-A4, Fisher Scientific, USA)—Reactive ingredients: malachite green in 100% methanol, 2 mg/ml

Leukostat solution 1 (CS#30 A4, Fisher Scientific)—Reactive ingredients: eosin Y, 0.1% w/v; formaldehyde, 0.1% w/v; sodium phosphate dibasic, 0.4% w/v; potassium phosphate monobasic, 0.5% w/v

Leukostat solution 2 (CS#30-C4, Fisher Scientific)—Reactive ingredients: methylene blue, 0.04% w/v; Azure A, 0.04% w/v; sodium phosphate dibasic, 0.4% w/v; potassium phosphate monobasic, 0.5% w/v

Mounting

Permount histological mounting medium (SP15-500, Fisher Scientific): typically a cell suspension of 3×10^5 cells/ml in a 100 μl volume should be cytospun onto glass slides for staining.

ii. Method
Cytocentrifugation

1. Load 100 μl cell suspension into a cytocentrifuge chamber and spin for 2 min at 500 rpm. Note that some cells are more sensitive to the centrifugation process than others. Providing a BSA buffer (1% BSA in PBS, 100 μl; spin for 2 min at 500 rpm before adding sample) prevents lysis of the cells by the centrifugal force. Typically HL-60, K562, U937, Daudi, Raji, Molt-3, Molt-4, CEM, H9, and Jurkat cells (as well as most other cell lines) can be spun directly onto the slide, but it is necessary to provide a buffer when using A1.1 cells.

Staining. After air-drying the slides for approximately 5 min, proceed with the staining process.

1. Fix the cells by staining in Leukostat fixative solution (10 dips); allow to drain.
2. Achieve nuclear staining by dipping the slide (10 dips) in Leukostat solution 1 and allowing to drain.
3. Counterstain the cytoplasm by dipping in Leukostat solution 2 (10 dips) and draining.

Mounting

1. After air-drying the slides, examine them under the light microscope to see whether the desired staining has been achieved.
2. If so, mount the slides with cover slips using an aqueous mountant.

Quantification

1. Examine the stained preparations under the 40× objective of a light microscope.
2. Score cells by their morphological characteristics as normal or apoptotic, by the criteria listed earlier.
3. Count a minimum of three fields of 100 cells per field, scoring normal and apoptotic cells and % apoptosis. Slides should be scored blindly.

Note. Apoptotic cells differ substantially from necrotic cells in their morphology. These necrotic cells must be identified as a distinct population in the cytospun preparations. Necrotic cells have several distinct features, including nuclear swelling, chromatin flocculation, and loss of nuclear basophilia (Duvall and Wyllie, 1986). Cells undergoing the final stages of necrosis burst and lyse their contents, resulting in "cell ghosts" that are identifiable on the cytospins.

b. Quantification of Cell Viability and Apoptotic Index by Acridine Orange/ Ethidium Bromide Uptake

This protocol employs the differential uptake of fluorescent DNA binding dyes acridine orange and ethidium bromide to determine viable and nonviable cells in a given population. These dyes can be used to determine which cells in the population have undergone apoptosis, and whether the cell is in the early or late states of apoptosis based on membrane integrity. Acridine orange intercalates into the DNA, giving it a green appearance. This dye also binds to RNA but because it cannot intercalate, the RNA stains red-orange. Thus a viable cell will have a bright green nucleus and red cytoplasm. Ethidium bromide is only taken up by nonviable cells. This dye also intercalates into DNA, making it appear orange, but only binds weakly to RNA, which may appear slightly red. Thus a dead cell will have a bright orange nucleus (the ethidium overwhelms the acridine) and its cytoplasm, if it has any contents remaining, will appear dark red. Both normal and apoptotic nuclei in live cells will fluoresce bright green (see Color Plate 4). In contrast, normal or apoptotic nuclei in dead cells will fluoresce bright orange. Thus, one can differentiate between early and late apoptotic cells in this system. Live cells with intact membranes will have a uniform green color in their nuclei. Early apoptotic cells whose membranes are still intact but have started to fragment their DNA will still have green nuclei since ethidium bromide cannot enter the cell, but chromatin condensation will become visible as bright green patches in the nuclei. As the cell progresses through the apoptotic pathway and membrane blebbing starts to occur, ethidium bromide may now enter the cell and stain them orange. Late apoptotic cells will have bright orange areas of condensed chromatin in the nucleus that will distinguish them from necrotic cells, which have a uniform orange color.

i. Materials

Dye mix: 100 μg/ml acridine orange (Cat. No. A-6014, Sigma St. Louis, MO) + 100 μg/ml ethidium bromide (Sigma Cat. No. E-8751), both prepared in PBS

Note.   Both acridine orange and ethidium bromide are mutagenic. Necessary precautions should be taken while handling.

Cell suspension: approx. 5×10^5–5×10^6 cells/ml; cell suspension concentration is not critical as long as a minimum of 200 cells per group are counted

ii. Method

1. Add 1 μl dye mix to 25 μl cell suspension. Cells may be in culture medium or PBS. Mix gently, swirling by hand. Dye uptake is practically instantaneous.

2. Place 10 μl cell suspension onto a clean microscope slide, cover, and examine under a $40\times$–$60\times$ objective with a filter combination suitable for reading fluorescien.

iii. Quantification

Count a minimum of 200 total cells, recording the number of cells in the following groups:

1. Live cells with normal nuclei (bright green chromatin with organized structure)

2. Early apoptotic (EA; bright green chromatin that is highly condensed or fragmented)

3. Late apoptotic (LA; bright orange chromatin that is highly condensed or fragmented)

4. Necrotic cells (N; bright orange chromatin with organized structure)

$$\% \text{ apoptotic cells} = \frac{\text{total number of apoptotic cells (EA + LA)}}{\text{total number of cells counted}} \times 100$$

$$\% \text{ necrotic cells} = \frac{N}{\text{total}} \times 100$$

Note.   If the cells have been dead for some period of time, or have begun to fragment into apoptotic bodies, they may have lost so much chromatin that the accurate determination of cell viability becomes difficult.

2. DNA Fragmentation Assays

The following protocols are dsigned to detect the results of the endonulease activity involved in the apoptotic process (see also chapter 3). To determine the optimum DNA fragmentation assay for a particular cell line, it is advisable to observe the morphological features of apoptosis in that cell line first, as determined by microscopic examination, and then to apply one or more of the following assays.

a. Agarose Gel Electrophoresis of Nucleosomal DNA Fragmentation in Total Genomic DNA

This procedure is based on the observation that DNA isolated from apoptotic cells is cleaved into a distinctive ladder pattern of approximately 200-bp integer multiples by activation of an endogenous endonuclease that cleaves the DNA in linker regions between nucleosome cores (Wyllie, 1980). Lysis of the DNA from apoptotic cells and subsequent agarose gel electrophoresis provides a reasonably good marker for apoptosis in most cell systems. However, this is not a definitive assay for reasons mentioned earlier. If the morphological features of apoptosis are apparent in a particular cell line in the absence of DNA fragmentation as visualized by agarose gel electrophoresis, we recommend that further analysis involving either nick end labeling or pulsed gel electrophoresis be carried out. For quantification of the extent of DNA fragmentation occurring during apoptosis, we refer the reader to quantitative assays such as the TUNEL assay or propidium iodide staining (see subsequent discussion.)

i. Materials

Lysis buffer: 20 mM EDTA, 100 mM Tris, pH 8.0, 0.8% (w/v) sodium lauryl sarcosine
RNase A/Tl cocktail, 1 mg/ml (Ambion Cat. No. 2286)
Proteinase K, 20 mg/ml (Ambion Cat. No. 2546).

ii. Method

1. Harvest 0.5×10^6 cells into 1.5-ml Eppendorf tubes (autoclaved). Do not use too many cells, or the digestion will not be completed and the resulting DNA solution will be too viscous. Centrifuge at 2000 rpm at 4°C and remove the supernatant.

2. Add 20 μl lysis buffer; mix the pellet and lysis buffer together by stirring with the pipet tip. Do not apply vigorous vortexing since this will shear the DNA. Add 10 μl 1 mg/ml RNase/Tl cocktail mix and mix well by flicking the tip of the tube. Incubate at 37°C for a minimum of 30 min but no more than 2 hr.

3. Add 10 μl proteinase K and incubate at 50°C for at least 1.5 hr. This step may be carried on overnight.

4. Add 5 μl loading buffer before loading sample onto agarose gel. Electrophoresis is carried out on a 1.0–1.5% agarose gel (ethidium bromide stained) in TAE buffer.

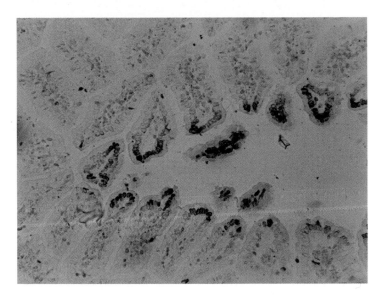

Color Plate 1 (Chapter 2) TUNEL staining in a cluster of villi from a paraffin section of rat small intestinal mucosa, demonstrating endogenous DNA fragmentation in nuclei at the villus tip. Magnification: 125×.

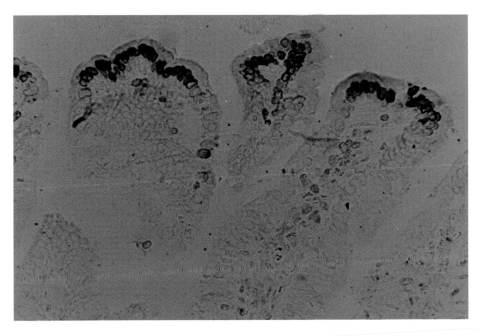

Color Plate 2 (Chapter 2) A higher magnification of TUNEL-positive villus tips of a rat small intestine, illustrating the normal morphology of PCD nuclei. Magnification: 500×.

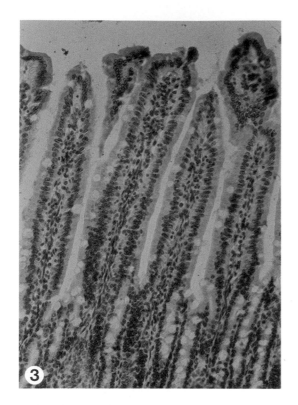

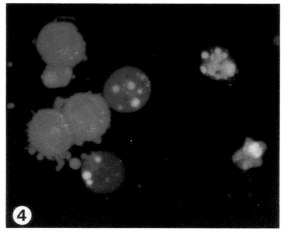

Color Plate 3 (Chapter 2) TUNEL staining of the entire tissue of the rodent small intestinal mucosa following a pretreatment with DNAase I (100 mg/ml). For more details, see Dvorak and Fallon (1991). Magnification: 250×.

Color Plate 4 (Chapter 9) Morphology of normal and apoptotic cells as assessed by acridine orange/ ethidium bromide staining. Live cells stain uniformly green apoptotic cells with intact plasma membranes appear green, with "dots" of condensed chromatin that are highly visible within, and apoptotic cells undergoing secondary necrosis are stained bright orange because of the entry of ethidium bromide into these cells.

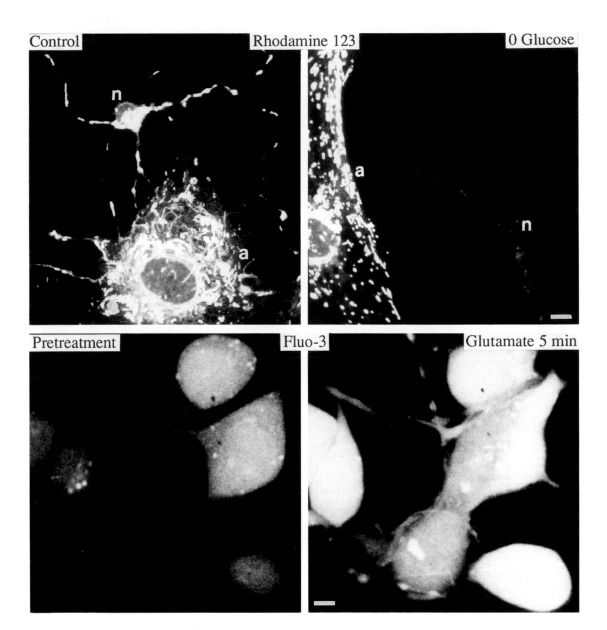

Color Plate 5 (Chapter 10, see also Fig. 6) Effects of metabolic and excitotoxic insults on mitochondrial transmembrane potential and intracellular free calcium levels assessed using the fluorescent dyes rhodamine 123 and fluo-3 and confocal laser scanning microscopy. *(Top left)* Rhodamine 123 fluorescence in a neuron (n) and an astrocyte (a) in a control rat hippocampal cell culture containing glucose; note that both the neuron and the astrocyte accumulated the dye in their mitochondria. *(Top right)* Rhodamine 123 fluorescence in a neuron (n) and an astrocyte (a) in a rat hippocampal cell culture that had been deprived of glucose for 14 hr; note that the mitochondria of the neuron accumulated little dye, indicating a reduction of transmembrane potential. Bar: 5 μm. *(Bottom)* Fluo-3 fluorescence in cultured human embryonic neocortical neurons in the resting state *(left)* and 5 min after exposure to 500 μ*M* glutamate *(right)*; note that glutamate caused a pronounced increase in fluorescence, indicating an elevation of intracellular free calcium levels. Bar: 1 μm.

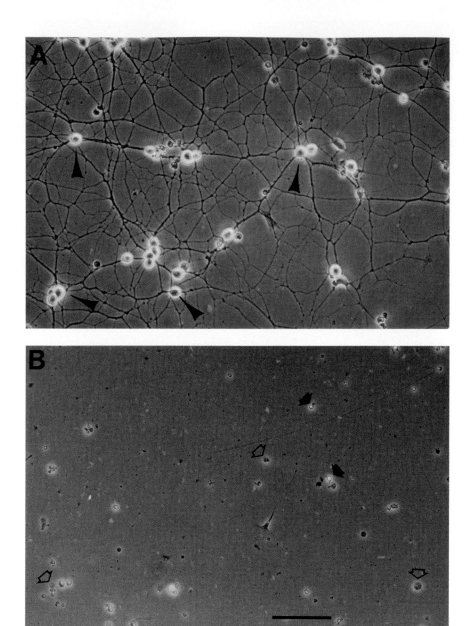

Color Plate 6 (Chapter 12)   Purified sensory neuron survival assay. Phase-contrast micrograph of purified sensory neurons in an assay for neurotrophic survival factors. DRG (dorsal root ganglion) cells are dissociated from E8 (embryonic day 8) sensory ganglia and nonneuronal cells are eliminated by differential preplating (see Section IV for details). Purified neurons are seeded on 35-mm laminin-coated dishes and individual cells are counted after 4 days in media treated with or without neuronal survival factors. (A) Survival and axonal growth from purified neurons cultured in media with 5 μg/ml BDNF (brain-derived neurotrophic factor). Arrow, phase bright neurons. (B) Debris from dead neuronal cells in identical sister cultures lacking neurotrophic factor treatment. Open arrows, membrane debris; closed arrows, nuclear debris from dead cells. Bar: 100 μm.

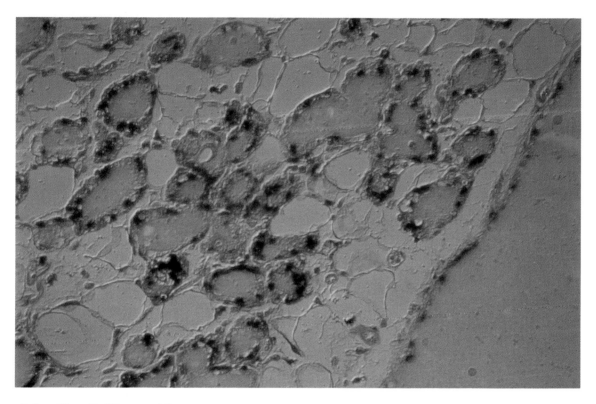

Color Plate 7 (Chapter 15) Photomicrograph showing TUNEL analysis of a 4-μm section from involuting mouse mammary gland 3.5 days after weaning.

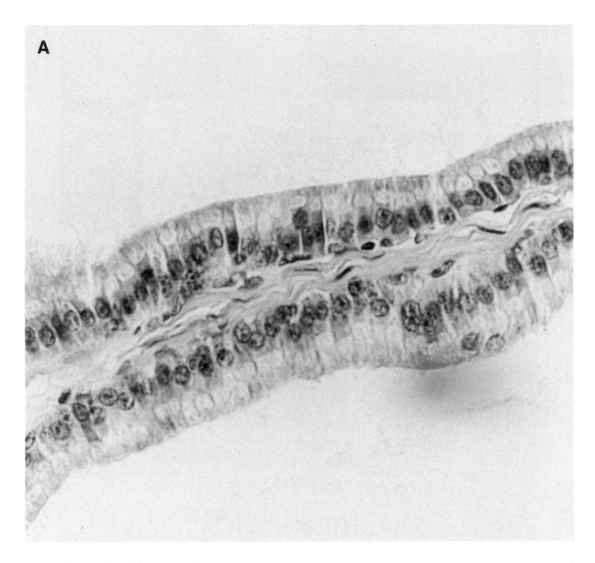

Color Plate 8 (Chapter 16) Histology of the normal and regressing rat ventral prostate gland. (A) Hematoxylin-stained thin section (200×) of a formaldehyde-fixed paraffin-embedded mature rat ventral prostate gland. The gland is a simple exocrine ductal system lined by a single layer of columnar secretory epithelial cells. (B) Higher magnification (400×) of a hematoxylin-stained thin section of a fixed embedded ventral prostate gland obtained from a 3-day castrated rat. Apoptotic bodies are easily identified within the prostatic epithelium by their characteristic morphology of shrunken cytoplasm and condensed pycnotic nuclei.

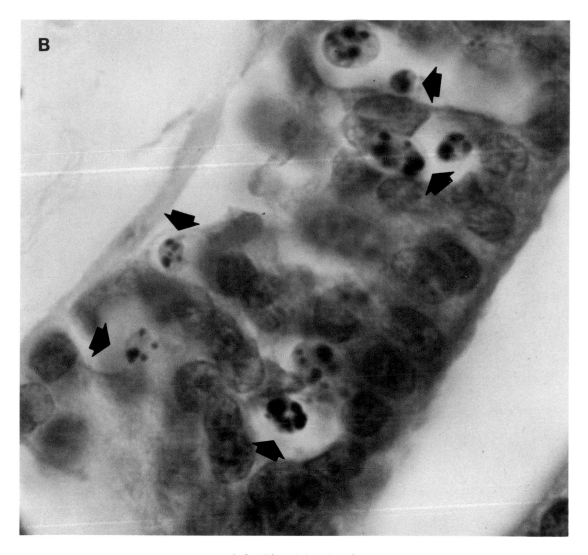

Color Plate 8 *(continued)*

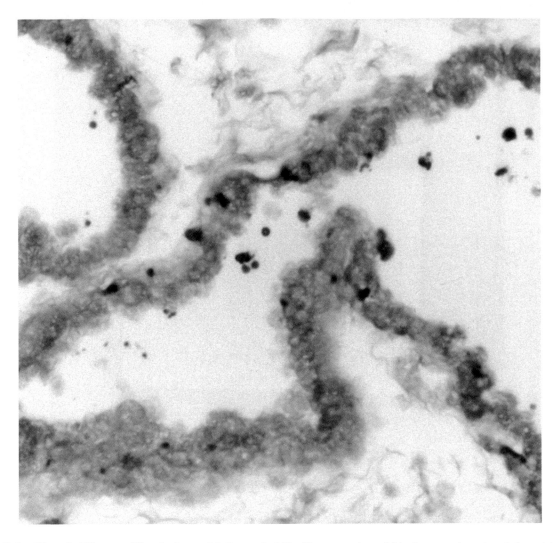

Color Plate 9 (Chapter 16) *In situ* gap-labeling method identifies apoptotic nuclei in the regressing rat ventral prostate gland. The ventral prostate gland from a 3-day castrated rat was fixed overnight in a fresh solution of 4% paraformaldehyde in phosphate-buffered saline. The tissue was dehydrated and embedded in paraffin using standard procedures. Thin sections (4 μM) were attached to poly-L-lysine coated slides, deparaffinized in xylene, and rehydrated through graded alcohols. Sections were treated with a 0.1% pepsin solution in water, pH 2.0, at 37°C for 15 min. Sections were rinsed in water and pre-incubated for 15 min in a buffer containing 50 mM Tris HCl, pH 7.5, 5 mM MgCl$_2$, 10 mM 2-mercaptoethanol, and 0.005% bovine serum albumin at room temperature. This buffer was replaced with an equivalent buffer solution containing 10 nM dATP, dCTP, and dGTP, 0.1 nM dig-dUTP, and 25 U/ml Klenow DNA polymerase. Parafilm strips were used as cover slips and incubation continued for 1 hr in a humidified chamber floating on a 15°C water bath. The slides were washed extensively in water, followed by 0.1 M sodium maleate, pH 7.4. Tissues were pre-incubated in blocking solution (from Boehringer Mannheim digoxigenin immunodetection kit) for 1 hr and subsequently incubated in blocking solution containing anti-digoxigenin monoclonal antibody (diluted 1:500) at 37°C for 45 min. Following sequential 5-min washes in sodium maleate (0.1 M, pH 7.4), were washed with 50 mM Tris-HCl, pH 9.5, 50 mM NcCl, and 10 mM MgCl$_2$, then incubated in this buffer containing NBT (nitroblue tetra zolium, final concentration 0.1%)/BCIP (5-bromo-4-chloro-3-indolyl phosphate, final concentration 0.25%) for 4 min. Slides were rinsed in water, counter-stained with 2% light green stain, dehydrated, and mounted for microscopy. Apoptotic nuclei with degraded DNA were identified by purple/black coloration.

5. Run the gel at a low voltage for better resolution (i.e., 35 V for approximately 4 hr or until the dye front has reached two-thirds of the way down the gel). Do not allow the gel front to run too far since this will reduce the resolution.

Note. When working with DNA samples, it is best to use wide-bore pipet tips to ensure that the DNA is not damaged by shearing. These tips can be obtained commercially or can be made simply by cutting off the ends of standard 200 μl pipet tips. Samples are best loaded into dry wells.

b. DNA Fragmentation Assayed by [^{3}H]Thymidine Release

Log phase cells are prelabeled with [^{3}H] thymidine and are subsequently subjected to treatments that induce apoptosis. Those cells that have undergone apoptosis fragment their DNA and, on harvesting and lysis of the cells, lose small DNA fragments into the supernatant. The loss of DNA from the cell as indicated by [^{3}H] thymidine loss may be quantified using standard scintillation counting techniques.

i. Materials

Lysis buffer A (1X): 20 mM Tris, pH 7.4, 4 mM EDTA, 0.4% Triton X-100
Lysis buffer B (1X): 20 mM Tris, pH 7.4, 4 mM EDTA, 1% SDS

ii. Procedure

1. Prelabel log phase cells with methyl-[^{3}H]thymidine at 10 μCi/ml for 1 hr at 37°C.
2. Remove the unincorporated radiolabel by washing the cells twice in PBS.
3. Incubate the radiolabeled cells for 1 hr in normal growth medium before exposing to the apoptosis-inducing conditions.
4. To quantify the fragmented DNA following different insults, remove triplicate 50-μl cell samples from each culture and plate into microtiter wells containing an equal volume of 2X lysis buffer A.
5. Allow the cells to lyse completely (10 min at room temperature) before centrifuging the lysates at 800 g to separate intact from fragmented DNA.
6. Measure the amount of [^{3}H] thymidine radioactivity associated with 50 μl of the supernatant using standard scintillation counting techniques.

[^{3}H]Thymidine available for release is determined by adding similar 50-μl aliquots of cells to an equal volume of 2X lysis buffer B to effect total lysis of cells; the whole lysate is then transferred to scintillation vials for counting. The percentage fragmented DNA may be calculated using the equation:

% fragmented DNA = (cpm in supernatant/cpm in total lysate) $\times$ 100

where CPM represents the level of radioactivity measured in counts per minute.

c. TUNEL Assay: **In Situ** *Nick End-Labeling Assay*

In this assay, individual cells are labeled with exogenous terminal deoxy-nucleotidyl transferase(TdT) (see also Chapter 2). TdT is a primer-dependent DNA polmerase that catalyzes the repetitive addition of deoxyribonucleotide from deoxynucleotide triphosphates to the terminal 3'-hydroxyl of a DNA or RNA strand with the release of inorganic pyrophosphate. Those cells with extensive DNA nicking incorporate b-dUTP to a greater extent than their unicked counterparts. The incorporated b-dUTP is detected by fluoresceinated avidin, and the cells that exhibit DNA strand breaks are identified by flow cytometry (see Fig. 2).

Note. By employing bivariate analysis (after staining with propidium iodide) and b-dUTP incorporation (detected with fluoresceinated avidin), it is possible to correlate the presence of DNA strand breaks with cell position in the cell cycle or DNA ploidy.

i. Materials

Cacodylate buffer: 0.2 *M* potassium cacodylate, 25 m*M* Tris-HCl, pH 6.6, 2.5 m*M* cobalt chloride, 0.25 mg/ml bovine serum albumin, 100 U/ml terminal deoxynucleotidyl transferase (Boehringer Mannheim, Cat. No. 220582), 0.5 n*M* biotin-16-dUTP (Boehringer Mannheim, Cat. No. 1093 070)

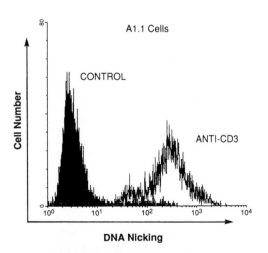

Fig. 2 *In situ* nick end-labeling of Al.1 cells. These cells have been induced to undergo apoptosis by cross-linking their CD3/TCR receptor complexes using anti-CD3 antibody. After an 18-hr incubation with anti-CD3 at 1 μg/ml, cells exhibiting DNA strand nicks incorporated biotin-dUTP, which can be detected by fluoresceinated avidin. Cells that have incorporated biotin-dUTP exhibit a log scale increase in fluorescence, in contrast to the normal population.

Saline Citrate Buffer: dilute 20X saline-sodium citrate buffer (Sigma Cat. No. S-6639) to 4X; add 2.5 mg/ml fluoresceinated avidin, 0.1% Triton X-100, 5% (w/v) non-fat dry milk, for final concentrations of 0.6 M NaCl and 0.06 M sodium citrate

Propidium iodide buffer: PBS, 5 μg/ml PI, 0.1% RNase A

Fixation buffer: 1% formaldehyde in PBS, pH 7.4, 70% ethanol

ii. Method
Cell fixation

1. Fix cells (0.5–1 $\times$ 10^6) in 1% formaldehyde for 15 min at 0–4°C.
2. To store cells before running the assay, wash the cells in PBS, resuspend in ice cold 70% ethanol, and keep at 0–4°C for up to 3 wk.

Elongation

1. Wash the fixed cells in PBS (3$\times$) and resuspend in 50 μl cacodylate buffer.
2. Incubate the cells for 30 min at 37°C.
3. Wash in PBS (3$\times$).

Staining

1. Resuspend the cells in 100 μl saline citrate buffer.
2. Incubate the cells in the saline citrate buffer at room temperature in the dark.

Propidium Iodide Staining

1. Rinse the cells in PBS containing 0.1% Triton X-100 and resuspend in 1 ml PI buffer.
2. Measure green fluorescence (labeled DNA strand breaks) and red fluorescence (position in the cell cycle) with excitation at 488 nm using 425 nm filters for green fluorescence and 620 nm LP for red fluorescence.

d. Propidium Iodide Staining for Cell Cycle Analysis

The observed reduced DNA stainability that is observed in apoptotic cells in a conventional cell cycle profile is thought to be a consequence of partial loss of DNA from these cells due to activation of an endogenous endonuclease and subsequent diffusion of the low molecular weight DNA products from the cell prior to measurement. Alternatively, marked condensation of the chromatin in apoptotic cells may render these areas of the DNA inaccessible to dyes. Therefore, the appearance of cells with low DNA stainability, lower than that of G_0/G_1 cells (sub-G_0/G_1 peak or A0 cells) in cultures treated with various drugs, is considered to be a marker of cell death by apoptosis (see Fig. 3).

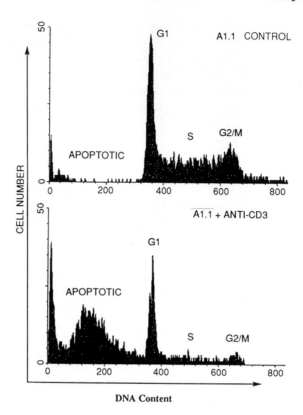

Fig. 3 Cell cycle profile of a healthy A1.1 T cell hybridoma cell population in log phase growth (*top*), exhibiting the normal distribution of cells between the G_1, S, and G_2/M phases of the cell cycle. A small proportion of the cells in this population has undergone spontaneous apoptosis and appears below the G_1 peak in the distribution. This result contrasts with the cell cycle profile of an A1.1 population that has been induced to undergo cell death by cross-linking their TCR/CD3 complexes by means of anti-CD3 antibody. Note the pronounced peak of cells with an apparent DNA content less than the normal diploid amount seen in the G_1 peak of cells; this peak corresponds to the population of apoptotic cells.

i. Materials

PBS, 5 mM EDTA
PBS, 5 mM EDTA plus 100 μg/ml propidium iodide
100% ethanol
RNase A/T1 cocktail (1 mg/ml RNaseA, T1)

ii. Method

1. Pellet cells from culture (10^6 cells total) and wash twice with cold PBS/ 5 mM EDTA.

2. Resuspend pellet in 1 ml PBS/5 mM EDTA. Fix/porate cells by slowly adding 1 ml 100% ethanol while vortexing. The fixed cells are stable and may be stored for some time at 4°C if necessary.

3. Incubate 30 min at room temperature.

4. Pellet cells. Resuspend in 0.5 ml PBS/5 mM EDTA. Add 20 μl RNaseA/T1 and mix. Incubate 30 min at room temperature. This step allows the soluble fragmented DNA to move out of the cells so that these cells then move to the left of the G_0/G_1 peak when analyzed.

5. Add 0.5 ml 100 μg/ml PI, vortex, and store at 4°C in the dark until analyzed (within 3–4 days).

6. Analyze using a FACS (FL2) in linear mode.

3. Cell Size/Granularity as Assessed by Flow Cytometry

Many workers have reported that changes in cell size and bouyant density accompany apoptotic cell death (Yamada and Ohyama, 1980; Wyllie and Morris, 1982; Martin *et al.*, 1990a). These changes are thought to be the consequence of extrusion of water from the cell due to vesicles budding from the endoplasmic reticulum, leading to a deflated ballon effect. Thus, cells in early apoptosis (i.e., prior to apoptotic body formation) will appear smaller and more dense than their normal counterparts. These changes can readily be detected in most cells by their light scattering properties on a flow cytometer (Fig. 4). The advantage of this assay is that it can deal with large numbers of cells and/or samples very rapidly and objectively. Note however that the precise light scattering changes that any given cell type will undergo during apoptosis will be cell type specific; therefore workers should check for a good correlation between the extent of apoptosis as assessed by morphological analysis and the extent of apoptosis as assessed by the changes in the light scattering properties of these cells in their particular system.

i. Method

1. Take cell from culture and analyze directly by flow cytometry. Cells should not be fixed.

IV. Conclusion

The use of cell lines in the study of apoptosis allows extensive analysis of a range of important cellular and molecular events. Cell lines permit manipulation of events in a manner that many other approaches will not allow. For example, the effects of expression or inhibition of particular genes can be accomplished more readily in cell lines than in other systems. Further, cell lines provide a uniformity of response that might not be seen *in vivo* or using cells *ex vivo*.

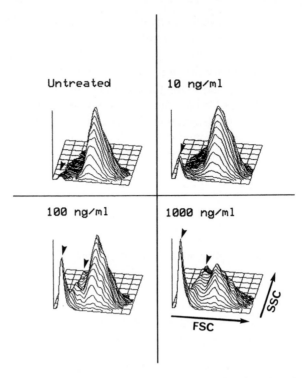

Fig. 4 Light scattering properties of CEM T lymphoblastoid cells undergoing anti-Fas-induced apoptosis. Cells were treated with the indicated concentrations of anti-Fas (CD95) IgM for 18 hr and were then analyzed on a FACScan flow cytometer without fixation. Note the appearance of dead cell peaks (arrows) with increasing amounts of anti-Fas antibody. Fas antigen-negative cells treated with the same concentrations of antibody do not undergo changes in their light scattering properties (not shown). FSC (low-angle forward light scatter) roughly equates with cell size, whereas SSC (90° or right angle side scatter) roughly equates with cell granularity or organelle density. In this type of analysis, apoptotic cells are observed to decrease in FSC (cell size) and increase in granularity (SSC), which is consistent with the idea that these cells condense in size because of extrusion of water rather than leakage of cellular contents. The tallest dead cell peak probably represents apoptotic bodies due to their small size and low granularity.

Of course, the use of cell lines also has inherent difficulties. The response of a particular cell line might not be representative of the normal tissue from which the line derived, and such responses might change with time in culture. Cells that have been cloned for any reason (e.g., following transfection thus might display variations that are not due to the manipulation (e.g., the action of the transfected gene product) and errors in interpretation may occur. Nevertheless, by attention to controls and reproducibility of effects, the information that can be garnered using cell lines for the study of apoptosis outweighs the difficulties.

As the preceding sections have shown, numerous methods are currently in use for determining cell viability. The particular methods chosen by an individual investigator will be dictated by a combination of personal preference and limitations imposed by the system under study, as well as by the availability of the specialized equipment required (such as a flow cytometer). Regardless of the exact methods chosen to measure cell death, it is important that conclusions are backed up by data from a combination of assay methods, since no single assay gives all the required information. Ideally, data should be gathered using a dye exclusion assay, a DNA fragmentation assay, and a morphological evaluation of stained cell preparations. New methods are constantly being developed. Hopefully it is only a matter of time before we have a simple and quantitative assay that can discriminate between apoptosis and necrosis.

Acknowledgments

The authors are indebted to the National Institutes of Health (AI31591) and the American Cancer Society (CB-82) for support. S. J. Martin is a Wellcome Trust International Travelling Fellow.

References

Alley, M. C., Scudiero, D. A., Monks, A., Hursey, M. L., Czerwinski, M. J., Fine, D. L., Abbott, B. J., Mayo, J. G., Shoemaker, R. H., and Boyd, M. R. (1988). Feasibility of drug screening with panels of human tumor cell lines using a microculture tetrazolium assay. *Cancer Res.* **48,** 589–601.

Askew, D. S., Ashmun, R. A., Simmons, B. C., and Cleveland, J. L. (1991). Constitutive c-myc expression in IL-3 dependent myeloid cell line suppresses cycle arrest and accelerates apoptosis. *Oncogene* **6,** 1915–1922.

Bell, P. A., and Jones, C. N. (1982). Cytotoxic effects of butyrate and other 'differentiation inducers' on immature lymphoid cells. *Biochem. Biophys. Res. Commun.* **104,** 1202–1208.

Bissonette, R. P., Echeverri, F., Mahboubi, A., and Green, D. R. (1992). Apoptotic cell death induced by c-myc is inhibited by bcl-2. *Nature* **359,** 552–554.

Clarke, A. R., Purdie, C. A., Harrison, D. J., Morris, R. G., Bird, C. C., Hooper, M. L., and Wyllie, A. H. (1993). Thymocyte apoptosis induced by p53-dependent and independent pathways. *Nature* **362,** 849–852.

Cohen, J. J. (1992). Morphological and biochemical assays of apoptosis. *Curr. Protocols Immunol.* 3.17–3.17.16.

Cohen, J. J., and Duke, R. C. (1984). Glucocortiocoid activation of a calcium-dependent endonuclease in thymocyte nuclei leads to cell death. *J. Immunol.* **132,** 38–42.

Cohen, J.J., Duke, R. C., Fadok, V. A., and Sellins, K. S. (1992). Apoptosis and programmed cell death in immunity. *Annu. Rev. Immunol.* **10** 267–293.

Collins, M. K. L., and Lopez-Rivas, A. (1993). The control of apoptosis in mammalian cells. *Trends Biochem. Sci.* **18,** 307–311.

Collins, R. J., Harmon, B. V., Gobe, G. C., and Kerr, J. F. (1991). Internucleosomal DNA cleavage should not be the sole criterion for identifying apoptosis. *Int. J. Radiat. Biol.* **61,** 451–453.

Cotter, T. G., Glynn, J. M., Echeverri, F., and Green, D. R. (1992). The induction of apoptosis by chemotherapeutic agents occurs in all phases of the cell cycle. *Anticancer Res.* **12,** 773–779.

D'Arpa, P., and Liu, L. F. (1989). Topoisomerase- targeting antitumor drugs. *Biochim. Biophys. Acta* **989,** 163–177.

Darzynkiewicz, Z., Bruno, S., Del-Bino, G., Gorczyca, W., Hotz, M. A., Lasspta, P., and Traganos, F. (1992). Features of apoptotic cells measured by flow cytometry. *Cytometry* **13**(8), 795–808.

Darzynkiewicz, Z. (1993). Induction of DNA strand breaks associated with apoptosis during treatment of leukemias. *Leukemia* **7,** 659–670.

Dive, C., and Hickman, J. A. (1991). Drug-target interactions: Only the first step in the commitment to a programmed cell death? *Br. J. Cancer* **64,** 192–196.

Dive, C., Evans, C. A., and Whetton, A. D. (1992). Induction of apoptosis—new targets for cancer chemotherapy. *Semin. Cancer. Biol.* **3**(6), 417–427.

Doherty, P. C. (1993). Cell mediated cytoxicity. *Cell* **75,** 607–612.

Duke, R. C., and Cohen, J. J. (1986). IL-2 addiction: Withdrawal of growth factor activates a suicide program in dependent T cells. *Lymphokine Res.* **5,** 289–299.

Duke, R. C., and Cohen, J. J. (1992). Morphological and biochemical assays of apoptosis. *Curr. Prot. Immunol.* **1,** 1–16.

Duke, R. C., Chervenak, R., and Cohen, J. J. (1983). Endogenous endonuclease-induced DNA fragmentation: An early event in cell-mediated cytolysis. *Proc. Natl. Acad. Sci. USA* **80,** 6361–6365.

Duke, R. C., Persechini, P. M., Chang, S., Liu, C. C., Cohen, J. J., and Young, J. D. E. (1989). Purified perforin induces target cell lysis but not DNA fragmentation. *J. Exp. Med.* **169,** 765–777.

Duvall, E., and Wyllie, A. H. (1986). Death and the cell. *Immunol. Today* **7,** 115–119.

Dyson, J. E., Simmons, D. M., Daniel, J., McLaughlin, J. M., Quirke, P., and Bird C. C. (1986). Kinetic and physical studies of cell death induced by chemotherapeutic agents or hyperthermia. *Cell Tissue Kinet.* **19,** 311–324.

Evan, G. I., Wyllie, A. H., Gilbert, C. S., Littlewood, T. D., Land, H., Brooks, M., Waters, C. M., Penn, L. Z., and Hancock, D. C. (1992). Induction of apoptosis in fibroblasts by c-myc protein. *Cell* **69,** 119–128.

Fanidi, A., Harrington, E. A., and Evan, G. I. (1992). Cooperative interaction between c-myc and bcl-2 proto-oncogenes. *Nature* **359,** 554–556.

Gagliardini, V., Fernandez, P. A., Lee, R. K. K., Drexler, H. C. A., Rotello, R. J., Fishman, M. C., and Yuan, J. (1994). Prevention of vertebrate neuronal death by the crmA gene. *Science* **263,** 826–828.

Gavrieli, Y., Sherman, Y., and Ben-Sasson, S. (1992). Identification of programmed cell death in situ via specific labeling of nuclear DNA fragmentation. *J. Cell Biol.* **119,** 493–501.

Gorczyca, W., Bruno, S., Darzynkiewicz, R. J., Gong, J., and Darzynkiewicz, Z. (1992a). DNA strand breaks occurring during apoptosis: Their early in situ detection by the terminal deoxynucleotidyl transferase and nick translation assays and prevention by serine protease inhibitors. *Int. J. Oncol.* **1,** 639–648.

Gorczyca, W., Bigman, K., Mittelman, A., Ahmed, T., Gong, J., Melamed, M. R., Lyons, A. B., Samuel, K., Sanderson, A., and Maddy, A. H. (1992b). Simultaneous analysis of Immunophenotype and apoptosis of murine thymocytes by single laser flow cytometry. *Cytometry* **13,** 809–821.

Green, D. R., and Cotter, T. G. (1992). Apoptosis in the immune system. *Sem. Immunol.* **4,** 355–362.

Green, D. R., Bissonette, R. P., Glynn, J. M., and Shi, Y. (1992). Activation-induced apoptosis in lymphoid systems. *Sem. Immunol.* **4,** 378–388.

Harmon, B. V., Bell, L., and Williams, L. (1984). An ultrastructural study on the "meconium corpuscles" in rat foetal intestinal epithelium with particular reference to apoptosis. *Anat. Embryol.* **169,** 119.

Harmon, B. V., Takano, Y. S., Winterford, C. M., and Gobe, G. C. (1991). The role of apoptosis in the response of cells and tumors to mild hyperthermia. *Int. J. Radiat. Biol.* **59**(2), 489–501.

Hendry, J. H., and Potten, C. S. (1982). Intestinal cell radiosensitivity: A comparison for cell death assayed by apoptosis or by a loss of clonogenicity. *Int. J. Radiat. Biol.* **42,** 621–628.

Henkart, P., Henkart, M., Millard, P., Frederikse, P., Bluestone, J., Blumenthal, R., Yue, C., and Reynolds, C. (1985). The role of cytoplasmic granules in cytotoxicity by large granular lymphocytes and cytotoxic T lymphocytes. *Adv. Exp. Med. Biol.* **184** 121–138.

Hinds, P. W., Dowdy, S. F., Eaton, E. N., Arnold, A., and Weinberg, R. A. (1994). Function of a human cyclin gene as an oncogene. *Proc. Natl. Acad. Sci. USA* **91,** 709–713.

Hockenbery, D., Nunez, G., Milliman, C., Schrieber, R. D., and Korsmeyer, S. J. (1990). Bcl-2 is an inner mitochondrial membrane protein that blocks programmed cell death. *Nature* **348**, 334–336.

Howell, D. M., and Martz, E. (1987). The degree of CTL-induced DNA solubilization is not determined by the human vs mouse origin of the target cell. *J. Immunol.* **138**, 3695–3698.

Iseki, R., Mukai, M., and Iwata, M. (1991). Regulation of T lymphocyte apoptosis. Signals for the antagonism between activation and glucocorticoid-induced death. *J. Immunol.* **147**, 4286–4292.

Ju, S. T., Ruddle, N. H., Strack, P., Dorf, M. E., and Dekruyff, R. H. (1990). Expression of two distinct cytolytic mechanicsms among murine CD4 subsets. *J. Immunol.* **144**, 23–31.

Kaufman, W. K., and Kaufman, D. G. (1993). Cell cycle control, DNA repair and the initiation of carcinogenesis. *FASEB J.* **7**, 1188–1191.

Kerr, J. F. R., Wyllie, A. H., and Currie, A. R. (1972). Apoptosis. A basic biological phenonemon with wide-ranging implications in tissue kinetics. *Br. J. Cancer* **26**, 239–257.

Kipreos, E. T., Lee, G. J., and Wang, J. Y. J. (1987). Isolation of temperature-sensitive tyrosine kinase mutants of v-abl oncogene by screening with antibodies for phosphotyrosine. *Proc. Natl. Acad. Sci. USA* **84**, 1345–1349.

Kizaki, H., Shimada, H., Ohsaka, F., and Sakurada, T. (1988). Adenosine, deoxyadenosine, and deoxyguanosine induce DNA cleavage in mouse thymocytes. *J. Immunol.* **141**, 1652–1657.

Kozopas, K. M., Yang, T., Buchan, H. L., Zhou, P., and Craig, R. W. (1993). MCL1, a gene expressed in programmed myeloid cell differentiation, has sequence similarity to BCL2. *Proc. Natl. Acad. Sci. USA* **90**, 3516–3520.

Lennon, S. V., Martin, S. J., and Cotter, T. G. (1991). Dose-dependent induction of apoptosis in human tumor cell lines by widely diverging stimuli. *Cell Prolif.* **24**, 203–214.

Lynch, M. P., Nawaz, S., and Gerschenson, L. E. (1986). Evidence for soluble factors regulating cell death and cell proliferation in primary cultures of rabbit endometrial cells grown on collagen. *Proc. Natl. Acad. Sci. USA* **83**, 4784–4788.

Martin, S. J. (1993). Apoptosis: Suicide, execution or murder? *Trends Cell Biol.* **3**, 141–144.

Martin, S. J., and Cotter, T. G. (1991). Ultraviolet B irradiation of human leukemia HL-60 cells in vitro induces apoptosis. *Int. J. Radiat. Biol.* **59**, 1001–1016.

Martin, S. J., Bradley, J. G. M., and Cotter, T. G. (1990a). Hl-60 cells induced to differentiate towards neutrophils subsequently undergo a programmed cell death (apoptosis). *Clin. Exp. Immunol.* **79**, 448–453.

Martin, S. J., Lennon, S. V., Bonham, A. M., and Cotter, T. G. (1990b). Induction of apoptosis (programmed cell death) in human leukemic HL-60 cells by inhibition of RNA or protein synthesis. *J. Immunol.* **145**, 1859–1867.

Martin, S. J., Green, D. R., and Cotter, T. G. (1994a). Dicing with death: Dissecting the components of the apoptosis machinery. *Tends Biochem. Sci.* **19**, 26–30.

Martin, S. J., Matear, P. M., and Vyakarnam, A. (1994b). HIV-1 infection of CD4+ T cells in vitro: Differential induction of apoptosis in these cells. *J. Immunol.* **152**, 330–342.

Masson, D., Corthesy, P., Nabholz, M., and Tschopp, J. (1985). Appearance of cytolytic granules upon induction of cytolytic activity in CTL-hybrids. *EMBO J.* **4**, 2533–2538.

McConkey, D. J., Hartzell, P., Jondal, M., and Orrenius, S. (1989). Calcium-dependent killing of immature thymocytes by stimulation via the CD3/T cell receptor complex. *J. Immunol.* **143**, 1801–1806.

McGahon, A., Bissonette, R., Schmitt, M., Cotter, K. M., Green, D. R., and Cotter, T. G. (1994). Bcr-abl maintains resistance of chronic myelogenous leukemia cells to apoptotic cell death. *Blood* **83**, 1179–1187.

Mercep, M., Bluestone, J. A., Noguchi, P. D., and Ashwell, J. D. (1988). Inhibition of transformed T cell growth in vitro by monoclonal antibodies directed against distinct activating molecules. *J. Immunol.* **140**, 324–335.

Miura, M., Zhu, H., Rotell, R., Hartwieg, E. A., and Yuan, J. (1993). Induction of apoptosis in fibroblasts by IL-1β converting enzyme, a mammalian honolog of the C. elegans cell death gene ced-3. *Cell* **75,** 653–660.

Miyashita, T., and Reed, J. C. (1993). Bcl-2 oncoprotein blocks chemotherapy-induced apoptosis in a human leukemia cell line. *Blood* **81,** 151–157.

Mosmann, T. (1983). Rapid colorimetric assay for cellular growth and survival: Application to proliferation and cytotoxicity assays. *J. Immunol. Meth.* **65,** 55–63.

Nagata, S. (1994). Apoptosis-mediating Fas antigen and its natural mutation. *In Curr. Commun. Cell Mol. Biol.* **8,** 313–326.

Nickas, G., Meyers, J., Hebshi, L. D., Ashwell, J. D., Gold, D. P., Sydora, B., and Ucker, D. S. (1992). Susceptibility to cell death is a dominant phenotype: Triggering of activation-driven T-cell death independent of the T-cell antigen receptor complex. *Mol. Cell. Biol.* **12,** 379–385.

Odaka, C., Kizaki, H., and Tadakuma, T. (1990). T cell receptor-mediated DNA fragmentation and cell death in T cell hybridomas. *J. Immunol.* **144,** 2096–2101.

Ostergaard, H. L., and Clark, W. R. (1989). Evidence for multiple lytic pathways used by cytotoxic T lymphocytes. *J. Immunol.* **143,** 2120–2126.

Pasternak, M. S., and Eisen, H. N. (1985). A novel serine esterase expressed by CTL. *Nature* **314,** 743–745.

Pierce, G. B., and Parchment, R. E. (1991). Hydrogen Peroxide as a mediator of programmed cell death in the blastocyst. *Differentiation* **46,** 181–186.

Podack, E. R., Hengartner, H., and Lichtenheld, M. G. (1991). A central role for perforin in cytolysis? *Ann. Rev. Immunol.* **9,** 129–157.

Prives, C., and Mandreidi, J. J. (1993). The p53 tumor suppressor protein: Meeting review. *Genes Dev.* **7,** 529–534.

Raff, M. C. (1992). Social controls on cell survival and cell death. *Nature* **356,** 397–400.

Russell, J. H., Masakowski, V. R., and Dobos, C. B. (1980). Mechanisms of immune lysis. I. Physiological distinction between target cell death mediated by cytotoxic T lymphocytes and antibody plus complement. *J. Immunol.* **124,** 1100–1105.

Savill, J., Fadok, V., Henson, P., and Haslett, C. (1993). Phagocyte recognition of cells undergoing apoptosis. *Immunol. Today* **14,** 131–136.

Scott, D. W., Livnat, D., Pennell, C. A., and Keng, P. (1986). Lymphoma models for B cell activation and tolerance III. Cell cycle dependence for negative signalling of WEHI-231 B lymphoma cells by anti-μ. *J. Exp. Med.* **164,** 156–164.

Sellins, K. S., and Cohen, J. J. (1987). Gene induction by gamma-irradiation leads to DNA fragmentation in lymphocytes. *J. Immunol.* **139,** 3199–3206.

Sellins, K. S., and Cohen, J. J. (1991). Hyperthermia induces apoptosis in thymocytes. *Radiat. Res.* **126,** 88–95.

Sentman, C. L., Shutter, J. R., Hockenbery, D., Kanagawa, O., and Korsmeyer, S. J. (1991). Bcl-2 inhibits multiple forms of apoptosis but not negative selection in thymocytes. *Cell* **67,** 879–888.

Shi, L., Kam, C. M., Powers, J. C., Aebersold, R., and Greenberg, A. H. (1992a). Purification of three cytotoxic lymphocyte granule serine proteases that induce apoptosis through distinct substrate and target cell interactions. *J. Exp. Med.* **176,** 1521–1529.

Shi, L., Kraut, R. P., Aebersold, R., and Greenberg, A. H. (1992b). A natural killer cell granule protein that induces DNA fragmentation and apoptosis. *J. Exp. Med.* **175,** 1521–1529.

Shi, L., Nishioka, W. K., Th'ng, J., Morton Bradbury, E., Litchfield, D. W., and Greenberg, A. H. (1994). Premature p34^{cdc2} activation required for apoptosis. *Science* **263,** 1143–1145.

Shi, Y. F., Sahai, B. M., and Green, D. R. (1989). Cyclosporin A inhibits activation-induced cell death in T-cell hybridomas and thymocytes. *Nature* **339,** 625–626.

Shi, Y. F., Bissonnette, R. P., Parfrey, N., Szalay, N., Kubo, M., and Green, D. R. (1991). In vivo administration of monoclonal antibodies to the CD3 T cell receptor complex induces cell death (apoptosis) in immature thymocytes. *J. Immunol.* **146,** 3340–3346.

Shi, Y, Glynn, J. M., Guilbert, L. J., Cotter, T. G., Bissonnette, R. P., and Green, D. R. (1992). Role for c-myc in activation-induced apoptotic cell death in T cell hybridomas. *Science* **257,** 212–214.

Sun, X.-M., Snowden, R. T., Skilleter, D. N., Dinsdale, D., Ormerod, M. G., and Cohen, G. M. (1992). A flow cytometric method for the separation and quantitation of normal and apoptotic thymocytes. *Anal. Biochem.* **204**, 351–356.

Th'ng, J. P., Wright, P. S., Hamaguchi, J., Lee, M. G., Norbury, C. J., Nurse, P., and Bradbury, E. M. (1990). The FT210 cell line is a mouse G2 phase mutant with a temperature-sensitive CDC2 gene product. *Cell* **63**, 313–324.

Takano, Y. S., Harmon, B. V., and Kerr, J. F. (1991). Apoptosis induced by mild hyperthermia in human and murine tumour cell lines: a study using electron microscopy and DNA gel electrophoresis. *J. Pathol.* **163**(4), 329–336.

Ucker, D. S., Ashwell, J. D., and Nickas, G. (1989). Activation-driven T cell death. I. Requirements for de novo transcription and translation and association with genome fragmentation. *J. Immunol.* **143**, 3461–3469.

Ueda, N., and Shah, S. V. (1992). Endonuclease-induced DNA damage and cell death in oxidant injury to renal tubular epithelial cells. *J. Clin. Invest.* **90**, 2593–2597.

Walker, R. P., Kokileva, L., LeBlanc, J., and Sikorska, M. (1993). Detection of the initial stages of DNA fragmentation in apoptosis. *BioTechniques* **15**, 1032–1036.

Waring, P. (1990). DNA fragmentation induced in macrophages by gliotoxin does not require protein synthesis and is preceded by raised inositol triphosphate levels. *J. Biol. Chem.* **265**, 14476–14480.

Warter, R. L. (1992). Radiation-induced apoptosis in a murine T-cell hybridoma. *Cancer Res.* **52**, 883–890.

White, E., Sabbitini, P., Debbas, M., Wold, W. S. M., Kusher, D. I., and Gooding, L. R. (1992). The 19 kilodalton adenovirus E1B transforming protein inhibits programmed cell death and prevents cytolysis by tumor necrosis factor α. *Mol. Cell. Biol.* **12**, 2570–2850.

Williams, G. T., and Smith, C. A. (1993). Molecular regulation of apoptosis: genetic controls on cell death. *Cell* **94**, 777–779.

Williams, G. T., Smith, C. A., Spooncer, E., Dexter, T. M., and Taylor, D. R. (1990). *Nature* **343**, 76–79.

Wyllie, A. H. (1980). Gluccocorticoid-induced thymocyte apoptosis is associated with endogenous endonuclease activation. *Nature* **284**, 555–556.

Wyllie, A. H., and Morris, R. G. (1982). Hormone-induced cell death: Purification and properties of thymocytes undergoing apoptosis after glucocorticoid treatment. *Am. J. Pathol.* **109**, 78–84.

Yamada, T., and Ohyama, H. (1980). Separation of the dead cell fraction from X-irradiated rat thymocyte suspensions by density gradient centrifugation. *Int. J. Radiat. Biol.* **37**, 695–705.

CHAPTER 10

Calcium, Free Radicals, and Excitotoxic Neuronal Death in Primary Cell Culture

Mark P. Mattson, Steven W. Barger, James G. Begley, and Robert J. Mark

Department of Anatomy and Neurobiology
Sanders-Brown Research Center on Aging
University of Kentucky
Lexington, Kentucky 40536

I. Introduction

The injury and death of neurons that occur as the result of both acute insults (e.g., stroke and head trauma) and chronic neurodegenerative disorders (e.g., Alzheimer's, Huntington's, and Parkinson's diseases) appear to involve disturbances in cellular calcium homeostasis and free radical metabolism (Choi, 1988; Jesberger and Richardson, 1991; Mattson *et al.*, 1993a; Mattson and Scheff, 1994). Dysregulation of calcium and free radical metabolism can be initiated by activation of receptors for glutamate, the major excitatory neurotransmitter in the mammalian brain. Neurons bearing glutamate receptors are particularly vulnerable when energy levels are diminished, as occurs in stroke and probably in age-related neurodegenerative disorders as well. Elucidation of cellular and molecular mechanisms of excitotoxic neuronal injury have been facilitated by the development of several cell biological technologies, including: (1) neural cell culture, (2) quantitative assays of neuronal injury and death, (3) fluorescence imaging of intracellular free calcium levels ($[Ca^{2+}]_i$), (4) evaluation of mitochondrial function, (5) quantification of reactive oxygen species, (6) assessment of cytoskeletal alterations elicited by excitotoxic/metabolic insults, and (7) detection of "stress response" proteins. This chapter presents protocols each of these technical approaches, as applied to embryonic rat and human brain neurons in dissociated cell culture. One strategy we have employed to study neuronal death is to elucidate the mechanisms that normally protect neurons from adverse environmental conditions. These studies have shown that many cellular signaling mechanisms exist that are designed to protect neurons against adverse environmental conditions such as metabolic and excitotoxic insults (Mattson *et al.*, 1993b). Indeed, many of the neuroprotective strategies acquired during evolution involve systems that control calcium and free radical metabolism.

Because this chapter focuses on the role of disturbances in calcium and free radical metabolism in neuronal injury, it is important to understand the cellular systems that normally regulate levels of calcium and free radicals (Fig. 1). In neurons, the resting intracellular free calcium concentration ($[Ca^{2+}]_i$) is approximately 100 nM, whereas the external concentration ($[Ca^{2+}]_o$) is approximately 1 mM. The low $[Ca^{2+}]_i$ is maintained by constant extrusion of Ca^{2+}, which is facilitated by ATP-dependent Ca^{2+} "pumps" in the plasma membrane and the endoplasmic reticulum (Carofoli, 1992). In addition, the plasma membrane Na^+/Ca^{2+} exchanger and cytosolic calcium-binding proteins appear to serve the important function of rapidly restoring $[Ca^{2+}]_i$ following a stimulated rise (Blaustein *et al.*, 1991; Mattson *et al.*, 1991). Calcium enters the cytoplasm through voltage-dependent and ligand-gated channels in the plasma membrane (Tsien *et al.*, 1991), as well as through channels in the endoplasmic reticulum that are activated by inositol triphosphate. Prominent among calcium-fluxing receptors are those for glutamate. Several distinct subtypes of glutamate receptor have been identified, including the metabotropic (G-protein-linked) and ionotropic subtypes (see Seeburg, 1993, for review). Ionotropic glutamate recep-

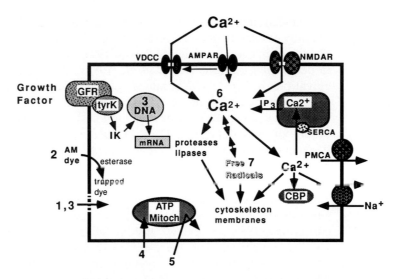

Fig. 1 Probes for assessing excitotoxic neuronal injury. Activation of glutamate receptors results in Ca^{2+} influx through N-methyl-D-aspartate (NMDA) receptors and voltage-dependent Ca^{2+} channels (VDCC), secondary to membrane depolarization mediated predominantly by non-NMDA ionotropic (AMPA) receptors. Elevation of cytosolic free Ca^{2+} levels is normally transient because of the activity of (i) the plasma membrane Ca^{2+}-ATPase (PMCA); (ii) the plasma membrane Na^+/Ca^{2+} exchanger; (iii) the endoplasmic reticulum Ca^{2+}-ATPase (SERCA); and (iv) the Ca^{2+} binding proteins (CBP). Overactivation of glutamate receptors results in sustained elevations of $[Ca^{2+}]_i$, which can directly activate proteases and can promote the formation of free radicals by activating lipases. Both Ca^{2+} and free radicals cause damage to structural proteins, enzymes, and membranes, which can lead to loss of ion homeostasis and cell degeneration. Probes for assessing cell viability include trypan blue (1), propidium iodide (3), and ethidium bromide (3), which are only taken up in cells with a damaged plasma membrane; the acetoxymethylester forms of calcein (2) and BCECF (2) which will only accumulate in healthy cells with esterase activity; rhodamine 123 (4) and JC-1 (4), whose uptake into mitochondria is dependent on transmembrane potential; MTT (5) and XTT (5), which are reduced in mitochondria and form a precipitate; and Alamar Blue (5), which is reduced by cell metabolites released into the culture medium. Intracellular free calcium levels can be assessed using the Ca^{2+} indicator dyes fura-2 and fluo-3 (6). Relative levels of reactive oxygen species can be assessed using 2,7-dichlorofluorescin (DCF; 7).

tors are subclassified as N-methyl-D-aspartate (NMDA) receptors and α-amino-3-hydroxy-5-methyl-4-isoxazolepropronic acid (AMPA)/kainate receptors. NMDA receptors possess an intrinsic Ca^{2+}-conducting pore that is activated by the combination of ligand binding and membrane depolarization. At resting membrane potential, ambient levels of Mg^{2+} block the calcium pore. AMPA/kainate receptors are highly permeable to Na^+ and play a major role in fast synaptic transmission. Membrane depolarization resulting from glutamate binding to AMPA/kainate receptors activates voltage-dependent calcium channels. Metabotropic glutamate receptors are linked via a G-protein to activation of phospholipase C, which initiates a cascade of phospholipid metabolism that leads to calcium release from internal stores (triggered by inositol triphosphate) and activation of protein kinase C by diacylglycerol.

Transient increases of $[Ca^{2+}]_i$ resulting from electrochemical stimulation and opening of ligand- and voltage-gated Ca^{2+} channels mediate information-coding processes in neural circuits (Malenka, 1991) and regulate growth cone behaviors in developing neurons (Kater *et al.*, 1988). However, uncontrolled prolonged elevations of $[Ca^{2+}]_i$ can result in neuronal degeneration and death (Mattson, 1992). Lethal elevations of $[Ca^{2+}]_i$ can be elicited by a variety of metabolic insults (e.g., hypoglycemia or mitochondrial poisons such as cyanide) that cause failure of ATP-dependent Ca^{2+} extrusion mechanisms. Calcium appears to damage cellular proteins and membranes by activating proteases such as the calpains (Siman and Noszek, 1988), and by promoting free radical production via activation of lipases (Verity, 1993) or nitric oxide synthase (Lipton *et al.*, 1993).

Free radicals (molecules with one unpaired electron or more) are constantly being produced in cells during oxidation–reduction reactions (see Halliwell and Gutteridge, 1985, for review). Free radicals in cells arise from several sources, but the oxygen used in the electron transport chain is a major source of a number of highly reactive oxygen species including the superoxide radical $(O_2\cdot^-)$, hydrogen peroxide (H_2O_2), and the hydroxyl radical $(OH\cdot)$. Transition metals promote production of the highly reactive and destructive hydroxyl radical by interacting with hydrogen peroxide and/or the superoxide radical. Additional cellular sources of free radicals include the arachidonic acid cascade of phospholipid metabolism and self-oxidation reactions involving flavins, catecholamines, and ferridoxins. The unsaturated bonds of fatty acids and cholesterol are particularly vulnerable to free radical attack, yielding lipid peroxides that are also free radicals. Thus, once initiated, lipid peroxidation becomes an autocatalytic phenomenon.

Excitatory amino acids promote free radical accumulation in neurons by several means, including activation of phospholipases that induce phospholipid hydrolysis, accumulation of arachidonic acid, and subsequent oxidation by cyclo-oxygenase and lipoxygenase (Verity, 1993); conversion of xanthine dehydrogenase to xanthine oxidase which catalyzes the oxidation of hypoxanthine or xanthine and generates the superoxide radical (McCord, 1985); and activation of nitric oxide synthase which generates nitric oxide, which reacts with superoxide anion resulting in formation of the highly destructive peroxynitrite radical (Lipton *et al.*, 1993).

This chapter presents approaches and protocols for studying various aspects of the role of calcium and free radicals in neuronal injury and death. We have not attempted a review of this area of investigation, and focus instead on methods used in this laboratory and the kinds of data we have obtained using these methods.

II. Brain Cell Culture Methods

The methods described in this section were initially developed for culture of neurons from a specific brain region, the hippocampus (Mattson and Kater,

1988; Mattson *et al.,* 1988a). However, we have successfully applied essentially identical procedures to a broad array of neuronal populations in embryonic rat and human mammalian central nervous systems (Mattson and Rychlik, 1990a; Cheng and Mattson, 1992). Each of the protocols described in subsequent sections has been applied to neurons from several regions of embryonic rat brain (hippocampus, septal area, neocortex) as well as to embryonic human neocortical neurons.

A. Procurement and Preparation of Tissue from Embryonic Rat and Human Brains

Timed pregnant female rats are purchased from Harlan Sprague Dawley (Indianapolis, IN) and used as the source of embryos (17 or 18 days of gestation). Dams are subjected to overdose of halothane anesthesia, and embryos are removed to sterile 100-mm plastic petri dishes containing cold 0.15 M saline with 10 μg/ml gentamicin sulfate. Heads are removed from the embryos and are transferred to 60-mm petri dishes containing cold HEPES-buffered (10 mM) Hank's balanced saline solution (HBSS) lacking Ca^{2+} and Mg^{2+} and containing 10 μg/ml gentamicin sulfate. Brains are removed and placed in HBSS-containing petri dishes where the meninges are removed. Brain regions of interest are removed by dissection with the aid of a dissecting microscope using transillumination (i.e., the light source is below the brain tissue). Brain tissue from all embryos (typically, 10–17 per pregnant rat) is pooled in 35-mm petri dishes containing HBSS. In preparation for enzymatic dissociation (see subsequent section), cerebral hemispheres are minced into pieces of approximately 1 mm^2 (each hemisphere is cut into 3–4 pieces) whereas hippocampi are left intact.

The relevance to humans of any studies done in lower species must ultimately be determined by performing similar studies in humans. This point is particularly salient when considering that our current understanding of mechanisms of neuronal death come almost entirely from animal studies. Therefore we have established methods for the procurement, cryopreservation, and cell culture of neurons from human embryonic cerebral cortex and hippocampus (Mattson and Rychlik, 1990a). Protocols for obtaining postmortem human fetal brain tissue from elective abortions must follow stringent federal and institutional guidelines. Our past studies have utilized brain tissue obtained from 12- to 16-wk-old fetuses obtained from elective abortions. Bottles containing sterile calcium- and magnesium-free HBSS (pH 7.2) are stored refrigerated at the surgical clinic, and fetal remains that include brain tissue are placed in the HBSS solution within 5–10 min of the surgery to be transported to the research laboratory. Cerebral cortices are identified based on their large size and presence of convolutions (immature gyri). The cortical tissue is removed to 60-mm petri dishes containing HBSS, and minced into pieces of approximately 2 mm^2.

Cell dissociation and culture procedures are essentially identical for rat and human brain tissues. The brain tissue is transferred to sterile 15-ml tubes; then the HBSS is removed and replaced with 3–5 ml HBSS containing 0.2% trypsin.

After a 15- to 20-min incubation in the trypsin solution, the tissue pieces are rinsed with fresh HBSS, and incubated for 5 min in 0.1% soybean trypsin inhibitor. After another wash in HBSS, cells are dissociated by triturating the tissue through the narrowed bore of a fire-polished pasteur pipet. Tissue is triturated until most tissue pieces of visible size have been disrupted. The ratio of tissue to volume of dissociation solution should be approximately: 8 rat hippocampi/1 ml; 6–8 rat cerebral hemispheres/1 ml; 1 human fetal cerebral hemisphere/30 ml. Aliquots of the cell suspension (typically 20–100 μl/35-mm diameter petri dish or 200–300 μl/60-mm petri dish) are added to culture dishes containing medium (see subsequent section). Cells are allowed to attach to the culture substrate (3–6 hr for rat hippocampal, septal, and cortical neurons; 24 hr for human cortical cells). Then the culture medium is replaced with fresh medium at a reduced volume (0.8 ml in 35-mm dishes and 1.6 ml in 60-mm dishes).

Cells from fetal rat (Mattson *et al.*, 1988b) or human (Mattson and Rychlik, 1990a) brain can be cryopreserved for culture at a later time. The method involves dissociation of cells in MEM + medium (for composition, see Section II,C) containing 8% dimethylsulfoxide. Aliquots (0.5 ml) of the cell suspension are distributed to cryovials, which are then slowly cooled (approximately 1°C/min) to −80°C and subsequently transferred to liquid nitrogen for long-term storage. A simple way of slowly freezing the cells is to sandwich the cryovials between two slabs of styrofoam (2-cm thick slabs) and place them in a −80°C freezer. Immediately before cell plating, the cryopreserved cells are rapidly thawed by placing the cryovials in a 37°C water bath. We typically find that 60–80% of the neurons survive the cryopreservation procedure. The cryopreservation methodology has proven particularly valuable for human fetal neocortical cells since stocks of frozen cells can be established and used over periods of months to years.

B. Culture Dishes and Growth Substrata

The choice of culture dish obviously depends on the nature of the experiments to be performed. For cell survival studies involving morphological evaluation and for immunocytochemical studies, 35-mm plastic dishes are useful. For fluorescence imaging studies involving calcium-, pH-, or oxidation-sensitive dyes, the cells must be grown on 0.17-mm thick glass. For this purpose we use 35-mm plastic dishes that have a 1-cm diameter hole drilled in the bottom and a glass coverslip affixed to the outer surface covering the hole. Such dishes can be purchased from Mat-Tek Inc. (part # P35G; Ashland, MA). For Western or Northern blot analyses, cells are cultured in 60-mm dishes at high density (200–500 cells/mm^2 of culture surface). For studies that employ a fluorescence plate reader, we typically use 24-, 48-, or 96-well plates.

Neuronal attachment and long-term cell survival are affected greatly by the culture substrate. Primary neurons do not attach or survive well on uncoated plastic. Two substrates that work well for primary neuronal cultures are poly-L-lysine and polyethyleneimine. Poly-L-lysine (#P1399; Sigma, St. Louis, MO) is prepared at a concentration of 10 μg/ml (for plastic dishes) or 1 mg/ml (for glass-bottom dishes) in borate buffer (borate buffer consists of 2.37 g borax and 1.55 g boric acid dissolved in 500 ml sterile water, pH 8.4). The culture surface is covered with the poly-L-lysine solution for 1 hr, and then rinsed thoroughly (typically 3–5 washes) with sterile, glass-distilled water. The dishes are allowed to dry and are then sterilized by exposure to UV light for 10 min. Although it is preferable to prepare the dishes within a few days of use, they can be stored at room temperature for up to 2 wk. Polyethyleneimine (50% solution, Sigma #P3143) is diluted 1 : 1000 in borate buffer. Dishes are incubated overnight in the polyethyleneimine solution (room temperature), washed 4 times with phosphate-buffered saline (PBS), and allowed to dry. The dishes are sterilized by exposure to UV light for 10 min, and culture medium is added to the dishes (polyethyleneimine-coated dishes cannot be stored dry). Although poly-L-lysine is the most commonly used substrate for primary neuronal cultures, it has been our experience that polyethyleneimine is consistently superior for maintenance of long-term cultures, particularly when the cells are grown on a glass surface.

C. Media for Culture Maintenance and Experimentation

Serum-supplemented media generally provide superior long-term neuronal survival in primary brain cell cultures. A maintenance medium we routinely use (referred to as MEM +) consists of Minimum Essential Medium with Earle's salts (prepared from powder; #410-1700EC; Gibco, Grand Island, NY) supplemented with 10 mM sodium bicarbonate, 2 mM L-glutamine, 1 mM pyruvate, 20 mM KCl, and 10% (v/v) heat-inactivated fetal bovine serum (FBS; Sigma). Using this medium we can routinely maintain dissociated neurons from embryonic rat hippocampus, cerebral cortex, or septal area for up to 3 wk, and human fetal cerebral cortical neurons for up to 2 mo. Medium containing serum promotes glial cell proliferation and may therefore not be suitable for some applications. In addition, certain manipulations including studies that involve exposure of cells to deoxyoligonucleotides require the absence of serum components, (Mattson *et al.*, 1993d). A serum-free defined medium that supports the survival of neurons consists of MEM supplemented with 5 μg/ml bovine insulin, 100 μg/ml human transferrin, 100 μg/ml bovine serum albumin (fraction V), 60 ng/ml progesterone, 16 μg/ml putrescine, 40 ng/ml sodium selenite, 42 ng/ml thyroxine, 33 ng/ml tri-iodo-L-thyronine, 2 mM L-glutamine, 1 mM sodium pyruvate, and 20 mM KCl (Sigma). An excellent commercially available defined medium is Neurobasal with B27 supplements (Gibco, Grand Island, NY). Glial cells will not proliferate in this serum-free medium. An alternative to serum-free medium that can be employed to suppress glial proliferation is the use of mitotic inhibitors such as cytosine arabinoside (10 μM), although in our experience such compounds reduce long-term neuronal survival.

For experiments that involve constant monitoring of cells in a room air atmosphere (e.g., $[Ca^{2+}]_i$ imaging studies), a medium buffered with HEPES rather than bicarbonate is required to maintain the proper pH. For short-term experiments (<4 hr), we routinely use an HBSS-based medium consisting of HBSS containing 2 mM $CaCl_2$, 1 mM $MgCl_2$, 10 mM glucose, 1 mM L-glutamine, and 10 mM HEPES (pH 7.2). For longer term experiments (>4 hr; e.g., glucose deprivation studies; Cheng and Mattson, 1991) cells fair better in a mixed buffer solution and a 6% CO_2 atmosphere; we employ Locke's solution which contains 154 mM NaCl, 5.6 mM KCl, 2.3 mM $CaCl_2$, 1.0 mM $MgCl_2$, 3.6 mM, $NaHCO_3$, 5 mM HEPES, and 10 mM glucose (glucose is 25 μM for glucose-deprivation studies).

III. Photomicrographic Assessment of Neuronal Injury

Cells are visualized and photographed at 100× magnification using Kodak Technical Plan (TP2415) 35-mm black-and-white film. We use a Nikon 2000 camera set at ASA 100 and use automatic exposure. The method relies on the ability to relocate the same microscope fields so they can be rephotographed at successive time points. This is accomplished by: (1) drawing a mark at one point on the edge of the bottom half of the culture dish, which is used to orient the dish on the microscope stage; (2) scratching a grid on the bottom (outer surface) of the 35-mm plastic culture dish using a single-edge razor blade (we typically use a 3 × 2 grid); and (3) locating microscope fields adjacent to cross-points of the grid lines; the field is arranged so that two intersecting lines form two sides of the microscope field. An example of an experiment in which glutamate-induced cell injury was examined with this method is shown in Fig. 2.

a. Advantages of This Method

(1) The very same cells are examined before and after experimental treatment; this establishes the morphological integrity of cells in each dish prior to experimentation. (2) Vulnerability of different types of cells in the culture (e.g., neurons and astrocytes) can be determined. (3) Interactive effects of different cell types may be discerned. For example, this method allowed us to discover that neurons contacting astrocytes are more resistant to excitotoxicity than are neurons not contacting astrocytes (Mattson and Rychlik, 1990b). (4) Since many morphological changes precede cell death, this method can provide a more sensitive index of neuronal injury than methods such as vital dye staining or measurement of lactate dehydrogenase (LDH) release (Monyer *et al.*, 1992; see below). For example, neurite fragmentation is an early sign of neuronal injury resulting from elevations of $[Ca^{2+}]_i$ and free radicals. The mechanistic basis of differential vulnerability of dendrites and axons to glutamate toxicity

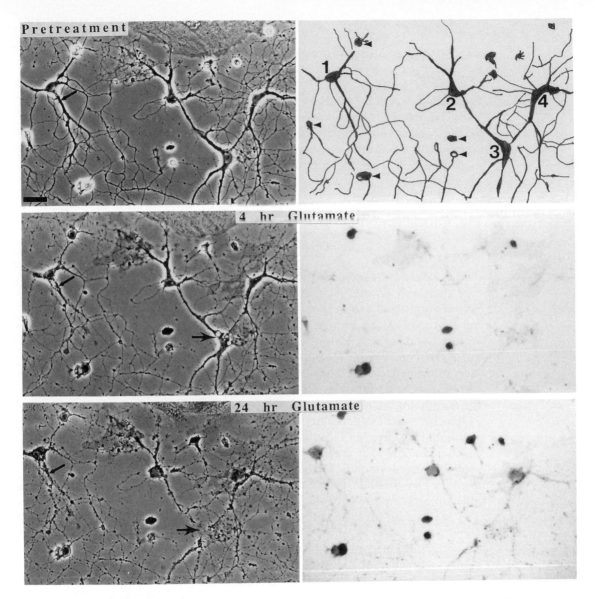

Fig. 2 Excitotoxicity in hippocampal cell culture. (*Top*) Phase-contrast image (*left*) and a tracing of the same (*right*) of embryonic hippocampal neurons (8 days in culture) prior to exposure to glutamate. Note that 4 cells with extensive neuritic arbors appear healthy. (*Middle*) A phase-contrast image of the same cells 4 hr after exposure to 200 μM glutamate (*left*). Note that some of the neurites are fragmented (e.g., arrow, neuron at left) and cell bodies appear vacuolated (e.g., arrow, neuron at lower right). A bright-field micrograph of the same cells following trypan blue staining (*right*). Note that all four neurons exclude trypan blue. (*Bottom*) All four neurons exhibit fragmented neurites and vacuolated cell bodies (*left*) and stain with trypan blue (*right*) 24 hr after exposure to glutamate.

was established using photomicrographic methods in cultured hippocampal pyramidal neurons (Mattson *et al.,* 1988a).

b. Disadvantages of This Method

(1) It is quite time consuming since it requires photographing cells, developing the film, and counting cells in each frame. (2) There is a subjective component in assessment of neuronal viability since the method involves determining whether individual cells meet the criteria of viable/undamaged.

IV. Colorimetric and Fluorescence Assays of Cell Survival

A. Colorimetric Dyes

Many dyes are excluded from cells with undamaged membranes, and therefore stain only severely damaged cells (Fig. 1; Table I). Trypan blue staining is commonly used to assess cell viability. This method simply involves incubating the cells in the presence of 0.4% trypan blue (Sigma #T8154) for 10–15 min, washing several times with saline, and then examining the cells using an inverted microscope with bright-field optics. Cell counts can be done immediately or the cells can be fixed (30 min in 4% paraformaldehyde/PBS), and counts can be done at a later time. This method can also be combined with the method for morphological assessment of cell survival (Fig. 2).

Another commonly used dye is 3-(4,5-dimethylthiazol-2-yl)-2,5-diphenyl tetrazolium bromide (MTT), introduced for viability assays by Mosmann (1983). In combination with a multiwell spectrophotometer, MTT can be used in a semiautomated viability assay. Spectrophotometric plate readers, commonly used for enzyme-linked immunosorbent assays (ELISAs), are generally available to more researchers than the fluorometric plate reader described later. The MTT assay takes advantage of the conversion of the yellow MTT to blue formazan crystals by mitochondrial succinate dehydrogenase in viable cells. The original assay had the disadvantages of being low in sensitivity and difficult to assay because of the insoluble nature of the crystalline reaction product. However, improvements (e.g., Denizot and Lang, 1986) have made the assay a reasonable alternative to other semiautomated methods. We have found consistent results using a kit available from Promega (Madison, WI). XTT is a tetrazolium compound similar to MTT that is also commonly used for cell viability assays (Parsons *et al.,* 1988; Scudiero *et al.,* 1988). (see Chapter 9, this volume)

Hippocampal neurons are plated in 96-well plates, coated as outlined earlier. To obtain consistent cell numbers, dissociated cells are diluted in MEM + medium to a concentration of approximately 10,000 cells/ml; 50-μl aliquots are added to wells containing 100 μl MEM + medium (placd in the wells for an overnight incubation before plating). After cell attachment, the medium in each well is replaced with 50 μl complete medium for sustained

Table I
Various Fluorescent Dyes of Utility in Studying Calcium- and Free Radical-Mediated Neuronal Injury

Dye	EX/EM[a]	Loading conditions[b]	Application	References
Calcein	485/530	1 μM, 30–45 min, RT	Live cell stain; membrane integrity	Moore et al. (1990)
BCECF	485/530	5 μM, 30 min, 37°C	Live cell stain; membrane integrity	Kolber et al. (1988)
Rhodamine 123	485/530	5 μM, 10 min, 37°C	Live cell stain; mitochondria potential	Lachowicz et al. (1989)
JC-1	485/530	10 μM, 10 min, 37°C	Live cell stain; mitochondria potential	Smiley et al. (1991)
Alamar Blue	530/590	10 μl per 100 μl medium, 1–6 hr, 37°C	Live cell stain; reduced by cell metabolite	Page et al. (1993)
Propidium iodide	485/645	2 μM, 30–45 min, RT	Dead cell stain; DNA	Tanke et al. (1982)
Ethidium bromide	485/645	2 μM, 30–45 min, RT	Dead cell stain; DNA	Beers and Wittliff (1975)
2,7-DCF	485/530	100 μM, 60 min, 37°C	Cellular oxidation	Heck et al. (1992)
Fluo-3	508/560	5 μM, 30 min, 37°C	Calcium concentration	Niggli and Lederer (1990)

[a] EX/EM, Excitation wavelength/emission wavelength (nm).
[b] RT, Room temperature.

maintenance. This procedure results in a cell density of approximately 250 cells/mm^2. Depending on the experimental objectives, glial growth can be inhibited with 10 μM cytosine arabinoside or substitution of the serum-containing medium with the defined medium described earlier. This is often desirable because glial cells may otherwise contribute to the background under conditions expected to specifically kill neurons. Two different approaches may be used to determine background. One is simply to kill all the cells in a few wells with a nonspecific toxin such as SDS immediately prior to addition of MTT. Alternatively, a few wells can be treated under conditions expected to kill neurons specifically, for example, 1 mM glutamate for 24 hr prior to addition of MTT. At the time viability is to be determined, 7 μl MTT dye (Solution A; Promega) is added to each well in a sterile manner and incubated under normal culture conditions for 1–3 hr, depending on the sensitivity required. Afterward, 50 μl solubilization cocktail (Solution B) is added to disperse the formazan reaction product. The cultures are typically left in Solution B overnight (at room temperature) to ensure adequate solubilization; in this case, the plate is sealed in an airtight chamber containing a moist paper towel to prevent evaporation.

The absorbance is then read at test and reference wavelengths. Although the optimal wavelengths are 570 nm for the test and 630 nm for the reference, we have had adequate results reading at 540 and 690 nm. The ratio of test to reference readings is calculated, and the ratio obtained from the background wells is subtracted from the other values. If necessary, the data thus obtained can be calibrated to cell number by performing the assay on cultures freshly plated (i.e., within 24 hr) at known numbers of viable cells to generate a standard curve. However, we usually express the data as percentages of the maximal expected signal (i.e., untreated wells). Because of the inherent variability in cell plating number, we use at least four wells per condition. Data obtained using this method, from an experiment in which glutamate toxicity was assessed in parallel cultures incubated in serum-containing or serum-free medium, are shown in Fig. 3.

B. Fluorescent Dyes

A large number of fluorescent dyes can be used to stain live or dead cells (Table I). These dyes can be used for cell counts under epifluorescence illumination and/or they can be used for analysis using a fluorescence plate reader. For fluorescence microscopy, cells should be grown in glass-bottom dishes. For the fluorescence plate reader, cells are grown in multiwell plastic plates (24-, 48-, or 96-well format). Dyes that are only taken up by viable cells include calcein-AM and BCECF-AM [2',7'-bis-(2-carboxyethyl)-5-(and-6)-carboxyfluorescein, acetoxymethyl ester] (Molecular Probes, Inc., Eugene, OR); these dyes are only trapped in healthy cells with active esterases that convert the dye from a lipophilic to a hydrophilic form. Dyes that stain dead

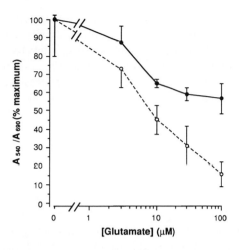

Fig. 3 MTT assay of glutamate toxicity. Rat hippocampal cells were plated in a 96-well plate and left in MEM + 10% fetal bovine serum or switched to serum-free defined medium at 3 days in culture. At 8 days in culture, the cells were exposed to the indicated concentrations of glutamate. Viability was assessed 24 hr later, as described in the text. The background formazan signal obtained in cultures exposed to 0.1% SDS was subtracted from the other values. Note that the minimal signal obtained in the serum-containing cultures (●) was greater than the minimal signal in the serum-free cultures (○) because of the increased glial cells that contribute to MTT conversion but are relatively resistant to glutamate.

cells include propidium iodide and ethidium bromide (these stains bind DNA). Following experimental treatment (e.g., 20 hr following exposure to excitatory amino acids), cells are incubated in the presence of a dye for a fixed time period (see Table I for details). Then the cultures are washed several times to remove extracellular dye. Cell counts can be made under epifluorescence illumination, and the percentage of viable (or nonviable) cells is determined. For cells grown in multiwell plates, an overall fluorescence/well can be determined using a fluorescence plate reader. We use a Millipore Cytofluor 2350 fluorescence plate reader, which provides rapid analysis of multiwell plates. Fluorescent dyes are generally excited maximally within a limited range of wavelengths, and the emitted fluorescence is also confined to a particular wavelength range. Therefore, it is necessary to use appropriate excitation and emission filters to optimize the specific fluorescence signal. The appropriate excitation and emission filters for each fluorescent dye discussed in this chapter are indicated in Table I.

In contrast to the dyes described so far, compounds have recently been developed that allow continuous monitoring of cell health. One such dye is Alamar Blue (Alamar, Inc., Sacramento, CA) (Page *et al.,* 1993). Alamar Blue is a nonfluorescent substrate which, after reduction by cell metabolites, becomes fluorescent. The reduced dye is soluble and will accumulate in the culture medium; the fluorescence intensity of the dye in the medium can be quantified with the fluorescence plate reader. Alamar Blue is nontoxic and extremely sensitive, and can be used to accurately assess viability of small populations of cells (as few as 200–500 cells/well). A caveat with the use of Alamar Blue is that, since the reduced dye continues to accumulate in the incubation medium, there is a time window (generally 1–6 hr) during which differences in cell viability are most readily detected; beyond this time window, dye accumulation approaches a saturation point (Fig. 4). (see Chapter 9, this volume)

V. Measurement and Manipulation of Intracellular Calcium Levels

Two major approaches have been applied to studies of the role of calcium in neuronal injury. One approach involves measuring intracellular free calcium levels [(Ca^{2+}]$_i$) and correlating [Ca^{2+}]$_i$ with various measures of cell injury. The other approach involves manipulating Ca^{2+} movements into and out of cells and subcellular compartments, and determining the effects such manipulations have on neuronal injury and viability. The methods described here have been used extensively in primary cultures of embryonic rat hippocampal, neocortical, and septal cells (Mattson *et al.*, 1988a,b, 1989a,b, 1993c,d; Cheng and Mattson, 1992).

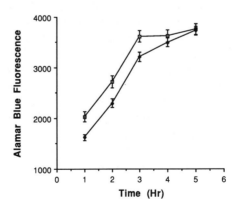

Fig. 4 Time course of change of Alamar Blue fluorescence in control (□) and glutamate-treated (◆) rat neocortical cell cultures. Cells in 24-well plates were left untreated (control) or were exposed to 500 μM glutamate for 20 hr. Alamar Blue was then added to the medium and the plate was scanned each hour for 5 hr. Note that during the first 3 hr the Alamar Blue fluorescence was lower in the medium of glutamate-treated wells, indicating lower cell viability. However, at 5 hr the difference between control and glutamate-treated wells was not apparent due to continued reduction of the dye by glia and by remaining viable neurons in the glutamate-treated cultures. Values are mean and SEM of 4 wells.

A. Calcium Indicator Dyes

The development of fluorescent indicators for the measurement of ion concentrations in living cells has revolutionized the study of signal transduction events. The use of such dyes for visualization of $[Ca^{2+}]_i$ has helped define the role of calcium in various aspects of neuronal morphology and viability (Mattson, 1992). Although the precise mechanisms through which calcium mediates these effects are still poorly defined, it is now evident that neuronal survival depends on an optimal range of $[Ca^{2+}]_i$.

Among the dyes most widely used for visualization and measurement of $[Ca^{2+}]_i$ are fura-2 and indo-1 (Grynkiewicz et al., 1985; Malgaroli et al., 1987). These dyes are based structurally on the calcium chelator [1,2-bis(2-aminophenoxy)ethane-N,N'-tetraacetic acid] (BAPTA) and are descendents of quin-2, the first dye used for such applications. Fura-2 and indo-1 have a greater energy yield per mole than quin-2; therefore, they can be used at much lower concentrations, reducing the undesirable buffering effects on $[Ca^{2+}]_i$ of the indicator itself. These dyes share with quin-2 the advantage of quantifying $[Ca^{2+}]_i$ in a reasonably accurate manner through the use of ratio measurements, which largely obviate artifacts due to variations in dye intensity. For instance, the binding of calcium to fura-2 increases the absorption at 340 nm and decreases the absorption at 380 nm. Therefore, a ratio of emission from 340-nm excitation to emission from 380-nm excitation is related to $[Ca^{2+}]_i$ and is independent of variations caused by differences in cell thickness, dye quantity, or photobleaching. This ratio can be used to determine $[Ca^{2+}]_i$ either by interpolation within a standard curve or by a calculation based on the two end-point ratios observed in the absence of calcium and in a saturating concentration of calcium (Grynkiewicz et al., 1985). Several imaging systems are now commercially available that automate the acquisition of images with synchronous control of excitation filter; most also perform calculation of $[Ca^{2+}]_i$ from the images acquired. In contrast to fura-2, indo-1 responds to calcium with a spectral shift in emission from excitation with a single wavelength and is thus more useful when an imaging system's versatility lies in detecting multiple emissions.

An additional dye frequently used is fluo-3. Fluo-3 differs from fura-2 and indo-1 in several ways. The excitation wavelengths for fluo-3 are longer and are thus compatible with a greater range of system specifications, including laser excitation and plastic cell substrates. In addition, fluo-3 exhibits a greater change (increase) in fluorescence intensity on Ca^{2+} binding than do fura-2 and indo-1. We commonly use this dye with confocal laser microscopy for subcellular localization of $[Ca^{2+}]_i$ and comparison with localization of organelles (identified with organelle-specific dyes) or specific proteins (localized by immunofluorescence). In addition, several investigators have used fluo-3 for relative $[Ca^{2+}]_i$ measurements in multiwell culture plates with the fluorescence plate reader already described. The disadvantage of fluo-3 is that it is unsuited to ratiometric quantification because of its monochromatic absorption and emission. With this dye, $[Ca^{2+}]_i$ can only be estimated from changes in intensity at a single emission wavelength from excitation at a single wavelength. Double-loading of cells with fluo-3 and SNARF (Molecular Probes, Eugene, OR), a dye commonly used for pH measurements, has been recommended for ratiometric quantification of $[Ca^{2+}]_i$ (Rijkers, 1990). Although this method might be a useful correction for differences in cell thickness and loading, it may not accurately correct differences in dye bleaching or leakage. Other long-wavelength dyes are now available that are related in structure to their prototype calcium green. These dyes are advantageous because of their rapid responses to calcium, making them suitable for use in conjunction with electrophysiological recordings.

1. Procedure for Imaging $[Ca^{2+}]_i$ in Primary Neuronal Cultures with Fura-2

A major consideration for fluorescent imaging is the choice of culture substrate. Fura-2 imaging is typically performed on an inverted fluorescence microscope, imaging cells that are attached to a thin layer of glass. For convenience, we plate cultures on commercially prepared 35-mm plates with a glass coverslip forming the bottom (MatTek, Inc.). However, one can easily plate cells on a coverslip (0.13–0.17 mm thick) lying in the bottom of a culture dish and then remove the coverslip and attach it to another chamber (using denture adhesive or a similar material) for imaging. For unknown reasons, glass coverslips of German origin provide better long-term attachment and survival of neurons. Any glass surface should be coated for optimal adherance of neurons. After etching the glass with 50% nitric acid for 16–20 hr, the coverslips are washed copiously with glass-distilled water and are coated with polyethyleneimine, as described. Neurons are plated on the coverslips and cultured as described. Note that prolonged storage of polyethyleneimine-coated glass in a dry state seems to compromise cell attachment and/or survival. We routinely coat dishes within 4 days of use, add MEM+ to the dishes immediately after coating, and then store them in a cell culture incubator.

Loading the cells with fura-2 (or most other indicator dyes) involves incubation with a membrane-permeable derivative that the cells are able to convert into a membrane-impermeable form in their cytosol. In most cases, membrane permeability is afforded through an acetoxymethyl ester (AM) linkage which is cleaved by ubiquitous cellular esterases. The optimum loading conditions vary with cell type and can depend on fura-2 concentration (usually 1–10 μM), solubility (often enhanced with mild dispersing agents such as Pluronic), and incubation time (usually 15–45 min). Although serum can contain esterases, which would obviously inhibit loading, many investigators have found that some amount of albumin aids loading; we load effectively in the presence of 10% heat-inactivated FBS. Our conditions are 3–6 μM fura-2/AM (Molecular Probes; dissolved as a 1 mM stock in DMSO) in complete cell culture medium for 30 min at 37°C (we have not found Pluronic necessary). This incubation is followed by a chase of 45 min to 2 hr in complete medium lacking fura-2 to allow for complete hydrolysis of the AM ester. Insufficient chase incubations can lead to incomplete hydrolysis, causing artificially low 340/380 ratios and muted responses; complete hydrolysis usually allows maximal ratios (i.e., in the presence of micromolar calcium ionophore in calcium-containing buffers) approaching 10. A chase period that is too long can result in leakage of the dye out of the cells or its compartmentalization into inappropriate organelles. (These problems seem to be minimal in primary neurons, but can pose a problem in cell lines.)

Imaging is typically performed in room air and therefore requires changing the buffer to one containing HEPES immediately before imaging. The buffer we use is HBSS containing 10 mM glucose and buffered to pH 7.2 with 10 mM HEPES. During imaging, several configurations can be used for reagent delivery and temperature maintenance (if necessary). Some investigators accomplish both by constant superfusion of the cells with warmed buffer. Multiple sources of the buffer (containing different treatments) can be connected via a manifold placed just before the inlet to the cell chamber, with effluent aspirated by an outlet vacuum line. Alternatively, conduction heating rings are commercially available that hold a 35-mm plate and have inlet tubing attachments for application of reagents. We find it reasonably easy and useful to perform treatments by direct addition to a stationary buffer, followed by gentle mixing with a pipetter. In any type of set-up, it is advisable to monitor the $[Ca^{2+}]_i$ in the cells of interest for 1–3 min before beginning the experimental treatments to ensure that buffer change, dye bleaching, and mock reagent additions do not significantly alter $[Ca^{2+}]_i$ from a stable baseline.

For most imaging systems, calibration of the ratios obtained from fura-2 imaging can be periodic and need not be performed after each experiment. Calibration can be performed by several means. The simplest method is to obtain ratios from two different solutions of fura-2, one containing millimolar Ca^{2+} and the other containing no added calcium and approximately 1 mM EGTA. Because of possible effects of other ions on the signal, we have attempted to use a buffer similar in cationic composition to the cytosol for these measurements: 130 mM KCl, 10 mM MgCl$_2$, 5 mM NaHCO$_3$, 10 mM HEPES (pH 7.0). Alternatively, one may measure the ratios in the cells being imaged under conditions of high and low Ca^{2+}. This is typically done by applying a calcium ionophore (ionomycin or 4-bromo A23187) in the presence of high extracellular calcium first, followed by a change to a buffer containing no added calcium and EGTA. The ratios obtained from either method can then be used in the following formula, first devised by Grynkiewicz et al. (1985).

$$[Ca^{2+}]_i = \frac{R - R_{min}}{R_{max} - R} (F_o/F_s) \, K_d$$

where R is the experimental ratio obtained, R_{max} is the ratio obtained from a solution of fura-2 containing calcium, R_{min} is the ratio obtained from a solution of fura-2 lacking calcium, F_o is the fluorescent emission detected from 380-nm excitation of the calcium-free solution, F_s is the fluorescent emission detected from 380-nm excitation of the calcium-containing solution, and K_d is the dissociation constant of fura-2 for Ca^{2+} (typically 224–228 nM).

An example of the application of Ca^{2+} imaging to the problem of excitotoxicity is shown in Fig. 5, where $[Ca^{2+}]_i$ (quantified using fura-2 imaging) following exposure to glutamate was correlated with subsequent neuronal degeneration. Figure 6 (see also Color Plate 5) shows an example of the application of confocal laser scanning microscopy to calcium imaging; in this case, cultured human embryonic neocortical neurons were loaded with fluo-3 and imaged before and after exposure to glutamate.

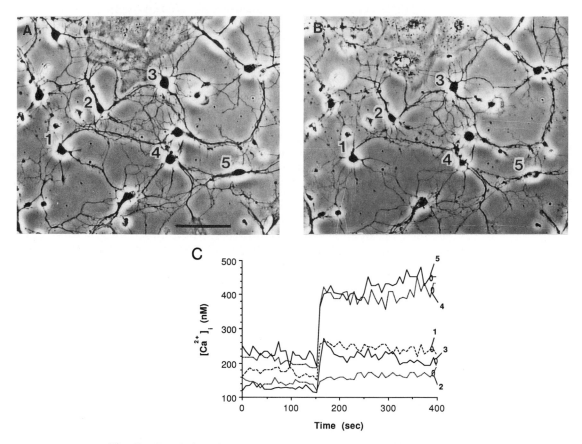

Fig. 5 Correlation of calcium responses to glutamate with subsequent neuronal degeneration. (A,B) Phase-contrast micrographs of a field of cultured hippocampal neurons prior to treatment (A) and 20 hr after exposure to 50 μM glutamate (B). Note that after exposure to glutamate, neurons 1, 2, and 3 appeared undamaged whereas neurons 4 and 5 exhibited vacuolated cell bodies and fragmented neurites. Bar: 70 μm. (C) Graph of intracellular free calcium levels (obtained by ratiometric imaging of the calcium indicator dye fura-2) in each of the five cells shown in the upper panel. Glutamate was added to the culture medium at $t = 150$ sec. Note that calcium levels rose considerably more in cells 4 and 5 (the cells that subsequently degenerated) than in cells 1–3.

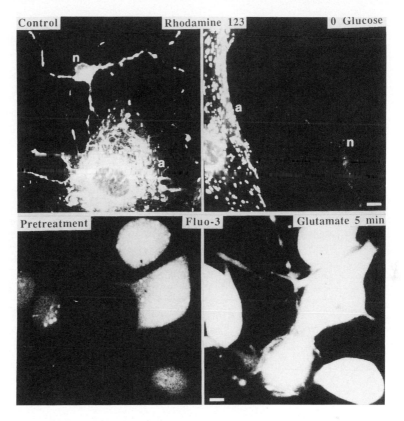

Fig. 6 (see also Color Plate 5) Effects of metabolic and excitotixic insults on mitochondrial transmembrane potential and intracellular free calcium levels assessed using the fluorescent dyes rhodamine 123 and fluo-3 and confocal laser scanning microscopy. (*Top left*) Rhodamine 123 fluorescence in a neuron (n) and an astrocyte (a) in a control rat hippocampal cell culture containing glucose; note that both the neuron and the astrocyte accumulated the dye in their mitochondria. (*Top right*) Rhodamine 123 fluorescence in a neuron (n) and an astrocyte (a) in a rat hippocampal cell culture that had been deprived of glucose for 14 hr; note that the mitochondria of the neuron accumulated little dye, indicating a reduction of transmembrane potential. Bar: 5 μm. (*Bottom*) Fluo-3 fluorescence in cultured human embryonic neocortical neurons in the resting state (*left*) and 5 min after exposure to 500 μM glutamate (*right*); note that glutamate caused a pronounced increase in fluroescence, indicating an elevation of intracellular free calcium levels. Bar: 1 μm.

B. Manipulation of Calcium Fluxes

In evaluating the involvement of Ca^{2+} in cell culture paradigms of excitotoxic/ metabolic neuronal injury, a variety of pharmacological manipulations can be employed that are based on the known cellular systems involved in the regulation of $[Ca^{2+}]_i$ (Table II). A common first step is to determine whether influx of extracellular Ca^{2+} is involved in the neurotoxicity. Varying the concentration of extracellular Ca^{2+} is very useful in determining whether Ca^{2+} influx is required for neurotoxicity. However, this change must be controlled carefully when studying relatively slow forms of neuronal degeneration that occur over

Table II
Agents that Affect $[Ca^{2+}]_i$ in Neurons and Their Mechanisms of Action

Agent	Effect on $[Ca^{2+}]_i$	Mechanism of action[a]
Glutamate	Increase	Ca^{2+} influx via NMDA channels and VDCC
NMDA	Increase	Ca^{2+} influx through NMDA receptor channels
Kainate/AMPA	Increase	Depolarizes PM; Ca^{2+} influx through VDCC
KCl	Increase	Depolarizes PM; Ca^{2+} influx through VDCC
Calcium ionophore A23187	Increase	Ca^{2+} influx through the ionophore
Thapsigargin	Increase	Blocks $ERCa^{2+}$ ATPase
Reduced $[Na^+]_o$	Increase	Blocks PM Na^+/Ca^{2+} exchange
Vanadate	Increase	Blocks PM Ca^{2+} ATPase
$0\ [Ca^{2+}]_o$ + Glutamate	Increase	Ca^{2+} release from ER stores
$0\ [Ca^{2+}]_o$ + A23187	Increase	Ca^{2+} release from ER stores
Reduced $[Ca^{2+}]_o$	Decrease	Reduction of Ca^{2+} influx through PM
Co^{2+}, La^{3+}	Decrease	Reduction of Ca^{2+} influx through VDCC
Nimodipine, nifedipine	Decrease	Reduction of Ca^{2+} influx through L channels
Omega conotoxin	Decrease	Reduction of Ca^{2+} influx through N channels
8-Bromo-cGMP	Decrease	Suppression of Ca^{2+} influx

[a] ER, Endoplasmic reticulum; PM, plasma membrane; VDCC, voltage-dependent calcium channels.

periods of days (e.g., growth factor deprivation or exposure to amyloid β-peptide; Cheng and Mattson, 1991; Mattson *et al.*, 1993e) since removal of extracellular Ca^{2+} itself can be detrimental during such prolonged incubations. Similarly, inorganic (Co^{2+}, Cd^{2+}, La^{3+}) and organic (e.g., nifedipine, nimodipine) blockers of voltage-dependent Ca^{2+} channels can also be neurotoxic during prolonged exposure periods. Glutamate receptor antagonists can be employed to reduce Ca^{2+} influx. For example, APV (DL-2-amino-5-phosphono-valeric acid) selectively blocks the NMDA-type glutamate receptor, CNQX (6-cyang-7-nitroquinoxaline-2,3-dione) blocks kainate/AMPA receptors, and MCPG (α-methyl-4-carboxyphenylglycine) blocks the metabotropic (G-protein-coupled) receptor (McCulloch, 1992). Compounds such as dantrolene sodium and ruthenium red are also available which block release of calcium from internal (endoplasmic reticulum) stores.

In addition to establishing whether Ca^{2+} influx is required for a particular insult to cause neuronal injury, it is important to establish whether elevation of $[Ca^{2+}]_i$ is sufficient to account for the neuronal injury. The latter possibility can be tested by stimulating elevation of $[Ca^{2+}]_i$ by means other than the proximate insult. For example, calcium ionophores such as A23187 and iono-mycin form Ca^{2+} "pores" in cell membranes resulting in Ca^{2+} movement into the cytoplasm (Mattson *et al.*, 1988b). The magnitude of the $[Ca^{2+}]_i$ rise induced can be controlled by varying the concentration of the ionophore and the concentration of extracellular Ca^{2+}. Selective release of Ca^{2+} from the endoplasmic reticulum can be induced with thapsigargin, a compound that inhibits SERCAs (Thastrup *et al.*, 1990). Another way in which $[Ca^{2+}]_i$ can be elevated is by incubating cells in medium with a low concentration of Na^+, which compromises the plasma membrane Na^+/Ca^{2+} exchange mechanism of Ca^{2+} removal; we substitute *N*-methyl-D-glucamine for NaCl (Mattson *et al.*, 1989b).

VI. Assessment of Cellular Oxidation

Several approaches are available for quantifying levels of free radicals generated in cultured neural cells. Relative levels of free radicals in living cells can be monitored using the oxidation-sensitive compound 2,7-dichlorofluorescin diacetate (DCF; Molecular Probes). Oxidation by peroxides converts DCF to 2,7-dichlorofluorescein which is fluorescent. The fluorescence can be quantified in neuronal cell cultures using imaging methods or a fluorescence multiwell plate reader (Goodman and Mattson, 1994). We have adapted methods previously described by Heck *et al.* (1992). Cells are loaded with DCF (50 μM, 50-min incubation) followed by 3 washes (2 ml/wash) in HBSS. Imaging studies employ a confocal laser scanning microscope system consisting of a Nikon Diaphot microscope coupled to a Sarastro 2000 system, which includes a Silicon Graphics personal IRIS workstation and "Imagespace" software (Molecular Dynamics, Sunnyvale, CA). The intensity of the laser beam and the sensitivity of the photodetector are held constant to allow quantitative comparisons between treatment groups of relative fluorescence intensity of cells. Cells are scanned only once with the laser because DCF is so sensitive to oxidation that exposure to the laser light induces photo-oxidation, resulting in increased fluorescence. Values for average staining intensity/cell are obtained using the Imagespace software supplied by the manufacturer (Molecular Dynamics). DCF fluorescence in multiwell plates can be quantified using the Cytofluor 2350 fluorescence plate reader.

Using these methods we have shown that excitatory amino acids induce free radical accumulation in cultured hippocampal and cortical neurons (Fig. 7A). The increased free radicals are part of the excitotoxic mechanism since agents such as vitamin E that scavenge free radicals can attenuate glutamate toxicity (Fig. 7B; see also Dykens *et al.*, 1987). Imaging of oxidative processes in neurons can provide an important complement to calcium imaging studies and, because the excitation and emission optima for DCF and some Ca^{2+} indicator dyes (e.g., fura-2) differ considerably, it is possible to image both $[Ca^{2+}]_i$ and $[peroxides]_i$ simultaneously.

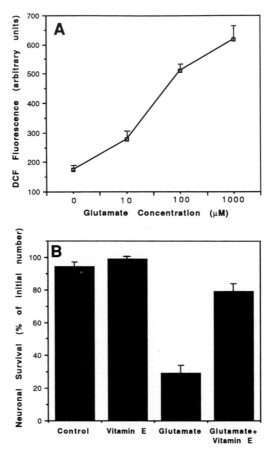

Fig. 7 Experiments to evaluate the role of free radicals in excitotoxicity. (A) Cultured rat hippocampal neurons grown in 24-well plates were exposed to the indicated concentrations of glutamate for 2 hr. The cells were then loaded with 2,7-dichlorofluorescin (100 μM for 60 min) and scanned using a fluorescence plate reader. Values represent the mean and SEM of determinations made in 4 separate wells. (B) Cultured rat hippocampal neurons were left untreated (control) or were exposed to vitamin E (50 μg/ml), glutamate (100 μM), or vitamin E + glutamate. Viable neurons in premarked microscope fields were counted prior to treatment and 24 hr after treatment. Survival is expressed as the percentage of pretreatment cells and values represent the mean and SEM of determinations made in 4 separate cultures.

VII. Evaluation of Mitochondrial Function

In addition to MTT and Alamar Blue (see preceding text), several fluorescent dyes are particularly sensitive to compromise of mitochondrial function. These dyes include rhodamine 123 (Mattson *et al.*, 1993f) and JC-1 (Smiley *et al.*, 1991). Both these reagents can be purchased from Molecular Probes. The uptake of these dyes by mitochondria is directly related to the transmembrane potential across the inner membrane (Johnson *et al.*, 1981). Stocks of rhodamine 123

(2.5 mM) and JC-1 (5 mM) are prepared in dimethylsulfoxide; 20-μl aliquots are stored at -20°C and are thawed only once. Neuronal cells are incubated for 10 min in the presence of 5 μM rhodamine 123 or 10 μM JC-1. Cells are then washed 3 times in HBSS and are imaged within 1 hr by confocal laser scanning microscopy. Images are acquired using a 60×, NA 1.3 oil immersion lens (Nikon). For comparison of relative levels of fluorescence/cell, images must be acquired using constant settings for laser intensity and detector gain. Average pixel intensity/cell can be determined using the Imagespace software. The fluorescence of mitochondria decreases with cell injury. An example of rhodamine 123 staining of neurons and astrocytes in control cultures and metabolically compromised cultures is shown in Fig. 6 (see also Color Plate 5).

VIII. Assessment of Cellular ATP Levels

Metabolic/excitotoxic insults usually involve energy failure, which may contribute to loss of calcium homeostasis and free radical production (Mattson *et al.*, 1993f). Therefore it is useful to monitor levels of cellular ATP. ATP levels in neuronal cell cultures can be quantified by a colorimetric assay similar to that described by Adams (1963). We have found that commercially available kits (e.g., Sigma kit #366) provide sufficient sensitivity to quantify ATP levels accurately in homogenates of cultured rat hippocampal and cortical cells (Mattson *et al.*, 1993f). The assay is based on the conversion of 3-phosphoglycerate to 1,3-diphosphoglycerate by phosphoglycerate phosphokinase and the subsequent conversion of 1,3-diphophosphoglycerate + NADH to glyceraldehyde-3-P + NAD + P, which is catalyzed by glyceraldehyde phosphate dehydrogenase. When NADH is oxidized to NAD, a decrease in absorbance at 340 nm occurs. Cells in 60-mm dishes (approximately 1 million cells/dish) are rinsed twice with cold PBS and then harvested in 0.5 ml of a solution consisting of a 1 : 1 mixture of 12% TCA and 0.2 M sodium citrate (pH 7.0). The lysate is centrifuged at 400 rpm for 10 min (4°C) and the supernatant is saved. A 0.4-ml aliquot of the supernatant is added to a solution containing 0.15 mg NADH (disodium salt), 0.5 ml PGA-buffered solution (18 mM 3-phosphoglyceric acid), and 0.75 ml water. The initial absorbance of the solution at 340 nm is determined; then 20 μl GAPD/PDK enzyme mixture (800 U/ml 3-phosphate dehydrogenase and 450 U/ml 3-phosphoglyceric phosphokinase) is added to the cuvette and the reduction in absorbance determined. Levels of ATP are calculated based on a millimolar absorptivity of NADH of 6.22.

IX. Cytoskeletal Markers of Neuronal Injury

The cytoskeleton of neurons has been the focus of much attention in the field of neurodegenerative disorders because characteristic alterations in many of its components are associated with the degenerative process (see Kosik, 1992,

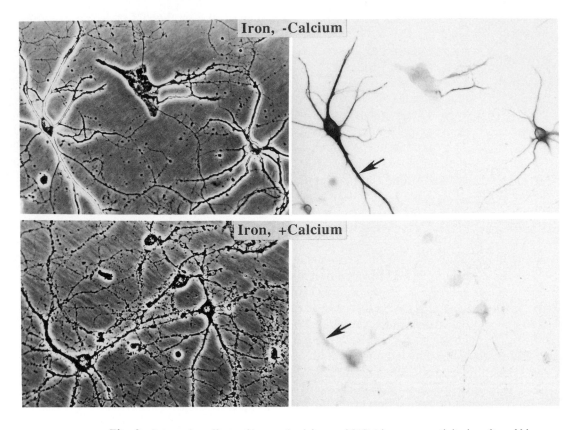

Fig. 8 Interactive effects of iron and calcium on MAP-2 immunoreactivity in cultured hippocampal neurons. Phase-contrast (*left*) and bright-field (*right*) micrographs of hippocampal cells (15 days in culture) that were exposed to 100 μM FeSO$_4$ for 2 hr in medium that either lacked calcium (no added calcium plus 1 mM EGTA; *top*) or contained 2 mM Ca^{2+} (*bottom*). The cells were then fixed and immunostained with an antibody against MAP-2 (see text). Note that dendrites (e.g., arrow) stain intensely in the cells exposed to iron in the absence of calcium, whereas MAP-2 immunoreactivity is lost in the dendrites of neurons exposed to iron in the presence of calcium (e.g., arrow).

for review). The structure of the cytoskeleton is normally highly dynamic, as microtubules and microfilaments assemble and disassemble in response to metabolic and environmental demands. Calcium is perhaps the major second messenger that dictates changes in the neuronal cytoskeleton in response to signals such as neurotransmitters and growth factors. By influencing polymerization/depolymerization of microtubules and microfilaments, calcium regulates behaviors of the neuronal growth cone and thereby plays a pivitol role in the formation and adaptive plasticity of neural circuits (Kater *et al.*, 1988; Mattson, 1992). Cytoskeletal components that are sensitive to calcium (directly or via the activities of calcium-activated kinases and proteases) include tubulin, actin, microtubule-associated protein 2 (MAP-2), tau, spectrin, and neurofilament proteins.

In mature neurons, the microtubule associated proteins MAP-2 and tau have distinct localizations. MAP-2 is present in dendrites but not axons, whereas tau is localized to axons; both MAPs are present at very low levels in the cell body (see Matus, 1988, for review). Elevation of $[Ca^{2+}]_i$ induced by glutamate or other means causes characteristic alterations in MAP-2 and tau immunoreactivity and localization. Antigenicity and localization of tau is altered by $[Ca^{2+}]_i$;

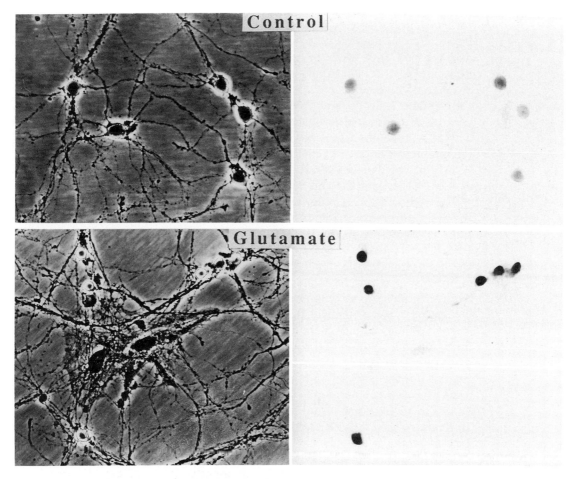

Fig. 9 Glutamate induces an increase in ubiquitin immunoreactivity in cultured hippocampal neurons. Phase-contrast (*left*) and bright-field (*right*) micrographs of hippocampal cells that were left untreated (control) or were exposed to 100 μM glutamate for 2 hr. The cells were then immunostained with a ubiquitin antibody (Sigma polyclonal rabbit; #U-5379) at a dilution of 1 : 500 and the immunoperoxidase method (Vector Laboratories, ABC kit). Note the marked increase in ubiquitin immunoreactivity in the cell bodies of the glutamate-treated neurons.

neurons become immunoreactive with a variety of antibodies (e.g., Alz-50 and 5E2) that do not stain them under basal conditions (Mattson, 1990). MAP-2 is highly sensitive to Ca^{2+}-dependent proteolysis that is mediated by calpain (Johnson *et al.*, 1991). In cultured hippocampal neurons, glutamate or Ca^{2+} ionophores cause a reduction in MAP-2 immunoreactivity in dendrites and accumulation in the cell body. These changes occur quite rapidly (min to hr) and precede cell death during exposure to toxic levels of excitatory amino acids. Subtoxic levels can also reduce MAP-2 immunoreactivity in dendrites, although the change is less robust. Figure 8 shows an example of the interactive effects of free radicals and calcium in altering MAP-2 immunoreactivity. Hippocampal neurons were exposed to iron in medium containing or lacking calcium. Iron caused a marked reduction in MAP-2 immunoreactivity and neuronal degeneration only when calcium was present in the medium. The effect of iron on MAP-2 was markedly suppressed when neurons were incubated in the absence of extracellular Ca^{2+}, indicating that the oxidative insult promoted Ca^{2+} influx. Indeed, direct measurement of $[Ca^{2+}]_i$ in cultured hippocampal neurons demonstrated an elevation of $[Ca^{2+}]_i$ in response to iron that is mediated by NMDA receptor activation (Zhang *et al.*, 1993). These observations illustrate the interactive nature of Ca^{2+} and free radicals in neuronal injury; elevation of $[Ca^{2+}]_i$ promotes free radical production and free radicals promote elevation of $[Ca^{2+}]_i$ (see Mattson and Scheff, 1994, for review).

The protocol for immunostaining cultured neurons with antibodies against the microtubule-associated proteins and spectrin follows (Mattson, 1990). Cultures are fixed for 30 min in a solution of 4% paraformaldehyde dissolved in PBS. The fixative is prepared by heating the PBS to boiling and then adding the paraformaldehyde; this solution is then cooled to 4°C prior to addition to cultures. It is important to prepare the fixative within 2–12 hr of use. The cold fixative is added to the cultures following removal of the culture medium, and the cultures are incubated at room temperature for 30 min. Cells are then washed 3 times with PBS and the membranes are permeabilized by incubating for 5 min in a solution of 0.2% Triton X-100 in PBS. Cells are then processed using an appropriate "ABC" kit from Vector Laboratories (a rabbit IgG kit is used when the primary antibody was raised in rabbit; a mouse IgG kit is used when the primary antibody is a mouse monoclonal). Following a 30-min exposure to blocking serum, the cells are incubated for 3–4 hr in primary antibody dissolved in PBS (plus blocking serum). Cells are then incubated for 1 hr in biotinylated secondary antibody followed by a 30-min incubation in ABC reagent (avidin–peroxidase complex). Cells are washed 3 times with PBS between all incubations. Finally, the cells are exposed for 5 min to diaminobenzidine/hydrogen peroxide solution, washed 3 times in water, and then wet-mounted by adding 5 drops of glycerol, which is then covered with a 22-mm^2 coverslip. The cells can be examined using an inverted microscope with phase-contrast and bright-field optics. Differences in cell staining intensity between control and treatment cultures can be determined using a semiquantitative method in which four categories of staining are used (Mattson, 1990).

The primary antibodies we have used, their sources, and recommended dilutions are Alz-50 (tau; Peter Davies, Albert Einstein College of Medicine; 1 : 10); 5E2 (tau; Kenneth Kosik, Brigham and Women's Hospital, Harvard Medical School; 1 : 200); and MAP-2 clone AP20 (Sigma; 1 : 1000).

X. Stress Response Proteins and Neuronal Injury

Various stressors including excitotoxic/ischemic insults induce the expression of a set of proteins that seem to play an important role in a protective response to cell injury (see Jaattela and Wissing, 1992, for review). These "stress response" proteins include hsp90, hsp72, hsp27, and ubiquitin. Levels of these proteins are increased in rodent models of ischemia *in vivo* (Nowak *et al.*, 1993). We have found that excitatory amino acids and Ca^{2+} influx can induce the expression of at least some of the heat-shock proteins in cultured hippocampal neurons (Mattson, 1990). An example is shown in Fig. 9; ubiquitin immunoreactivity was markedly increased in neurons exposed to glutamate for 2 hr over ubiquitin immunoreactivity in untreated neurons. The protocol for immunostaining for the various heat-shock proteins is essentially identical to that used for immunostaining with antibodies against cytoskeletal proteins (see Section IX). In our experience, the increased heat-shock protein levels appear prior to morphological signs of neuronal degeneration, and then remain elevated as the neurons degenerate or recover.

XI. Conclusions

In this chapter we have presented an array of technical approaches that can be applied to studies of excitotoxic neuronal injury in cell culture. Technological developments in imaging and automated quantification of fluorescence in multiwell cultures have provided the means to evaluate parameters of neuronal injury at the subcellular level and to quantify neuronal injury rapidly. Calcium and free radicals, two principle players in the process of nerve cell injury and death, can now be quantified with high spatial resolution in primary cell cultures. In addition, the array of pharmacological and molecular tools with which to manipulate calcium and free radical production is rapidly expanding. We have presented examples of protocols that can be routinely applied to studies of calcium and free radical metabolism as they relate to neuronal injury. It is hoped that this information will provide a basis for expanding knowledge of mechanisms of neuronal injury and plasticity.

Acknowledgements

We thank M. Barger, S. Bose, Y. Goodman, S. E. Laughran, and N. McCants for technical assistance. We thank R. E. Rydel for providing information on polyethyleneimine as a growth substrate and on the use of Alamar Blue. Original research in M.P.M.'s laboratory was supported by the National Institutes of Health, the Alzheimer's Association, the Metropolitan Life Foundation, and the French Foundation for Alzheimer's Research.

References

Adams, H. (1963). Adenosine 5′-triphosphate determination with phosphoglycerate kinase. *In* "Methods of Enzymatic Analysis" (H. U. Bergmeyer, ed.), pp. 539–543. Academic Press, New York.

Beers, P. C., and Wittliff, J. L. (1975). Measurement of DNA and RNA in mammary gland homogenates by the ethidium bromide technique. *Anal. Biochem.* **63**, 433–443.

Blaustein, M. P., Goldman, W. F., Fontana, G., Krueger, B. K., Santiago, E. M., Steele, T. D., Weiss, D. N., and Yarowsky, P. J. (1991). Physiological roles of the sodium–calcium exchanger in nerve and muscle. *Ann. N.Y. Acad. Sci.* **639**, 254–274.

Carafoli, E. (1992). The Ca^{2+} pump of the plasma membrane. *J. Biol. Chem.* **4**, 2115–2118.

Cheng, B., and Mattson, M. P. (1991). NGF and bFGF protect rat hippocampal and human cortical neurons against hypoglycemic damage by stabilizing calcium homeostasis. *Neuron* **7**, 1031–1041.

Cheng, B., and Mattson, M. P. (1992). IGF-I and IGF-II protect cultured hippocampal and septal neurons against calcium-mediated hypoglycemic damage. *J. Neurosci.* **12**, 1558–1566.

Choi, D. W. (1988). Glutamate neurotoxicity and diseases of the nervous system. *Neuron* **1**, 623–634.

Denizot, F., and Lang, R. (1986). Rapid colorimetric assay for cell growth and survival. Modifications to the tetrazolium dye procedure giving improved sensitivity and reliability. *J. Immunol. Meth.* **89**, 271–277.

Dykens, J. A., Stern, A., and Trenkner, E. (1987). Mechanism of kainate toxicity to cerebellar neurons in vitro is analogous to reperfusion tissue injury. *J. Neurochem.* **49**, 1222–1228.

Goodman, Y., and Mattson, M. P. (1994). Secreted forms of β-amyloid precursor protein protect hippocampal neurons against amyloid β-peptide induced oxidative injury. *Exp. Neurol.* **128**, 1–12.

Grynkiewicz, G., Poenie, M., and Tsien, R. Y. (1985). A new generation of calcium indicators with greatly improved fluorescence properties. *J. Biol. Chem.* **260**, 3440–3450.

Halliwell, B., and Gutteridge, J. M. (1985). "Free Radicals in Biology and Medicine." Clarendon Press, Oxford.

Heck, D. E., Laskin, D. L., Gardner, C. R., and Laskin, J. D. (1992). Epidermal growth factor suppressses nitric oxide and hydrogen peroxide production by keratinocytes. *J. Biol. Chem.* **267**, 21277–21280.

Jaattela, M., and Wissing, D. (1992). Emerging role of heat shock proteins in biology and medicine. *Ann. Med.* **24**, 249–258.

Jesberger, J. A., and Richardson, J. S. (1991). Oxygen free radicals and brain dysfunction. *Int. J. Neurosci.* **57**, 1–17.

Johnson, G. V. W., Litersky, J. M., and Jope, R. S. (1991). Degradation of microtubule-associated protein 2 and brain spectrin by calpain: A comparative study. *J. Neurochem.* **56**, 1630–1638.

Johnson, L. V., Walsh, M. L., Bokus, B. J., and Chen, L. B. (1981). Monitoring relative mitochondrial transmembrane potential in living cells by fluorescence microscopy. *J. Cell Biol.* **88**, 526–532.

Kater, S. B., Mattson, M. P., Cohan, C. S., and Connor, J. A. (1988). Calcium regulation of the neuronal growth cone. *Trends Neurosci.* **11**, 315–321.

Kolber, M. A., Quinones, R. R., Gress, R. E., and Henkart, P. A. (1988). Measurement of cytotoxicity by target cell release and retention of the fluorescent dye bis-carboxyethyl-carboxyfluorescein (BCECF). *J. Immunol. Meth.* **108**, 255–264.

Kosik, K. S. (1992). Cellular aspects of Alzheimer neurofibrillary pathology. *Prog. Clin. Biol. Res.* **379**, 183–193.

Lachowicz, R. M., Clayton, B., Thallman, K., Dix, J., and VanBuskirk, R. C. (1989). Rhodamine 123 as a probe of in vitro toxicity in MDCK cells. *Cytotechnol.* **2**, 203–211.

Lipton, S. A., Choi, Y. B., Pan, Z. H., Lei, S. Z., Chen, H. S., Sucher, N. J., Loscalzo, J.,

Singel, D. J., and Stamler, J. S. (1993). A redox-based mechanism for the neuroprotective and neurodestructive effects of nitric oxide and related nitroso-compounds. *Nature* **364**, 626–632.

Malenka, R. C. (1991). The role of postsynaptic calcium in the induction of long-term potentiation. *Mol. Neurobiol.* **5**, 289–295.

Malgaroli, A., Milani, D., Meldolesi, J., and Pozzan, T. (1987). Fura-2 measurement of cytosolic free Ca^{2+} in monolayers and suspensions of various types of animal cells. *J. Cell. Biol.* **105**, 2145–2155.

Mattson, M. P. (1990). Antigenic changes similar to those seen in neurofibrillary tangles are elicited by glutamate and calcium influx in cultured hippocampal neurons. *Neuron* **4**, 105–117.

Mattson, M. P. (1992). Calcium as sculptor and destroyer of neural circuitry. *Exp. Gerontol.* **27**, 29–49.

Mattson, M. P., and Kater, S. B. (1988). Isolated hippocampal neurons in cryopreserved long-term cultures: Development of neuroarchitecture and sensitivity to NMDA. *Int. J. Dev. Neurosci.* **6**, 439–452.

Mattson, M. P., and Rychlik, B. (1990a). Cell culture of cryopreserved human fetal cerebral cortical and hippocampal neurons: Neuronal development and responses to trophic factors. *Brain Res.* **522**, 204–214.

Mattson, M. P., and Rychlik, B. (1990b). Glia protect hippocampal neurons against excitatory amino acid-induced degeneration: Involvement of fibroblast growth factor. *Int. J. Dev. Neurosci.* **8**, 399–415.

Mattson, M. P., and Scheff, S. W. (1994). Endogenous neuroprotection factors and traumatic brain injury: Mechanisms of action and implications for therapy. *J. Neurotrauma* **11**, 3–33.

Mattson, M. P., Dou, P., and Kater, S. B. (1988a). Outgrowth-regulating actions of glutamate in isolated hippocampal pyramidal neurons. *J. Neurosci.* **8**, 2087–2100.

Mattson, M. P., Guthrie, P. B., and Kater, S. B. (1988b). Intracellular messengers in the generation and degeneration of hippocampal neuroarchitecture. *J. Neurosci. Res.* **21**, 447–464.

Mattson, M. P., Guthrie, P. B., and Kater, S. B. (1989a). A role for Na^+-dependent Ca^{2+} extrusion in protection against neuronal excitotoxicity. *FASEB J.* **3**, 2519–2526.

Mattson, M. P., Murrain, M., Guthrie, P. B., and Kater, S. B. (1989b). Fibroblast growth factor and glutamate: Opposing roles in the generation and degeneration of hippocampal neuroarchitecture. *J. Neurosci.* **9**, 3728–3740.

Mattson, M. P., Rychlik, B., Chu, C., and Christakos, S. (1991). Evidence for calcium-reducing and excitoprotective roles for the calcium binding protein (calbindin-D28k) in cultured hippocampal neurons. *Neuron* **6**, 41–51.

Mattson, M. P., Barger, S. W., Cheng, B., Lieberburg, I., Smith-Swintosky, V. L., and Rydel, R. E. (1993a). Amyloid precursor protein metabolites and loss of neuronal calcium homeostasis in Alzheimer's disease. *Trends Neurosci.* **16**, 409–415.

Mattson, M. P., Cheng, B., and Smith-Swintosky, V. L. (1993b). Growth factor-mediated protection from excitotoxicity and disturbances in calcium and free radical metabolism. *Sem. Neurosci.* **5**, 295–307.

Mattson, M. P., Cheng, B., Culwell, A. R., Esch, F. S., Lieberburg, I., and Rydel, R. E. (1993c). Evidence for excitoprotective and intraneuronal calcium-regulating roles for secreted forms of β-amyloid precursor protein. *Neuron* **10**, 243–254.

Mattson, M. P., Kumar, K., Wang, H., Cheng, B., and Michaelis, E. K. (1993d). Basic FGF regulates the expression of a functional 71 kDa NMDA receptor protein that mediates calcium influx and neurotoxicity in hippocampal neurons. *J. Neurosci.* **13**, 4575–4588.

Mattson, M. P., Tomaselli, K., and Rydel, R. E. (1993e). Calcium-destabilizing and neurodegenerative effects of aggregated β-amyloid peptide are attenuated by basic FGF. *Brain Res.* **621**, 35–49.

Mattson, M. P., Zhang, Y., and Bose, S. (1993f). Growth factors prevent mitochondrial dysfunction, loss of calcium homeostasis and cell injury, but not ATP depletion in hippocampal neurons deprived of glucose. *Exp. Neurol.* **121**, 1–13.

Matus, A. (1988). Microtubule-associated proteins. *Annu. Rev. Neurosci.* **11**, 29–41.

McCord, J. (1985). Oxygen-derived free radicals in postischemic tissue injury. *N. Engl. J. Med.* **312**, 159–163.

McCulloch, J. (1992). Excitatory amino acid antagonists and their potential for the treatment of ischaemic brain damage in man. *Br. J. Clin. Pharmacol.* **34,** 106–114.

Monyer, H., Giffard, R. G., Hartley, D. M., Dugan, L. L., Goldberg, M. P., and Choi, D. W. (1992). Oxygen or glucose deprivation-induced neuronal injury in cortical cell cultures is reduced by tetanus toxin. *Neuron* **8,** 967–973.

Moore, P. L., MacCoubrey, I. C., and Haugland, R. P. (1990). A rapid pH insensitive two color fluorescence viability (cytotoxicity) assay. *J. Cell Biol.* **111,** 58a.

Mosmann, T. (1983). Rapid colorimetric assay for cellular growth and survival: Application to proliferation and cytotoxicity assays. *J. Immunol. Meth.* **65,** 55–63.

Niggli, E., and Lederer, W. J. (1990). Real-time confocal microscopy and calcium measurements in heart muscle cells: Towards the development of a fluorescence microscope with high temporal and spatial resolution. *Cell Calcium* **11,** 121–130.

Nowak, T. S. Jr., Osborne, O. C., and Suga, S. (1993). Stress protein and proto-oncogene expression as indicators of neuronal pathophysiology after ischemia. *Prog. Brain Res.* **96,** 195–208.

Page, B., Page, M., and Noel, C. (1993). A new fluorometric assay for cytotoxicity measurements in vitro. *Int. J. Oncol.* **3,** 473–476.

Parsons, J. L., Risbood, P. A., Barbera, W. A., and Sharman, M. N. (1988). The synthesis of XTT: A new tetrazolium reagent that is bioreducible to a water soluble formazan. *J. Heterocyclic Chem.* **25,** 911–914.

Rijkers, G. T. (1990). Improved method for measuring intracellular Ca^{2+} with fluo-3. *Cytometry* **11,** 923.

Scudiero, D. A., Shoemaker, R. H., Paull, K. D., Monkr, A., Tierney, S., Nofziger, T. H., Currens, J. J., Seniff, D., and Boyd, M. R. (1988). Evaluation of a soluble tetrazolium/formazan assay for cell growth and drug sensitivity in culture using human and other tumor cell lines. *Cancer Res.* **48,** 4827–4833.

Seeburg, P. H. (1993). The molecular biology of mammalian glutamate receptor channels. *Trends Neurosci.* **16,** 359–365.

Siman, R., and Noszek, J. C. (1988). Excitatory amino acids activate calpain I and induce structural protein breakdown in vivo. *Neuron* **1,** 279–287.

Smiley, S. T., Reers, M., Mottoloa-Hartshorn, C., Lin, M., Chen, A., Smith, T. W., Steele, G. D., and Chen, L. B. (1991). Intracellular heterogeneity in mitochondrial membrane potentials revealed by a J-aggregate-forming lipophilic cation JC-1. *Proc. Natl. Acad. Sci. USA* **88,** 3671–3675.

Tanke, H. J., Van Der Linden, P. W. G., and Langerak, J. (1982). Alternative fluorochromes to ethidium bromide for automated read out of cytotoxicity tests. *J. Immunol. Meth.* **52,** 91–96.

Thastrup, O., Cullen, P. J., Drobak, B. K., Hanley, M. R., and Dawson, A. P. (1990). Thapsigargin, a tumor promoter, discharges intracellular Ca^{2+} stores by specific inhibition of the endoplasmic reticulum Ca^{2+}-ATPase. *Proc. Natl. Acad. Sci. USA* **87,** 2466–2470.

Tsien, R. W., Ellinor, P. T., and Horne, W. A. (1991). Molecular diversity of voltage-dependent Ca^{2+} channels. *Trends Pharmacol. Sci.* **12,** 349–354.

Verity, M. A. (1993). Mechanisms of phospholipase A2 activation and neuronal injury. *Ann. N. Y. Acad. Sci.* **679,** 110–120.

Zhang, Y., Tatsuno, T., Carney, J. M., and Mattson, M. P. (1993). Basic FGF, NGF, and IGFs protect hippocampal and cortical neurons against iron-induced degeneration. *J. Cereb. Blood Flow Metab.* **13,** 378–388.

CHAPTER 11

Use of Cultured Neurons and Neuronal Cell Lines to Study Morphological, Biochemical, and Molecular Changes Occurring in Cell Death

Jason C. Mills, ⋆ **Songli Wang,**† **Maria Erecińska,**⋆,† **and Randall N. Pittman**⋆,†

⋆ Cell Biology Graduate Group and
† Department of Pharmacology
School of Medicine, University of Pennsylvania
Philadelphia, Pennsylvania 19104

I. Introduction

The phenomenon of neuronal programmed cell death has drawn considerable attention in recent years. Most of what has been learned, however, is descriptive in nature: which cells die and when, which growth factors are involved, and what type of pathology the tissue assumes. Only very recently have advances been made in understanding the actual process of death on a cellular level. Much of this recent progress is due to the advent of various means of studying the death process in isolated neurons in tissue culture, where pharmacological and environmental variables can be controlled and where only one cell type is involved (Martin *et al.*, 1988; Scott and Davies, 1990). Even more progress is likely to occur as a result of the study of various differentiated neuronal cell lines that have been developed (Ruckenstein *et al.*, 1991; Jensen *et al.*, 1992; Koike, 1992; Mesner *et al.*, 1992; Pittman *et al.*, 1993). Although cell lines carry the risk of being less representative of *in vivo* conditions, they have several significant advantages over primary cultures as model systems: (1) they are a stable, dependable source of nearly identical, pure cells; (2) large numbers of cells can be grown; (3) death in the same cell type at different phases of differentiation can be examined; and (4) the cells can be manipulated genetically to study the effects on death.

We have used a subline of the pheochromocytoma cell line PC12 (Greene and Tischler, 1976) to study various aspects of neuronal cell death (Mills *et al.*, 1993; Wang and Pittman, 1993). This chapter will describe how this cell line (termed PC6-3) is maintained and differentiated, as well as how it is used to characterize morphological, biochemical, and molecular events in cell death. In addition, since differentiated PC6-3 cells can be used as a model system for sympathetic neurons, a protocol for growing primary cultures of sympathetic neurons from the superior cervical ganglia is also outlined.

II. Culturing Sympathetic Neurons

A. Background

Primary cultures of sympathetic neurons undergo transcription- and translation-dependent cell death after removing nerve growth factor (Martin *et al.*, 1988). The advantage of primary cultures of neurons over neuronal cell lines is that they probably more closely reflect events occurring *in vivo;* however, the disadvantage is that only a limited number of rather heterogeneous cells can be obtained, making biochemical and molecular studies difficult. Sympathetic neurons are grown in a large number of laboratories; many different culture conditions will support long term growth (see Hawrot and Patterson, 1979; Johnson and Argiro, 1983; Kessler, 1985). The following section presents one of those methods.

B. Materials

rat tail collagen (Collaborative Research; collagen can be made in the laboratory, but over the last 2 yr, for some unknown reason, a number of batches of collagen made in our laboratory have been somewhat toxic)

24-well tissue culture dishes (Costar, Cambridge, MA)

collagenase (Type 1A; Sigma, St. Louis, MO)

dispase (Boehringer-Mannheim, Indianapolis, IN)

calcium- and magnesium-free Hank's balanced salt solution (HBSS, pH 7.35): 8 g/liter NaCl, 0.4 g/liter KCl, 0.48 g/liter Na_2HPO_4, 0.06 g/liter KH_2PO_4, 1 g/liter glucose, 0.01 g/liter phenol red

DME medium (high glucose; JRH Biosciences, Lenexa, KS)

Ham's F12 medium (JRH)

fetal bovine serum (FBS; HyClone, Logan, VT)

rat serum (prepared in the laboratory from clotted blood)

penicillin/streptomycin (Sigma), prepared as a $100\times$ stock (10,000 units/ml penicillin and 10 mg/ml streptomycin) in HBSS, filter sterilized, and stored at $-20°C$

N2 serum-free medium components (see Section III,B)

35-mm Petri dishes (Falcon)

nerve growth factor (NGF; see Section III,B)

anti-NGF IgG (the IgG fraction of rabbit antiserum isolated in our laboratory by protein A-Sepharose column chromatography)

C. Procedure

1. Coat tissue culture wells (24- or 48-well dishes are good for growing neurons) with a thin film of rat-tail collagen (1–2 mg/ml); allow to dry at room temperature overnight. Generate the thin film either by adding excess collagen solution to each well and tipping the plate at a 45° angle to remove all excess collagen, or by adding a single drop to each well and using a pasteur pipet with the small end sealed and bent to "rake" the collagen over the well. Wash the dried collagen-coated wells one time with sterile water and one time with sterile HBSS prior to use.

2. Sympathetic ganglia are situated at the bifurcation of the carotid in the neck and are easy to recognize and remove; however, the best way to learn

the dissection is not to read it in a book, but to watch someone who is familiar with the procedure. Many laboratories grow sympathetic neurons. Visiting one of them to learn how to do the dissections properly is worthwhile. E20-P1 rat pups should be used as a source of ganglia.

3. After removing excess debris (e.g., carotid, fat, and nerve roots) from the ganglia, cut them into 2–3 pieces (to increase access by enzymes) and place them in a 35-mm petri dish in a sterile solution of HBSS containing 5 mg/ml dispase and 1 mg/ml collagenase. Use about 1 ml enzyme solution per 20 ganglia, and incubate the dish for 1–1.5 hr at 37°C in an air incubator, or seal tightly and incubate in a 6% CO_2 incubator at the same temperature.

4. "Coat" a 3-ml syringe fitted with an 18-gauge needle prior to use by drawing up and expelling serum-containing medium several times (remove excess serum by drawing up and expelling HBSS). Using the needle and syringe, transfer the ganglia to a sterile 15-ml polypropylene tube and triturate about 10 times gently, being careful to avoid generating excess air bubbles. If ganglia are not easily dissociated, add 1 ml fresh enzyme per 20 ganglia and incubate ganglia an additional 30 min at 37°C.

5. Centrifuge the solution of cells at 500 *g* for 8–10 min and suspend the pellet in serum-containing medium; centrifuge. Resuspend the pellet with serum-containing medium and centrifuge again for a total of two washes.

6. Resuspend the cell pellet in DME/F12 medium containing 3% rat serum (10% FBS works about as well), 100 ng/ml 2.5S NGF, 100 units penicillin, and 0.1 mg/ml streptomycin; plate onto collagen-coated wells. The number of cells plated per well depends on whether cells will be used for morphology and cell counts (low density) or for biochemistry (high density). In general, plate the equivalent of 2 ganglia per well of a 24-well dish for biochemistry experiments and 0.4–0.5 ganglia per well for cell counts or morphological studies.

7. Nonneuronal cells can be killed by treating cultures with 20 μM fluorodeoxyuridine and 20 μM uridine during the first week of culture. Cytosine arabinoside should not be used because it kills neurons in a manner very similar to that seen following NGF removal (Martin *et al.,* 1990). Alternatively, the number of nonneuronal cells can be reduced by preplating cells on tissue culture plastic (not coated with a substrate) for 1 hr, and then plating the cells that remain in suspension onto collagen-coated wells.

8. Feed cultures every 2–3 days. These cells can be used for experiments after 5 days, but routinely are used after 7–10 days in culture.

9. Initiate cell death by removing medium and replacing it with medium without NGF and containing 30–75 μg/ml anti-NGF IgG (the concentration depends on the titer of the antibody). Death of neurons occurs 18–36 hr later.

D. Notes

1. Ganglia can also be mechanically dissociated in a 35-mm petri dish using a pair of fine #5 jeweler's forceps. The sheath is removed (although it is often

not possible to remove the *entire* sheath), and the two forceps are used to tear and tug at the pieces of ganglion [two dogs (i.e., the forceps) fighting over a sock (i.e., the ganglion) is a fairly accurate analogy of how to mechanically dissociate ganglia]. The solution containing dissociated cells, small chunks of ganglia, and debris is allowed to settle for 3–5 min, after which the cell suspension (without the settled debris) is centrifuged at 500 *g* for 8–10 min; then the cell pellet is resuspended in growth medium and plated. The advantage of mechanical dissociation is that practically all nonneuronal cells remain associated with the "chunks" of ganglia; therefore, cultures are almost pure neurons. The disadvantage is that only about 30% of the total number of neurons can be obtained compared with the number obtained with enzymatic dissociation.

2. Neurons can be grown in serum-free medium using N2 components (Bottenstein *et al.*, 1980; see Section III,B for components). If grown on a collagen film substrate, cells should be cultured for 2 days in serum-containing medium before shifting into serum-free medium. If cells are grown on a laminin substrate or on an absorbed collagen substrate (generated by incubating wells with a 100 μg/ml collagen solution for 2 hr followed by rinsing wells, but not allowing them to dry), then neurons may be plated initially in serum-free medium. In general, laminin substrates and absorbed collagen substrates "roll up" after 4–6 days but are good for short-term cultures.

III. Maintenance and Differentiation of PC6–3 Cells

A. Background

PC6-3 cells were subcloned in this laboratory from the PC12 cell line (Pittman *et al.*, 1993). In their naive state, they grow as single, serum-dependent cells that can be passaged repeatedly. When treated with NGF, however, the cells undergo a slow, but relatively steady, differentiation into a neuronal phenotype. Within 6–8 days, the cells have developed neurites, have become nonmitotic, and can be subcultured out of the standard PC12 medium (RPMI-based, horse-serum containing) into medium designed to support primary cultures of neurons (DMEM/F12/N2 with or without FBS). After subculturing into DME/F12-based medium, the cells assume a more rounded neuronal morphology, grow a very extensive neurite network, and will undergo dramatic, transcription-dependent cell death on withdrawal of NGF (for more details, see Pittman *et al.*, 1993).

B. Materials

100-mm Corning (or Costar) tissue culture dishes
RPMI 1640 medium (JRH)
equine serum (HyClone)
fetal bovine serum (HyClone)
penicillin/streptomycin (see Section II,B)

trypsin/EDTA (GIBCO, Grand Island, NY, 0.05% trypsin and 0.53 m*M* EDTA)

freezing medium [made with 7% DMSO and 93% calf serum (HyClone)]

rat-tail collagen solution (Collaborative Research), bought in bulk, stored at 4°C

NGF (2.5 S, bought in bulk from Collaborative Research in lyophilized form, suspended in 100 mg/ml BSA, aliquotted, dried in a Speed-Vac concentrator, stored in dried form at 4°C)

DMEM (high glucose, without pyruvate; JRH)

Ham's F12 (JRH)

6-well tissue culture plates (Corning, 35-mm well diameter)

N2 components (listed as final, diluted concentrations): 5 μg/ml insulin (kept as a 500× stock in 5 m*M* HCl at 4°C), 10 μg/ml transferrin (kept as a 1000× stock in PBS at −20°C), 30 n*M* sodium selenate (kept as a 100 μM stock in dH$_2$O, 4°C), 20 n*M* progesterone (kept as a 1000× stock in PBS at 4°C), 100 μM putrescine (kept as a 100× stock in PBS, 4°C), 100 μg/ml BSA (from 1000× stock in PBS, aliquotted, −20°C)

anti-NGF antibody (see Section II,B)

C. Procedure

The goal of this technique is to take immortalized, nonneuronal, serum-dependent cells and make them terminally differentiated, nonmitotic, and NGF-dependent cells to serve as models of neuronal cell death (Fig. 1). Thus, the two facets to PC6-3 cell culture are (1) general maintenance of the naive line and (2) differentiation of the naive line into the postmitotic neuronal phenotype.

1. Naive Cell Maintenance

Naive cells are typically maintained on 100-mm tissue culture dishes and fed every other day (or Monday, Wednesday, Friday) with RPMI medium supplemented with 10% equine serum, 5% FBS, 100 units/ml penicillin, and 0.1 mg/ml streptomycin. Every 5–7 days, naive cells are split at a 1:6 dilution by treating with trypsin/EDTA for 5 min. After 5–6 passages, that particular line of naive cells is discarded and new experiments are conducted using a fresh tube of cells from working stocks kept at −80°C in freezing medium. The original lines are kept in liquid nitrogen.

2. Differentiation of PC6-3 Cells

1. Adsorb a 50 μg/ml solution of rat-tail collagen (stock collagen diluted in dH$_2$O) to 100-mm tissue culture dishes (5 ml/dish) for 2 hr.

2. Treat naive cells with trypsin, wash them off the plate, pellet in a table-

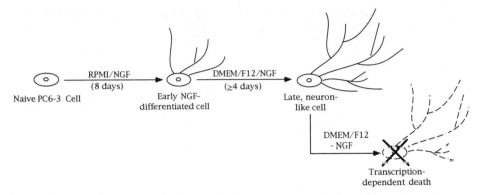

Fig. 1 Schematic of PC6-3 differentiation protocol. Naive cells are fed RPMI-based medium in the presence of horse and fetal bovine serum and NGF for ~8 days. They are then subcultured into DMEM/F12-based, serum-free medium with added N2 components and cultured ≥4 days until death is induced by replacing the medium with DMEM/F12/N2 lacking NGF.

top centrifuge, and resuspend in the same RPMI-based medium used for the naive cells, but with 100 ng/ml added NGF.

3. Aspirate the collagen solution from the dishes and distribute the cell suspension to each of the dishes to make a 1 : 8 or 1 : 10 split of the naive cells.

4. Incubate cells at 37°C, 6% CO_2, and feed with RPMI/NGF medium (described in Step 2) exactly every 2 days until day 7–9, when the cells are ready for subculture into DME/F12-based medium. Only about half the medium is changed at each feeding so there is always some conditioned medium left in the dish.

5. At least 2 hr prior to subculture, coat 6-well plates with 150 μg/ml collagen (~1.5 ml/well).

6. Subculture cells by removing RPMI/NGF medium from the 100-mm dishes and using DME/F12-based serum-free medium to triturate the cells off the dish. Transfer the suspended cells to a tissue culture flask or tube (wide-mouth, disposable glass pipets work best for this step). After the cells have been suspended in DME/F12 medium, replace the collagen solution in the 6-well plates with the cell suspension (2 ml suspension/well). The DME/F12 medium is a 1 : 1 mix of DMEM and Ham's F12 with added N2 components, penicillin/ streptomycin, and 100 ng/ml NGF. Typically, before adding the DME/F12 medium to the 100-mm dishes, calculate the total amount needed to fill all the 6-well plates. Then split that amount (plus a little extra for safety) equally into each of the 100-mm dishes so no additional dilution is necessary and so the cell suspension can later be accurately and evenly distributed into each well of the 6-well plates. The final dilution is usually about 1.5 to 2 100-mm plates for every 6-well plate (i.e., ~1.5 : 1).

7. Keep cells at 37°C, 6% CO_2 and feed on day 10, replacing about half the

medium in each well. If the experiments are to be performed on day 13 or later, feed the cells on day 12 as well.

3. NGF Removal

On day 12, apoptosis is induced by replacing the DME/F12/NGF medium described earlier with the same medium containing anti-NGF antibody instead of NGF. Once NGF is removed, the cells will die over a very reproducible time course (as assayed by counting cells in the presence of trypan blue, which only stains nonviable cells). At 8 hr, about 10% of the cells are dead. By 16 hr, the percentage of dead cells reaches about 35%, and by 24 hr, it is up to 70%. Note that death is a *process,* and many of the cells that are still trypan blue negative at a given time point have, nevertheless, passed an irreversible "commitment" point after which they can no longer be rescued by NGF. For example, even though at 12 hr following NGF removal only 20% of the cells have actually died, an additional 30% have committed themselves to die within 36 hr, even if NGF is added back to the medium. Since the cell population clearly goes through a dying *process* that evolves over time, the biochemical and molecular studies (outlined in Sections V and VI) are always conducted at several representative time points during the course of death.

D. Notes

1. Collagen from a number of sources (both rat-tail and calf-skin) has been tried. The most consistent results have been obtained using rat-tail collagen from Collaborative Research. Technically, collagen is more soluble in an acidic solvent, but for the times and dilutions described here, sterile distilled H_2O works well. The advantage of using H_2O is that there is no need to rinse the plates before adding cells.

2. There is some latitude with the timing of the system. Cells can be subcultured into DMEM/F12 on day 7–9 with little difference. Day 8 is standard. Death can be induced on day 11–14, but the results seem most dramatic on day 12.

3. NGF is a sticky molecule. Anti-NGF is used primarily to inactivate NGF stuck to the matrix and tissue culture dish. If the antibody is left out, the cells will still die, but they will take longer to do so.

4. Most of the N2 component stocks can be kept for several months. One exception is insulin, which should be remade every 4–6 wk. Once diluted into the medium at a neutral pH, insulin is even more labile. NGF is not stable for long in solution either, so generally medium—once it is mixed—is not kept for more than ~2 wk.

5. The final phase of differentiation (i.e., that which occurs in DME/F12-based medium) will also occur in DMEM/F12 + 5–10% FBS. The cells will

still die on NGF removal. However, for biochemical and molecular studies, a serum-free system has obvious advantages.

6. The cells are usually not grown on 100-mm plates when subcultured into DME/F12 because researchers have found that, at this stage, the cell monolayer is much more sensitive to shear force and, in the larger plates, has a tendency to detach from the surface in large sheets. This phenomenon may be related to the fact that, as the cells become more neuronal, their very extensive neurite networks increase tension on the collagen substrate, which eventually weakens its adhesion to the surface of the plate.

IV. Time-Lapse Videomicroscopy Studies

A. Background

A reliable tissue culture system makes a more in depth examination of how individual cells die possible. For example, with the aid of a long-term time-lapse video system, we have used PC6-3 cells to perform a detailed analysis of the morphological changes that occur during death. The chief advantage of this particular video system is that the cells are grown in exactly the same medium, at the same CO_2 concentrations, for the same amount of time as their sister cells in the incubator, which are used for other experiments. Thus, it is easy to correlate morphological changes as seen on video with biochemical and molecular ones being studied in sister cultures.

B. Materials

cylinder of compressed 5 or 10% CO_2 with balance air (Airco, Huntingdon Valley, PA)

35-mm glass-bottom tissue culture dishes (Mat-Tek, Ashland, MA)

collagen solution (see Section III,B)

150-mm and 60-mm plastic petri or tissue culture dish

Repel-Silane (LKB, Gaithersburg, MD)

flexible tubing

C. Procedure

1. Cell Culture

1. Grow cells in RPMI/NGF medium as described earlier until they are ready to be subcultured into DME/F12/NGF medium.
2. The night before subculturing, add a 450 μg/ml solution of collagen to the glass well in the bottom of the glass-bottom dish.
3. The next day, when sister cells are being subcultured into 6-well dishes for biochemical or molecular studies, add a ~1 : 4 dilution of the cell suspen-

sion to the glass well (after the collagen is removed). Then, after ~5 min, add 2 ml fresh medium to fill the rest of the dish.

4. Feed the glass-bottom dishes every 2 days; replace about half the medium each time (carefully!).

5. On days 11–14, use the cells for morphological studies. These studies are performed by placing a glass-bottom dish in a specially designed incubation chamber (Fig. 2) on the stage of a Nikon Diaphot inverted microscope in a 37°C warm room. The microscope is fitted with a Newvicon video camera (Panasonic), and the image is recorded on a JVC (BR-9000) time-lapse video recorder.

2. Incubation Chamber Construction

The incubation chamber is constructed using inexpensive, readily available materials and is designed to keep the cells at the proper humidity and CO_2 concentration with minimal optical interference.

1. Cut a circular hole ~55 mm in diameter in the center of a 150-mm dish using a heated scalpel.

2. Cut off the rim of the top of a 60-mm dish and affix using melted paraffin wax to form a concentric, water-tight ring within the 150-mm dish.

3. Cut a hole slightly smaller than a large round coverslip in the center of the lid of the 150-mm dish. Affix a round, silanized 25-mm coverslip with either wax or Vaseline to cover the underside of the hole.

4. Puncture two small holes in the 150-mm chamber using heated needles. Make one on the side of the dish about two-thirds of the way up from the bottom and about the diameter of a 14-g needle. Make the other just at the periphery of the coverslip on the lid and about the diameter of an 18-g needle. Thread flexible tubing of a size that will make a very tight fit through each of the holes. The tubing on the side carries the humidified CO_2 into the chamber and over the 60-mm inner wall to where the glass-bottom dish will be. The tubing on the top serves as a port for the removal and addition of medium to the glass-bottom dish.

3. Preparation of the Chamber for a Video Study

1. Immediately prior to every use of the incubation chamber, cut a piece of plastic wrap (e.g., Saran® Wrap) that will eventually be used to cover the outside bottom and sides of the chamber and to suspend the 35-mm dish.

2. Cut a small, coverslip-sized hole in the center of the plastic wrap.

3. Transport the 35-mm dish into a 37°C warm room and place it in the center of the incubation chamber, on the plastic wrap, with the glass bottom squarely on the coverslip-sized hole.

A

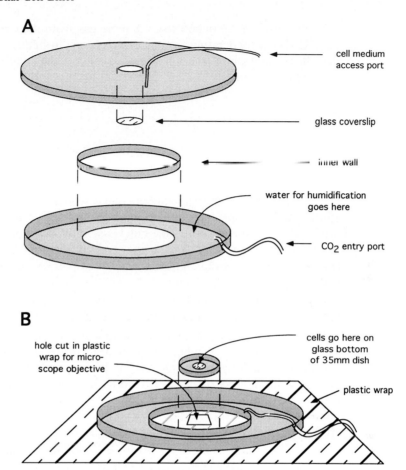

cell medium
access port

glass coverslip

inner wall

water for humidification
goes here

CO_2 entry port

B

hole cut in plastic
wrap for micro-
scope objective

cells go here on
glass bottom
of 35mm dish

plastic wrap

Fig. 2 Incubation chamber. (A) Schematic of entire chamber. (B) Bottom of chamber as it looks immediately prior to a video study.

4. Remove the lid from the glass-bottom dish. Cover the portion of the 150-mm dish that surrounds the 60-mm inner wall with a thin layer (4–5 mm) of water to provide humidity.
5. Replace the incubation chamber lid with the hole in the top of the chamber positioned directly over the glass bottom well (to allow for maximum light passage into the chamber).
6. Bring the plastic wrap up around the edge of the chamber to make a semi-gas-tight seal. Place the whole chamber on the microscope stage with light from the condenser passing through the coverslip on the lid of the chamber into the medium in the glass-bottom dish, through the cells, out the glass bottom, and into the objective.

7. At a rate of ~3 bubbles/sec, bubble gas from the cylinder containing 5 or 10% CO_2 (depending on choice of base medium) through dH_2O in a vacuum flask, and direct the humidified CO_2 into the incubation chamber using flexible tubing.

D. Notes

1. Cells can be videotaped using this system for at least 2–3 days. The only limitation seems to be the nonsterility of the system. Occasionally, after 48–72 hr, some growth of mold is evident.

2. The collagen does not adhere very well to the glass coverslips in the 35-mm dishes, especially after 5 days or so. Thus, there is often some "lift off" of the adsorbed collagen, but since only a small area of the coverslip is observed anyway, a field with good attachment can almost always be found. Caution should be exercised, however, whenever transporting the cells.

3. The plastic wrap in the incubation chamber is important since it increases retention of moisture and CO_2 in the system.

4. Note that this system has two significant advantages over typical systems for time-lapse video: (1) there is no need to make any special modifications of the culture medium for proper pH buffering and (2) since the cells can be observed for 2 days or more, the entire course of death of *all* the cells in a given field can be studied. In short, this system allows for more-or-less direct comparisons between studies of changes in individual cell morphology and studies of biochemical changes in populations of cells.

5. A 37°C warm room is not essential. The incubation chamber should also be effective in a system that uses an air-curtain stage heater to maintain proper temperature.

V. Methods for Studying Biochemical and Physiological Changes in Cell Death

A. Background

Large numbers of homogeneous cells in identical states of differentiation permit easy population studies of intracellular changes that occur during death (see Chapters 3–7 and 9–11, this volume). Protein synthetic rate, glucose utilization, and lactate production have all been studied in PC6-3 cells at various times after NGF removal. The goal of these studies is achieving a better understanding of the overall metabolic changes that occur in a dying cell, which might provide insight into understanding *how* cells manage to kill themselves.

B. Determining the Rate of Protein Synthesis

1. Materials

Labeling medium: DMEM/F12 (in powder form from Sigma, comes with 15 mM HEPES; lacks L-leucine, L-lysine, L-methionine, $CaCl_2$, $MgCl_2$, $MgSO_4$, $NaHCO_3$, and phenol red; all the missing components are added back once the stock is prepared, except phenol red and L-methionine), N2 components, and 100 $\mu Ci/ml[^{35}S]$methionine (in the form of Tran-S-label, ICN, Costa Mesa, CA)

Sample electrophoresis buffer: either modified O'Farrell's (9.5 M urea, 2% ampholines, 2% NP-40, 0.04 M DTT with 0.2% SDS; see O'Farrell, 1975) or Laemmli's (4×: 4 ml 0.5 M Tris, pH 6.8; 2 ml 20% SDS; 1.6 ml 0.2% bromophenol blue, 0.4 ml 2-mercaptoethanol, 2.0 ml glycerol; see Laemmli, 1970)

6 and 12% trichloroacetic acid (TCA)

Millipore vacuum filtration apparatus

glass fiber filters (Schleicher & Schuell, Keene, NH)

Coomassie stain: 0.4% Serva Blue G (Serva, Paramus, NJ) in 50% methanol, 10% acetic acid

5 mg/ml BSA stock for making protein standards

2. Procedure

a. Experimental Design

The goal of this study is to estimate the rate at which dying cells—relative to controls—are synthesizing new protein. The approach is to treat both populations with medium containing a radiolabeled amino acid, which is taken up by cells and incorporated into the protein that is actively being translated. Thus, when the cells are harvested and the protein is extracted, the quantity of radiolabeled protein will be proportional to the amount of new protein synthesized during the time the cells were exposed to the labeled amino acid.

b. Quantification of Incorporated Label

1. Culture PC6-3 cells for 12 days using the differentiation protocol described already.

2. Replace NGF-containing medium with NGF-antibody-containing medium in half the wells on each plate (−NGF group) and with more NGF containing medium in the other half (+NGF group).

3. Culture the cells until exactly 4 hr before they are to be harvested; at this point, replace the medium with labeling medium.

4. After 4 hr of labeling, place the 6-well plate in an ice-water bath. Completely remove the radioactive medium from each of the three wells of one of the two groups (+NGF or −NGF). Divide exactly 750 μl ice cold PBS nonquantitatively into each of those three wells; leave the largest fraction in

the first well and give the other two wells just enough PBS to keep from drying out. Then remove the labeling medium from the three wells of the other group and replace with PBS in the same fashion. Next, use a cell scraper or rubber policeman to detach the cells from the first well of the first group. Transfer the resulting cell suspension to the next well of the same group until all three wells are suspended and combined in 750 μl PBS, which is then transferred to a microcentrifuge tube on ice. Then scrape and suspend the wells from the same sample similarly, this time using 500 μl PBS. Repeat the entire scraping and suspension process with the three wells of the other experimental group, and transfer the cell suspension from that group to a different tube.

5. After harvesting the two samples from each time point, you should have two microcentrifuge tubes—one from the three +NGF wells, the other from the three −NGF. Centrifuge each tube (containing 1.25 ml PBS/cell mixture) for approximately 30 sec at 4°C in a microcentrifuge at 13,000 g. Then remove the supernatant very carefully and completely with a pipetter. Invert the tube and blot the remaining liquid with the corner of a Kimwipe.®

6. Add 150–300 μl sample buffer to each pellet (both Laemmli's and O'Farrell's have been used, depending on whether the sample is to be electrophoresed later on a 1- or 2-dimensional gel, respectively). Triturate the pellet vigorously using a 26-g 1/2″ needle and a 1-cc syringe (this step is designed to shear the DNA in the sample).

7. Boil samples in Laemmli's buffer for 5 min. Incubate those in O'Farrell's at room temperature for 1–2 hr (to ensure solubilization of proteins). After that, the samples can be snap-frozen and stored at −80° C for later use.

8. To determine the amount of incorporated label, transfer 3- to 5-μl aliquots of each sample, in triplicate, to microcentrifuge tubes containing 300 μl ice-cold 12% TCA. Precipitate on ice for 5–10 min.

9. To trap the protein precipitated by the TCA, prepare a Millipore vacuum filtration apparatus with glass fiber filters pre-wetted in 12% TCA. Then transfer the samples onto the filters (1 sample/filter, of course) and aspirate the liquid through the filter using a vacuum pump. Wash each filter 3–4× in ice-cold 6% TCA. After the last wash, dry the filters containing the precipitated protein under vacuum, transfer to scintillation vials, add scintillation cocktail, and determine radioactivity using a scintillation counter.

c. Standardization of Labeled Protein Relative to Total Protein
Since the goal of this study is to determine the rate of protein synthesis in cells that are actively dying during the course of the experiment, it is essential to use some means of standardizing the raw ^{35}S incorporation data. One way is to perform cell counts using trypan blue exclusion (see Section III,C,3) on cells grown in parallel, separate plates and then to express the data in terms of labeled methionine incorporation per viable cell. Another method is simply to use the same protein extracts already obtained for measurement of [^{35}S]methionine incorporation and to determine the concentration of *total*—labeled and

unlabeled—protein. Since dead cells tend to be removed from the wells when the labeling medium is discarded, total protein concentration in each sample gives a good estimate of how much protein could potentially have been labeled and, thus, can be used to determine a "percentage incorporation" in each sample. The "percentage incorporation" can then be used to compare relative rates of protein synthesis of cells during any stage in the death process.

To measure protein concentrations in samples that contain large amounts of either SDS or NP-40 and 2-mercaptoethanol, which cannot be measured accurately by common protein assays, we use a version of Esen's protein quantification procedure (Esen, 1978; Marder *et al.*, 1986).

1. Blot samples (and standards consisting of various concentrations of BSA from 0.5 to 2.5 mg/ml) in triplicate 5-μl drops on Whatman paper, dry in a 60°C oven for a few minutes, stain in 0.4% Coomassie for 5 min, wash repeatedly in dH$_2$O (until background has been greatly reduced), and dry again in the oven.

2. Using a paper punch, cut out blotted samples and incubate in 1 ml 1% SDS for at least 1 hr at 65°C in glass test tubes in a water bath (unblotted areas of the Whatman paper are used as blanks).

3. Measure the optical density at 600 nm in a spectrophotometer. Standardize samples relative to the BSA standard concentrations.

4. Once total protein is known, protein synthetic rate can be expressed as incorporated label/total protein for each sample.

3. Notes

1. A curve of data that might be expected when using this protocol to quantify protein synthesis in this model system is depicted in Fig. 3.

2. The same samples that are collected for determining the rate of protein synthesis can also be used in either 1-dimensional (for samples solubilized in Laemmli's) or 2-dimensional (for those in modified O'Farrell's) gel electrophoresis. If the gels are subsequently exposed to X-ray film, those proteins whose expression changes during death can be observed and potentially identified.

C. Glucose Utilization

1. Materials

Earle's basic salt solution (IX): 117.2 mM NaCl, 5.4 mM KCl, 1.8 mM CaCl$_2$, 0.8 mM MgSO$_4$, 1.0 mM NaH$_2$PO$_4$, 26.2 mM NaHCO$_3$, 5.6 mM glucose (can also be made as 10× stock, although, in this case, the NaHCO$_3$ and glucose are added only) when smaller aliquots are diluted to 1×

[^{3}H]deoxy-D-glucose ([^{3}H]2-deoxyglucose), 30.6 Ci/mmol (NEN/DuPont, Boston, MA)

4% Nonidet P 40 (NP-40) in dH$_2$O

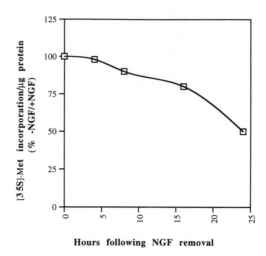

Fig. 3 Hypothetical plot of protein synthetic rate after NGF removal, based on results typically obtained in this laboratory using the methods described in the text. The data points represent [35S] methionine incorporation into total protein at each time point following NGF withdrawal, expressed as a percentage of control (+ NGF) [35S]methionine incorporation at the same time point. Note that typically, cell protein from 3–6 separate wells is pooled to generate each data point.

2. Procedure

a. Experimental Design

Radioactive 2-deoxyglucose (2-DG) uptake into cells is used to compare the rate of glucose utilization and, by inference, the metabolic rate in dying and control cell populations. Before such a comparison is possible, however, two experimental conditions must be met: the rate of 2-DG uptake must be relatively constant over the time course of the experiment and extracellular glucose must be at a concentration sufficient to saturate its cellular transporter (see Brookes and Yarowsky, 1985). When the rate of 2-DG uptake is constant, the *net* 2-DG uptake measured in the experiment reflects the *overall rate* of uptake. When the transporter is saturated, the rate of glucose uptake (as measured by the 2-DG analog) will directly reflect the rate glucose is actually being used by the cell.

b. Measuring Net Uptake

The first step before establishing a paradigm to estimate glucose utlization with 2-DG should be to perform a saturation study with various concentrations of extracellular glucose (from ~2 to 20 mM). A time course of 2-DG uptake should also be performed at a constant extracellular glucose concentration (a convenient time range is from 5 to 30 min).

The paradigm established for the PC6-3 system—based on our saturation and time course studies—follows.

1. Culture and differentiate cells, as described earlier, for 12 days.

2. For each time point, remove the medium from one 6-well plate. Add 1 ml fresh NGF-containing medium to each of 3 wells (NGF + samples); add the same volume of NGF-antibody-containing medium to the other 3 wells (NGF − samples).

3. Incubate plates for exact times (usually 8, 16, and 24 hr; see Section III,C,3 for more details). Then discard the medium or save for lactate measurements (see Section V,D).

4. Add 750 μl Earle's salt solution with the standard 5.6 mM glucose (a concentration determined to be at saturation by our preliminary studies) and 40 nM 2-DG to all the wells on the plate. Pre incubate the Earle's solution at 7.5% CO_2 for several hours to equilibrate it to the proper pH. Conduct the entire experiment in a cell culture incubator at 37°C and 7.5% CO_2.

5. After a 20-min incubation, rapidly aspirate the medium (using a vacuum line). Wash the cells 3 times quickly in ice cold PBS.

6. Add 500 μl 4% NP-40 to each well and leave at room temperature for at least 30 min. Then transfer the cell extract to scintillation vials, fill with scintillation cocktail, and determine radioactivity using a scintillation counter.

7. The measurement of interest is glucose utilization in cells that are in the process of dying and not yet dead; therefore, the number of living cells at each time point is determined using trypan blue in sister plates cultured in parallel to the 2-DG plates.

3. Notes

1. It is important to estimate the amount of 2-DG that is simply "trapped" in extracellular space at the various time points, since total counts may not exactly reflect intracellular radioactivity and cells change shape during the course of dying. Such an estimate can be obtained by measuring radioactivity in extracts prepared from parallel wells that have been treated with [^{3}H]2-DG for only a very short time (a few seconds) and then washed normally.

2. It is important to run control, NGF-containing cultures for each time point because there tends to be some variability in glucose usage depending on the amount of time since the cells were last fed.

3. A graph based on hypothetical values for 2-DG uptake, depicting the kinds of results that can be expected from this protocol, is shown in Fig. 4.

D. Lactate Accumulation

1. Materials

40% perchloric acid (PCA)

2.5 M KHCO$_3$

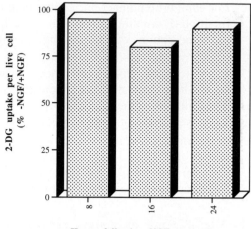

Fig. 4 Hypothetical graph of 2-deoxyglucose (2-DG) uptake after NGF removal, based on typical results. The columns represent 2-DG incorporation (standardized to the number of *living* cells and not total uptake) at each time point, expressed as a percentage of control (+ NGF) uptake at the corresponding time. Cell counts (to determine the number of *living* cells) and scintillation counts (to determine 2-DG uptake) are typically performed in triplicate.

lactate dehydrogenase (5 mg/ml; Boehringer-Mannheim)
reaction buffer (0.5 *M* glycine/0.4 *M* hydrazine, pH 9)
NAD^+ (32 m*M* stock in dH_2O, stored in aliquots at $-20°C$)

2. Procedure

a. Experimental Design

For cells in tissue culture, which are very glycolytically active, lactate production—like glucose uptake—is a good indication of metabolic activity. However, since lactate is measured as an accumulation over time, whereas glucose uptake is measured at a given moment, the two assays are complementary rather than redundant.

b. Measuring Lactate Accumulation

1. Culture and differentiate cells as previously described. Replace the medium in each 6-well plate with NGF-containing and NGF-antibody-containing medium, as for the glucose utilization studies (Section V,C,2,b, Step 2).

All of the following steps are performed at 4°C.

2. At exact time points (usually anywhere from 4 to 36 hr), remove 750 μl

       medium from each well and immediately acidify in tubes containing 100 μl 40% PCA.

3. Perform trypan blue cell counts either in the same wells from which medium is taken for lactate determination or in parallel ones.

4. Centrifuge the acidified medium samples at 8000 g for 10 min to remove precipitated debris. Neutralize the supernatant with 2.5 M KHCO$_3$ to pH 6–7. (Measure pH using a pH meter.) Record the exact volume of dilution since it will eventually be needed to determine concentration in the original sample.

5. Pellet the PCA precipitated after neutralization with KHCO$_3$ by centrifuging at 3000 g for 10 min. Discard the pellet.

6. Dilute 25- to 50-μl aliquots of the samples (after clearing) with dH$_2$O up to 200 μl. To this, add 800 μl reaction buffer plus 50 μl NAD$^+$ (to make a final concentration of 1.6 mM).

7. Take an optical density (OD) reading at 340 nm to establish baseline reaction values.

8. Start the reaction by adding 10 μl LDH (final concentration: 5 μg/ml) and continue until the OD stabilizes (10–15 min under these conditions).

9. Take OD readings again of samples and blanks (either buffer and enzymes alone or DMEM/F12/N2 medium treated the same way as the samples—there should be no difference).

10. Determine lactate concentrations in the original samples by multiplying the OD reading by the dilution factor (due to the PCA and KHCO$_3$) and dividing this number by the molecular extinction coefficient (6.23) and by the dilution factor due to the size of the aliquot used in the LDH reaction.

3. Notes

1. Several kits are commercially available for determining lactate levels.

2. Although it is commonly believed that the reaction buffer is stable for at least 1 mo at 4°C, we always make it fresh.

3. Under the conditions just outlined, 1 nmol lactate in 1 ml (i.e., 1 μM) will give an OD reading of 0.006 at 340 nm. Since most spectrophotometers are reasonably linear up to ~2.0, the range of measurable lactate concentrations is from about 10 μM to at least 300 μM. Optimal readings are probably between 30 and 100, which will give ODs of 0.180 to 0.600. In our system, measured values of lactate concentration in 1 ml medium, produced by ~300,000 cells, are in the range of 1–6 mM. The dilution introduced by the acidification and neutralization steps is about twofold. Thus, a 25- to 50-μl aliquot of the sample in Step 6, providing an additional 20- to 40-fold dilution, gives OD values in the optimal range.

4. As in the other biochemical assays, the measurement of interest is lactate production in dying, but not yet dead, cells; therefore, some estimate of lactate production per live cell is needed. Unfortunately, since lactate accumulation over time is measured, a count of living cells at any given time point cannot be used. Thus to standardize for living cells at each specific time point, the history of cell death up to that point must be known. Since cell counts from multiple time points exist for each experiment, a viability curve can be constructed from the ratio of −NGF to +NGF cell counts at each time point. The area under this curve can be used to estimate "total living cell time," and a factor can be obtained to standardize the −NGF raw data to account for already dead cells. In practice in this laboratory, the −NGF cell counts at various times are ratioed relative to controls at each point, a second order polynomial curve is fitted using a computer graphing program, and the equation corresponding to that curve is integrated from 0 to each time point. This integral is then divided into the control integral for the same period (which will, by definition, always be 1.00 times the number of hours) to achieve the "dead cell" correction factor.

5. A hypothetical graph, based on typical values for lactate acquired in this laboratory, is depicted in Fig. 5.

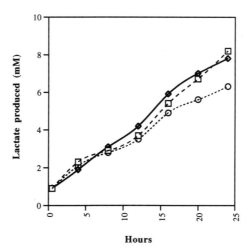

Fig. 5 Hypothetical plot of lactate production in +NGF (◇) and −NGF (○,□) populations based on typical results. Data points represent the cumulative concentration of lactate achieved over the given time period in the standard 1 ml of medium in each well. For the "−NGF" cultures, "raw" data points are plotted (○) along with corrected ones (□); the correction factor is determined as described in the text. Note that typically, lactate concentrations are determined in each of 3 separate wells and then averaged to generate each data point.

════════════   **VI. Electrophoretic Mobility Shift Assays**

A. Background

A cell line such as PC6-3 cells, that undergoes apoptosis, allows investigators to characterize changes in binding of nuclear proteins to specific DNA sequences during various stages of cell death relatively easily (Wang and Pittman, 1993). Eventually, studies along these lines may help elucidate the molecular cascade of transcription factors that is likely to be associated with death.

Gel shift assays consist of two phases: (1) preparation of nuclear extracts from dying and control cells and (2) reaction of those extracts with specific radiolabeled double-stranded oligonucleotides. During the reaction, specific DNA binding proteins (many of which are presumably transcription factors) will bind to the radiolabeled oligonucleotides that they recognize. When the reaction mixture is run on a low percentage polyacrylamide gel, the bound proteins will retard the migration of the oligonucleotides and will form a high molecular weight band. The patterns of DNA–protein binding can then be compared between control and dying cells.

B. Nuclear Extract Preparation

1. Materials

DTT (2000× stock, 1.0 M in dH$_2$O)

protease inhibitors (Boehringer-Mannheim; all stocks stored in aliquots at −20°C): PMSF (300× stock, 0.15 M in isopropyl alcohol), pepstatin (1000× stock, 1 mg/ml in methanol), TPCK (285× stock, 20 mg/ml in ethanol), leupeptin (10,000× stock, 5 mg/ml in dH$_2$O), trypsin inhibitor (200× stock, 2 mg/ml in dH$_2$O)

PBS: 0.15 M NaCl, 0.01 M NaH$_2$PO$_4$, 0.005 M Na$_2$HPO$_4$, pH 7.2, 0.5 mM PMSF (added immediately before use)

hypotonic buffer: 10 mM HEPES (pH 7.9, 2.4 g/liter), 1.5 mM MgCl$_2$ (0.3 g/liter), 10 mM KCl (0.75 g/liter), full complement of protease inhibitors (added immediately before use)

extraction buffer: 20 mM HEPES (pH 7.9), 20% glycerol, 0.42 M NaCl, 1.5 mM MgCl, 0.2 mM EDTA, full complement of protease inhibitors (added immediately before use)

dialysis and storage buffer: 20 mM HEPES (pH 7.9), 20% glycerol, 100 mM KCl, 0.2 mM EDTA, 0.5 mM PMSF, 0.5 mM DTT (added immediately before use)

dounce all-glass homogenizer

micro stirbar

2000-dalton cut-off dialysis tubing (7.6-mm diameter; Spectrum, Houston, TX)

2. Procedure

Start with 5×10^7 to 1×10^8 cells. Keep cell pellets, lysate, nuclear pellets, and extracts on ice water at all times.

1. Grow cells for 12 days in NGF-containing medium as described earlier. Replace medium in half the wells with NGF-antibody-containing medium and, after various times, remove cells from the plate by triturating with ice cold PBS. Centrifuge the cell suspension at 1000 g for 10 min at 4°C.

2. Resuspend the cell pellet in 3 ml hypotonic buffer, and transfer to a chilled homogenizer; incubate on ice water for 10 min. Lyse cells by homogenizing with 10–20 strokes until the pellet is disrupted. Lysis can be checked using a phase contrast microscope. Keep the homogenizer on ice to minimize heat build up.

3. Transfer the cell lysate to a 15-ml tissue culture tube and centrifuge at 1000 g for 10 min.

4. Save the supernatant at this step to study cytoplasmic proteins. For nuclear extraction, however, resuspend the pellet in 200–300 μl (~3 volumes) of hypotonic buffer. Then transfer the nuclear suspension to a microfuge tube and centrifuge at 13000 g for 1 min to remove residual cytoplasmic impurities.

5. Resuspend the pellet, containing nuclei, in 0.5 ml extraction buffer and homogenize with ~40 strokes until complete lysis is observed (this can be checked under the microscope). Transfer the homogenate to a glass or polypropylene scintillation vial containing a micro stirbar. Rinse the pestle and homogenizer with an additional 0.25 ml extraction buffer, and add to the vial.

6. Stir the extract for 30 min in an ice water bath.

7. Transfer the extract to a microcentrifuge tube and centrifuge for 30 min at 13000 g to remove membranous material.

8. Dialyze the supernatant from this spin against 300 ml dialysis buffer for 2 hr with a buffer change after 1 hr.

9. Centrifuge the dialysate at 13000 g for 15 min. Discard the pellet.

10. Aliquot the supernatant into ~20-μl portions, freeze in liquid nitrogen, and store at −80°C (prior to use in DNA binding reactions).

3. Notes

1. We cannot overemphasize that everything the extract comes in contact with during the assay should be at 0–4°C.

2. Nuclear extracts are best if frozen only once. Even when stored at −80°C, extracts are best if used within 2–3 mo.

3. A protein assay should be performed on the extracts to determine the volume that should be used during the DNA binding reaction described next.

4. See Dignam *et al.* (1983) for a detailed analysis of how various conditions affect nuclear extraction.

C. Gel Shift Assay

1. Materials

G-25 quick spin column (In Vitrogen, San Diego, CA)

running buffer for 20% gel: 1× TBE

gel matrix (one 20 × 16 × 0.15 cm 20% gel): 33.3 ml 30% acrylamide/0.8% bis-acrylamide stock, 11.6 ml dH$_2$O, 5 ml 10× TBE, 0.4 ml 10% ammonium persulfate, 40 μl TEMED

1 liter 10× TBE: 108 g Tris base, 55 g boric acid, 40 ml 0.5 M EDTA (no need for pH titration)

4 liters running buffer for 4.5% gel (5×): 570.8 g glycine (380 mM), 15.68 g EDTA (2 mM), 123.2 g Tris (50 mM)

gel matrix (two 20 × 16 × 0.15 cm 4.5% gels): 14.8 ml 30% acrylamide/0.8% bis-acrylamide stock, 5 ml 50% glycerol, 20 ml 5× running buffer, 61.5 ml dH$_2$O, 0.7 ml ammonium persulfate, 35 μl TEMED

eluting buffer: TE (10 mM Tris, pH 8, 1 mM EDTA)

reaction buffer: 10 mM Tris, pH 7.5, 50 mM NaCl, 1 mM DTT, 1 mM EDTA, 5% glycerol, 1.0 μg/ml poly dI-dC (Boehringer-Mannheim; this buffer is made fresh immediately before use)

2. Procedure

1. Purchase two complementary strands of DNA—in single-stranded, lyophilized form—encoding a consensus protein-binding site of interest. Reconstitute each strand in dH$_2$O to a concentration of 150 ng/ml. Combine the two strands in one tube at room temperature, boil for 5 min (to ensure denaturation), and then immediately place on ice for 3 min. End label the two complementary single strands with [^{32}P]ATP using T4 polynucleotide kinase for 30 min at room temperature (consult any standard molecular biology reference for a labeling protocol). Next, incubated the strands at 65°C for 45 min and allow to hybridize by slow cooling back to room temperature (cooling should take less than 1 hr).

2. Eliminate free nucleotides by spin column. Add 1/10 volume of 80% glycerol to the labeled double-stranded oligonucleotides. Electrophorese on a 16 × 20 cm 20% polyacrylamide gel (25 mAmp/gel, ~6 hr).

3. Wrap the gel (while still wet) in plastic wrap and expose on X-ray film for ~30 min.

4. With the aid of the exposed X-ray film, excise the band corresponding to the pure double-stranded oligonucleotides. Elute in 200–300 μl eluting buffer at room temperature for 2 hr.

5. Count the eluent in a scintillation counter for quantification; then use in the DNA binding reaction.

6. Begin the DNA binding reaction by pre-incubating the reaction buffer with 5 μg nuclear extract (5 min, room temperature). This step allows the poly dI-dC oligonucleotides in the reaction buffer to react with any nonspecific nucleotide binding activity in the extract before the labeled oligonucleotides are introduced. Then add 5000 cpm labeled oligonucleotide and incubate for at least an additional 30 min. The reaction is optimal in a total volume of between 15 and 20 μl.

7. After the reaction is complete, add 1/10 volume 80% glycerol and tracer bromophenol blue to the sample. Electrophorese on a 4.5% polyacrylamide gel (25 mAmp/gel, ~2 hr).

8. Vacuum dry the gel under heat and expose to X-ray film.

3. Notes

1. For some reason, the concentration of DTT in the reaction buffer seems to be critical, affecting both signal and background noise. In this laboratory, 1 mM was found to be optimal for maximizing the signal-to-noise ratio.

2. Undoubtedly, when attempting to establish a mobility shift assay system, it is best to use standard extracts with known activity for a specific probe first to determine an optimal concentration of poly dI-dC and an optimal reaction time for reducing nonspecific binding. For example, in this laboratory, HeLa cell nuclear extracts were used as standards because of their well-characterized binding to oligonucleotides with CRE (cyclic AMP response enhancer) consensus sequences (e.g., Verri *et al.*, 1990)

3. On each 4.5% gel, at least one lane should contain sample from a reaction conducted in the presence of 50–100× excess unlabeled probe to ensure the specificity of the binding reaction. Also, labeled probe alone (with no extract) should be run in one lane as a negative control.

4. Mutant oligonucleotide sequences with very similar sequences to those of interest, (usually 2 or 3 nucleotides are transposed) but known not to have any binding activity, can be run as additional controls for specificity. Of course, since a reasonable amount of variability in the consensus protein binding sites often occurs, mutant sequences might still be expected to have some binding activity. Thus it is essential that they be well-characterized before use.

Acknowledgment

The authors thank Jeff Ware for his invaluable tutoring in the use of Macintosh drawing software.

References

Bottenstein, J. E., Skaper, S. D., Varon, S. S., and Sato, G. H. (1980). Selective survival of neurons from chick embryo sensory ganglionic dissociates utilizing serum-free supplemented medium. *Exp. Cell Res.* **125,** 183–190.

Brookes, N., and Yarowsky, P. J. (1985). Determinants of deoxyglucose uptake in cultured astrocytes: The role of the sodium pump. *J. Neurochem.* **44,** 473–479.

Dignam, J. D., Liebovitz, R. M., and Roeder, R. G. (1983). Accurate transcription initiation by RNA polymerase II in a soluble extract from isolated mammalian nuclei. *Nucleic Acids Res.* **11,** 1475–1489.

Esen, A. (1978). A simple method for quantitative, semiquantitative, and qualitative assay of protein. *Anal. Biochem.* **89,** 264–273.

Greene, L., and Tischler, A. (1976). Establishment of a noradrenergic clonal line of rat adrenal pheochromocytoma cells which respond to nerve growth factor. *Proc. Natl. Acad. Sci. USA* **73,** 2424–2428.

Hawrot, E., and Patterson, P. H. (1979). Long-term culture of dissociated sympathetic neurons. *Meth. Enzymol.* **58,** 574–584.

Jensen, L. M., Zhang, Y., and Shooter, E. M. (1992). Steady-state polypeptide modulations associated with nerve growth factor (NGF)-induced terminal differentiation and NGF deprivation-induced apoptosis in human neuroblastoma cells. *J. Biol. Chem.* **267,** 19325–19333.

Johnson, M. I., and Argiro, V. (1983). Techniques in the tissue culture of rat sympathetic neurons. *Meth. Enzymol.* **103,** 334–347.

Kessler, J. A. (1985). Differential regulation of peptide and catecholamine characters in cultured sympathetic neurons. *Neuroscience* **15,** 827–839.

Koike, T. (1992). Molecular and cellular mechanism of neuronal degeneration caused by nerve growth factor deprivation approached through PC12 cell culture. *Prog. Neuropsychopharmacol. Biol. Psychiatry* **16,** 95–106.

Laemmli, U. K. (1970). Cleavage of structural proteins during the assembly of the head of bacteriophage T4. *Nature (London)* **227,** 680–681.

Marder, J. B., Mattoo, A. K., and Edelman, M. (1986). Identification and characterization of the psbA gene product: The 32-kDa chloroplast membrane protein. *Meth. Enzymol.* **118,** 384–395.

Martin, D. P., Schmidt, R. E., DiStefano, P. S., Lowry, O. H., Carter, J. G., and Johnson, E. M. (1988). Inhibitors of protein synthesis and RNA synthesis prevent neuronal death caused by nerve growth factor deprivation. *J. Cell Biol.* **106,** 829–844.

Martin, D. P., Wallace, T. L., and Johnson, E. M. (1990). Cytosine arabinoside kills postmitotic neurons in a fashion resembling trophic factor deprivation: Evidence that a deoxycytidine-dependent process may be required for nerve growth factor signal transduction. *J. Neurosci.* **10,** 184–193.

Mesner, P. W., Winters, T. R., and Green, S. H. (1992). Nerve growth factor withdrawal-induced cell death in neuronal PC12 cells resembles that in sympathetic neurons. *J. Cell Biol.* **119,** 1669–1680.

Mills, J. C., Wang, S. L., and Pittman, R. N. (1993). Timelapse videomicroscopic studies of morphological changes during apoptotic and necrotic cell death. *Mol. Biol. Cell.* **4,** 2140.

O'Farrell, P. H. (1975). High resolution two-dimensional electrophoresis of proteins. *J. Biol. Chem.* **250,** 4007–4021.

Pittman, R. N., Wang, S. L., DiBenedetto, A. J., and Mills, J. C. (1993). A system for characterizing cellular and molecular events in programmed neuronal death. *J. Neurosci.* **13,** 3669–3680.

Ruckenstein, A., Rydel, R., and Greene, L. (1991). Multiple agents rescue PC12 cells from serum-free cell death by translation- and transcription-independent mechanisms. *J. Neurosci.* **11,** 2552–2563.

Scott, S. A., and Davies, A. M. (1990). Inhibition of protein synthesis prevents cell death in sensory and parasympathetic neurons deprived of neurotrophic factor *in vivo. J. Neurobiol.* **21,** 630–638.

Verri, A., Mazzarrello, P., Biamonti, G., Spadari, S., and Focher, F. (1990). The specific binding of nuclear protein(s) to the cAMP responsive element (CRE) sequence (TGACGTCA) is reduced by the misincorporation of U and increased by the deamination of C. *Nucleic Acids Res.* **18,** 5775–5780.

Wang, S. L., and Pittman, R. N. (1993). Altered protein binding to the octamer motif appears to be an early event in programmed neuronal cell death. *Proc. Natl. Acad. Sci. USA* **90,** 10385–10389.

CHAPTER 12

Methods for Studying Cell Death and Viability in Primary Neuronal Cultures

James E. Johnson

Bowman Gray School of Medicine
Department of Neurobiology and Anatomy
Program in Neuroscience
Winston-Salem, North Carolina 27157-1010

I. Advantages and Caveats of *in Vitro* Assays for Neuronal Cell Death

The cellular origin of neuronal fibers was very controversial at the beginning of the 20th century. Whereas one model favored the development of nerve fibers via a syncytium of protoplasm derived from migrating cells, another conflicting model favored the outgrowth of protoplasmic extensions directly

Copyright © 1995 by Academic Press, Inc. All rights of reproduction in any form reserved.

from neuronal cells. A novel experimental approach was taken by the pioneer neuroembryologist Ross Harrison to answer this question. This approach required the development of early tissue culture methods to isolate and observe neurons when they were maintained away from other cellular components of the developing embryo. Harrison removed the neural tube from developing frog and chick embryos and maintained the dissected tissue in culture with drops of clotted lymph or plasma (Harrison, 1910, 1912, 1914). Observations from these early experiments provided direct evidence that axons could arise from nerve cells alone rather than from the fusion of migrating cells in the forming peripheral nerve. Harrisons' technical ingenuity not only made a major contribution to the foundation of the neuron doctrine, but also demonstrated some of the unique experimental advantages of cell culture techniques that subsequently were to be exploited by other investigators. Primary neuronal cultures allow the investigator to define the biological components of an experiment arbitrarily and to observe the response of living cells to controlled environmental stimuli directly. When experiments utilize neurons isolated away from other cells, a unique opportunity exists to access the direct effects of a defined molecule on neuronal behavior without the intervention or mediation of surrounding glial components, presynaptic input, or feedback from synaptic targets. For this reason, experiments with primary neuronal cultures have been especially useful for observing the direct effects of putative neurotrophic factors on cell survival and differentiation. *In vitro* experiments have also been useful for investigations of events that regulate cell death or cytotoxicity, including studies of cellular mechanisms that normally result in neuronal cell death (or apoptosis) during development. Primary cell cultures have been frequently used to test the pathological events that result in cytotoxicity (including ischemia, excitatory neurotoxicity, and many types of environmental neurotoxicity). Experiments with sample populations of identified neurons have also allowed the development of rapid assays for neuronal survival and cytotoxicity. These assays have made primary cell culture a useful tool in screening unknown compounds for putative cell survival or toxic activity. Primary neuronal cultures have been the traditional system of choice to screen protein purification fractions for neurotrophic activity. These cultures permit direct dose–response assays, synergistic and additive assays with multiple known factors, and investigations of the intracellular regulatory mechanisms for neuronal viability and cytotoxicity. Both primary and secondary neuronal cultures have been useful for the investigation of receptor binding, transduction, and intracellular changes associated with neuronal survival or cell death. These studies include characterizing intracellular second messengers, the regulation of transcriptional and translational events, and other biochemical, morphological, and behavioral responses by neurons to cell survival or cytotoxic factors.

A primary culture assay can reduce the components of an experiment to a dissected tissue explant or isolated cells observed and manipulated in a defined environment. Although enhanced experimental control is the principal advantage of primary culture experiments, it is also the major limitation and caveat when interpreting observations. The artificial nature of *in vitro* assays requires that conclusions regarding normal developmental or physiological principles be further examined *in vivo* (see Chapters 8–11, 13, and 15). For example, primary neuronal cultures are often prepared from the developing nervous system for pragmatic reasons. The long-term survival and maintenance of neurons isolated from the adult nervous system is frequently more difficult. Although neurons are always damaged by their removal and dissociation, the isolation of more mature neurons, with better developed dendrites and axons, is likely to be more traumatic than the damage inflicted on immature cells. Investigations of developmental events *in vitro* must therefore utilize neurons that are traumatized by dissociation. This procedure requires isolated cells to regenerate and adapt to their new culture environment. Cells in culture may therefore be selected for either survival or cell death by criteria that do not correspond to similar events in the normal developing embryo. In this context, it is important to distinguish *in vitro* events that mediate cell repair, attachment, and neurite extension from the regulation of survival and cytoxocity *in vivo*.

This chapter will review some of the advantages and pitfalls of neuronal cell death assays *in vitro*. In particular, the relative merits and risks of methods to identify and isolate neurons will be examined, as will the utility of methods to distinguish cell survival from changes in cell attachment, proliferation, or adaptation *in vitro*. Protocols are provided in Section IV for dissociating cells from the peripheral and central nervous system, identifying specific cell populations, isolating subgroups of cells, and measuring cell viability and cytoxocity. In addition, a variety of alternative cytotoxicity and viability assays will be reviewed for their capacity to measure cell death *in vitro* accurately or rapidly. Many good manuals are available for general procedures in the preparation of cell cultures as well as for specialized protocols for particular populations of nerve or glial cells. The protocols described in Section IV provide only some of many alternative methods for culture preparation and analysis.

II. Neuronal Cultures

Once cells are dissociated from neuronal tissues, they must be identified to determine the yield of viable cells and the proportion of cells that are neuronal or glial. Cell identification can be difficult given the diversity of cell phenotypes that often populate a small region of the nervous system. Neighboring cells can differ greatly in their neuronal birthdates, anatomical connections, cellular geometry, function, electrophysiology, and receptor and neurotransmitter phenotypes. It is not surprising then that neuronal subpopulations often differ in their dependence on specific neurotrophic factors for survival or their sensitivity to excitatory neurotoxins (see Table I).

The response of a numerically small subpopulation of neurons to survival or

Table I
Antigens for Cell Identification in Mixed Neuronal Cultures

Antigen marker	Cellular expression	Antibody localization	Comments	Reference/commercial source[a]
Neurofilament	Neurons	Cytoskeletal type IV intermediate filament; labels entire cell	Onset of subunit (70, 140, and 200 kDa) expression is variable	Debus (1983), Franke (1991); BM#814326, SIG#N5139, CHEM#AB1983
13H9	Neurons	Intracellular, very specific to only growth cones and lamellipodia	Antibody binds to ezrin, a cytoskeletal link between actin and microtubules	Birgbauer and Solomon (1989)
Class III β tubulin	Neurons	Cytoskeletal marker	Isotype of tubulin expressed in neurons and some tumor cells; not expressed in nonneuronal cells or glia	Sullivan et al. (1986), Burgoyne et al. (1988)
MAP2ab	Neurons	Only dendrites and cell soma are strongly labeled	Expressed in growing axons and may sometimes persist within axons of cultured neurons; reported in reactive astrocytes	Banker and Waxman (1988), Higgins et al. (1988); SIG#M1406
TAU	Neurons	Axons strongly labeled	Soma and dendrites also labeled in some (hippocampal and cortical) neuronal cultures	Kossik and Finch (1987); SIG#T5530, CHEM#MAB370
Gap-43	Developing neurons	Found on membranes of growing axons during development	Also expressed by adult regenerating axons; antibodies cross-reactive for many species	Skene (1989). Goslin et al. (1988); SIG# G9264
Tetanus toxin binding/anti-tetanus toxin labeling	Neurons; also a subpopulation of astrocytes (fibrous or type II)	Surface membrane ganglioside receptor for toxin makes binding suitable for live cell labeling; same as A2B5	Antibody is against nontoxic proteolytic fragment C; toxin binding detected with antibody	BM#131621
Neuron-specific enolase	Neurons	Cytoplasmic glycolytic enzyme	One of several related glycolytic enzymes; may be absent in young neurons	CHEM#MAP314, DAK#A589
A2B5 (also called neuron surface antigen)	Neurons; also a subpopulation of astrocytes (fibrous or type II)	Tetra-sialo-gangliosides on the cell surface	Found on neurons as well as precursors for glia	ATTCC#A2B5, CHEM#MAB312
Synaptophysin	Neurons, neuromuscular junctions, adrenal cells, islet cells, hypophysis	38-kDa presynaptic vesicle protein	Suitable for labeling in culture	Fletcher et al. (1990), Wiedenmann and Franke (1985); BM#902314, SIG#S5768. CHEM#MAB372
Parvalbumin	Some neuronal subpopulations, muscle cells	Soma, abundant cytosolic marker	Ca^{2+} binding protein	Celio et al. (1988), SIG#P6038
Calbindin	Some neuronal subpopulations, muscle	Soma, abundant cytosolic marker	Ca^{2+} binding protein	Celio et al. (1900): SIG# C8666
Thy1.1	Variety of neuronal subpopulations, thymocytes, fibroblasts	Surface membrane glycolipid, will cap with antibody binding on live cells, enhances axonal growth	Shows specificity in many regions for cells with long axons (i.e., retinal ganglion cell); also on fibroblast	HAR #MCA47
Choline acetyltransferase (ChAT)	Neuronal subpopulations	Soma, cytosolic; phenotypes can be induced or selected by trophic environment	May not be expressed early; may require trophic stimuli; rate limiting production of ACh	Martinou et al. (1989); CHEM#AB143, CHEM#MAB305
Acetylcholinesterase (AChe)	Neuronal subpopulations	Soma, cytosolic	Enzymatic histochemistry can also be used for detection	CHEM#MAB303, DAK#A032

Antigen/Marker	Cell type	Localization/description	Comments	References
Tyrosine hydroxylase (TH)	Neuronal subpopulations	Soma, cytosolic; phenotypes can be induced or selected by trophic environment	May require trophic stimuli during development	Won et al. (1989); CHEM#AB151
Phenyl N-methyltransferase (PNMT)	Neuronal subpopulations	Soma, cytosolic; phenotypes can be induced or selected by trophic environment	May require trophic stimuli during development	CHEM#AB110
GAD	Neuronal subpopulations	Soma, cytosolic; phenotypes can be induced or selected by trophic environment	May require trophic stimuli during development; yields GABA	Hoch and Dinledine (1986); CHEM#AB108
GABA	Neuronal subpopulations	Soma, cytosolic; phenotypes can be induced or selected by trophic environment	May require trophic stimuli during development	Neal et al. (1983); CHEM#AB131
Serotonin	Neuronal subpopulations	Soma, cytosolic; phenotypes can be induced or selected by trophic environment	May require trophic stimuli during development	CHEM#AB125
Neuropeptides (substance P, CGRP, CCK, enkephalin)	Neuronal subpopulations	Soma, cytosolic; phenotypes can be induced or selected by trophic environment	Found in sensory ganglia and a variety of CNS neurons; may require trophic stimuli during development	Huettner and Baughman (1986); SP—CHEM#AB1977, CHEM#MAB356; CGRP—CHEM#AB1971; LENKEPH—CHEM#AB1975; CCK—CHEM#AB1972
Fibronectin	Fibroblasts	Cell surface extracellular matrix/membrane protein	Sometimes found on astrocytes	CHEM#AB1942. DAK#A245
Glial fibrillary acidic protein (GFAP)	Astrocytes (Type I and Type II)	Cytoskeletal 51-kDa intermediate filament protein	Increased expression in reactive cells	Debus (1983), Franke (1991); SIG#G9269, CHEM#MAB049, DAK#M761
S-100	Astrocytes, ependymal cells, Schwann cells, melanocytes	Cytosolic; mechanism of cellular release not clear	Contains reported neurotrophic activity for many populations	SIG# S2644, DAK#Z311
Galactocerebroside	Oligodendrocytes	Glycolipid on membrane surface	Suitable for labeling living cells	Benjamin et al. (1987); SIG# G9152, CHEM#SB142
Cyclic-nucleotide phosphodiesterase (CNPase)	Oligodendrocytes, Schwann cells	Cytosolic marker shared by both central and peripheral myelinating cells	Neurons and astrocytes are not labeled	Sprinkle et al. (1987); SIG# C5922
Vimentin	Astrocytes, fibroblasts, endothelial cells, melanocytes	Cytoskeletal 53-kDa type 3 intermediate filament	Transient expression in immature neurons	Osborn et al. (1984); SIG#V6630, CHEM#MAB0413, DAK#M725
Nonspecific esterase	Microglia	Cytoplasmic; not expressed by other glia or neurons	May not be expressed early	
Nonneuronal enolase	Variety of nonneuronal cells	Cytoplasmic; distinct from neuronal isotype	Transient expression by early neurons	
Griffonia simplicifolia agglutinin (GsA)	Endothelial cells	Lectin binding marker	Found within brain blood vessels	
Hoechst 33258 or 33342 Bisbenzimide dyes	All cells	Cell-permeant water-soluble DNA stains: not antibodies but very useful as double labels	Fluorescent dye useful for obtaining total cell number for immunofluorescent studies	MOLP#H3569, MOLP#3570
BrdU incorporation/anti-BrdU labeling	Incorporated by dividing cells	Nuclear stain	Living cells incorporate nucleotide analog during S phase, followed by antibody labeling	

[a] There are many commercial sources for these antibodies and reagents. For a current list of companies and prices, consult Linscott's Directory of Immunological and Biological Reagents, 40 Glen Drive, Mill Valley, CA 9494 . SIG, SIGMA Chemical Co.; CHEM, Chemicon; BM, Boehringer Mannheim; ATTCC, American Type Tissue Culture Collection; HAR, Harland/Serotech; DAK, DAKO Corp.; MOLP, Molecular Probes.

cytotoxic factors can often be masked by the presence of overwhelming numbers of nonresponsive cells. This signal-to-noise ratio problem can be eliminated by a sensitive and selective assay that can amplify the signal from responsive cells. Cell markers can be used to identify specific subpopulations of neurons and to quantify changes in survival. In other instances, *contaminating* cells cocultured with responsive neurons may either mask or mediate a response to neurotrophic factors. The interpretation of observations can be difficult when other cells affect the capacity of cells to respond in mixed cultures. This sometimes occurs with the high density plating or *in vitro* proliferation of cells that adapt to an artificial culture environment by the induction and secretion of paracrine survival factors. For instance, developing sensory dorsal root ganglion (DRG) neurons normally die *in vivo* when they are experimentally deprived of trophic support from their synaptic targets in the skin or spinal cord. However, target-deprived DRG neurons can survive *in vitro* when cocultured with nonneuronal cells from the ganglia that rapidly proliferate and colonize the dish. Therefore, an *in vitro* response of DRG neurons to putative survival factors cannot be detected when *contaminating* nonneuronal cells mask the response by the induction of compensating trophic support.

A wide variety of strategies is available to identify cells by molecular or anatomical labeling methods and thereby study the survival of a specific neuronal population in heterogeneous cultures. Once they have been identified, responsive populations can sometimes be further enriched in the preparation of purified cell cultures. Under these circumstances it is often possible to unmask the direct effects of a particular survival or cytotoxic stimulus on the cells *in vitro*.

A. Cell Identification *in Vitro*

From a practical point of view, the primary strategies that carefully consider the dissection of selected neuronal populations are the most important. The known regional and temporal distribution of cells in the developing embryo can be exploited to facilitate their identification after isolation in culture. Peripheral sensory and autonomic neurons have historically been preferred in many neuronal culture preparations because of their segregation within distinct ganglia that are easily harvested from the embryo. The general identification and isolation of these cells is simplified by their anatomical and temporal distribution relative to many central nervous system (CNS) populations. Nevertheless, within the brain and spinal cord, the collection and identification of neurons can also exploit known temporal and anatomical patterns of cell production and migration.

A number of useful dissection guides for the selection of different neuronal or glial populations from the developing peripheral and central nervous system have been published (Banks, 1979; Davies, 1989; Shahar, *et al.,* 1989; Banker and Goslin, 1990, 1991; Fedoroff and Richardson, 1992). See Table II for guidelines on cell culture establishment.

Table II
Culture Dishes and Density: Approximate Cell Numbers for Primary Cultures and Cell Lines

Vessel	Well #/diameter (mm^2)	Growing surface area (mm^2)	Maintenance media volume (ml)	Primary cells[a]			Cell lines confluent cells
				Low density	Intermediate density	High density	
Microtitration plate	96 wells × 6.4	32	0.1–0.2	0.7–3.5×10^2	1.1–1.4×10^3	3.5–5.6×10^3	10^5
Greiner dish	4 wells × 10 or 1 dish × 35	75	0.08/well or 1.5–2/dish	0.4–1.9×10^3	6–7.5×10^3	1.9–3×10^4	1.9×10^5
Multiwell plates	24 wells × 16	201	0.5–1	1–5×10^3	1.6–2×10^4	5–8×10^4	5×10^5
35-mm Dish	1 dish × 35	804	1.5–2	0.4–2×10^4	$6.4,$–8×10^4	2–3.2×10^5	2×10^6
60-mm Dish	1 dish × 60	2,124	5	1.1–5.3×10^4	1.7–2.1×10^5	5.3–8.5×10^5	5.2×10^6
T25 Flask	1 flask	2,500	5–10	1.3–6.3×10^4	2–2.5×10^5	0.6–1×10^6	5×10^6
T75 Flask	1 flask	7,500	20–30	0.4–1.9×10^5	6–7.5×10^5	1.9–3×10^6	2×10^7

[a] Low density, 5–25/mm^2; intermediate, 80–100 cells/mm^2; high density, 250–400 cells/mm^2.

At the onset of the experiment, it is important to estimate an anticipated yield of neurons that may be harvested from a given region of the nervous system at the given stage of development. This estimate is especially useful in the evaluation of methods that ensure optimal yield and minimal loss or damage during isolation. The potential cell number is best derived from estimates made from tissue sections. Virtually all the methods that have been used to characterize the diversity of neuronal phenotypes *in vivo* have also been used in the analysis of neuronal differentiation and identification *in vitro*. These include microscopic, ultrastructural, biochemical, electrophysiological, pharmacological, and molecular criteria. The identification and selection of a specified neuronal phenotype is enhanced by the use of multiple requisite properties or indicators of differentiation. Cultured neurons have been visualized with all the techniques that are also used *in vivo*. A sampling of the many available protocols for staining and characterizing neurons *in vitro* includes thionin staining, silver staining (Toran-Allerand, 1976), scanning electron microscopy, and transmission electron microscopy (Fedoroff and Richardson, 1992).

1. Immunostaining Advantages and Caveats

Methods that selectively immunostain cells *in vitro* have proven to be especially useful in identifying different subpopulations in mixed cultures (for a sampling of *in vitro* immunostaining protocols, see Johnson, 1989; Barnstable, 1980; Banker and Goslin, 1991). A number of antibodies have proven useful for the identification of neuronal phenotypes, including antibodies against the three cytoskeletal subunits of the neurofilament protein and the cytosolic enzyme neuron-specific enolase. Antibodies and these antigens are commercially available from many vendors and are widely employed as general neuronal markers *in vitro*. Another cytoskeletal protein, MAP2 (microtubule associated protein 2), is expressed by cultured neurons in both the soma and the dendrites (Caceres *et al.*, 1984; Matus *et al.*, 1986; Higgins *et al.*, 1988). An antibody against this protein has been used to identify and characterize details of the phenotypic structure of dendrites from hippocampal neurons in culture (Banker and Waxman, 1988). Aside from the aforementioned neurofilament proteins, antibodies against two additional markers have been localized to axons growing from cultured hippocampal or cortical neurons. These are antibodies against GAP43 (growth associated protein) (Skene, 1989) and the microtubule associated protein tau (Kosik and Finch, 1987; Goslin *et al.*, 1988, 1990). The *in vitro* expression of several of these phenotypic labels is sometimes correlated with maturation and may not be early enough to make them useful indicators for determining the initial number of neurons at the onset of cultures. In some cases, their expression has been reported to be delayed for long periods of time in culture (Shaw *et al.*, 1985; Marangos and Schmechel, 1987). Neuronal surface markers are particularly advantageous because they can be visualized in living cells. Extracellular plasma membrane gangliosides that bind tetanus toxin (Dimpfel et al., 1977; Mirsky et al., 1978) or the monoclonal antibody A2B5 (Eisenbach et al., 1979) have been used as neuronal surface markers. However, note that in some mixed CNS cultures these antigens may also be found on a subpopulation of glial cells (Raff et al., 1983a,b). A number of other surface antigens, including L1 (also known as NILE or NG-CAM), have been useful in identifying neurons in vitro, but they are also strongly expressed by many other cell types (Jessell, 1988; Rutishauser and Jessell, 1988). The developing nervous system can provide regional variations that help the investigator identify cultured neurons. Specific subpopulations of neurons can sometimes be identified in vitro by the presence of phenotypic markers unique to their region. For instance, the antibodies SC1 or Islet 1 have been used to identify selectively, or in some cases purify, chick motoneurons from other spinal cord interneurons (although SC1 is only transiently expressed in vitro and in vivo; Bloch-Gallego et al., 1991). Similarly, several different antigens appear to be unique to Purkinje cells when compared locally with other cerebellar neurons. Antibodies against these regionally specific markers (cGMP-dependent protein kinase, calbindin, PEP-19) have proven useful in the identification of dissociated Purkinje cells in cerebellar cultures (Hockberger et al., 1989). Other regionally specific antigens have been used to identify or purify dissociated retinal ganglion cells from other retinal cells (for a review, see Johnson, 1989).

Relying on cell specific markers for neuronal survival assays in vitro has inherent weaknesses. For example, differences in the numbers of a labeled subpopulation of cultured neurons may also arise from an increase in the proliferation of neuronal proecursors or an increase in the expression of the label used to identify neurons. Note that under some circumstances cultured neuronal precursors can divide in vitro to increase neuronal number (Rohrer and Thoenen, 1987) and undifferentiated cells can be recruited to acquire phenotypic neuronal labels (Rohrer et al., 1985). In addition, some neurotrophic factors are clearly pleiotropic with multiple effects on both survival and neuronal differentiation. Therefore, when cell-specific markers are required to identify neuronal subpopulations in mixed cultures, both cell proliferation and cell differentiation effects can be misinterpreted as effects on cell survival. Assays that monitor the incorporation of [^{3}H]thymidine or 5-bromodeoxyuridine (BrdU) in labeled cells can be used to assess the effects of proliferation in these survival assays (see Johnson, 1989, for a protocol to double label cells with immunofluorescence and [^{3}H]thymidine). At least three strategies are available to determine if an increase in label expression can confound subpopulation cell counts with neuronal survival assays in vitro: (1) Determine if the expression of the subpopulation markers can be induced following a period of survival factor deprivation in vitro. (2) Identify the same subpopulation of cells by labeling neurons with stable anatomical retrograde axonal tracers in vivo prior to the preparation of cultures. (3) Purify live cells (using either immunofluorescent markers or anatomical labels) at the onset to test survival in homogeneously labeled cells (Johnson *et al.*, 1986).

2. Anatomical Labeling

A reliable technique for the identification of neuronal subpopulations exploits the anatomical specificity of long axon connections *in vivo*. Retrograde axonal tracers can label specific neuronal populations when placed in their respective synaptic targets 24–48 hr prior to the preparation of primary cultures. For instance, this method has been used to label definitively subpopulations of spinal motoneurons (Dorhman *et al.*, 1986; Honig and Hume, 1986), retinal ganglion cells (Nurcombe and Bennett, 1981; Armson and Bennett, 1983; Sarthy *et al.*, 1983; Johnson *et al.*, 1986; Johnson, 1989), and neurons from the visual cortex *in vitro* (Huettner and Baughman, 1990). Chick embryos are advantageous for this method since anatomical tracers can often be placed with greater ease *in ovo* than *in utero*. However, neonatal rodents have also been used in these experiments. Labeling tracers have included fluorescent latex microspheres, the carbocyanine dye DiI, true blue, and horse radish peroxidase (HRP). However, several drawbacks exist for the routine use of this method in neuronal survival assays. Anatomical tracers frequently fail to label the majority of neurons in a given subpopulation since only a portion of the developing cells that have established axonal growth in target areas is typically labeled. As a result, the yield of labeled cells is often smaller than by specific immunostaining methods. Second, the method is labor intensive and therefore better suited for the development and validation of experiments with other labeling methods and for cell purification techniques (Johnson, 1989).

B. Purifying Neuronal Subpopulations

Identified populations of neurons can often be separated from other dissociated cells in suspension at the onset of an *in vitro* survival assay. Primary neuronal cultures that are homogeneous (for a particular phenotype) are especially advantagenous for testing the direct effects of a given factor on the survival of isolated cells in low density cultures. These cultures also have the added advantage of accounting for the identity and proportion of cell types prior to the treatment of cells *in vitro*. Purified cells or their conditioned media can also be combined in experiments that address cell-to-cell interactions *in vitro*. Protocols are available for the purification or enrichment of specific cell populations throughout the peripheral and central nervous system (a small sampling of protocols is provided for the purification of the following cells: spinal, cranial, and mesencephalic sensory neurons—Davies, 1989; peripheral sympathetic and parasympathetic neurons—Higgins *et al.*, 1988; adrenal chromaffin cells—Acheson and Thoenen, 1989; retinal ganglion cells—Sarthy *et al.*, 1983; Barres *et al.*, 1988; Johnson, 1986, 1989; retinal photoreceptors—Adler, 1992; cerebellar granule cells—Levi *et al.,* 1989; cerebellar Purkinje cells—Messer *et al.,* 1989; hippocampal neurons—Goslin *et al.,* 1988, 1990; CNS astrocytes, oligodendrocytes; and microglia—Young and Antel, 1990; Cole and de Velis, 1992; Hao *et al.,* 1992; peripheral Schwann cells—Johnson and Bunge, 1990; spinal motoneurons—Dorchman *et al.,* 1986; Schaffner *et al.,* 1987; Bloch-Gallego, 1990; spinal preganglionic sympathetic neurons—Clendening and Hume, 1990; *Aplysia* neurons—Goldberg, 1986; insect neurons—Levi-Montalcini and Aloe, 1989). Generally, neurons from discrete sensory and autonomic ganglia are easier to purify than neuronal subpopulations in the CNS. The most simple and rapid method of purifying neurons from peripheral ganglia is by a brief pre-plating step in which nonneuronal cells differentially adhere to a plastic substrate (see Davies, 1989, for a detailed protocol). Neurons are then harvested from the supernatant and cultured in dishes with a more favorable substratum for neuronal attachment and axonal growth. The original number of neurons and their subsequent survival or cytotoxicity can be easily followed in low density cultures. Other methods have been used to eliminate populations selectively, including the transient treatment of cultures with mitotic inhibitors, complement-mediated cytotoxicity when coupled with cell-specific surface antibodies (Rohr and Thoenen, 1987), defined media selective for the maintenance of subpopulations (Bottenstein and Sato, 1985), photo-induced cytotoxicity when coupled with cell-specific 5-brU and Hoechst dye 33258 (Shine, 1989), and the differential sedimentation of neurons (Davies, 1989). Purifying specific CNS neurons can sometimes create enormous problems in both the purification and the yield of cells. For instance, many large projection neurons (i.e., retinal ganglion cells, Purkinje cells, and motoneurons) with long axons make up only a very small proportion (1%) of the total number of cells in their region. However, by taking advantage of their relatively earlier birthdates and differentiation, many investigators have been able to exploit the relative buoyant properties of these cells by differential centrifugation on metrizamide (Serva) or Percoll (Pharmacia) gradients (Sarthy *et al.,* 1983; Dohrman *et al.,* 1986; Johnson *et al.,* 1986; for a protocol, see Johnson, 1989). Other techniques, including the use of argon lasers to rapidly identify and select (10^3–10^4 cells/second) immunofluorescent neurons or neurons retrogradely labeled with fluorescent tracers by FACS (fluorescent activated cell sorting) (Abney *et al.,* 1983; Schaffner *et al.,* 1987; Maxwell *et al.,* 1988; Barald, 1989, have exploited the specific immunological properties of specific CNS neurons. Unfortunately, although this technology has proven to be especially useful for sorting many humoral cells (i.e., lymphocyte subpopulations) and isolated cell nuclei, the parameters required for the separation of specific neuronal populations typically require extensive experimentation. The yield, purity, and viability of separated neurons can be more problematic because of difficulties preventing stirred cell suspensions of mutually adhesive neurons from aggregating prior to separation. The cell panning technique is another method that exploits surface immunostaining to achieve remarkable levels of CNS neuron purification. Labeled cells bind with high affinity to substrate-bound secondary antibodies in an immunoadsorbent assay. This technique has been used to purify numerically rare subpopulations of neurons. For instance, although retinal ganglion cells make up less than 1% of retinal cells, they can be harvested to homogeneity (>99%) with impressive yields of up to 50% (Barres *et al.*, 1988). Finally, a very creative method of cell blotting has been used to allow the natural survival requirements of neurons to select putative or known neurotrophic factors transferred to nitrocellulose (Manthorpe, 1989). Cells survive only when bound to selective regions of the paper that contain the specific proteins capable of providing neurotrophic support.

III. General Cytotoxicity and Viability Assays

A. Cytotoxicity Assays

Cytotoxicity assays utilize techniques to identify dying or dead cells *in vitro* rapidly and selectively. These assays have been applied to investigations of cell death with secondary neuronal and glial cell lines or with primary cultures. Many of the assays can be used for dissociated cells in suspension, as well as for neurons adherent to the dish substrate with extensive neurite outgrowth. The term *cytotoxicity* suggests that these assays can detect stimuli that evoke a response detrimental to cell survival. Each assay measures an abnormal change in cell physiology as an indirect measure of cell toxicity. Therefore it is important to select an assay that is likely to be useful in detecting the specific effects of a harmful stimulus or the deprivation of a component required for survival. Common features of many of these assays are methods designed to detect an abnormal increase in the permeability of the plasma membrane correlated with cell pathology and subsequent death. Some assays are impermeant dyes for cytosolic components whereas others employ fluorescent impermeant dyes that intercalate with DNA only in the nuclei of cells with poor membrane integrity. Finally, cytotoxic changes in membrane permeability can be detected by the abnormal leakage of cytosolic enzymes into the culture media during cell death. DNA fragmentation has been used as an alternative to cell permeability assays for cell death. Indicators of DNA damage within dying or dead cells have been used in both biochemical and *in situ* anatomical methods. In general, multiple methods that rely on indicators of cytotoxicity or cell viability may be helpful in the validation or interpretation of any one cytotoxicity assay.

1. Membrane Permeability Assays

These assays assume that (1) the abnormal passage of normally impermeant markers that are either in colloidal suspension or too polar or too large for passage through the plasma membrane can be used as indicators of cell damage and (2) that leakage of impermeant extracellular dyes into the cell or the leakage of normal cellular components into the media can be correlated with dead or dying cells. The relative size of the molecular probe applied in membrane permeability assays has been used as an indicator of the size or quantity of damage to the membrane (Thelestam and Mollby, 1976, 1980). In some experiments, the rate at which the same neuronal cultures are stained with different cytotoxic indicators has been demonstrated to vary (Juurlink et al., 1991b, Juurlink and Hertz, 1993). Events that cause cell death may differ greatly in different experiments (i.e., damage during cell dissociation, metabolic effects of specific cellular or neurotoxins, oxygen and nutrient deprivation used for in vitro models of ischemia, antibody- or complement-mediated cell death, survival factor deprivation and apoptosis experiments, and excitatory amino acid toxicity). The choice of cytotoxicity indicator may depend on the mechanism of cell death in the experimental paradigm, for example, necrosis vs apoptosis. An increase in membrane permeability that accompanies cell death can be indicated with impermeable colloidal stains, DNA intercalating dyes, and the leakage of normal cytosolic components.

a. Impermeant Colloidal Dyes

Most living cells will actively exclude the colloidal dyes trypan blue, nigrosin, eosin, and erythrosin B (Sigma Chemical Co., St. Louis, MO) (Kaltenbach et al., 1958; Melamed et al., 1969; Walum et al., 1985; Fedoroff and Richardson, 1992). Dansyl lysate, a green fluorescent dye, is also used as a dye exclusion marker for cytotoxicity. Cells stained with any of these impermeable dyes are assumed to have lost their vital capacity to maintain plasma membrane integrity (Tennant, 1964). However, not all unstained cells may be assumed to be viable or to remain viable for very long. Some cells may be in the early or minor stages of membrane leakage whereas other dead cells may be poorly indicated due to an extensive loss of the cytosolic proteins stained by these dyes. Despite these and other difficulties (Melamed et al., 1969), colloidal dye exclusion has proven to be useful for identifying dead cells. The dyes have been especially useful for accessing cell death from traumatic damage to membranes during the dissociation of tissues for primary culture and from the cytotoxic actions of antiserum and complement (Saijo, 1973). Trypan blue and nigrosin have been widely used for hemocytometer cell counts of freshly dissociated cells (a description of the use of these dyes for this method is provided by Johnson, 1989, and Fedoroff and Richardson, 1992). Cells are counted only if they exclude the dye, whereas damaged or dying cells that are stained are not included in calculations for cell density. Dansyl lysine (DL) is a fluorescent dye that is also membrane impermeant. DL is excited in the UV spectrum, allowing for its combined use with other common fluorescent probes including rhodamine- and fluorescein-labeled antibodies for cell identification.

b. Impermeant DNA Stains

Ethidium bromide (EB) and propidium iodide (PI) are impermeant fluorescent stains that bind DNA without apparent basepair preference. Their fluorescence is enhanced by DNA binding and their red emission allows the simultaneousdetection of fluorescein-labeled probes or antibodies (see Chapters 5 and 9; Jones and Senft, 1985). Both fixed and dead cells will readily stain with PI and EB, whereas living cells exclude the dyes. In a study on the effects of ischemic conditions on cultured neurons, viability indicators of mitochondrial activity disappeared immediately before the onset of PI fluorescence in the nucleus (Juurlink and Hertz, 1993). Although both stains are structurally similar phenanthridinium DNA intercalators, they differ slightly in their routine usage. EB is often used to detect DNA in gels; it is slightly less water soluble and more membrane permeable, and therefore is less suitable than PI for cytotoxicity assays. PI has been used frequently to access cell death via membrane damage in a variety of experiments including apoptosis and electrofusion studies. A homodimer of EB, ethidium homodimer-1 (EH 1) is reported to have a 1000-fold higher DNA binding affinity as well as a lower membrane permeability than either EB or PI. The red fluorescence of this dye is enhanced 40-fold with DNA binding. All three intercalating dyes can be excited by the mercury or xenon lamps used with fluorescent microscopy and by fluorescent microplate readers. In addition, they all are excited by argon lasers and are therefore suitable for flow cytometry and FACS (fluorescence-activated cell sorter) technology (see Chapter 4). A new series of ultrasensitive dimeric analogs of intercalating DNA (PO-PRO, BO-PRO, YO-PRO, TO-PRO; Molecular Probes, Eugene, OR) has exhibited great promise as fluorescent dyes for cytotoxicity. These dyes are virtually nonfluorescent unless bound to nuclei acids. They are very impermeant to live cells, but stain dead cells with a very high affinity for DNA. The increased sensitivity of these dimeric analogs is reported to permit the detection of only 4 pg double-stranded DNA in 5-mm wide bands on vertical agarose gels (1 mm thick). Ethidium monoazide is an impermeant stain that has been used to label DNA covalently in dead cells. Both live and dead cells incubated with ethidium monoazide are illuminated with UV light. This exposure produces a highly reactive form of the dye that is covalently bound to the DNA. The cells can be processed or fixed without fear of dilution or loss of staining on subsequent treatment.

2. Cytoplasm Leakage Assays

Some cytotoxicity assays attempt to measure the leakage of normal components of the cytoplasm into culture media. These assays are typically biochemical measurements that rely on the detection of intracellular soluble enzymes or their reaction products in culture media (Walum, 1983; Walum and Peterson, 1982). The release of lactate dehydrogenase (LDH) into culture media is perhaps the most frequently used leakage assay. LDH activity is relatively easy to measure by the conversion of NAD to NADH in the presence of excess lactate substrate (Wahlefeld, 1983). Released LDH activity in culture media has been used to access total accumulated cell death over a period of time in culture. This has proven to be a rapid screening assay for many toxicity assays and many different cell types (Mitchell et al., 1980; Walum et al., 1990; Atterwill et al., 1992; Honegger and Schilter, 1992). Release of LDH activity has been used to measure neuronal death from excitatory amino acids in enriched neuronal cultures (Frandsen and Schousboe, 1987; Frandsen et al., 1989) and in mixed neuron–glial cultures (Ko and Choi, 1987, 1988). Quantitative measures of cell death in culture have frequently used LDH cytotoxicity assays (Yu et al., 1989). In glial cultures, the presence of stains for viable cell activity (R123; Johnson et al., 1980; see Section III,B) have been correlated with the retention of LDH. Conversely, the loss of viability staining and the onset of cytoxicity-associated nuclear staining (PI) has been correlated with the release of LDH (Juurlink and Hertz, 1993). However, there are caveats regarding the use of LDH leakage as a measure of cell death in vitro. The first is the latency of enzyme leakage. The gradual release of this high molecular weight enzyme by dying cultured neurons is relatively slow when compared with the rapid PI nuclear staining at the onset of cell death in the same cells (Juurlink and Hertz, 1993). Second, measurement of LDH release provides an indicator of cell death but in mixed neuronal and glial cultures, this assay provides little information about the identity of the population of cells that are dying. This is true of even highly enriched neuronal cultures with few glial contaminants. Apparently some neurons and glia contain and release unequal amounts of LDH during cell death. For example, in enriched primary cortical cultures containing 90% GABAergic neurons, only 60% of the total LDH activity is released after complete neuronal destruction with excitoxocity (Frandsen and Schousboe, 1987; Frandsen et al., 1989). In enriched (90%) glutamatergic cerebellar granule cells mixed with only 10% glia, nearly half the total LDH activity in the culture can be accounted for by the small subpopulation of glial cells (Juurlink and Hertz, 1993). A more sensitive assay has been widely used in cytotoxicity studies with a radioactive isotope of chromium (^{51}Cr). The reduced form of the isotope, ^{51}Cr^{3+}, is easily taken up by living cells in the salt Na$_2$^{51}CrO$_4$ and is oxidized to the membrane impermeable ^{51}Cr^{2+} form (Zawydiwski and Duncan, 1978). The membrane leakage of oxidized chromium has proven to be a very sensitive assay for cell death. To measure cytotoxicity, living cells are preloaded for 1 hr with Na$_2$^{51}CrO$_4$ and thoroughly washed to remove reduced chromium from the medium. The release of oxidized chromium into medium is then compared in control and treated cultures. The retention of oxidized chromium isotopes by living cells can also be used as a viability assay. Following treatment, the amount of retained radioactivity can be recovered from cell lysates with Triton X-100 (Thelestam and Mollby, 1976). A common feature of observations of apoptosis in vitro is the presence of DNA fragmentation in dying cells. This process has been perhaps best characterized in cultured lymphocytes, in which researchers have demonstrated that endogenous endonuclease activity can be associated with normal apoptosis. Analysis of extracted genomic DNA from cultures containing a sufficient number of dying cells can sometimes reveal a ladder of nucleosomal DNA fragments in agarose gels as an indicator of cell death (see Chapter 3). Methods to identify individual dying cells by ISEL (*in situ* end labeling) of nuclei have been developed using the TUNEL technique [terminal deoxynucleotidyl transferase (TdT)-mediated dUTP-biotin end labeling].

B. Viability Assays

Viability assays attempt to distinguish living from pathological cells by using indicators of vital cell functions. Some viability assays exploit normally functioning cytosolic enzymes to produce easily detected fluorescent or colored products within the living cell. Other assays utilize probes that are sensitive to the maintenance of vital ion gradients required to maintain organelle membrane potentials or pH differences. Finally, more traditional viability assays depend on the incorporation of labeled substrates into normal cellular metabolism. At best, any viability assay is an indirect and relative measure of the current status of the cell relative to neighboring cells or in response to specific pharmacological or environmental stimulus *in vitro*. (See Fig. 1 for a summary of the viability and cytotoxicity assays.)

1. Enzymatic Assays

a. Esterase Products Used to Check Membrane Integrity

As do some of the cytoxocity assays previously described, these tests make two assumptions: (1) that the leakage of impermeant cytosolic dyes out of the cell can be used as an indicator of plasma membrane damage and cell pathology and (2) that the retention of impermeant intracellular dyes is an indicator of cell viability. Whereas staining with an extracellular impermeant dye (i.e., trypan blue or PI) is an indicator of cytoxocity, cellular retention of an impermeant dye produced by cytosolic enzymes is an indicator of cell viability. These assays rely on relatively neutral or nonpolar acetoxymethyl (AM) esters that readily and passively diffuse across cell membranes. Once inside the cell, these nonfluorescent esters become substrates for endogenous intracellular esterase enzymes. The product of esterase activity is a fluorescent and polar product that is rendered relatively impermeant to the plasma membrane. Live cells in culture retain fluorescence since the polar product cannot easily exit through the membrane. Dead or dying cells with a damaged membrane will rapidly leak the dye, even if they have residual esterase activity. Experiments using cells stained by this assay have demonstrated that live cells will rapidly lose their fluorescent staining with the onset of cytotoxicity induced by antibodies, killer cells, toxins, or detergents. A number of AM esterase probes have been used

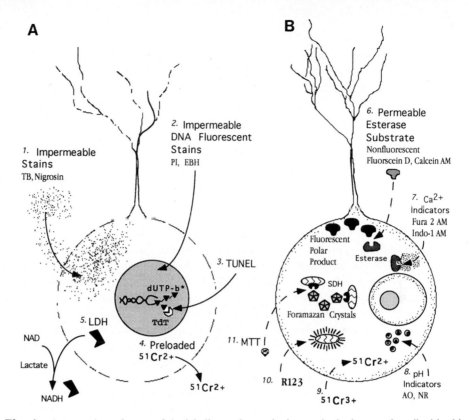

Fig. 1 An overview of some of the labeling and quantitative methods that are described in this chapter for assays of cell death and viability *in vitro*. (A) Cytotoxicity Assays. 1. Membrane impermeable colloidal stains that are normally excluded by living cells can readily pass through damaged membranes of dead cells to stain cytosolic proteins. TB (Trypan blue) and Nigrosin are examples. 2. Fluorescent DNA intercalating dyes that are normally membrane impermeable can rapidly stain the nuclei of dead and dying cells. PI (propidium iodine) and EBH (ethidium homodimer 1) are examples. 3. DNA fragmentation can sometimes be detected during apoptosis by *in situ* end-labeling with biotinylated nucleotides. TUNEL [terminal deoxynucleotidyl transferase (TdT)-mediated dUTPO–biotin end-labeling] can detect these events. 4. Cultured cells can be preloaded with salts containing reduced radioactive chromium. The oxidation of chromium yields a normally membrane impermeable form retained by living cells. However, during cell death there is an increased leakage of radioactivity into the media. 5. Cell death can be monitored with automated assays that detect the presence of normal cytosolic enzymes released into the culture media. LDH (lactate dehydrogenase) is one such marker. (B) Viability assays. 6. Nonfluorescent analogs of fluorescein are membrane permeable. They can be converted by cytosolic esterase enzymes to polar fluorescent compounds that are retained by living cells. Fluorescein diacetate and calcein AM (acetoxymethyl esters) are examples. 7. Changes in intracellular free calcium can be measured in living cells with permeable, AM (acetoxymethyl ester) analogs of the calcium ion indicators fura-2 and indo-1. 8. Proton pumps maintain the acidic environment of some organelles (i.e., lysosomes) in living cells. Their activity can be labeled with pH-sensitive fluorescent dyes and stains such as AO (acridine orange) and NR (neutral red). 9. Membrane permeable salts of reduced chromium can be used to assay the retention of radioactivity following chromium oxidation by living cells. Leakage of oxidized chromium can also be used as an assay of cell death (see 4). 10. R123 (Rhodamine 123) is a potential sensitive fluorescent dye specific for mitochondria in living cells. 11. The mitochondrial enzyme SDH (succinate dehydrogenase) reduces MTT (a soluble tetrazolium salt) into water insoluble blue formazan crystals in living cells.

for viability tests. Products of these AM esters vary in fluorescence (blue, green, yellow–green, or red), polarity, cell retention, and pH sensitivity. Fluorescein diacetate, an analog of fluorescein, was among the first of these compounds used for viability staining. Intracellular esterases cleave fluorescein diacetate, producing fluorescein (Rotman and Papermaster, 1966). This compound has since been replaced in many experiments by one of two analogs with enhanced cellular retention—bis-carboxyethyl-carboxyfluorescein (BCECEF) AM or calcein AM. Calcein AM has more widespread application because it has longer cellular retention times and is less pH sensitive than other fluorescein analogs.

b. MTT Microassays

The MTT assay was developed by Mosmann (1983) for rapid measurement of lymphocyte viability and proliferation in 96-well microplate cultures. A tetrazolium salt—3-(4,5-dimethylthiazol-2-yl)-2,5-diphenyl tetrazolium bromide (MTT; Sigma Cat. No. M-2128)—is reduced within the active mitochondria of living cells by the enzyme succinate dehydrogenase. The salt is reduced to an insoluble blue formazan product in living cells, but apparently not in the mitochondria or cellular debris of dead cells (Carmichael, 1987). The assay has been adapted as a viability assay of neuronal cultures for counting cells and for automated colorimetric methods (Manthorpe et al., 1986). Cultures of isolated and purified chick ciliary ganglion neurons have been compared for direct counts of viable cells containing blue formazan crystal with automated colorimetry assays (Manthorpe, 1986; Manthorpe and Varon, 1989). These studies have demonstrated that the colorimetric assay of purified neurons in microplate cultures is both reliable and sensitive (500 ciliary ganglion neurons/well). The colorimetric assay has been used for the analysis of the survival effects of putative neurotrophic factors on both peripheral and central neurons as well as on glial cells in vitro (Manthorpe et al., 1986; Zhou et al., 1991; Barres et al., 1993; Bozycko-Coyne et al., 1993). Several caveats must be kept in mind when using the colorimetric assay for MTT in neuronal cultures. The microplate assay has been reported to be accurate and sensitive with low density cultures of large peripheral neurons (sensory and parasympathetic), whereas higher density cultures may be required for smaller CNS cortical neurons (Bozycko-Coyne et al., 1993). This may be due in part to the relative differences in the quantity of succinate dehydrogenase activity per cell. All automated assays yield data for the entire culture well or dish, and are therefore best suited for primary cultures composed of identified and purified neurons or glia instead of secondary cell lines.

2. Ion Pump Assays

The maintenance of cellular and organelle compartments is essential for the special environmental needs of vital enzymes in living cells. A number of viability assays have been developed to monitor the maintenance of ion or proton gates in living cells. Some chromophores and fluorescent dyes are sensitive to minor pH changes. Acridine orange and neutral red have both been used to detect the maintenance of acidic organelles such as lysosomes. Similarly, R123 (rhodamine 123) has been used as a membrane-potential-sensitive dye that is selective for mitochondria in living cells. One advantage of these membrane potential indicators is that the validity of their use for a particular experimental paradigm can be checked with other viability or cytotoxicity indicators.

For instance, as previously mentioned, the loss of R123 fluorescence in living cells has been demonstrated to precede immediately the onset of propidium iodide staining of the nucleus (a cytotoxicity marker). Other ion indicators can be used to monitor and measure dynamic changes in intracellular concentrations of calcium associated with the capacity of living cells to respond to receptor-mediated events. Tween, fura 2, fura 2 AM, indo-1, indo-1 AM, and indo-2 are modified fluorescent calcium chelators that have all been used to monitor intracellular free calcium. The intensity of fluorescence increases with calcium concentration so real time measurements of intracellular calcium levels can frequently be made. However, the investigator must carefully select indicators to avoid problems with compartmentalization and the binding of indicators to protein-bound calcium. The AM analogs of calcium indicators are nonfluorescent and relatively nonpolar. They can readily penetrate cells and the endogenous esterases can liberate the indicator for calcium fluorescence. By combining multiple indicators, it is possible to identify and measure both living and dead cells (at different fluorescent wavelengths) in the same culture. A number of these assays have been developed with fluorescein diacetate/PI; EH-1/calcein AM, and R123/PI. The use of multiple live cell indicators has also been done (i.e., NR/BCECF AM). In addition to the manual counting of labeled cells, a number of colorimetric or fluorimetric assays have been used. Multiplate readers have been used to automate the MTT microassay as well as the calcein AM assay. Electronic cell counting instruments, fluorescent cytometry, and FACS have all been adapted to measure standard cytotoxic or viability indicators. Perhaps the most common automated assays use LDH release or chromium isotope detection in media or cell lysates. However, all automated methods assume relative cellular homogeneity and must be carefully controlled, as previously described.

IV. Protocols

A. Dissociating and Purifying Sensory Neurons from Chick Dorsal Root Ganglia for Survival or Axonal Growth Assays

Dorsal root ganglia (DRG) are dissected from E8-E11 chick embryos (see Color Plate 6). Optimal conditions for ganglia dissection and neuronal purification are with E9 embryos. DRG are exposed by a ventral dissection and harvested by the removal of their peripheral and central nerve roots with #5 fine forceps (Dumont). Care should be taken to avoid direct forceps pressure or handling of the ganglia. Instead they should be collected by manipulating the short amputated stump of the dissected central or peripheral roots. Ganglia are typically collected from 3–5 embryos for a period of not more than 1 hr. With practice, 40–60 ganglia can be rapidly removed from each embryo, depending on the stage of development. Ganglia are stored in ice cold Ca^{2+}/Mg^{2+}-free PBS during the collection process.

Harvested ganglia are placed in a fresh proteolytic solution containing 0.01% DNase and 0.1% trypsin (Worthington) in 2 ml Ca^{2+}/Mg^{2+}-free PBS. The digestion solution is maintained for 20 min at 37°C in a warm water bath. The ganglia are then gently centrifuged at 300 g for 2 min at 4°C. The proteolytic solution is removed and the ganglia are gently washed twice with 4 ml F14 media containing 10% horse serum (F14HS) to block further digestion.

Ganglia are centrifuged at 300 g for 2 min at 4°C between and following F14HS washes. They are then gently dissociated in 2 ml fresh F14HS with a 1-ml Gelman pipetman. Five repeated smooth strokes are applied with the pipetman in an effort to dissociate cells mechanically from the digested ganglia. This is the most critical step of the procedure. Care should be taken to avoid the introduction of air bubbles while firmly and smoothly manipulating the ganglia in and out of the pipet tip. At the end of five strokes, all undissociated ganglia are allowed to sediment rapidly to the bottom of the 15-ml sterile tube and 1.5-ml dissociated cell suspension is collected. A fresh 1.5 ml F14HS is added and the remaining ganglia are further dissociated by five additional smooth strokes. This dissociation process is repeated a third time, after which any undissociated ganglia are discarded. The combined 5-ml solution of dissociated cells is then filtered with a 45-μm sterile nylon mesh into a fresh 15-ml sterile centrifuge tube. The filter is gently rinsed with 0.5 ml fresh media. This filtration step removes small aggregates of undissociated cells.

The 6 ml F14HS solution containing dissociated cells is added to a 60-mm plastic culture dish (Nunc) and cultured for 210 min to pre-plate nonneuronal cells. During this time only nonneuronal cells adhere to the surface of the plastic culture dish. Neurons should initially contain amputated axons if correctly dissociated. During the pre-plating period, the dishes should not be disturbed in the incubator. At the end of the period, the dish should be inspected to ensure nonneuronal pre-plating. When gently moving the microscope stage, the phase-bright round neurons should roll along the surface by the movement of the media while the flat nonneuronal cells should remain adherent to the surface of the dish.

Neurons should be harvested from the pre-plating dish by carefully tilting the side of the dish and collecting the 6 ml suspended neurons. To ensure that all the suspended neurons are harvested, gently rinse the surface of the dish twice with the cell suspension to force all the cells to one edge for collection. Care should be taken not to dislodge nonneuronal cells by rinsing the dish gently with media without unnecessary pressure.

Harvested neurons should be seeded on precoated culture dishes or wells. Dishes are coated overnight with 0.5 mg/ml poly-DL-ornithine (PORN; Sigma) in borate buffer (pH 8.4) followed by 3 washes with sterile dH_2O and air drying in a sterile hood. Dry dishes coated with PORN are then coated with laminin (GIBCO; 12–20 μg/ml PBS) for a minimum of 4 hr. Laminin-coated dishes are washed twice with PBS and once with F14HS immediately before use. Dishes precoated with laminin should not be allowed to dry between washes.

A cell density count should be made of harvested neurons using the vital stain trypan blue (Sigma). The stain is diluted (5 μl stain : 45 μl suspended cells) before hemocytometer counting. Cells are seeded to dishes at a density equivalent to 30 neurons/mm². The cell density in the suspension should be adjusted so that it is roughly equal to 2.3×10^3 cells in 80 μl/well for each of the 4 wells (9-mm) in a Greiner dish, 6×10^3 cells in 0.5 ml/well for each of the 24 wells (16-mm) in a multiwell plate, or 2.4×10^4 cells in 1 ml/dish for a 35-mm dish. Although brief crosswise horizontal and vertical movements may be used to distribute cells evenly over the surface, care should be taken to avoid circular motions that concentrate cells within the dish. The dish should not be disturbed during the first 2 hr of plating to promote neuronal attachment. Additional media can be added to Greiner dishes at the end of 2 hr.

Cell survival or neurite outgrowth can be measured after 48 hr in culture. Assays of unknown should also include cultures treated as both positive controls (i.e., with 5 ng/ml NGF or BDNF) and negative controls (no added factors). Positive controls are useful to determine the quality of harvested cells and culture conditions for each assay. At least 20% of the total surface area should be measured for each assay.

B. Preparing Retinal Cultures

1. Eye Collection

Embryos are collected from timed pregnant rats following anesthesia with methoxyflurane. Only animals weighing 600 mg +/−150 mg are used for experiments requiring E17 embryos. Anesthetized early postnatal pups (P0–P4) are also used for the purification of ganglion cells from retinas or experiments requiring prior labeling *in vivo*. Retinas harvested with ganglion cells anatomically labeled by midbrain injections of retrograde tracers are dissected 24 hr after HRP injection or 48 hr after injection with true blue. Retinas harvested from the eyes of embryonic animals are capable of longer and greater ganglion cell survival than older retinas.

2. Retinal Dissections

With the aid of a dissecting microscope, retinas are carefully dissected by the removal of the eye sclera, lens, choroid, pigmented epithelium, and iris tissues with #5 Dumont forceps in sterile cold Ca^{2+}/Mg^{2+}-free PBs. Once the retinas are dissected free, any adhering vitreous humor and vitreal blood vessels are removed. Care is taken not to directly handle or damage retinas with forceps. Harvested retinas are collected in cold F14 medium (Irvine Scientific modified F12) on ice. Approximately 8–12 retinas are collected for each experiment investigating ganglion cell survival in dissociated (E17) retinal cell cultures. More retinas (16–20) are required for the purification of ganglion cells harvested from postnatal animals. All retinas are collected within a 60-min period prior to proteolytic digestion.

3. Proteolytic Digestion

Harvested retinas collected in ice cold F14 medium are pelleted by gentle centrifugation (60 g for 3 min) and placed in a pretreatment solution at 37°C. Pretreatment to remove any remaining vitreous humor adherent to the retina is achieved with 0.01% hyaluronidase in 10 ml Ca^{2+}/Mg^{2+}-free PBS (5 min, 37°C). This solution is removed and embryonic retinas are digested for 15 min at 37°C in 10 ml Ca^{2+}/Mg^{2+}-free PBS containing 0.05% trypsin (Worthington) and 0.01% DNase I (Sigma). Postnatal retinas require a more extensive digestion, which is accomplished by increasing the trypsin concentration to 0.1% and the trypsin digestion period to 30 min. Proteolytic digestion is stopped by the gentle centrifugation of retinas (60 g for 3 min) and the replacement of the trypsin solution with a wash consisting of F14 medium containing 10% (v/v) heat-inactivated horse serum (F14HS). Embryonic retinas are gently repelleted (60 g for 3 min) and placed in a dissociation solution of F14HS and 0.01% DNase. Postnatal retinas must be washed twice with F14HS tissue dissociation.

4. Dissociation

Digested retinas are then mechanically dissociated into individual cells by carefully pipetting retinas in 2 ml fresh F14HS containing 0.01% DNase. Five smooth strokes through a 1-ml sterile pipet tip are made with a 1-ml Gelman pipetman. Extreme care should be taken to avoid air mixing or air bubbles. Large undissociated pieces of retina are then allowed to sediment, the suspended cells in the supernatant are removed, and the remaining undissociated retinal tissue is pipetted with five smooth strokes in 2 ml fresh F14HS containing 0.01% DNase. This process is repeated a third time to achieve a final volume of 6 ml media containing dissociated cells. This technique reduces the mechanical damage by removing free cells as they are dissociated.

5. Cell Plating

A hemocytometer count of cell suspensions is made after trypan blue staining to detect an increase of membrane permeability in damaged or dying cells. At

E17, 2–3 $\times$ 10^7 total cells containing 2–3 $\times$ 10^5 ganglion cells can be harvested per retina. Virtually all the suspended retinal cells should be trypan blue negative from embryonic retinas, whereas some of the larger cells from postnatal retinas will be damaged and stained with trypan blue. Under optimal conditions, dissociated neurons should retain the amputated stump of a neurite immediately following dissociation. When neurons attach, these cut ends are rapidly absorbed and new outgrowth is generated. Two hundred thousand trypan blue-negative retinal cells are added to each 35-mm dish coated with poly-DL-ornithine and laminin. The homogeneous distribution of cells is achieved by an initial plating in 1 ml to allow the rapid attachment of cells to the surface.

C. Retinal Cell Survival Assays

The initial media is very gently exchanged with 2 ml fresh F14HS, with or without the desired test concentration of putative neurotrophic substances 210 min after plating (Fig. 2). This step requires the careful and gradual addition of new media along the wall of the culture dish at the 12 o'clock position in every dish. Care should also be taken not to allow the surface of the dish to dessicate in the laminar flow hood by completing the treatment media exchange one dish at a time. This time point allows cell attachment to be completed prior to the onset of the survival assay. An initial measurement of cell number is made on identical sister cultures immediately following media exchange with control media. This permits an analysis of the number and seeding density of all cells, neurons, and retinal ganglion cells at the onset of the experiment. Cells are identified with immunocytochemical, fluorescent, and retrograde markers. The measurement of the number and density of identified cells is repeated after 24 hr, 2 days, 4 days, 7 days, and 12 days *in vitro*. During this time period, cultures are maintained by exchanging 1 ml conditioned media with the same volume of fresh media containing appropriate experimental supplements every 2 days. Care is taken to add media gently along the side of the dish at the 12 o'clock position in every dish. Several control experiments are required to determine whether the number of identified cells measured in long term cultures represents the survival of originally plated cells. The addition of [^{3}H]thymidine or 5-BrdU in sister cultures is used to determine whether any identified cells arise from populations that incorporate the labeled nucleotides during the S phase of cell proliferation *in vitro*. Using this control and a variety of growth factor treatments, we have always observed that many nonneuronal cells in mixed retinal cell cultures (but no dissociated neurons) arise from proliferating cells *in vitro*. A change in cell number due to the induction of immunocytochemical labels used to identify cells can be controlled by using alternative retrograde tracers to follow the fate of identified cells unequivocally after plating. In addition, potential problems with label induction can be controlled by isolating the labeled population before cell plating and treating only cells in purified cultures.

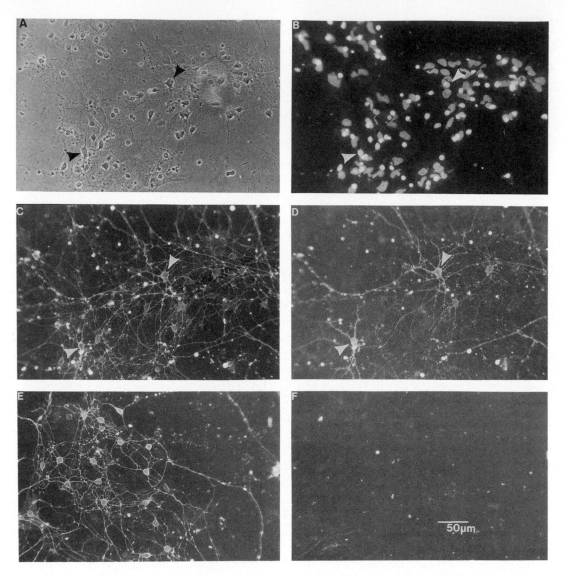

Fig. 2 Cell identification in central nervous system (CNS) survival assays. Dissociated cells from the embryonic day 17 (E17) rat retina. (A–D), Dissociated cells grown for 4 days in media containing 5 μg/ml BDNF (brain-derived neurotrophic factor). The same microscopic field is photographed to demonstrate the requirement of cell-specific markers in survival assays using mixed CNS cultures. (A) Phase contrast microscopy. (B) Blue bisbenzimide fluorescence of all cell nuclei. (C) Green FITC immunofluorescence of tetanus toxin binding to the membranes of all neurons but not to nonneuronal cells. (D) Red RITC immunofluorescence with anti-Thy1.1, a surface glycoprotein found only on a numerically small subpopulation of retinal neurons, predominantly ganglion cells. Note that a population of retinal cells is labeled by methods specific for neurons and the neuronal subpopulation with the ganglion cell marker (white arrows; C and D). These cells can also be seen in phase contrast and bisbenzimide fluorescent staining (black arrows, A; white arrows, B). (E,F) Sister cultures grown without the survival factor have approximately the same total number of neuronal cells (E); however, the specific subpopulation of cells with the ganglion cell marker is absent in cultures without BDNF (F).

D. Triple Fluorescent Labeling for Total Cells/Total Neurons/Ganglion Cells

1. Cell Culture Preparation

Living retinal cultures grown in Greiner 35-mm dishes are gently washed with Krebs buffer containing BSA (K-BSA). All fluid exchanges are made very carefully at the 12 o'clock rim position of each well on a draft-free table top on black plastic. Fluid within each well is removed with a Rainin microliter pipet tip connected to a gentle water vacuum equipped with a trap flask. Special care is taken to remove the contents from each well gently to avoid cell disruption. Exchanges are made by the removal and immediate addition of 80 μl for each individual well. Dessication during the transfer is eliminated by using the left hand for the vacuum removal pipet tip and the right hand for the immediate micropipet addition of the next solution.

2. Labeling Cell Surface Markers on Live Cells

Each well is washed three times with K-BSA to block nonspecific antibody and toxin binding. The first antibody/toxin solution consists of K-BSA containing tetanus toxin (25 μg/ml) for toxin binding site localization on all retinal neurons and mouse monoclonal antibody OX-7 against the Thy 1.1 glycoprotein found predominantly on a subpopulation of retinal neurons—the ganglion cells (Serotech/Harland; ascites diluted 1 : 20). Alternatively, antibody A2B5 (Chemicon) may also be used for labeling these surface gangliosides on all retinal neurons. All antibody solutions are prepared by diluting frozen concentrated aliquots. Diluted solutions are gently mixed and centrifuged. Then the supernatant is used to avoid the addition of any protein precipitated during the thawing process. Labeling surface antigens is greatly improved with live cell preparations. Cells are incubated for 40 min at room temperature. Cells are then gently washed three times with K-BSA blocking buffer before treatment with the second solution.

The second solution consist of K-BSA blocking buffer containing RITC-conjugated anti-mouse IgG (1 : 50 in K-BSA; Dapatts, Denmark) as a secondary antibody to detect OX-7 anti-Thy-l. This solution also contains a human primary antibody against tetanus toxin (1 : 50 in K-BSA; Behring-Werke, Marburg, Germany). Cells are incubated for 20 min at room temperature under a black plastic cover to avoid extended light exposure). Following this treatment, cells are washed gently twice with K-BSA blocking buffer.

3. Fixation

Excess Krebs buffer without BSA (2 ml) is added to each dish to dilute and remove serum proteins in the blocking buffer prior to fixation. Cultures are then fixed by the addition of ice cold 4% paraformaldehyde in PBS. Cultures are fixed for 20 min at room temperature and then washed twice with excess

Krebs buffer without BSA (2 ml) to remove all fixative. This is followed by two washes with K-BSA blocking solution.

The third solution consists of K-BSA blocking buffer containing FITC-labeled goat anti-human IgG (1 : 50 in K-BSA; Behring-Werke). This solution is prepared immediately prior to incubation. Fixed cells are incubated in this solution for 20 min at room temperature. The labeled secondary antibody in this solution is used to detect primary antibodies against tetanus toxin. Cultures are washed twice with K-BSA following this treatment.

4. Fluorescent Labeling of All Cell Nuclei

Cultures are finally treated with Hoechst bisbenzimide nuclear stain (1 μg/ml K-BSA, Hoechst No. 33258; Molecular Probes). Cells are treated for 20 min at room temperature and then washed twice with K-BSA buffer.

5. Maintenance and Microscopy

Labeled cultures are then washed in PBS and wells are covered with a cover slip mounting medium that retards dessication (PBS/glycerol, 1 : 1). Care must be taken to avoid the addition of excess cover slip medium by adding only 30 μl/well. Each well is then carefully covered with a 9-mm circular glass cover slip (EM Corp.). Cells can be observed immediately with an inverted phase-contrast microscope equipped for epifluorescence. The release of fluorescent light can be quenched by the plastic bottom of the dish. To eliminate this problem, the wall of the 35-mm Greiner dish is removed with pliers to permit the use of short working distant fluorescence optics directly against the glass cover slip. When sealed in a dark humidified environment, labeled dishes can be maintained for up to 2 wk (care should be taken to avoid dehydration during storage and fluorescence quenching during prolonged observations or photography). Each well may be viewed using four methods. All the normal features of the cells may be viewed by bright-field phase-contrast microscopy. Blue bisbenzimide fluorescence can be used for measurements of total cell number by labeling all cell nuclei (excitation wavelength, 340–380 nm). The nuclei of neurons are typically condensed spheres emitting very bright blue fluorescence whereas the same amount of DNA stain is diluted in the larger irregular nonneuronal nuclei, forming a pale blue stain. Yellow/green FITC fluorescence can be used to localize tetanus toxin binding on all neurons (excitation wavelength, 450–490 nm). Bright RITC fluorescence is used for Thy-1 localization on ganglion cells (excitation wavelength, 530–560 nm).

E. Solutions

1. Phosphate Buffered Saline (PBS): Ca^{2+} and Mg^{2+} free

Ingredients	1L	2L	3L
NaCl	8 g	16 g	40 g
KCl	200 mg	400 mg	1 g
$Na_2HPO_4 \cdot 12H_2O$	2.889g	5.778 g	14.445 g
(for $\cdot 2H_2O$)	(1.411 g)	(2.882 g)	(7.205 g)

Stir gently, check pH, and adjust to 7.2.

2. F 14 Medium

Prepare from powdered modified Coons F-12 (Irvine Scientific, Cat. No. 9409).

Ingredient	500 ml	5L
$NaHCO_3$	987 mg	9.87 g
Penicillin 'G	30 mg	300 mg
Streptomycin sulfate	50 ml	500 ml

Add powdered ingredients to highly purified distilled deionized water (i.e., MilliQ purified water to 18 MOhm resistance). Initially add ingredients to 90% of the final volume and stir gently (avoid heat or excessive stirring). After all solutes are added, bring to the desired final volume and adjust pH to 7.1 (Note: pH will rise 0.1–0.2 units during sterile filtration). Sterilize with 0.22-μm sterile filter and use or discard within 30 days.

3. Krebs Ringer Solution

125 mM NaCl (7300 mg/L)

1.2 mM $Mg_2SO_4 \cdot 7H_2O$ (296 mg/L)

4.8 mM KCl (358 mg/L)

1.2 mM KH_2PO_4 (H_2O) (163 mg/L)

1.3 mM $CaCl \cdot 2H_2O$ (191 mg/L)

5.6 mM glucose (1.1 mg/L)

25 mM HEPES (5960 mg/L)

mM NaAscorbate (176 mg/L)

Adjust to pH 7.3; sterilize by 0.22 μm filtration. For K-BSA blocking solution, add 100 μl BSA (Sigma) before filtration.

4. 4% Paraformaldehyde Fixative in PBS

To avoid solute precipitation, use the following two stock solutions:

Stock Solution A: 8% paraformaldehyde

Heat 100 ml H_2O to 75–80°C in a covered flask (use a fume hood to avoid toxic vapor). Remove flask from heat, stir, and add 8 g paraformaldehyde (Note: handle with care). Add 2 drops 5 N NaOH until the milky solution is clear. This solution is volatile and should be stored sealed at 4°C.

Stock Solution B: $2\times$ PBS solution

Add cool equal volumes of stock solutions A and B, stir, and adjust pH to 7.4 immediately prior to use.

References

Abney, E. R., Williams, B. P., and Raff, M. C. (1983). Tracing the development of oligodendrocytes from precursor cells using monoclonal antibodies, fluorescence-activated cell sorting and cell culture. *Dev. Biol.* **100,** 166–177.

Adams, R. L. P. (1980). Cell culture for biochemists. *Lab. Tech. Biochem. Mol. Biol.* **8,** 11–22.

Arakawa, Y., Sendtner, M., and Thoenen, H. (1990). Survival effect of ciliary neurotrophic factor (CNTF) on chick embryonic motoneurons in culture: Comparison with other neurotrophic factors and cytokines. *J. Neurosci.* **10,** 3507–3515.

Armson, P., and Bennett, M. (1983). Neonatal retinal ganglion cell cultures of high purity: Effect of superior colliculus on their survival. *Neurosci. Lett.* **38,** 181–186.

Atterwill, C. K., Hillier, G., Johnston, H., and Thomas, S. M. (1992). A tiered system for *in vitro* neurotoxicity testing: A place for neural cell line and organotypic cultures. 83–108.

Bainbridge, D. R., and Macey, M. M. (1983). Hoechst 33258: A fluorescent nuclear counterstain suitable for double-labelling immunofluorescence. *J. Immunol. Meth.* **62,** 193–195.

Banker, G. (1980). Trophic interactions between glia and hippocampal neurons in culture. *Science* **209,** 809–810.

Banker, G. A., and Cowan, W. M. (1977). Rat hippocampal neurons in dispersed cell culture. *Brain Res.* **126,** 397–425.

Banker, G. A., and Cowan, W. M. (1979). Further observations on hippocampal neurons in dispersed cell cultures. *J. Comp. Neurol.* **187,** 469.

Banker, G., and Goslin, K. (1988). Developments in neuronal cell culture. *Nature* **336,** 185–186.

Banker, G., and Goslin, K., eds. (1991). "Culturing Nerve Cells." MIT Press, Cambridge, Massachusetts.

Banker, G. A., and Waxman, A. B. (1988). Hippocampal neurons generate natural shape in culture. *In* "Intrinsic Determinants of Neuronal Form and Function" (R. Lasek and M. M. Black, eds.), pp. 61–82. Liss, New York.

Banks, M. M. (1979). Basic methods/dispersion and disruption of tissues. *Meth. Enzymol.* **57,** 119–131.

Barald, K. F. (1989). Culture conditions effect the cholinergic development of an isolated subpopulation of chick mesencephalic neuronal crest cells. *Dev. Biol.* **135,** 349–366.

Barnstable, J. C. (1980). Monoclonal antibodies which recognize different cell types in the rat retina. *Nature* **286,** 231–235.

Barres, B. A., Silverstein, B. E., Corey, D. P., and Chun, L. L. Y. (1988). Immunological, morphological and electrophysiological variation among retinal ganglion cells purified by panning. *Neuron* **1,** 791–793.

Barres, B. A., Schmid, R., Sendnter, M., and Raff, M. C. (1993). Multiple extracellular signals are required for long-term oligo dendrocyte survival. *Development* **118**, 283–295.

Bloch-Gallego, E., Huchet, M., El M'Hamdi, H., Xie, F. K., Tanaka, H., and Henderson, C. E. (1991). Survival *in vitro* of motoneurons identified or purified by novel antibody-based methods is selectively enhanced by muscle-derived factors. *Development* **111**, 221–332.

Borenfreund, E., and Puerner, J. A. (1985). Toxicity determined in vitro by morphological alterations and neutral red absorption. *Toxicol. Lett.* **24**, 119–124.

Borenfreund, E., Babich, H., and Martin-Alguacil, N. (1988). Comparison of two in vitro cytotoxicity assays—The neutral red (NR) and tetrazolium MTT tests. *Toxic In Vitro* **2**, 1–6.

Bottenstein, J. E., and Sato, G. (1985). "Cell Culture in the Neurosciences." Plenum Press, New York.

Doulton, A. A., Baker, G. B., and Walz, eds. (1992) "Practical Cell Culture Techniques," Neuromethods, Vol. 23. Humana Press, Clifton, New Jersey.

Bozycko-Coyne, D., McKenna, B. W., Connors, T. S., and Neff, N. T. (1993). A rapid fluorometric assay to measure neuronal survival in vitro. *J. Neurosci. Meth.* **50**, 205–216.

Bruinink, A. (1992). Serum-free monolayer cultures of embryonic chick brain and retina: Immunoassays of developmental markers, mathematical data analysis and establishment of optimal culture conditions. *In* "The Brain in Bits and Pieces" (G. Zbinden, ed.). M. T. C. Verlag, Zollikon, Switzerland.

Burgoyne, R. D., Cambray-Deakin, M. A., Lewis, S. A., Sarkar, S., and Cowan, N. I. (1988). Differential distribution of beta-tubulin isotypes in cerebellum. *EMBO J.* **7**, 2311–2319.

Caceres, A., Banker, G. A., Steward, O., Binder, L., and Payne, M. (1984). MAPZ is localized to the dendrites of hippocampal neurons which develop in culture. *Dev. Brain Res.* **13**, 314–318.

Carnow, T. B., Manthorpe, M., Davis, G. E., and Varon, S. (1985). Localized survival of ciliary ganglion neurons identifies neurotrophic factor bands on nitrocellulose blots. *J. Neurosci.* **5**, 1965–1971.

Chen, L. B. (1988). Mitochondrial potential in living cells. *Annu. Rev. Cell Biol.* **4**, 151–181.

Clendening, B., and Hume, R. I. (1990). Cell interactions regulate dendrite morphology and responses to neurotransmitters in embryonic chick sympathetic preganglionic neurons *in vitro*. *J. Neurosci.* **10**, 3992–4005.

Conn, P. M., ed. (1990). "Cell Culture," Methods in Neurosciences, Vol. 2. Academic Press, New York.

Corey, D. P. (1983). Patchclamp: Current excitement in membrane physiology. *Neurosci. Comm.* **1**, 99–110.

Cowan, W. M., Fawcett, J. W., O'Leary, D. D., and Stanfield, B. B. (1984). Regressive events neurogenesis. *Science* **225**, 1258–1265.

Davies, A. M. (1989). Neurotrophic factor bioassays using dissociated neurons. *In* "Nerve Growth Factors" (R. A. Rush, ed.), IBRO Methods in the Neurosciences, Vol. 12, p. 95–110. John Wiley and Sons, New York.

deErausquin, G. A. (1990). Gangliosides normalize distorted single cell intracellular free Ca^{2+} dynamics after toxic doses of glutamate in cerebellar granule cells. *Proc. Natl. Acad. Sci. USA* **87**, 8017.

Dimpfel, W., Huang, R. T. C. and Habermann, E. (1977). Gangliosides in nervous tissue cultures and binding of ^{125}I-labeled tetanus toxin. *J. Neurochem.* **29**, 329–334.

Dohrmann, U., Edgar, D., Sendtner, M., and Thoenen, H. (1986). Muscle derived factors that support survival and promote fiber outgrowth from embryonic chick spinal motor neurons in culture. *Dev. Biol.* **118**, 209–221.

Dohrmann, U., Edgar, D., and Thoenen, H. (1987). Distinct neurotrophic factors from skeletal muscle and the central nervous system interact synergically to support the survival of cultured embryonic spinal motor neurons. *Dev. Brain Res.* **124**, 145–152.

Dyer, S., Derby, M., Cole, G., and Glaser, L. (1983). Identification of subpopulations of chick neural retinas by monoclonal antibodies: A fluorescence activation cell sorter screening technique. *Dev. Brain Res.* **9,** 197–203.

Eisenbarth, G. S., Walsh, F. S., and Nirenberg, M. (1979). Monoclonal antibody to a plasma membrane antigen of neurons. *Proc. Natl. Acad. Sci. USA* **76,** 4913–4917.

Eriksson, H., and Heilbronn, E. (1989). Extracellularly applied ATP alters the calcium flux through dihydropyridine-sensitive channels in cultured chick myotubes. *Biochem. Biophys. Res. Commun.* **159,** 878–885.

Essig-Marcello, J. S., and VanBuskirk, R. G. (1990). A double-label in situ cytotoxicity assay using fluorescent probes neutral red and BCECF-AM. *In Vitro Toxicol.* **3,** 219–227.

Fedoroff, S., and Richardson, A. (1992). "Protocols for Neural Cell Cultures." Humana Press, Totowa, New Jersey.

Fields, K. L. (1979). Cell type-specific antigens of cells of the central and peripheral nervous system. *Curr. Top. Dev. Biol.* **13,** 237–257.

Fields, K. L. (1985). Neuronal and glial surface antigens on cells in culture. *In* "Cell Culture in the Neurosciences" (J. E. Bottenstein and G. Sato, eds.), pp. 45–93. Plenum Press, New York.

Frandsen, A., and Schousboe, A. (1987). Time and concentration dependency of the toxicity of excitory amino acids on cerebral neurones in primary culture neurochem. *Int.* **10,** 583–591.

Frandsen, A., Frejer, J., and Schousboe, A. (1989). Direct evidence that excitotoxicity in cultured neurons in mediated via N-methyl-D-aspartate (NMDA) as well as non-NMDA receptors. *J. Neurochem.* **53,** 297–299.

Goldberg, D. J., and Burmeister, D. W. (1986). Stages in axon formation: Observations of growth of *Aplysia* axons in culture using video-enhanced contrast–differential interference contrast microscopy. *J. Cell Biol.* **103,** 1921–1931.

Goslin, K., Scyreyer, D. J., Skene, J. H. P., and Banker, G. (1988). Development of neuronal polarity: GAP-43 distinguishes axonal from dendritic growth cones. *Nature* **326,** 672–674.

Goslin, K., Schreyer, D. J., Skene, J. H. P., and Banker, G. (1990). Changes in the distribution of GAP-43 during the development of neuronal polarity. *Neurosci.* **10,** 588–602.

Grynkyewicz, G. M., Poenie, G. M., and Tsien, R. Y. (1985). A new generation of Ca^{2+} indicator with greatly improved fluorescence properties. *J. Biol. Chem.* **260,** 3440–3450.

Harrison, R. G. (1970). Observations on the living developing nerve fiber. *Anat. Rec.* **1,** 116–118.

Harrison, R. G. (1910). The outgrowth of the nerve fiber as a mode of protoplasmic movement. *J. Exp. Zool.* **9,** 797–846.

Harrison, R. G. (1912). The cultivation of tissues in extraneous media as a method of morphogenetic study. *Anat. Rec.* **6,** 181–193.

Harrison, R. G. (1914). The reaction of embryonic cells to solid structures. *J. Exp. Zool.* **17,** 521–544.

Higgins, D., Waxman, A., and Banker, G. A. (1988). The distribution of microtubule-associated protein 2 changes when dendritic growth is induced *in vitro*. *Neuroscience* **24,** 583–592.

Hilwig, I., and Gropp, A. (1972). Staining of constitutive heterochromatin in mammalian chromosomes with a new fluorochrome. *Exp. Cell Res.* **75,** 122–126.

Hirschberg, H., Skere, H., and Thorsby, E. (1977). *J. Immunol. Meth.* **16,** 131.

Hockberger, P. E., Tseng, H. Y., and Connor, J. A. (1989). Development of rat cerebellar Purkinje cells: Electrophysiological properties following acute isolation and in long-term culture. *J. Neurosci.* **9,** 2258–2271.

Honegger, P., and Schilter, B. (1992). Serum-free aggregate cultures of fetal rat brain and liver cells: Methodology and some practical applications in neurotoxicology. 51–79.

Honig, M. G., and Hume, R. I. (1986). Fluorescent carbocyanine dyes allow living neurons of identified origin to be studied in long-term cultures. *J. Cell Biol.* **103,** 171–187.

Huettner, J. E., and Baughman, R. W. (1990). Primary culture of identified neurons from the visual cortex of postnatal rats. *J. Neurosci.* **6,** 3044–3060.

Huschtscha, L. I., Lucibello, F. C., and Bodmer, W. F. (1989). A rapid micro method for counting cells *in situ* using a fluorogenic alkaline phosphatase enzyme assay. *In Vitro Cell Dev. Biol.* **25**, 105–108.

Jensen, L. M., Zhang, Y., and Shouter, E. M. (1992). Steady-state polypeptide modulations associated with nerve growth factor (NGF)-induced terminal differentiation and NGF deprivation-induced apotheosis in human neuroblastoma cells. *J. Cell Biol. Chem.* **267**, 19325–19333.

Jessell, T. M. (1988). Adhesion molecules and the hierarchy of neural development. *Neuron* **1**, 3–13.

Johnson, J. E. (1989). Retinal cultures used to assay CNS neurotrophic factors. *In* "Nerve Growth Factors" (R. A. Rush, ed.), IBRO Methods in the Neurosciences, Vol. 12, p. 111–137. John Wiley and Sons, New York.

Johnson, J. E., Barde, Y.-A., Schwab, M., and Thoenen, H. (1986). Brain derived neurotrophic factor supports the survival of cultured rat retinal ganglion cells. *J. Neurosci.* **6(10)**, 3031–3038.

Johnson, L. V., Walsh, M. L., and Chen, L. B. (1980). Localization of mitochondria with rhodamine 123. *Proc. Natl. Acad. Sci. USA* **77**, 990–994.

Jones, K. H., and Senft, J. A. (1985). An improved method to determine cell viability by simultaneous staining with fluorescein diacetate propidium iodide. *J. Histol. Cytol.* **33**, 77–79.

Juurlink, B. H., and Hertz, L. (1993). Ichemia-induced death of astrocytes and neurons in primary culture: pitfalls in quantifying neuronal cell death. *Dev. Brain Res.* **71**, 239–246.

Juurlink, B. H. J., Munoz, D. G., and Ang, L. C. (1991a). Motoneuron survival in vitro: Effects of pyruvate, α-ketoglutarate, gangliosides and potassium. *Neurosci. Lett.* **133**, 25–28.

Juurlink, B. H. J., Sochocka, E., Hertz, L., Shuaib, A., and Code, W. E. (1991b). Ischemic-hypoxic injury in cultured neurons and astrocytes measured by a dye technique. *Anesthesiology* **75**, A610.

Kaltenback, J. P., Kaltenback, M. H., and Lyons, W. B. (1958). Nigrosin as a dye for differentiating live and dead ascites cells. *Exp. Cell Res.* **15**, 112–117.

Katz, L. C., Burkhalter, A., and Dreyer, W. J. (1984). Fluorescent latex microspheres as a retrograde neuronal marker for *in vivo* and *in vitro* studies of visual cortex. *Nature* **310**, 498–500.

Koh, J. Y., and Choi, D. W. (1987). Quantitative determination of glutamate mediated cortical neuron injury in cell culture by lactate dehydrogenase efflux assay. *J. Neurosci. Meth.* **20**, 83–90.

Koh, J. Y., and Choi, D. W. (1988). Vulnerability of cultured cortical neurons to damage by excitotoxins: differential susceptibility of neurons containing NADPH-diaphorase. *J. Neurosci.* **8**, 2153–2163.

Koike, T. (1992). Molecular and cellular mechanism of neural degeneration caused by nerve growth factor deprivation approached through PC12 cell culture. *Prog. Neuropsychopharmacol. Biol. Psychiatry* **16**, 95–106.

Kosik, K. S. and E. A. Finch (1987), MAP2 and tau segregate into dendritic and axonal domains after the elaboration of morphologically distinct neurites: an immunochemical study of cultured rat cerebrum *J. Neurosci.* **7**, 3142–3153.

Krishnan, A. (1975). Rapid flow cytofluorometric analysis of mammalian cell cycle by propidium iodide staining. *J. Cell Biol.* **66**, 188–193.

Kudo, Y., Takeda, K., and Yamazaki, K. (1990). Quin 2 protects against neuronal cell death due to Ca^{2+} overload. *Brain Res.* **528**, 48.

Laughton, C. (1984). Quantification of attached cells in microtiter plates based on Coomassie brilliant blue G-250 staining of total cellular protein. *Anal. Biochem.* **140**, 417–423.

Lee, M. K., Tuttle, J. B., Rebhun, L. I., Cleveland, D. W., and Frankfurter, A. (1990). The expression and posttranslational modification of a neuron-specific beta-tubulin isotype during chick embryogenesis. *Cell Motil. Cytoskel.* **17**, 118–132.

Malgarol, A., Milani, D., Meldolesi, J., and Pozzan, T. (1987). Fura-2 measurements of cytosolic free Ca^{2+} in monolayers and suspensions of various types of animal cells. *J. Cell Biol.* **105**, 2145–2155.

Manthorpe, M., and Varon, S. (1988). Use of chick embryo ciliary ganglion neurons for the *in vitro* assay of neuronotrophic factors. *In* "A Dissection and Tissue Culture Manual for the Nervous System" (A. Shakar and B. Haber, eds.), pp. 317–321. Liss, New York.

Manthorpe, M., Skaper, S. D., and Varon, S. (1981). Neuronotrophic factors and their antibodies: *In vitro* microassays for titrating and screening. *Brain Res.* **230**, 295–306.

Manthorpe, M., Fagnani, R., Skaper, S. D., and Varon, S. (1986). An automated calorimetric microassay for neurotrophic factors. *Dev. Brain Res.* **25**, 191–198.

Manthorpe, M., Ray, J., Pettmann, B., and Varon, S. (1989). Ciliary neurotrophic factors. *In* "Nerve Growth Factors" (R. A. Rush, ed.), IBRO Methods in the Neurosciences, Vol. 12, p. 31–56. John Wiley and Sons, New York.

Marangos, P., and Schmechel, D. E. (1970). Neuron Specific enolase, a clinically useful marker for neurons and neuroendocrine cells. *Annu. Rev. Neurosci.* **10**, 269–295.

Marangos, P. J., and Schmechel, D. E. (1987). Neuron specific enolase, a clinically useful marker for neurons and neuroendocrine cells. *Annu. Rev. Neurosci.* **10**, 269–295.

Martin, D. P., Schmidt, R. E., DiStefano, P. S., Lowig, O. H., Carter, J. G., and Johnson, E. M. (1988). Inhibitors of protein synthesis and RNA synthesis prevent neuronal death caused by nerve growth factor deprivation. *J. Cell Biol.* **106**, 829–844.

Martin, D. P., Wallace, T. L., and Johnson, E. M. (1990). Cytosine arabinoside skills postmitotic neurons in a fashion resembling trophic factor deprivation: Evidence that a deoxycytidine-dependent process may be required for nerve growth factor signal transduction. *J. Neurosci.* **10**, 184–193.

Matus, A., Bernhardt, R., Bodmer, R., and Alaimo, D. (1986). Microtubule-associated protein 2 and tubulin are differentially distributed in the dendrites of developing neurons. *Neuroscience* **17**, 371–389.

Maxwell, G. D., Forbes, M. E., and Christie, D. S. (1988). Analysis of the development of cellular subsets present in the neural crest using cell sorting and cell culture. *Neuron* **1**, 557–568.

Melamed, M. R., Kemetsky, L. A., and Bouse, E. A. (1969). Cytotoxic test automation: A live–dead cell differential counter. *Science* **163**, 285–286.

Mesner, P. W., Winters, T. R., and Green, S. H. (1992). Nerve growth factor withdrawal-induced cell death in neuronal PC12 cells resembles that in sympathetic neurons. *J. Cell Biol.* **119**, 1169–1680.

Mirsky, R., Wendon, L. M. B., Black, P., Stolkin, C., and Bray, D. (1978). Tetanus toxin: A cell surface marker for neurons in culture. *Brain Res.* **148**, 251–259.

Mitchell, D. B., Santone, K. S., and Acosta, D. (1980). Evaluation of cytotoxicity in cultured cells by enzyme leakage. *J. Tissue Cult. Meth.* **6**, 113–121.

Moore, P. L., MacCoubrey, I. C., and Haugland, R. P. (1990). A rapid, pH insensitive, two color fluorescence viability (cytotoxicity) assay. *J. Cell Biol.* **304**, 58a.

Mosmann, T. (1983). Rapid calorimetric for cellular growth and survival: Application to proliferation and cytotoxic assays. *J. Immunol. Meth.* **65**, 55–63.

Neher, E., and Sakmann, B. (1976). Single channel currents recorded from membranes of denervated frog muscle fibers. *Nature* **260**, 779–802.

Neher, E., Sakmann, B., and Stienback, J. H. (1978). The extracellular patch clamp: A method for resolving currents through individual open channels in biological membranes. *Pflugers Arch.* **375**, 219–228.

Nieto-Sampredro, M., Needels, D. L., and Cotman, C. W. (1985). A simple objective method to measure the activity of factors that promote neuronal survival. *J. Neurosci. Meth.* **15**, 37–48.

Nurcombe, V., and Bennett, M. (1981). Embryonic chick retinal ganglion cells identified in vitro. *Exp. Brain Res.* **44**, 249–258.

Oliver, M. H., Harrison, N. K., Bishop, J. E., Cole, P. J., and Laurent, G. J. (1989). A rapid convenient assay for counting cells cultured in microwell plates: Application for assessment of growth factors. *J. Cell Sci.* **92**, 513–518.

Paul, J. (1975). "Cell and Tissue Culture." Churchill Livingstone, New York.

Pettmann, B., Manthorpe, M., Powell, J. A., and Varon, S. (1988). Biological activities of nerve growth factor bound to nitrocellulose paper by Western blotting. *J. Neurosci.* **8,** 3624–3632.

Pittman, R. N., Wang, S. L., DiBenedetto, A. J., and Mills, J. C. (1993). A system for characterizing cellular and molecular events in programmed neuronal death. *J. Neurosci.* **13,** 3669–3680.

Raff, M. C. (1989). Glial cell diversification in the optic nerve. *Science* **243,** 1450–1455.

Raff, M. C., Mirsky, R., Fields, K. L., Lisak, R. P., Sidman, S., Silberberg, D. N., Gregson, N. A., Leibowitz, S., and Kennedy, M. C. (1978). Galacto-cerebroside is a specific cell-surface antigen marker for oligodendrocytes in culture. *Nature* **274,** 813–816.

Raff, M. C., Abney, E. R., Cohen, J., Lindsay, R., and Noble, M. (1983a). Two types of astrocytes in cultures of developing rat white matter: differences in morphology, surface gangliosides and growth characteristics. *J. Neurosci,* **3,** 1289–1300.

Raff, M. C., Miller, R. H., and Noble, M. (1983b). A progenitor cell that develops in vitro astrocyte or an oligodendrocyte depending on culture medium. *Nature* **303,** 390–396.

Rago, R., Mitchen, J., and Wilding, G. (1990). DNA fluorometric assay in 96-well tissue culture plates using Hoechst 33258 after cell lysis by freezing in distilled water. *Anal. Biochem.* **191,** 31.

Rohrer, H., and Thoenen, H. (1987). Relationship between differentiation and terminal mitosis: Chick sensory and ciliary neurons differentiate after terminal mitosis of precursor cells, whereas sympathetic neurons continue to divide after differentiation. *J. Neurosci.* **7,** 3739–3748.

Rohrer, H., Henke-Fahle, S., El-Shark, T., Lux, H. D., and Thoenen, H. (1985). Progenitor cells from embryonic chick dorsal root ganglia differentiate *in vitro* in neurons: Biochemical and electrophysiological evidence. *EMBO J* **4,** 1709–1714.

Rosner, H., Greis, C., and Henke-Fahle, S. (1988). Developmental expression in embryonic rat and chicken brain of a polysialoganglioside-antigen reacting with the monoclonal antibody Q211. *Dev. Brain Res.* **42,** 161–171.

Rotman, B., and Papermaster, B. W. (1966). Membrane properties of living cells as studied by enzymatic hydrolysis of fluorogenic esters. *Proc. Natl. Acad. Sci. UA* **55,** 134–141.

Rudge, J., Davis, G. E., Manthorpe, M., and Varon, S. (1987). An examination of ciliary neuronotrophic factors from avian and rodent tissues using a blot and culture technique. *Dev. Brain Res.* **32,** 103–110.

Rutishauser, U., and Jesse, T. J. (1988). Cell adhesion molecules in vertebrate neural development. *Physiol. Rev.* **68,** 819–857.

Rutishauser, U., Acheson, A., Hall, A. K., Mann, D. M., and Sunshine, J. (198B). The neural cell adhesion molecule (NCAM) as a regulator a cell interactions. *Science* **240,** 53–57.

Saijo, N. (1973). A spectroscopic quantitation of cytotoxic action of antiserum and compliment by trypan blue. *Immunology* **24,** 683–690.

Sarthy, P., Curtis, B., and Catterall, W. (1983). Retrograde labeling, enrichment and characterization of retinal ganglion cells from the neonatal rat. *J. Neurosci.* **3(12),** 2532–2544.

Schaffner, A. C., St. John, P. A., and Barker, J. F. (1987). Fluorescence-activated cell sorting of embryonic mouse and rat motoneurons and their long-term survival *in vitro. J. Neurosci.* **7,** 3088–3104.

Scott, S. A., and Davies, A. M. (1990). Inhibition of protein synthesis prevents cell death in sensory and parasympathetic neurons deprived of neurotrophic factor in vitro. *J. Neurobiol* **21,** 630–638.

Shahar, A., deVellis, J., Vernadakis, A., and Haber, B., eds. (1989). "A Dissection and Tissue Culture Manual of the Nervous System." Liss, New York

Shaw, G., Danker, G. A., and Weber, K. (1985). An immunofluorescence study of neurofilament protein expression by developing hippocampal neurons in tissue culture. *Eur. J. Cell Biol.* **39,** 205–216.

Silverstein, B. E., and Chu, L. L. Y. (1987). Purification of rat retinal ganglion cells by panning. *Soc. Neurosci. Abstr.* **13,** 294.1.

Skehan, P. (1990). New calorimetric cytotoxicity assay for anticancer-drug screening. *J. Natl. Cancer Inst.* **82,** 11047.

James E. Johnson

Skene, J. H. P. (1989). Axonal growth associated proteins. *Annu. Rev. Neurosci.* **12,** 127–156.

Sullivan, K. F., Havercroh, J. C., Machlin, P. S., and Cleveland, D. W. (1986). Sequence and expression of the chicken beta 5- and beta-tubulin genes define a pair of divergent beta-tubulins with complementary patterns of expression. *Mol. Cell Biol.* **6,** 4409–4418.

Tennant, J. R. (1964). Evaluation of the trypan blue technique for determination of cell viability. *Transplantation* **2,** 685–694.

Thelestam, M., and Mollby, R. (1976). Cytotoxic effects on the plasma membrane of human diploid fibroblasts—A comparative study of leakage tests. *Med. Biol.* **54,** 39–49.

Thelestam, M., and Mollby, R. (1980). Screening and characterization of membrane damaging effects in tissue culture. *Toxicology* **17,** 189–193.

Thoenen, H., Barde, Y. A., Davies, A. M., and Johnson, J. E. (1987). Neurotrophic factors and neuron death. *In* "Selective Neuronal Cell Death" (G. Bock and M. O'Connor, eds.), Ciba Symposium 126, pp. 82–95. John Wiley & Sons, Sussex.

Tsien, R. (1981). A non-disruptive technique for loading calcium buffers and indicators into cells. *Nature* **290,** 527–528.

Tsien, R. Y. (1988). Fluorescence measurement and photochemical manipulation of cytosolic free calcium. *Trends Neurosci.* **11,** 419.

Tsien, R. Y. (1989). Fluorescent probes of cell signaling. *Annu. Rev. Neurosci.* **12,** 227.

Tsien, R. Y., Pozzan, T., and Rink, T. J. (1982). Calcium homeostasis in intact lymphocytes: Cytoplasmic free calcium monitored by a new intracellular trapped indicator. *J. Cell Biol.* **94,** 325–334.

Toran-Allerand, C. D. (1976). Golgi-cox modifications for the impregnation of organotypic cultures of the CNS. *Br. Res.* **118,** 293–298.

Wahlefeld, A. W. (1983). UV-method with L-lactate and NAD. *In* "Methods of Enzymatic Analysis," Vol. III. Verlag-Chemie, Deerfield Beach, Florida.

Walum, E., and Peterson, E. (1982). Tritiated 2-deoxy-D-glucose as a probe for cell membrane permeability studies. *Anal. Biochem.* **120,** 8–11.

Walum, E., Peterson, A., and Erkell, L. J. (1985). Photometric recording of cell viability using trypan blue in perfused cell cultures. *Xenobiotica* **15,** 701–704.

Walum, E., Stenberg, K., and Jenssen, D. (1990). "Understanding Cell Toxicology: Principles and Practice." Ellis Harwood, New York.

Walum, E., Wang, L., Jones, K., Nordin, M., Clemedson, C., and Varnbo, I. (1992). Cellular neuronal development *in vitro*—Neurobiological and neurotoxicological studies in cultured model systems. 116–130.

Wigzell, H. (1965). Quantitative titrations of mouse H-2 antibodies using ^{51}Cr-labelled target cells. *Transplantation* **3,** 423.

Yu, A. C. HJ., Gregory, G. A., and Chan, P. H. (1989). Hypoxia-induced dysfunctions and injury of astrocytes in primary cell cultures. *J. Cereb. Blood Flow Metab.* **9,** 20–28.

Zawydiwski, R., and Duncan, G. R. (1978). Spontaneous ^{51}Cr release by isolated rat hepatocytes: An indicator of membrane damage. *In Vitro* **14,** 707–714.

Zbinden, G., ed. (1992). "The Brain in Bits and Pieces: In Vitro Techniques in Neurobiology, Neuropharmacology, and Neurotoxicology." M. T. C. Verlag, Zollikow, Switzerland.

Zhou, M. H., Zhao, L. P., and Ren, L. S. (1991). Effects of a 30-kd protein from tectal extract of rat on cultured retinal neurons. *Science* **34,** 908–915.

CHAPTER 13

Neuron Death in Vertebrate Development: *In Vivo* Methods

Peter G. H. Clarke* and Ronald W. Oppenheim†

* Institut d'Anatomie, Faculté de Médecine
Université de Lausanne
1005 Lausanne, Switzerland

† Department of Neurobiology and Anatomy and
Neuroscience Program
Bowman Gray School of Medicine
Wake Forest University
Winston-Salem, North Carolina 27157

I. Determination of Neuron Death *in Vivo*

A. General Considerations

    Neuronal death is one of the main events in the making of the nervous system, involving the loss from different neuronal populations of varying proportions

Copyright © 1995 by Academic Press, Inc. All rights of reproduction in any form reserved.

of neurons, ranging from 0 to 75% or more and averaging about 50% (Oppenheim, 1991). The study of neuronal death involves demonstrating that it occurs in a given situation, estimating the magnitude and timing of the loss, evaluating which particular neurons die, analyzing why and how they die, and understanding the role or purpose of the loss. A wide range of methods is available for achieving these ends; most involve the use of histological sections, although biochemical analysis of homogenized tissue can also provide useful information. In this chapter we attempt to give a broad overview of the whole range of methods rather than concentrating on a given one. Hence, it is beyond our scope to provide detailed protocols or recipes, but we systematically cite references in which these can be obtained. We deliberately avoid duplicating the description of methods that are covered in depth in other chapters of the present volume (e.g., methods for examining DNA fragmentation; see Chapters 2 and 3).

B. Counting Healthy Neurons in Histological Sections

Counting healthy neurons in histological sections is the most direct and widely used method for estimating the number of neurons that die, and the timing of their loss. Since most cases of neuronal death occur in postmitotic populations, there is rarely any need to consider complex tissue kinetics; the number of neurons lost is simply the initial number minus the final number *in a defined population*. However, the fact that subtraction is used makes the final estimation of neuronal death highly sensitive to errors in the estimations of total neuronal number. For example, in a situation where 30% of the neurons die, a 10% underestimation of the pre-death number and a 10% overestimation of the post-death number will cause more than 50% error in neuronal loss, which will be calculated as 13% instead of 30%.

To meet the high standards of accuracy that are required in counting healthy neurons, it is essential to avoid two main sources of error: (1) those due to inadequate definition of the population to be counted and (2) those in counting.

1. Defining the Population to Be Counted

Ideally, an unambiguous boundary should be specified, and it should be proved that neurons do not migrate in or out of this region. This task may be easy in peripheral ganglia but more difficult in the spinal cord or brain. It is dangerous to rely on a cytoarchitectonic boundary, especially when this is defined mainly by a change in neuronal density, as in some layers in the cerebral cortex and nuclei in the brainstem. Here, a small displacement of neurons from a low-density region might cause them to pile up along the border of an adjacent high-density one (or vice versa), leading to a change in the apparent boundary. Boundaries defined by changes in the size of neurons or in their intensity of Nissl staining should also be used with caution. For example, large seasonal volume changes in the canary high vocal center (HVC) nucleus were initially reported on the basis of these two criteria, but researchers subsequently found that the volume of HVC remained constant if it was defined on the basis of estrogen receptor immunoreactivity or by retrograde labeling from its axon target region (Gahr, 1990; Kirn *et al.,* 1991). Apparently the "boundaries" detected on the basis of the Nissl criteria did not reliably define HVC, because of seasonal changes in the sizes and staining intensities of neurons in the deeper parts of HVC. In contrast, avian spinal motoneurons are the largest and most intensely Nissl-stained neurons in the ventral horn at stages before, during, and after cell death (Fig. 1). Thus, as discussed in more detail later, these criteria can be used to define this population of cells reliably.

Logically, a fundamental difference exists between identifying a particular group of neurons that undergo cell death based on its *inherent characteristics* (size, shape, expression of antigens, axonal projections, etc.) and defining the group by *long-term exogenous labeling*. The disadvantage of the former approach is that the characteristics may change with time. The disadvantage of the latter is that the labeling may disrupt development.

When exogenous labels are used, in some cases the counts will need to be performed on the labeled neurons. In others, it may be sufficient to use the labels only to check the validity of other spatial and morphological criteria, after which the counts can be performed in unlabeled material; this approach is not only more convenient but also more rigorous, because it avoids the possibility that the labeling may modify the pattern of neuron death. For example, retrograde labeling of chick isthmo-optic neurons followed by varying survival times was used to show that reductions in their numbers during embryogenesis were not due to the outward migration of the neurons (Clarke, 1982; see also O'Leary and Cowan, 1982). In the same study, differences in the "birthdates" (as judged by thymidine autoradiography) of the isthmo-optic neurons relative to the neurons in surrounding regions served as further proof against outward migration (Clarke, 1982). Once outward migration had been ruled out, counts were performed in Nissl-stained sections without additional labeling. Similarly, the regions containing limb innervating motoneurons in the mouse embryo have been identified by retrograde labeling from limb muscles (e.g., Lance-Jones, 1982), after which motoneuron counts could be carried out accurately in the same region using only Nissl-stained material (e.g., Oppenheim, 1986; Oppenheim *et al.,* 1986).

It is essential to check that exogenous labels remain detectable in the neurons for a sufficiently long period of time and do not affect neuronal survival. For example, a more complicated experimental paradigm had to be employed in at least one study because retrogradely transported horseradish peroxidase disappeared from neurons after about 3 days (Clarke, 1982). Certain fluorescent labels such as the organic DNA-binding dye *True Blue* have the advantage of persisting longer in neurons (O'Leary and Cowan, 1982), but here, too, care is required because intraocular injection of a moderately high dose (0.4 mg) of the chemically related dye *Fast Blue* is toxic to the retina in E13 chick embryos

(P. G. H. Clarke, unpublished data). The main danger in experiments on cell death seems not to be a direct toxic effect on the retrogradely labeled cells, but their retrograde degeneration as a result of dye toxicity at the injection site. The same result has been found in the brains of adult rats following labeling with another organic fluorescent dye, *Fluoro-Gold* (hydroxystilbamidine), which is a member of the aromatic diamidines, a class of therapeutic agents. Small volumes of pressure-injected Fluoro-Gold at concentrations of 2.5% or greater killed neurons at the injection site, whereas remote retrogradely labeled neurons survived and retained their label for at least 6 wk (Schmued *et al.*, 1993). Other investigators have found that retrogradely transported Fluoro-Gold remained in neurons for more than 1 yr without noticeable toxicity or loss of fluorescence (Divac and Morgenson, 1990). Because of the known pharmacological properties of this class of dyes, in some cases affecting neuronal activity (see Schmued *et al.*, 1993, for discussion), one must be wary of other abnormal effects, even in the absence of toxicity.

Another potential disadvantage of identifying cells by retrograde labeling from targets is that the uptake, retrograde transport, or visualization of the tracer may change over the cell death period so researchers may conclude that significant cell death occurs when, in fact, based on other more reliable criteria, little if any cell loss can be observed (for discussion, see Hardman and Brown, 1985; Oppenheim, 1986). Additionally, because some neurons may project an axon into a peripheral nerve while undergoing migration, caution must be exercised in concluding that all retrogradely labeled cells are confined to adult boundaries such as the ventral horn (Levi-Montalcini, 1950; Chu-Wang *et al.*, 1981; Moody and Heaton, 1981).

The problem of toxicity must also be considered when [^{3}H]thymidine is used (as a birthdate-specific label) in heavy doses to assess neuronal loss during development. However, it is nontoxic and nonteratogenic at doses more than an order of magnitude above those required for adequate (light-microscopic) labeling. For example, 160 μCi [^{3}H]thymidine given systemically to 5-day chick embryos produces no detectable brain abnormality, even with long survival (P. G. H. Clarke, unpublished data), whereas a mere 10 μCi yields adequate labeling in the brain (Clarke, 1982). The problem of [^{3}H]thymidine toxicity in fetal mice is addressed by Mareš *et al.* (1974). When 5'-bromodeoxyuridine (BrdU) is used instead of [^{3}H]thymidine for birthdate-specific labeling, an analogous danger of toxicity or abnormal DNA expression exists, in principle. Although one report has been made of BrdU toxicity to neurons in the chick embryo (Bannigan, 1987), this result may be accounted for by the relatively large dose of *tritiated* BrdU that was used, since other researchers have failed to observe BrdU toxicity (Miller and Nowakowsi, 1988; Oppenheim *et al.*, 1992, 1993; J. Caldero and R. Oppenheim, unpublished data). BrdU has also been reported to inhibit the death of mesenchymal cells in the limb of the chick embryo (Toné *et al.*, 1983).

Distinguishing Neurons from Glia

Not only must the spatial limits of a neuronal population be defined and supplemented with other criteria, but the neurons to be counted must also be distinguished from glia (see Chapter 12, this volume). Inclusion of the latter might confound the neuronal counts, either because of glial proliferation or because of the massive cell death that can occur in glial populations (Barres *et al.*, 1992). Neurons are frequently recognized purely on the basis of their morphology in Nissl-stained sections, notably from their large size, large unstained nucleus, and abundant Nissl substance. Even in favorable situations, where the neurons are large and uniform, this approach may not be reliable. For example, although spinal ganglion cells are easily distinguished from satellite cells, there is currently debate over whether late-generated or late-differentiated cells in the spinal ganglia of rats differentiate long after birth to become neurons (Devor and Govrin-Lippmann, 1991; La Forte *et al.*, 1991). Evidence also suggests the late differentiation of motoneurons (Farel *et al.*, 1993) and sensory neurons (St. Wecker and Farel, 1994). Although these particular examples are not directly relevant to the issue of neuronal death, they remind us of the logical possibility that late differentiation into neurons of cells not initially recognizable as such could mask the simultaneous death of other neurons. This problem is often difficult to solve by immunocytochemistry because even the best neuronal and glial markers are rarely expressed strongly in 100% of a given population, even in adults (e.g., Kitamura *et al.*, 1987); during development this problem is exacerbated because ''neuron-specific'' and ''glia-specific'' antigens are often not expressed by immature cells of these classes. In at least one case, however, a neuron-specific marker has been shown to be a reliable means for selectively identifying all spinal motoneurons in the chick embryo spinal cord, even at the earliest stages of cell death (Ericson *et al.*, 1992; Yamada *et al.*, 1993; H. Yaginuma, S. Homma, and R. Oppenheim, unpublished data). In some cases, neurons can also be distinguished more reliably using the earlier birthdates of neurons as the criterion (Clarke, 1982; Oh *et al.*, 1991). This can be achieved by [^{3}H]thymidine labeling and autoradiography or by 5'-BrdU labeling and immunohistochemistry, but the stereological problems in counting labeled cells are greater, and less well worked out, with the autoradiographic method than with other forms of labeling. Furthermore, there is likely to be some overlap, however small, in neuronal and glial birthdates within a nucleus or ganglion.

A related problem is that the shrinkage of neurons at a given stage in development, or after an experimental intervention, might cause them to be misidentified as glia or as belonging to a different neuronal class at later stages (e.g., injured motoneurons may atrophy and resemble smaller interneurons), causing an overestimation of neuronal death. Neuron shrinkage has not been documented to date in normal development and has been ruled out in the case of the isthmooptic nucleus of chick embryos, both by long-term retrograde labeling and by [^{3}H]thymidine labeling (Clarke, 1982). However, shrinkage can certainly occur after de-afferentation or de-efferentation (Clarke, 1991), and appears to be one of the factors contributing to controversy in attempts to estimate cell death after such procedures (e.g., Barron et al., 1988, 1989).

2. Accurate Counting

The principles of neuronal counting are the same as those of counting any kind of particle. To provide a full review of such methods is beyond the scope of this chapter, so we give here merely a brief introduction to the most widely used approaches for counting neurons.

The traditional approach is to count all neurons for which a part (or all) of the cell (or of the nucleus, or of the nucleolus) lies within the section. This method is easy and quick, but introduces the problem that, in most cases, sectioned neurons will be counted more than once, leading to a spuriously high count. Of course one can ensure that no neuron is counted more than once merely by counting sufficiently spaced sections (every nth) and multiplying by n; this approach does not, however, solve the problem because the *probability* of a given neuron being counted twice is still high. The traditional and widely adopted solution is to multiple the raw counts by the so-called Abercrombie (1946) correction factor, which was actually first proposed by Linderstrøm-Lang et al. (1934) and again, in a slightly modified version, by Floderus (1944). The latter version is $C = T/(T + \overline{h} - 2R)$, where T is the section thickness, $\overline{h}$ the mean height of the particles to be counted, and R is the height of "lost caps," that is, the minimum distance that the upper or lower extreme of a particle must protrude into a section to be counted. The other versions differed only in that R was considered negligible and was set equal to 0.

The correction factor approach has been heavily criticized in recent years on the grounds that it involves several restrictive assumptions that render it invalid in most real situations, in which the assumptions are untrue (e.g., Williams and Rakic, 1988; Coggeshall, 1992).

In our opinion, the correction factor approach can be vindicated, especially when the heights of the particles to be counted are less than the section thickness (Clarke, 1992, 1993). Papers attributing enormous errors to Abercrombie-corrected counts contradict each other with respect to the direction of the error, and themselves suffer from significant faults. Coggeshall et al. (1990) found that Abercrombie-corrected counts of neuronal *nucleoli* exceeded the true number of *neurons* by 16–40%, but part of this excess can be explained simply by the observed fact that some neurons contained several nucleoli. In contrast, Pakkenberg et al. (1991) found their Abercrombie-corrected counts to be *too low* (by 16% for Parkinson-diseased brains and by 40% for controls), but their paper contains mistakes of calculation and other defects that undermine credibility. In their Table 2, the raw counts ("$Q_A \times 40$") for the 80- and 84-year-old Parkinsonian females are only half the values required to yield the "corrected" counts; also, the same correction factor is used for all brains, whereas it should have been calculated independently for each brain. According to the critics of the correction factor approach, the two most restrictive assumptions are that the particles to be counted are spherical and of uniform size. However, the assumption of spherical shape is only needed when the particle heights are derived from tangential measurements. In sufficiently thick sections, the heights can be measured directly by focusing up and down in the z plane of the microscope; the section thickness should then be measured in the same way. This approach requires the use of an oil immersion objective of high numerical aperture and magnification (see Clarke, 1993, for further details). The error due to the uniform size assumption can be calculated for theoretical situations; ultimately, in sections whose thickness exceeds the greatest nuclear height by a factor of more than about 1.5, the error will always be negligible (Clarke, 1992). For cases in which this latter criterion is not quite met, a more complicated correction factor is available that is theoretically exact (regardless of section thickness) but whose practical use still requires that the nuclear heights be less than the section thickness so they can be measured. Using the preceding terminology, this correction factor is

$$C = \Sigma_{i=1}^{m} T/(T - h_i)/\Sigma_{i=1}^{m} (T + h_i - 2R)/(T - h_i),$$

where h_i is the height of particle i and m is the number of (whole) particles sampled (randomly) for measuring (Clarke, 1993). In all correction factors, the sampling of neurons for measurement of nuclear (or nucleolar) height must be unbiased and uniform; this can be achieved by the method of Gundersen (1977).

When the heights of the particles exceed the section thickness, the assumption of sphericity becomes necessary since the particle heights will now have to be deduced from measurements in the x and y planes. In such cases, one should take the nominal value of section thickness rather than measuring it optically, because sections may shrink in thickness as they dry on the slide, thereby flattening particles that had previously been spherical. Deduction of the heights from the x and y measurements can be a complicated matter requiring the uniform size assumption in most cases as well (e.g., Wimer, 1977).

To avoid these problems, it is obviously desirable to use relatively thick sections, although it is then necessary to focus up and down carefully in order not to miss cells piled on top of one another. Also, one should generally choose as "particle" the smallest possible easily visualized and unique part of the neuron: single nulceoli (if there is one per neuron) in preference to nuclei, nuclei in preference to cells. If single nucleoli are counted in relatively thick sections, it is often not necessary to use a correction factor at all. When individual neurons contain several nucleoli, one can consider a "particle" to be the "nucleolar ensemble," that is, the abstract body formed by joining the centers of all the nucleoli in a nucleus (e.g., Clarke, 1985).

The logical extension of this method is to try to count objects of *zero* height. This is a new approach to stereology, as discussed subsequently.

The "Disector" and Its Descendants

In a seminal paper (Sterio, 1984), the "disector" method was proposed for counting the *tops* (or *bottoms*) of particles by comparing two different parallel planes of known separation (less than the particle height), designated the *reference plane* and the *look-up plane*. Particles present in one and absent in the other are counted. These clearly have a top lying between the two planes, yielding the number of particle tops and, hence, particles in the known volume between the two planes. In other words, a density is obtained. This density can be measured at uniformly spaced points throughout the region, and the mean density can be multipled by the volume of the tissue to give the number of particles contained therein. The great theoretical elegance of this method resides in its conceptual simplicity and in its lack of assumptions. It requires no measurements; there is no need to take account of errors due to "lost caps" because these will be equal, on average, in the two planes and will tend to cancel out. Moreover, this original version of the disector, now called the "physical disector," is suitable in practice for counts of neuronal nuclei in semi-thin sections, the very situation that is most difficult to treat using the correction factor approach. However, the practical implementation of the physical disector is relatively difficult. A way must be found to compare the two planes; this involves the use of two microscopes and should ideally include a means of superimposing the two images. Even then, in regions of high cell density, it may be difficult to decide whether profiles in the two planes correspond to the same particle. Another source of error in the disector method is that it is highly sensitive to imprecision in the measurement of the distance between the two planes, but this problem is avoided in a related method called the *fractionator*. These methods are thoroughly reviewed by Gundersen (1986) and Gundersen *et al.* (1988). More general reviews on stereology applied specifically to the nervous system are provided by Royet (1991) and Mayhew (1992).

In relatively thick sections, an alternative version of the disector is available that is easier to use: the *optical disector*. As its name suggests, this method involves defining the reference and look-up planes by optical sectioning (Gundersen, 1986; Gundersen *et al.*, 1988). This approach has all the theoretical advantages of the physical disector and is easier to use. One simply changes the plane of focus by a defined amount, watching constantly to count how many of the cell nuclei in the field of view disappear. Although this technique does not inherently require special equipment, in practice one needs some kind of video imaging system or some automatic means of displacing the plane of focus by a predefined distance (e.g., Williams and Rakic, 1988; West and Gundersen, 1990), because otherwise one loses track of individual cells while shifting one's eyes to the z control.

A related approach is the "Counting Box" method of Howard *et al.* (1985), which has been adapted for neurobiological use by Williams and Rakic (1988). In this case, particles are counted between two planes whose separation is *greater* than the particle height. This technique again requires special equipment in the version proposed by the authors. However, a simplified modification can be used without special equipment (Clarke, 1993): one simply counts all the neuronal nuclei, sectioned or not, in the whole thickness of the section; then one counts the number of these that intersect the upper (or lower) surface. The latter count is subtracted from the former to yield the number of nuclear bottoms (or tops). In trials, this theoretically unbiased approach gives the same results (within 1%) as the modified version of the Abercrombie method (Clarke, 1993). The total time needed using the two methods is about the same.

As noted earlier, the reason for using stereological methods or correction factors in estimating neuronal numbers in a population is to avoid or minimize errors from counting the same cell more than once. However, Oppenheim (Oppenheim *et al.*, 1982, 1989) has consistently used another cell counting method for almost 20 years that does not require either stereology or correction factors to generate reliable estimates of cell numbers in the nervous system. As illustrated in Fig. 1, in this method neurons are counted only if they exhibit all of the following criteria: large soma, a clear nucleus with intact nuclear membrane, and at least one large clump of nucleolar material. By following identified neurons in serial sections (8–12 μm thick, paraffin embedded), we have found that individual neurons exhibit all these criteria less than 2% of the time in adjacent sections (Fig. 1). As indicated in Table I, this has been shown for a variety of different types of neurons in the developing chick and mouse. Furthermore, comparison of cell numbers in some of these populations (e.g., avian motoneurons and dorsal root ganglion neurons), using this method or the unbiased correction factor described earlier (Clarke, 1993), yields cell numbers that differ by no more than 5%. Although a potential disadvantage of this method is that it may be slower when used by a naive individual, a trained person can complete cell counts in a population of neurons containing 15,000 cells by counting every 10th or 20th section in less than 30 min.[1]

3. Counts of Axons

Because of the difficulty of counting neuronal cell bodies, it may be useful in some cases to check whether the results match the number of axons. Clearly the latter cannot always be relied on for estimating neuron death, because axons may be lost without death of the parent cell, a single neuron may give several axonal branches into a single nerve or fiber tract, or new axons may grow. Nevertheless, a close correlation between counts of neurons and of their axons is grounds for confidence in the results. For example, Chu-Wang and Oppenheim

[1] It is perhaps worth noting that, because this method of counting neurons was taught to Oppenheim by two of the pioneers in neuronal death research (Viktor Hamburger and Rita Levi-Montalcini), the cell counts published in their early classic papers are also likely to be valid and reliable estimates of neuronal numbers and of the extent of cell loss in the populations they examined.

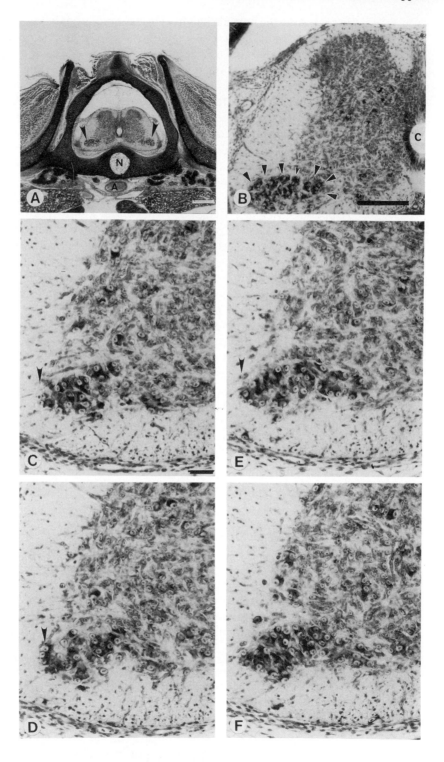

Table I
Different Neuronal Populations Examined in Serial Sections[a]

Cell type	Number of cells	Species
Spinal motoneurons	125	Chick
Cranial (trochlear) motoneurons	75	Chick
Sensory (DRG) neurons	60	Chick
Spinal motoneurons	100	Mouse
Sensory (DRG) neurons	50	Mouse
Isthmo-optic neurons	65	Chick
Sympathetic ganglion	40	Chick
Ciliary ganglion	60	Chick

[a] In all these cell types, less than 2% of the cells examined met all of the criteria for inclusion in cell counts (see text) in two adjacent sections (paraffin embedded, 8 to 12-μm sections).

(1978b) found that the number of motoneurons in a given spinal cord segment was approximately equal to the number of axons in the corresponding ventral root at all times during the period of neuron death, confirming the validity of the cell counts. Similar results from comparisons of axon and neuron counts have been reported by others (Prestige and Wilson, 1972; Schmalbruch, 1987). Axon and neuron numbers have also been shown to coincide after various experimental manipulations of normal cell death (Oppenheim and Chu-Wang, 1983).

C. Identifying and Counting Dying Neurons in Histological Sections

As noted in Sections I, A and B, the *apparent* loss of neurons during development by cell death may actually reflect events that do not involve degeneration (e.g., selective migration of neurons away from a region, phenotypic changes, inaccurate counting methods, misidentification of boundaries). In other words, a decrease in healthy cells in a population during development does not necessarily indicate that cell death has occurred. For this reason, many investigators have also attempted to relate the appearance and rate of occurrence of degener-

Fig. 1 Illustrations of motoneurons in the developing chick embryo lumbar spinal cord on embryonic day 9. (A) Low power view of the entire spinal cord. The distinctly stained motor nucleus is indicated by arrowheads. N, Notochord; A, dorsal aorta. Bar: 150 μm. (B) A higher magnification of the motor nucleus. The boundary is indicated by arrowheads. c, Central canal. Bar: 100 μm. (C–F) Serial sections illustrating that motoneurons that meet the criteria for inclusion in the cell counts (see text) do not exhibit all these criteria in adjacent sections. For instance, the neuron indicated by the arrowhead in D is absent in C and E. Bar: 30 μm. All the sections shown in Figs. 1–3 and 5 are from paraffin-embedded blocks sectioned at 8–12 μm and stained with the Nissl stain thionin.

ating neurons with decreases in numbers of healthy cells. Only in a few favorable cases is it possible to account quantitatively for the total loss of healthy neurons by the estimated number of dying ones (e.g., Oppenheim *et al.,* 1978; Wong and Hughes, 1987), because this requires knowledge of the time taken to clear away the dying or dead neurons. Even when this is not possible, however, the mere presence of significant numbers of degenerating neurons during the same period when healthy neurons are declining in number provides strong evidence that at least some of the cell loss is due to active degeneration. Accordingly, it is quite clear that the identification and quantitative analysis of degenerating neurons in a specific region or population during development provides a valuable and often critical means for verifying that neuronal death is occurring. The major issues to consider in such an approach are: (1) the identification of degenerating cells; (2) confirmation that these are, in fact, dying *neurons* (rather than glia or other nonneuronal cells); and (3) a means of accurately estimating (counting) the number of dying neurons in a population.

Although our focus here is on the examination of degenerating cells in histological sections stained with dyes selective for Nissl substance, in some favorable cases it is possible to identify dying cells on the surface of tissues and organs by vital staining of whole mounts (e.g., Hamburger and Levi-Montalcini, 1949; Saunders, 1962; Toné *et al.,* 1983; Jeffs *et al.,* 1992; Abrams *et al.,* 1993). However, for nervous tissue, this technique is only useful at very early stages of development and, even then, stained cells primarily reflect debris-filled phagocytic cells such as macrophages.

In addition to using basic aniline dyes for staining the Nissl substance and vital dyes, an increasing number of investigators are using the nuclear DNA stain propidium iodide to stain dead and degenerating cells (e.g., Barres *et al.,* 1992). It is virtually impossible, with a nuclear stain such as this, to distinguish whether the degenerating cells are neurons, but the problem can be solved in some cases by prior retrograde labeling of the neurons *in vivo* with retrogradely transported nuclear stains, as has been demonstrated with Diamidino Yellow (Harvey *et al.,* 1990). With these techniques, dying cells can be identified in frozen sections with fluorescence optics by the presence of a condensed and fragmented nucleus. Despite their usefulness in screening or sampling a structure of cell population for cell death that is accompanied by nuclear breakdown, fluorescent nuclear stains are not well suited to detailed serial analysis of cell death in an entire structure or cell population because of the need for frozen sections.

As discussed in more detail in the following section, the most reliable means for determining that cellular profiles observed with the light microscope (LM) are, in fact, degenerating neurons is to re-embed sections and examine them with the electron microscope (EM). Alternatively, adajcent LM sections can be labeled with immunological markers for neurons; sometimes researchers can show that a dying cell expresses neuronal markers (e.g., Barres *et al.,* 1992; Soriano *et al.,* 1993). However, a potential problem with this approach is that by the time a neuron is recognizable in the LM as degenerating, it may have lost the appropriate antigenic markers. Another useful approach is to label neurons retrogradely from their targets and demonstrate that degenerating cells contain the label (e.g., Oppenheim and Chu-Wang, 1977). In this case, however, one must show that the label itself is not toxic.

Because none of these methods was used in many past studies in which degenerating cells were identified as dying neurons, one can legitimately raise questions about the reliability of these techniques. Although we do not wish to underestimate the potential problem of interpretating neuronal death in such studies, in general we feel that in most cases investigators have, in fact, accurately identified degenerating neurons. Many of these previous studies provide excellent light micrographs of putative degenerating profiles that have been shown in other neuronal populations to be dying neurons when examined with EM, or in which the same neurons have been identified as pyknotic by other investigators using the EM (e.g., Hamburger and Levi-Montalcini, 1949; Hamburger, 1958; Hughes, 1961; Pannese, 1974; Hamburger *et al.*, 1981; Oppenheim *et al.*, 1982,1986; Dunlop and Beazley, 1987; Horsburgh and Sefton, 1987; Harman, 1991). Moreover, in several cases, investigators have specifically controlled against the inclusion of mitotic cells or red blood cells (two prominent cell types that may superficially resemble dying cells) in their analysis of neuron degeneration (e.g., Hughes, 1961; Oppenheim, 1986; Kirn and DeVoogd, 1989; Harman, 1991; McKay and Oppenheim, 1991; Johnson and Bottjer, 1993). Some authors have also included phagocytic cells containing degenerating neuronal debris in their counts of dying cells (e.g., Oppenheim *et al.*, 1978; Hamburger *et al.*, 1981). The problem here is that it is not possible to determine whether such profiles contain the debris of one of several dead cells. Accordingly, whenever possible, a deliberate attempt should be made to exclude such profiles from counts of dead and dying cells. In Table II, we list the various criteria that have been used to identify and characterize dying neurons with the LM; examples of some of these features are given in Fig. 2.

The possibility that many earlier studies of degenerating neurons during development may have inadvertently included red blood cells in their estimates of the number of dying neurons was raised in a paper by Coggeshall *et al.* (1993). These authors have examined cell death in the developing rat dorsal root ganglion (DRG) and claim that a significant number of apparent degenerating neuronal profiles are actually nuclei of immature red blood cells. From this result, they conclude that previous studies of cell death in the developing rat DRG are very likely incorrect and must be reexamined; by inference they cast doubt on studies relating to other species (they cite papers on the avian DRG, e.g., Hamburger and Levi-Montalcini, 1949; Carr and Simpson, 1978; Hamburger *et al.*, 1981) and different neuronal populations.

For the reasons delineated here, we feel that this conclusion is not justified. (1) These authors fail to use electron microscopy (or histochemical or immunocytochemical methods) to support their claim that the profiles they see are, in fact, nuclei of intact erythrocytes. This step is important not only for accurate identification of red blood cells, but also because red blood cells themselves

Table II
Some Common Descriptions of Dying and Degenerating Neurons from Observations Using the Light Microscope

Description	Species/cell type	Source
Morphological characteristics vary considerably; in most instances they are spherical in shape and consist of a large, deeply stained spherical central part which is surrounded by a thin hyaline surface layer; in other instances, one observes small deeply stained particles in groups of three or more without a distinct cellular boundary	Chicken/DRG	Hamburger and Levi-Montalcini (1949)
First phase resembles a normal prophase; later, the nucleus becomes pyknotic with a large clump of basophilic material and the cytoplasm shrinks and the nuclear membrane contracts; finally, the nucleus and cytoplasm fragment and are phagocytized	Frog/motoneurons	Hughes (1961)
There is a breakdown of nuclear elements into one or several large darkly stained elements in a clear cytoplasm; cells later fragment into small clusters	Chicken/motoneurons	Oppenheim et al. (1978)
The nucleus and cytoplasm are uniformly hyperchromatic; the nucleoplasm may contain small areas of condensed chromatin	Rat/superior colliculus	Giordano et al. (1980)
Pyknosis is by no means uniform; rather is assumes a variety of forms characterized by very deep stain of the basophilic nuclear material	Chicken/motoneurons	Hamburger (1975)
Degenerating cells appear in a variety of forms, ranging from multiple basophilic small spheres in an otherwise intact nucleus and cytoplasm to unrecognizable fragments; frequently, the cell mass is condensed in a deeply stained homogeneous sphere; in other instances, the cell becomes vacuolated or ballooned	Chicken/DRG	Hamburger et al. (1981)
Darkly stained nuclei with intensely stained chromatin distributed in large clumps	Rat/Photoreceptors	Spira et al. (1984)
Darkly stained and homogeneously stained, shrunken and sometimes fragmented nuclei and eosinophilic and dense or pale cytoplasm	Rat/cortex	Ferrer et al. (1990)
Condensed, darkly stained nucleus often haloed by clear cytoplasm and sometimes fragmented nuclei	Wallaby/thalamus	Harman (1991)
Pyknotic nuclei are intensely basophilic and vitrified in appearance with one or several large or small spheres surrounded by a halo of unstained cytoplasm	Rat/retina	Horsburgh and Sefton (1987)
Shrunken, darkly stained liquefied appearance	Finch/forebrain	Kirn and DeVoogd (1989)
Loss of internal detail, reduction of nuclear volume and nuclear fragmentation; nuclei also stain darkly and evenly and have a liquid-like appearance; pale or absent cytoplasm	Rat/retina	Sengelaub and Finlay (1982)

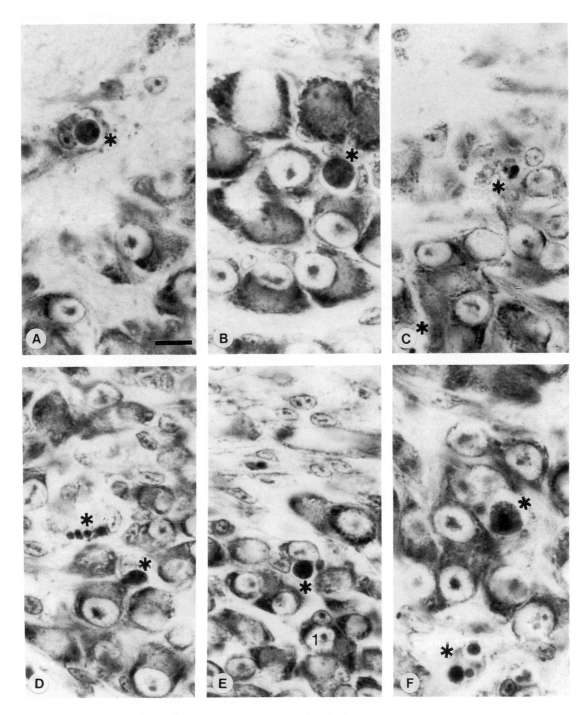

Fig. 2 Illustrations of different profiles of degenerating neurons (asterisks) in the dorsal root ganglia (A,B,D,E) and motor nucleus (C,F) in the developing chick embryo lumbar spinal cord on embryonic day 8.5. As discussed in the text, profiles such as those shown in A do not typically exhibit the appropriate criteria for inclusion in cell counts in adjacent sections (see Fig. 5). Bar: 8μm (for A–F).

may undergo programmed cell death (Sabin, 1921) and because other blood-borne cells actively phagocytize degenerating debris and may appear as degenerating profiles within blood vessels (Sabin, 1921; Kirn and DeVoogd, 1989; McKay and Oppenheim, 1991). (2) According to Bauer *et al.* 1993), red blood cells in rodent embryos become enucleated at a stage prior to the time when Coggeshall *et al.* (1993) claim to see immature red blood cells with nuclei in the rat DRG. Furthermore, apparent pyknotic neurons can be observed at ages when mammalian red blood cells are no longer nucleated (Oppenheim, 1986; Sengelaub and Arnold, 1989). (3) Previous EM studies of cell death in the DRG of avian and rodent embryos have consistently failed to report that any of the degenerating profiles are red blood cells (e.g., Pannese, 1974,1976; Carr and Simpson, 1978; Scaravilli and Duchen, 1980). (4) Although red blood cells in avian species retain their nuclei (Sabin, 1921; Romanoff, 1960; Edmonds, 1966), the morphology of these cells is quite distinguishable from that of dying neurons (see preceding discussion and Fig. 3). (5) Labeling and counting red blood cells in the avian embryo, based on a stain selective for hemoglobin, shows that their numbers in the DRG are not correlated with the number of degenerating cells (Fig. 4). Furthermore, even assuming that previous investigators mistook *all* red blood cells for dying neurons—which is highly unlikely—this would still only account for 1/4 of the total number of dying cells derived from counts of Nissl-stained ganglia. (6) In contrast, counts of dying neurons correlate well with the rate of neuronal loss. Logically, the curve of dying neuron number against time should have the same time course as the rate of change of total neuron number. We have tested this possibility using the counts of healthy and degenerating chick embryo motoneurons published by Hamburger (1975) and Oppenheim and Chu-Wang (1983); we find that the prediction is strikingly fulfilled (P. G. H. Clarke, unpublished data). Likewise, preventing neuron death in the avian DRG by treatment with nerve growth factor (NGF) reduces the numbers of dying neurons to almost zero (Hamburger *et al.,* 1981; Hamburger and Yip, 1984). It seems unlikely that NGF would reduce the number of red blood cells. (7) Finally, if a significant number of the profiles identified in the avian DRG as dying neurons are red blood cells, it is difficult to understand why these would exhibit such precise spatio-temporal developmental regula-

Fig. 3 Illustrations of mitotic cells, degenerating cells, and red blood cells in the lumbar spinal cord of the chick embryo on embryonic day 8.5. (A) A mitotic cell in telophase in the ventricular zone. Bar: 10 μm (for A–E). Two degenerating neurons in the dorsal root ganglion (asterisks). The arrow indicates two daughter cells of a recent mitotic division. (C) Two degenerating neurons (asterisks) in the dorsal root ganglion. (D,E) Red blood cells (arrows) in blood vessels (dashed lines) in the motor nucleus (D) and dorsal root ganglion (E). Note the granulated appearance of the nucleus and the clear cytoplasm of red blood cells, and their distinctly different morphology (and size) compared with degenerating cells in B and C. (F) Red blood cells in a blood vessel (bv). Bar: 15 μm (F).

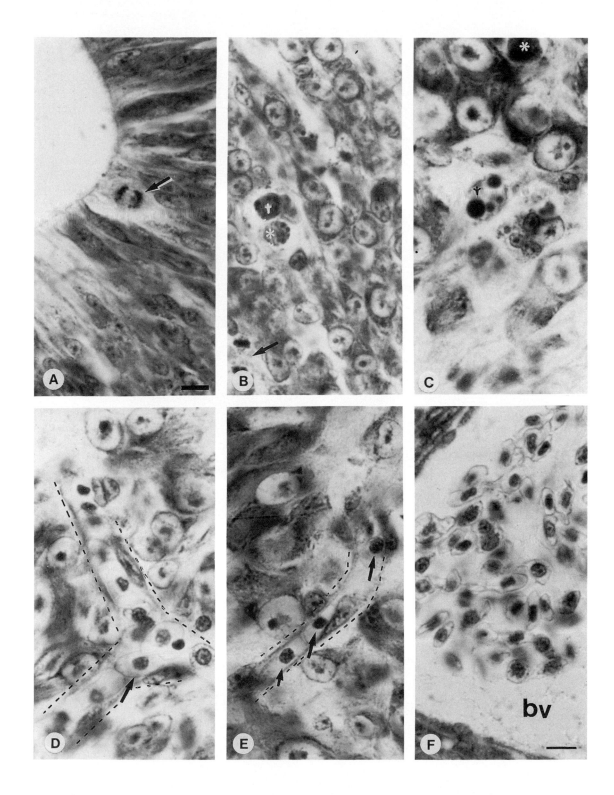

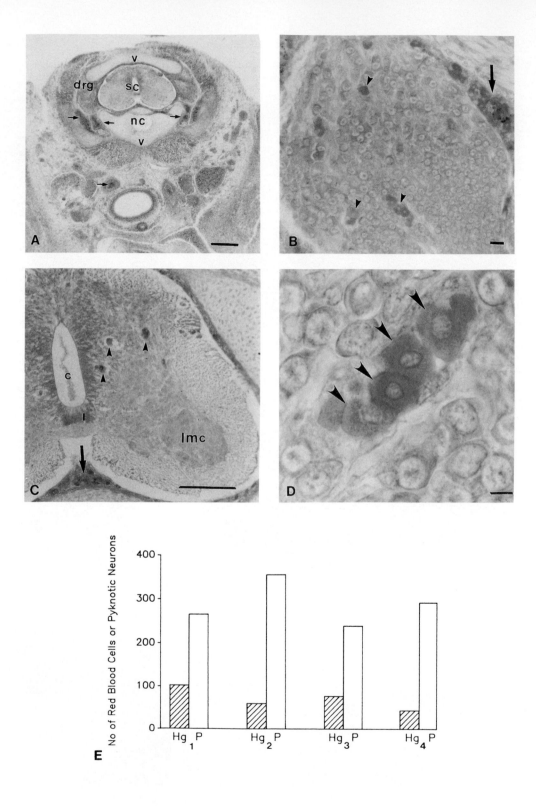

tions (e.g., Hamburger and Levi-Montalcini, 1949; Hamburger *et al.*, 1981; Hamburger and Yip, 1984). For example, significantly fewer such profiles are present in limb than in nonlimb ganglia, yet ganglia at both levels appear to be vascularized to the same extent (Feeney and Watterson, 1946). As Hamburger argued, it seems considerably more likely that this result reflects differences in target regulation of neuronal survival in limb and nonlimb ganglia (Hamburger and Yip, 1984).

Despite the weight of evidence against the claims of Coggeshall *et al.* (1993), we do not wish to give the impression that one should entirely dismiss their arguments. For example, we cannot exclude the possibility that previous investigators have included a few red blood cells in their estimates of apparent dying neurons in the rat DRG. However, we would argue that even if this is shown to be the case, it is unlikely that one can generalize this to avian DRG, and certainly not to other neuronal populations in birds of mammals, based solely on the evidence presented by Coggeshall *et al.* (1993). The available evidence strongly supports the conclusion that, in the vast majority of cases, the cells that have been reported as dying neurons are, indeed, dying neurons.

The final issue regarding the use of data on degenerating neurons as an assay for neuronal cell death concerns the methodology for accurately counting dying cells. In most studies, the number of dying cells is counted in thick (8–12 μm) serial sections of paraffin-embedded material without the use of either stereological methods or correction factors to eliminate the probability of counting the same profile twice. Despite this failure, it seems likely (depending on the morphological criteria used to identify a dying cell) that in most cases this method is valid and reliable. For example, in our own studies of cell death in a variety of neuronal populations in the chick embryo (e.g., motoneurons and sensory, sympathetic, and ciliary ganglion neurons), using the specific criteria described in Table II and illustrated in Fig. 2, we have found that only

---

Fig. 4 (A–D) Transverse sections (8–12 μm) of brachial spinal cord and DRG from an E8 chick embryo. These paraffin-embedded sections have been processed with a stain selective for hemoglobin content in red blood cells (Puchtler *et al.*, 1964). (A) A low power photomicrograph of the entire trunk in the brachial region. Arrows indicate blood vessels containing darkly staining red blood cells. Bar: 150 μm. (B,D) Low (B) and higher (D) power photomicrographs of red blood cells (arrowheads) in brachial ganglion 15. Compare the red blood cells in D with the pyknotic neurons and red blood cells shown in Fig. 3, in which a Nissl stain was employed. Arrows in B indicate blood vessels outside the DRG. Bar: 20 μm (B); 10 μm (D). (C) Brachial spinal cord. Arrows indicate blood vessels around the periphery of the cord. Arrowheads indicate a few red blood cells in the gray matter. Bar: 25 μm. Abbreviations: c, central canal; sc, spinal cord; drg, dorsal root ganglion; nc, notochord; v, vertebra; lmc, lateral motor column. (E) Comparisons of the number of red blood cells stained for hemoglobin (Hg) with the number of pyknotic neurons (P) in Nissl-stained sections (see Figs. 1–3) in brachial ganglion 15 of an E8 embryo. Cells were counted in every 5th section (paraffin-embedded sections, 8–12 μm). The numbers 1–4 indicate data from four separate ganglia. The numbers of pyknotic DRG cells are comparable to the control data reported by Hamburger *et al.* (1981).

1–2% of such degenerating profiles are present in adjacent serial sections (Fig. 5). Because most other studies in the literature have used rather similar criteria for identifying and counting dying neurons, we conclude that published data on dying cell numbers are probably reasonably accurate estimates of their true numbers.

Another potentially more serious but related problem is the issue of the clearance or turnover time for dying neurons, that is, the time required for a healthy appearing neuron to degenerate and disappear. Although most estimates of clearance times for dying developing neurons indicate that the turnover time is rapid (e.g., <5 hr), a potential problem here is in making comparisons of the numbers of dying neurons between control and experimental conditions. For example, a reduction in the number of dying neurons following treatment with a neurotrophic factor could indicate either inhibition of cell death or a faster rate of clearance of the dying cells. However, in practice, as long as one also assesses the number of surviving neurons after treatment and shows that they are increased, it seems reasonable to conclude that the reduced numbers of dying cells do, in fact, reflect suppression of cell death. Accordingly, although we do not wish to underestimate the potential problems of interpretation in this situation, we would argue that by assessing both dying and surviving neurons one can effectively eliminate this concern.

A final problem related to counting of degenerating cells is the manner of expressing the data one obtains. The possible alternatives are: the number of dying cells per section; the number for the whole population (e.g., per ganglia or per lumbar ventral horn); or the number per unit of healthy cells (e.g., dying cells per 1000 healthy cells in a population). Although it probably makes little difference which one of these measures is used in normal control animals, problems could arise from using the first two when comparing control and experimentally manipulated animals in which the manipulation is expected to alter neuronal survival. For example, if an experimental perturbation increases cell numbers in a population from 10,000 (control) to 20,000 (experimental), expressing only the total number of dying cells in the entire population (e.g., if there are a total of 1,000 dying cells in both groups) might suggest that a reduction in cell death did not contribute to this difference, when, in fact, if the data are expressed as dying cells per 1,000 healthy cells, one would conclude just the opposite, since this method now indicates a 50% reduction in dying cells (i.e., 100 vs 50 per 1,000 healthy cells).

D. Dying and Healthy Neurons Examined by Electron Microscopy

Ultrastructural investigation is relevant for several reasons. First, since neuronal survival is dependent on the making and receiving of connections, it is important to study the process of synaptogenesis, which requires electron microscopy. Second, electron microscopy may clarify the neuronal identity of dying and healthy cells, removing the ambiguity of counts performed using the

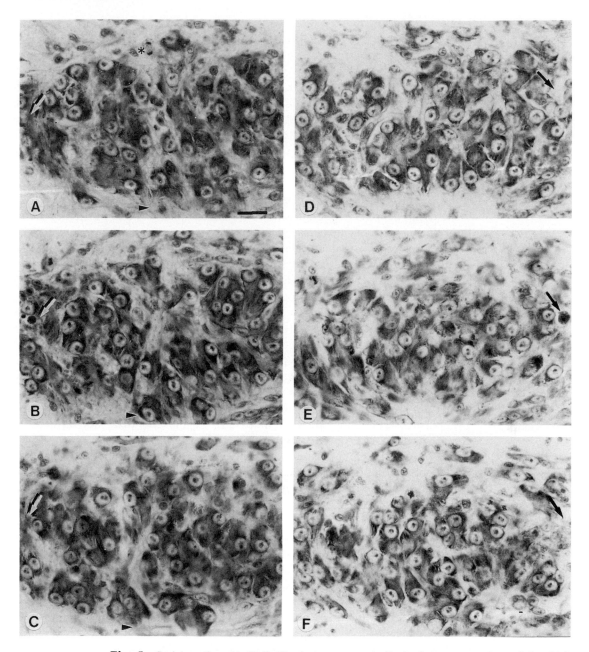

Fig. 5 Serial sections (A–C, D–F) of motoneurons in the lumbar motor nucleus of the chick embryo on embryonic day 9. The arrows in B and E indicate degenerating motoneurons which are absent from adjacent serial sections rostral (A,D) and caudal (C,F) to B and E. The asterisk in A indicates a mitotic cell in telophase. The arrowhead in B indicates a healthy motoneuron that is not found in serial sections rostral (A) or caudal (C) to B. Bar: 16 μm.

LM. Third, ultrastructural studies can help determine the state of differentiation of neurons when the cell death process begins. Fourth, and most important, electron microscopy provides esential information concerning the actual process of cell destruction.

In embryogenesis, the well-known apoptosis–necrosis distinction (Kerr *et al.*, 1972) seems too narrow; at least three morphologically distinct kinds of cell death occur (Schweichel and Merker, 1973; Clarke, 1990). This classification applies to neurons as well as to other cells. The first type of neuron death, described as the *nuclear type* by Pilar and Landmesser (1976) or *type 1* by Chu-Wang and Oppenheim (1978a), does resemble apoptosis because both exhibit clumping of nuclear chromatin, blebbing of the cell membrane, and loss of ribosomes from rough endoplasmic reticulum and polysomes. However, in these dying neurons, the clumped chromatin is in a few balls in the middle of the nucleus whereas in classical apoptosis it is at the nuclear margin. The second type of neuron death, described as the *autophagic type* (Clarke, 1990), apparently involves the destruction of most of the cell in its own secondary lysosomes. The cytoplasm is destroyed by enhanced autophagy, the plasma membrane is channeled into the autophagic vacuoles following enhanced endocytosis, and the nucleus is reduced in size by a largely mysterious process that involves the transfer of DNA (or its breakdown products) into the largest of these vacuoles (Clarke and Hornung, 1989; Hornung *et al.*, 1989; González-Martín *et al.*, 1992). The third type of Schweichel and Merker (1973) and Clarke (1990), described in the context of neurobiology as the *cytoplasmic type* by Pilar and Landmesser (1976) or *type 2* by Chu-Wang and Oppenheim (1978a), involves the dilation of the nuclear envelope, the Golgi apparatus, and the endoplasmic reticulum, and sometimes also of mitochondria, but little or no change in the nucleus until very late in the degenerative process.

The methods for identifying these types of neuronal death involved electron microscopy, combined in some cases with other techniques, which required resolving a number of compatibility problems (see Chapter 1, this volume). For example, Clarke and Hornung (1989) employed thymidine autoradiography, for which fixation with glutaraldehyde is usually avoided because it tends to trap free [^{3}H]thymidine or small molecules containing it, whereas one wants to detect only [^{3}H]thymidine incorporated to DNA. However, glutaraldehyde is required for good ultrastructural preservation. This problem was solved by perfusing cells first with buffered paraformaldehyde for 5 min to wash out the small labeled molecules, and then adding glutaraldehyde for a further 15 min of perfusion (Clarke and Hornung, 1989). In other cases, electron microscopy was combined with enzyme histochemistry for various endogenous phosphatases and for exogenous horseradish peroxidase (Decker, 1978; Hornung *et al.*, 1989) and with immunocytochemistry. (González-Martín *et al.*, 1992); full details are given in the methods sections of these papers.

As noted earlier, the use of electron microscopy in cell death research is also valuable for determining the state of differentiation of neurons at the time that cell death occurs (also see Section II,B,1). For example, in avian motoneurons and ciliary neurons researchers have shown that cells initiating cell death do so at a relatively advanced stage of differentiation after synaptic contacts with their targets and afferent synapses begin to form (Pilar and Landmesser, 1976; Oppenheim and Chu-Wang, 1977; Chu-Wang and Oppenheim, 1978a,b; Furber *et al.,* 1987; Okada *et al.,* 1989). In contrast, in other populations, immature neurons, and precursor cells may undergo cell death at much earlier stages (Levi-Montalcini, 1950; O'Connor and Wyttenbach, 1974; Pannese, 1976; Carr and Simpson, 1982; Martin-Partido and Navascués, 1990; von Bartheld and Bothwell, 1993; Homma *et al.,* 1994).

Finally, as discussed in detail in Section I,C, electron microscopy is also often required to demonstrate that healthy cells assumed to be neurons are in fact neurons and not glia, and that profiles identified as dying or degenerating neurons at the light microscopic level are not glia, mitotic cells, or red blood cells. Although not all studies of cell death in the nervous system have used electron microscopy to determine that dying, degenerating, pyknotic, or apoptotic cells seen with the LM are dying neurons, many have; in virtually all such cases, the dying cells have been confirmed to be neurons (e.g., O'Connor and Wyttenbach, 1974; Pannese, 1976; Pilar and Landmesser, 1976; Oppenheim and Chu-Wang, 1977,1978a; Scaravilli and Duchen, 1980; Furber *et al.,* 1987; Homma *et al.,* 1994). In none of these studies were degenerating cells that were first identified as neurons in the LM found to be glia, mitotic cells, red blood cells, or other nonneuronal phenotypes when later examined with the EM. Therefore, it seems reasonable to conclude that the identification and assessment of neuronal death by the identification of dying cells in the LM is a valid and reliable method as long as the precautions discussed here and in the preceding sections are considered.

E. Estimating Neuron Death in Subpopulations by Selective Labeling

When different subpopulations of neurons are intermingled in the same region, the problems of recognizing and quantifying neuronal death in a given subpopulation are considerable. When exogenous labels are used, they must reveal the *entire* subpopulation, which is not necessary when using the labels merely to supplement other criteria for recognizing a spatial boundary. Moreover, when long-term labeling is required (as it usually is), it is difficult to avoid the problem that the label may change the pattern of cell death. To illustrate the problems involved, we shall limit discussion to two related examples concerning the elimination of transient ipsilateral projections between the eye and the brain in chick embryos. These examples have been chosen deliberately because the ipsilaterally projecting neurons were reported to have died in one case but not in the other. We describe the results of this research only to the extent necessary to clarify the methodological difficulties.

Although the retinofugal and retinopetal projections in hatched chicks and mature chickens are entirely crossed, in chick embryos a weak ipsilateral component occurs transiently in both projections (Clarke and Cowan, 1975; McLoon and Lund, 1982). The occurrence of such transient pathways was demonstrated by, respectively, retrograde or anterograde tracing following the intraocular injection of horseradish peroxidase. This technique posed a few problems of interpretation. The possibility had to be considered that peroxidase might diffuse into the optic chiasm in the immature optic nerve, where it might conceivably label axons of passage or, in the retinopetal case, late-arriving growth cones. However, this possibility could be ruled out directly by the fact that no tracer was detectable in the chiasm. Further confirmation was later obtained in the retinopetal case, because retinal activity blockade caused the ipsilateral axons to persist beyond the end of their normal period of occurrence (Péquignot and Clarke, 1992b).

Nonetheless, it was, and is, much harder to show that this axon elimination involves the death of the parent neurons. This interpretation was initially suggested by both groups because the ipsilateral components were lost during the periods when their neuronal populations of origin were undergoing phases of cell death, but direct evidence was needed.

In the case of the isthmo-optic projection, this evidence was provided by O'Leary and Cowan (1982) by means of long-term retrograde labeling with True Blue. They injected True Blue intraocularly just before the start of the neuronal death period of the isthmo-optic nucleus, but fixed the embryos 1 wk later when the transient projection has disappeared. Could True Blue have reduced the chances of survival of the ipsilaterally projecting neurons, leading to a spurious conclusion? Apparently not, because any toxic effects of the dye on the retina or on the retrogradely labeled neurons would have been expected to affect cells labeled in *both* isthmo-optic nuclei, but the contralateral nucleus seemed normal. This argument carries weight but is not absolutely rigorous, because neurons with aberrant axons might be more vulnerable than others to dye toxicity. (See preceding text for further discussion of dye toxicity.)

In contrast to these results on the isthmo-optic system, Williams and McLoon (1991) reported that the loss of the chick's ipsilateral retinotectal projection is not due solely to neuronal death. They showed this by two kinds of long-term retrograde labeling; the two were complementary because they were subject to different problems. The first approach was to make large injections of Fast Blue into the optic tectum at E9, and to compare the numbers of retinal ganglion cells labeled ipsilaterally at E11 (before the period when the ipsilateral projection is eliminated) or at E18 (after the elimination). The loss was about 60%, only slightly greater than for contralaterally projecting ganglion cells. However, this was not decisive evidence for the survival of neurons contributing to the transient ipsilateral injection, because Fast Blue probably diffused into nontectal visual nuclei known to receive a permanent ipsilateral projection from the retina, and neurons labeled at E18 may have belonged to the latter group. To solve this problem, the authors relied mainly on a second series of experiments involving very localized injection of the carbocyanine dye diI (1,1'-dioctodecyl-

3,3,3',3'-tetramethylindocarbocyanine) or of fluorescent microspheres into the tectum, using the same temporal paradigm used with Fast Blue. Significant numbers of ipsilaterally projecting ganglion cells were found at E18. This modification seemed to have solved the diffusion problem, but another problem remained that was essentially the converse of the toxicity problem discussed earlier. Neuron death in development is active, but can be prevented or delayed by the inhibition of macromolecular synthesis (Martin *et al.*, 1988; Oppenheim *et al.*, 1990) or by axotomy (Farel, 1989), and requires electrical activity in the target region (reviewed by Oppenheim, 1991; see also Péquignot and Clarke, 1992a,b). The chemical properties of the tracers, or mechanical damage associated with their injection, may have affected the labeled neurons or their targets in such a way that it permitted survival of ipsilaterally projecting neurons that are normally fated to die. Similar problems arise in other systems as well.

F. Biochemical Estimation of Neuron Death

Various methods have been used to count neurons and, hence, deduce neuron death without the tedium of histology.

a. Counts in a Hemacytometer

Zamenhof (1976) removed the cerebella of newly hatched chicks, fixed them by immersion in a formaldehyde solution, dissociated them into a suspension, and stained their Nissl material. The large cell types of the cerebellum (Purkinje, Golgi, and deep nuclear) were then counted in a hemacytometer. This method of counting cells avoids the tedium of histology, but is limited to brain regions that can be dissected. More seriously, large numbers of cells might be ruptured by the dissociation procedure, which involves mechanical disintegration followed by sonication. This concern is supported by the fact that Zamenhof's total of large cells (in normal chicks) was 1.75×10^5 on average, whereas counts in histological sections of only the Purkinje cells in the cerebellum of a 2-wk-old chick came to 2.62×10^5 (Armstrong and Clarke, 1979).

b. Flow Cytometry

A more modern approach to counting dissociated cells is available in the sophisticated and highly automated technique of flow cytometry (see Chapters 4–6 and 9 this volume; for review, see Darzynkiewicz and Crissman, 1990), which enabled one to count cells with a common labelling characteristic rapidly. For example, from a population of cells stained for DNA with a fluorescent dye, separate counts can be obtained of diploid cells and tetraploid cells. However, the previously mentioned problem of cell rupture during the dissociation procedure remains. This technique has not, to our knowledge, been used for counting neurons *in vivo*, but it was used by Morris and Taylor (1985) to estimate the proportion of nonproliferating cells in the neural retinae of chick embryos, and it is useful for counting neurons *in vitro*.

c. Biochemical Measurement of Total DNA

A standard method of measuring the total number of cells in a given piece of tissue is to determine the total mass of DNA biochemically and divide it by the known mass of DNA per diploid cell. This method does not lend itself easily to the study of cell death, since the loss of some cells may be masked by the generation of others. Moreover, it fails to distinguish neuronal from nonneuronal cell types.

d. Measurement of Labeled DNA

Several authors have attempted to improve on this approach by studying the loss of radioactivity from a tissue in which a group of cells was selectively labeled with radioactive exogenous thymidine. Thymidine is known to be incorporated specifically and permanently into self-replicating DNA (Mareš et al., 1974; but for a discussion of possible sources of error, see Maurer, 1981). The basic approach is to deliver [3H]thymidine to several animals or embryos at a single precise age when the molecule is believed to label mainly neuronal precursor cells, to sacrifice the animals at subsequent ages, and to measure the radioactivity incorporated into DNA. This approach has revealed little of any cell death in the rat external granular layer (EGL) during natural development, but substantial EGL cell death in X-irradiated (Griffin et al., 1978) or thyroid-deficient rat pups (Patel and Rabié, 1980). The method has also revealed natural cell death in the cerebral hemispheres of developing chick embryos (Hyndman and Zamenhof, 1978), and abnormally high levels of cell death in the optic tectum of chick embryos following early destruction of the contralateral eye (Bondy et al., 1978).

The main disadvantages of this approach are threefold. First, it provides only minimal information on the location of the cell death. Second, it is not specific to a given class of cells. Even when the time of [3H]thymidine injection is optimally chosen to label mainly the precursors of neurons (or a specific class of neurons), other cell types are also labeled. In autoradiographic experiments this is not a major problem, because additional criteria are available. However, when the sole measure is the readout of a scintillation counter, even weak labeling of many nonneuronal cells (or neurons of classes not being studied) can provide serious contamination. Likewise, the very weak labeling that occurs through unscheduled DNA synthesis in all nonproliferating cells, including neurons (Cameron et al., 1979; Korr and Schultze, 1989), is a further source of contamination. Third, the radioactivity in the DNA of cells that degenerate can be reutilized and re-incorporated into DNA in substantial amounts, as has long been known to occur in many tissues (e.g., Maruyama, 1964) including brain (Schelper and Adrian, 1980; Korr et al., 1984). In some cases it may be possible to circumvent the problem of reutilization by administering very large doses of nonradioactive thymidine (e.g., daily intraperitoneal injections of 2.4 mg thymidine in mice; Maruyama, 1964) during the suspected cell death period. However, once the blood–brain barrier has developed, this method may be difficult to use in the central nervous system because of the limited penetration of thymidine. In adults, [³H]thymidine labeling in the brain is about 10-fold weaker than in other tissues (Mareš *et al.,* 1974). Although reutilization of thymidine can be blocked with reserpine (Patel *et al.,* 1979), side effects of the drug may complicate data interpretation. A less problematic alternative is to label with a thymidine analog that has a lower rate of reutilization (e.g., Schelper and Adrian, 1980). In the final analysis, however, with currently available methods we cannot recommend any of these techniques.

II. Methods Involved in the *in Vivo* Testing of Hypotheses Concerning the Signals that Regulate Neuron Death

A. Testing Hypotheses

Neuron death must be understood at three different levels. At the level of the whole network of neurons, we need to understand the role of neuron death in producing an optimally adaptive nervous system. At the level of cell-to-cell communication, we need to know what signals determine whether or not a neuron will survive. At the level of individual cells, we need to understand the chain of events by which neurons are destroyed. Although essential information can be obtained from descriptive studies of normal development, hypotheses must ultimately be tested experimentally at each of these levels.

However, we will discuss *in vivo* methods related only to the second level—cell-to-cell signaling. Our understanding at the network level is not yet advanced enough to be tested *in vivo;* the mechanisms of neuronal destruction are mostly analyzed *in vitro* (see Chapters 10, 11 and 12).

B. Methods Relating to the Regulation of Cell Death in Peripheral Projections

Neuronal populations projecting to peripheral targets (i.e., motoneurons and autonomic and sensory ganglion cells) have been the most extensively studied cell groups with respect to the cellular and molecular mechanisms that mediate survival. The reasons for this are twofold. First, historically, naturally occurring cell death was first recognized as being a normal part of neuronal development in these populations (for reviews, see Oppenheim, 1981; Hamburger, 1992). Second, the properties of these cell groups make them more amenable to experimental approaches than most central nervous system groups. Accordingly, most of our understanding about the mechanisms regulating cell death has come from studies of motoneurons, sensory neurons, and autonomic neurons. In the following discussion, we shall focus mainly on the methods used to examine four fundamental questions concerning mechanisms of cell death in one of these populations: spinal cord motoneurons in avian embryos. The rationale for concentrating on this single population in only one species is that, in many respects, more is known about these neurons than about the others, and much (although by no means all) of what has been learned about the survival of avian spinal motoneurons can be generalized to the other populations.

1. Differentiation, Target Projections, and Afferent Inputs prior to the Onset of Cell Death

Although programmed cell death in the nervous system can involve proliferating precursor cells, early postmitotic neurons, migrating neurons, and differentiating neurons (Glücksmann, 1951; Silver, 1978; Carr and Simpson, 1982; Homma et al., 1994), the massive death of avian spinal motoneurons occurs at a relatively advanced stage of differentiation (with one exception; see Levi-Montalcini, 1950; Yaginuma et al., 1994) following the projection of axons to targets and the initiation of afferent synaptic inputs. The evidence in support of this claim comes from several studies using a variety of different methods, as described here.

Although researchers had long suspected that all avian spinal motoneurons, even the half that will subsequently die, project axons to their peripheral targets prior to the onset of cell death (e.g., Hamburger, 1958), the advent of new and better methods was required to demonstrate this unequivocally. First, investigators showed by axon counts in the ventral root using the EM that the number of axons prior to as well as during and after the cell death period matched the number of motoneurons present in the ventral horn (Chu-Wang and Oppenheim, 1978a). Although this result was sufficent to demonstrate that motoneurons fated to die project axons out of the spinal cord, it did not exclude the possibility that the axons of cells that later die were unable actually to reach their targets in the limb. To demonstrate this, it was necessary to use retrograde labeling of motoneurons from the limb prior to the onset of cell death. By injecting massive amounts of the retrograde tracer horseradish peroxidase into the limb bud on E5 (cell death begins on E6) it was possible to show that more than 90% of the motoneurons were labeled, demonstrating that virtually all motoneurons, even those that would later die, had innervated the limb. Subsequent studies have shown that actual synapse formation begins at the onset of the cell death period and continues for several days (Dahm and Landmesser, 1991).

Both electrophysiological (e.g., Provine, 1972; Bekoff, 1976) and ultrastructural (Oppenheim et al., 1974) methods have been used to demonstrate that afferent synapses begin to form on spinal motoneurons just prior to the onset of cell death on E5–6. Although these studies were unable to determine whether synaptic input was present on those specific motoneurons that will later degenerate, in a subsequent study using EM, researchers showed that synapses were, in fact, present on motoneurons undergoing active degeneration (Okada et al., 1989).

Despite having shown that motoneurons fated to die send axons to their targets and receive afferent input prior to their degeneration, these studies failed to address the question of whether the population of motoneurons that later die differs in other aspects of early differentiation. To address this question, two different approaches were used. First, motoneurons in the ventral horn of normal embryos were examined in E5–6 with the EM to determine whether the cells could be separated into two groups that exhibited different stages of differentiation that might reflect cells fated to die and cells that would survive (Chu-Wang and Oppenheim, 1978a,b). Although no differences could be detected, these observations suffered from the fact that, because motoneurons fated to die are randomly distributed in the ventral horn, it was not possible to identify specific cells with known fates. Thus, to address this issue more adequately, a second experimental approach was employed. This technique involved the unilateral removal of the limb bud anlage at a stage (E2) prior to cell death and before axonogenesis. Studies going back to the turn of the century (e.g., Shorey, 1909; Hamburger, 1934, 1958) had shown that, in this situation, virtually all motoneurons deprived of their targets undergo cell death. Accordingly, with this model it was possible to compare the state of differentiation of a population of neurons prior to degeneration in which all the cells were fated to die. Once again, comparisons of differentiation in the target-deprived motoneurons and in contralateral (normal, control) motoneurons in the same embryo, prior to the onset of cell death, failed to reveal any differences between the two populations (Oppenheim *et al.*, 1978). In addition to examining cytological differentiation with the LM and EM, this study also employed biochemical assays of motoneuron-specific enzymes; these also failed to reveal any differences.

Although collectively these studies provide strong evidence for the argument that, prior to the actual onset of degeneration, motoneurons fated to die or to survive undergo comparable levels of morphological differentiation, we cannot exclude the possibility that other more subtle aspects of differentiation (biochemical, molecular) may differ in these groups. This remains an important issue because, if such early differences do exist, cell death may not be probabilistic or stochastic, but may be genetically predetermined in a subset of motoneurons, despite its apparently random distribution in the motor column. At the moment, however, the available evidence does not support this idea.

2. Regulation of Survival by Targets and Afferents

The idea that the synaptic targets of spinal motoneurons are a major source of epigenetic signals that regulate survival and differentiation grew out of studies done earlier in this century on the general problem of neuron–target (or central–peripheral) interactions (Oppenheim, 1981; Jacobson, 1991). By the 1950s the available evidence strongly supported the notion that targets regulate the survival rather than the proliferation, migration, or early differentiation of motoneurons (e.g., Hamburger and Levi-Montalcini, 1949). More recent experimental studies are entirely consistent with this idea (e.g., Hamburger, 1958, 1975; Hollyday and Hamburger, 1976; Oppenheim *et al.*, 1978; O'Brien and Oppenheim, 1990). Perhaps the strongest support for this possibility comes from observations that the removal or transplantation of supernumerary limb bud anlagen does not alter the pre-cell-death numbers of motoneurons that assemble in the ventral horn. Rather, these manipulations increase (limb removal) or reduce (supernumerary limb) the extent of motoneuron death. This result suggests that, in avian spinal motoneurons, the targets control final cell numbers only to the extent that they are able to regulate the number of cells that survive.

Because of the historical interest in issues related to neuron–target (or central–peripheral) interactions, and the considerable evidence indicating that targets can modulate survival, researchers long believed that the targets of motoneurons were the major, if not the sole, source of survival-promoting signals. However, more recent evidence suggests that the existence of other sources of trophic support, including nonneuronal cells and afferent inputs, is very likely (Oppenheim, 1991; Korsching, 1993). The first evidence for the role of afferents in motoneuron survival came from experiments in which intraspinal, supraspinal, and sensory afferent inputs were deleted early in development (prior to the onset of cell death) and were found to increase the normal death of motoneurons, even in the presence of normal motoneuron–target interactions (Okado and Oppenheim, 1984). Because the numbers of motoneurons were normal in this situation prior to cell death, and because there were increased numbers of dying cells during the period when healthy cell numbers were declining, researchers concluded that afferents normally influence motoneuron survival. Although it is not possible to exclude entirely the possibility that deafferentation may also perturb the phenotype of motoneurons so they would no longer be included as motoneurons in the cell counts, we consider this highly unlikely since motoneuron properties are established before the onset of cell death (see preceding discussion).

In summary, considerable evidence shows that both efferent contacts and afferent inputs begin to develop prior to, and continuously during, the period of motoneuron cell death, and that perturbations of these interactions alter motoneuron survival in a predictable manner, consistent with their role as putative sources of survival-promoting signals. The precise mechanisms by which motoneuron interactions with targets and afferents regulate survival have been the focus of many recent studies, but a discussion of these is beyond the scope of this chapter.

C. Methods Relating to the Regulation of Cell Death in Central Projections

Although the greater complexity and inaccessibility of the central nervous system tends to make neuronal death there more difficult to study, the appropriate choice of system can largely offset the difficulties. The central nervous system even has certain advantages, notably the availability of the eye as an *in vivo* closed chamber into which one can inject pharmacological agents for the retina, and the accessibility of axon-sparing neurotoxins for central but not peripheral use.

1. Axotomy: Change in Size of Input Field or Target

Experiments showing the importance of afferents and efferents for a central neuron's survival date back to the 19th century, when neuronal shrinkage or death was found to be provoked by cutting the input or output axons (reviewed by Clarke, 1991). Somewhat later, the rise of experimental embryology provided techniques that made it possible to vary the size of the input field to a group of central neurons, although the potential relevance of such manipulations to naturally occurring neuron death was not realized at the time. The early experiments were always performed on amphibian embryos, and included the excision of an eye primordium (Larsell, 1929), the grafting of an additional nasal placode (Burr, 1924) or optic vesicle (May and Detwiler, 1925), and the heteroplastic transplantation of the primordium of an abnormally large eye (Twitty, 1932).

Later, similar experiments were performed in chick embryos, including destruction of the cerebellar primordium to show the dependence of cerebellar afferents on their target (Harkmark, 1956); destruction of the optic primordium to show the dependence of primary visual nuclei on their optic afferents (Filogamo, 1950), of the isthmo-optic nucleus on its axonal target, the retina (Cowan and Wenger, 1968), or of the trochlear nucleus on its axonal target, the superior oblique muscle (Cowan and Wenger, 1967); and grafting of additional eye primordium to show that this reduces neuron loss in the isthmo-optic and trochlear nuclei (Boydston and Sohal, 1979).

In view of the wide variety of surgical techniques used in these experiments and the classical nature of the procedures, we shall not attempt to describe them here. Details are available in the previously cited papers and in a monograph devoted to these techniques (Hamburger, 1960).

In mammals, the difficulties of performing such experiments have been overcome in a number of ways. Some researchers have chosen to study species that are born so precociously that much of their neuron death is postnatal, for example, the hamster (Finlay *et al.*, 1986), the ferret (Thompson *et al.*, 1993), or various marsupials (Crewther *et al.*, 1988; Coleman and Beazley, 1989). In such cases, the surgery (axotomy or lesion) is performed postnatally. Other investigators have developed techniques for operating prenatally; for example, Rakic and Riley (1983) removed one eye from rhesus monkey fetuses after 63–65 days of gestation, and then returned them to the womb to survive 100 more days until full term, when they were delivered by caesarean section and allowed to survive for several more months. Another approach has been to use genetic mutants in which the source of input or the axonal target of a group of neurons is affected selectively (Caddy and Biscoe, 1979). A refinement of this genetic approach is to make aggregation chimeras between the mutant and wild-type strains, to vary the number of neurons in the affected population over a wide range (Herrup and Sunter, 1987). Finally, when two groups of neurons with distinctly different periods of neurogenesis are connected unidirectionally, their mutual dependence can be analyzed using an antimitotic drug. Thus, Chen and Hillman (1989) injected methylazoxymethanol acetate into pregnant rats, thereby reducing either the number of cerebellar granule cells or the number of Purkinje cells selectively in the offspring, depending on the time of injection. When the granule cells were depleted by this means there was no secondary loss of Purkinje cells, but primary depletion of Purkinje cells caused a secondary loss of granule cells, suggesting that neuron number in the cerebellar cortex is regulated by output but not by input.

The dependence of neurons on their afferent and efferent connections is greatest during a critical period in development. In the case of afferents, this dependence can be analyzed simply by varying the moment of the lesion, but in the case of efferents, once these projections have reached the target a means of destroying the target neurons without directly affecting the innervating axons is required. One approach to solving this problem is to inject the supposedly axon-sparing neurotoxin kainate into the target (Carpenter et al., 1986; Catsicas and Clarke, 1987). Both studies indicated that the target dependence of the innervating neurons was maximal during their period of naturally occurring neuron death and ended abruptly soon afterward. Although there is no evidence for direct toxicity of kainate on axon terminals, it is now well known that metabotropic kainate receptors occur frequently on presynaptic axons. There-fore, it may be preferable in future experiments to use N-methyl-D-aspartate, which appears to act only rarely presynaptically.

2. Modifying Levels of Trophic Factor

a. Systemic Delivery

As in the periphery, one means of evaluating the role of trophic factors in the central nervous system is to deliver trophic molecules, or antibodies against them, systemically; the additional problem exists, however, that they may not penetrate the blood–brain barrier. According to most authors, the barrier to intravascularly injected tracers increases during development (e.g., Wakai and Hirokawa, 1978; Clarke, 1984), but there is controversy over whether its early "openness" is physiological or is caused by an increase in plasma concentration or blood volume as a result of tracer injection (Saunders, 1992). Therefore, when systemic delivery is used, the crossing of trophic molecules and antibodies into neural parenchyme must be monitored; when moderately large quantities have been injected to raise the plasma concentration or blood volume by more than about 5%, controls may be needed for possible side effects of disrupting the blood–brain barrier (e.g., increased penetration of hormones into the brain).

A further problem of systemic delivery is that it is hard to tell whether any effects observed are due to a direct action of the exogenous molecules on the neurons being studied or whether the effects are mediated indirectly via a chain of interactions involving other cells (neurons, glia, endocrine cells). One also cannot infer whether the effects (direct or indirect) on a neuron are mediated via its axon, cell body, or dendrites.

b. Localized Injections

It is desirable to inject trophic molecules (or antibodies against them) directly into a chosen region of neural parenchyme. Techniques are available for chronic administration of chosen substances by the implantation of micropellets of impregnated Silastic® (Heaton, 1977) or by single microinjections (Heaton and Kosier, 1978). However, if such techniques are used to inject trophic factors, it will be necessary but very difficult to monitor (and, ideally, control) the changing spatiotemporal gradients of trophic factor as it diffuses through the tissue.

One region in which such problems are relatively slight is the eye. For example, in chick embryos during the last 12 days of incubation, tracer molecules (radioactive amino acids, peroxidase, etc.) injected in a volume of 2–6 μl into the vitreous body spread through it rapidly to produce almost uniform labeling of the retina within less than 1 hr (P. G. H. Clarke, unpublished data). Subsequent leakage from the eye is generally slow; after an intravitreal injection of peroxidase, the amount remaining the the eye halves every 24 hr (Clarke, 1982). Moreover, in chick embryos it is easy to make intravitreal injections at all stages of development. Hence, by means of the intravitreal injection of trophic factor (or antibodies), concentrations of trophic factors available to the (centrifugal) isthmo-optic fibers can be modulated in a precise and controlled way. However, careful experimental designs are needed to distinguish a direct effect of trophic factor on the isthmo-optic terminals from an indirect one due to the action of trophic factor on the retinal target cells of the isthmo-optic fibers.

Such problems of interpretation may be still greater when analyzing antero-grade trophic effects, because it will be very difficult to modify the anterograde transport of a particular trophic molecule in a group of axons without affecting these axons and their parent cells in other ways as well. In view of recent evidence that intravitreally injected neurotrophins are transported *antero-gradely* to the optic tectum in chick embryos (von Bartheld *et al.*, 1993), one could attempt to modify the anterograde transport of a particular trophic factor in retino-tectal fibers by injecting it intraocularly. Indeed, evidence in this system suggests both activity-independent and -dependent components of anterograde support, of unknown molecular basis (Catsicas *et al.*, 1992). However, any effect on cell survival in the tectum would be difficult to interpret because the injected trophic factor might also act on retinal ganglion cells, causing them to modify their production and anterograde transport of other trophic molecules, or it might affect their electrical activity.

Another potential means of modulating trophic support in the retina is the intravitreal injection of antisense oligodeoxynucleotides. This has not been done to date, but the feasibility of the antisense approach *in vivo* is clear from the experiments of Osen-Sand *et al.* (1993), who used it to reduce the expression of SNAP-25 in the embryonic chick retina. For analysis of retrograde trophic support, this sophisticated approach offers no clear advantage over the intravitreal injection of antibodies against trophic factors, and suffers from the disadvantage that more than 1 day may be needed for the injected antisense nucleotides to affect molecular levels in the retina significantly. However, to reduce trophic factor levels selectively in the anterograde pathway, the antisense approach appears to be the only method available, since antibodies would not be expected to penetrate into cells and would therefore not affect the production and anterograde transport of trophic factors. Again, a problem will be to distinguish a direct effect due to trophic factor reduction in ganglion cells from an indirect one due to perturbed intraretinal trophic interactions.

c. Localized Modulation of Trophic Factor Production
Molecular techniques may ultimately provide a means, not limited to the optic system, for modifying trophic factor production in particular groups of neurons. This has not yet been achieved, but eventually may be possible by using genetically engineered embryos in which the inserted trophic factor cDNA is linked to a promoter that is induced by a molecule present in the nuclei of only a defined population of cells.

3. Monitoring and Modifying Electrical Activity

Electrical activity is known to affect neuron survival in the developing central nervous system (Oppenheim, 1991; Péquignot and Clarke, 1992a,b; Catsicas *et al.,* 1992; Galli-Resta *et al.,* 1993). Activity-mediated regulation of the production of trophic factors may contribute to this phenomenon but does not fully explain it, because neuron survival depends not just on the amount of electrical activity but on its pattern. Current evidence suggest that Hebb-like mechanisms sensitive to the synchronicity of incoming action potentials may determine the survival of synapses and, hence, of neurons (Clarke, 1991).

Therefore, it is important to monitor the pattern of electrical activity during the period when neuron death occurs naturally, and to modify it in a controlled way to observe how this change affects neuron survival. These aims are more difficult to achieve in the early developmental stages involved than in adults.

a. Monitoring Activity
The only current method useful for monitoring activity in relation to neuronal death is that of electrophysiological recording. Alternative methods such as deoxyglucose autoradiography or immunocytochemistry of c-*fos* expression are insufficiently sensitive to reveal the low levels of activity that occur when

neuron death is occurring. Moreover, the latter methods fail to reveal the *timing* of the activity, which is so important in the present context.

In chick embryos, spontaneous or evoked electrical activity has been monitored with macro- or microelectrodes in various parts of the central nervous system, especially in the motor system (Provine, 1972) and in the visual system (retina and optic tectum: Blozovski, 1971; Sedlácek, 1972; Rager, 1976a,b). At the time of naturally occurring neuron death (E10–E16 in the chick visual system), spontaneous activity exists but is low. The chick embryos must be kept in a moist, warm chamber because electrophysiological recording from their brains or eyes involves lifting the head out of the egg. Because of the flexibility of the skull, it is difficult to hold the head rigidly; for embryos younger than E14, Rager (1976a,b) made special plastic head molds to hold the head without causing injury.

In mammals, prenatal electrophysiology has been attempted *in vitro* and *in vivo*. Although the *in vitro* approach provides useful information on the development of functioning synapses, it cannot be relied on to reproduce perfectly the intensity and spatio-temporal pattern of firing of action potentials that occur in normal development. Therefore, it is preferable, where possible, to remove the fetus from the uterus and record from it while it is still attached to the mother (Naka, 1964). One useful technique for immobilizing the fetus while maintaining its temperature and humidity is to place it on the mother in a dish with a slot in the bottom for the umbilical cord, and to cover the fetus with low melting point wax maintained at 38°C, a hole being made in the wax to uncover regions destined for stimulation or recording (Fitzgerald, 1987). This technique was used by Maffei and Galli-Resta (1990) to demonstrate correlated firing between neighboring retinal ganglion cells in late fetal rats. However, to study the overall pattern of action potentials and their synchrony across a wide extent of retina, Meister *et al.* (1991) found it necessary to resort to an *in vitro* approach. These researchers isolated retinas from fetal, neonatal, or adult ferrets or cats and placed a piece of retina, about 3 mm in diameter, in a dish containing aerated Ringer's solution so the ganglion cell layer was facing downward and in close proximity to an array of 61 electrodes incorporated into the dish. By this means, these investigators were able to demonstrate bursts of action potentials that swept across the retina at about 100 μm per sec. They argued that these *in vitro* results reflect the *in vivo* situation.

b. Modifying Electrical Activity

Since the period of naturally occurring neuron death in the central nervous system almost always occurs before the neurons become responsive to sensory stimulation, the only methods available for modifying activity during this period are electrical (or perhaps magnetic) stimulation or pharmacological intervention. Until now, only the latter approach has been used. In most cases, tetrodotoxin (a specific, high-affinity blocker of voltage-dependent sodium channels) has been injected into an eye to block action potentials in the retina. In neonatal rodents this changes the pattern of death among the retinal ganglion cells without affecting the total amount of death (O'Leary *et al.*, 1986), but it does lead to a rapid increase in cell death in the superior colliculus (Galli-Resta *et al.*, 1993). In chick embryos, this manipulation rapidly provokes neuron death in the optic tectum (Catsicas *et al.*, 1992), whereas in the isthmo-optic nucleus neuron death is reduced during the period when it occurs naturally but is provoked subsequently (Péquignot and Clarke, 1992a,b). Unfortunately, tetrodotoxin leaves the eye very quickly. Thus, Dubin *et al.* (1986) estimated, from experiments in kittens, involving different doses of tetrodotoxin, that half the injected dose leaves the eye in 7 hr. In chick embryos, recent experiments by M.-P. Primi (unpublished data) using tritiated saxitoxin (whose properties are similar to those of tetrodotoxin) indicate a still more rapid loss from the eye, although the rate of proportional loss depends on the dose injected. These experiments suggest that the rapid loss is due to the pharmacological action of the toxin, perhaps through its blockade of action potentials in the axons controlling the blood vessels that drain the eye. Because of the rapid loss from the eye, repeated heavy doses of tetrodotoxin (or saxitoxin) are required, leading to the possibility that the high percentage that leaks into the circulation will cause nonspecific effects. Since at low concentrations tetrodotoxin affects only neurons and striated muscle fibers, the most likely nonspecific effects would be caused by its direct action on the brain regions being studied and their connections by indirect interference with the nervous control of the lungs, heart, or glands. Indeed, Péquignot and Clarke (1992a) found a major reduction of neuron death in the isthmo-optic nucleus ipsilateral to the injected eye, despite the fact that the isthmo-optic projection is more than 99% crossed (in chick embryos; 100% in adults).

Attempts have also been made to infuse tetrodotoxin into the brain. For example, Friedman and Shatz (1990) used osmotic minipumps to infuse tetrodotoxin into the brains of cat fetuses for up to 15 days, and found no effect on retinal ganglion cell death. The high systemic levels produced are inevitably a problem in such experiments. In this particular case, the cerebrospinal fluid concentration of tetrodotoxin appears to have been between 0.1 and 1.0 μM, about 100 times greater than the equilibrium dissociation constant of tetrodotoxin binding in nervous tissue. The authors failed to find tetrodotoxin in the vitreous humor by means of a bioassay of unstated sensitivity.

Although not used to date to investigate neuron death, many other pharmacological agents are available for modifying central neural activity. Kobayashi *et al.* (1990) compared the effects of intraocularly injected grayanotoxin I or tetrodotoxin on the refinement of optic terminals in the tectum, finding similar effects in both cases. In contrast to tetrodotoxin, which blocks sodium channels, grayanotoxin I is known to open them, but the consequences of this action in these experiments were not investigated physiologically. This is regrettable, because one must know whether grayanotoxin blocked action potentials and, if so, for how long after the injection to interpret the experiments. Reiter and

Stryker (1988) used osmotic minipumps to infuse muscimol, a blocker of GABA-A receptors, into the visual cortex of kittens in which one eye had been sutured. They found that the cortical neurons became more readily excited, and drew the important conclusion that (among other things) cortical plasticity depends on postsynaptic membrane conductance or polarization, because they believed the ionotropic GABA-A receptors, unlike the metabotropic GABA-B, to occur only presynaptically. Extending this approach of selective postsynaptic blockade to analyze the activity-dependent mechanisms that regulate neuron death will be interesting. In addition to GABA-A receptors, it may be useful to block or activate glycine ionotropic glutamate receptors, which also appear to be mostly postsynaptic. However, it will be difficult in most situations to be certain that a given receptor is purely postsynaptic, because this has been evaluated in relatively few situations. GABA-A receptors are, in fact, known to occur presynaptically in the spinal cord (Eccles *et al.*, 1963; Stuart and Redman, 1992), although apparently not in the cerebral cortex or hippocampus (Waldmeier and Baumann, 1990).

References

Abercrombie, M. (1946). Estimation of nuclear populations from microtome sections. *Anat. Rec.* **94**, 239–247.

Abrams, J. M., White, K., Fessler, L. I., and Steller, H. (1993). Programmed cell death during *Drosophila* embryogenesis. *Development* **117**, 29–43.

Armstrong, R. C., and Clarke, P. G. H. (1979). Neuronal death and the development of the pontine nuclei and inferior olive in the chick. *Neuroscience* **4**, 1635–1647.

Bannigan, J. G. (1987). Autoradiographic analysis of 5-bromodeoxy-uridine on neurogenesis in the chick embryo spinal cord. *Dev. Brain Res.* **36**, 161–170.

Barres, B. A., Hart, I. K., Coles, H. S. R., Burne, J. F., Voyoodie, J. T., Richardson, W. D., and Raff, M. C. (1992). Cell death and control of cell survival in the oligodendrocyte lineage. *Cell* **70**, 31–46.

Barron, K. D., Dentinger, M. P., Popp, A. J., and Mankes, R. (1988). Neurons of layer Vb of rat sensorimotor cortex atrophy but do not die after thoracic cord transection. *J. Neuropathol. Exp. Neurol.* **47**, 62–74.

Barron, K. D., Banerjee, M., Dentinger, M. P., Scheibly, M. E., and Mankes, R. (1989). Cytological and cytochemical (RNA) studies on rubral neurons after unilateral rubrospinal tractotomy: The impact of GM_1 ganglioside administration. *J. Neurosci. Res.* **22**, 331–337.

Bauer, H. C., Bauer, H., Lametschwandtner, A., Amberger, A., Ruiz, P., and Steiner, M. (1993). Neorvascularization and the appearance of morphological characteristics of the blood–brain barrier in the embryonic mouse central nervous system. *Dev. Brain Res.* **75**, 269–278.

Bekoff, A. (1976). Ontogeny of leg motor output in the chick embryo: A neural analysis. *Brain Res.* **106**, 271–291.

Diozovski, D. (1971). Maturation des réponses visuelles évoquées par la stimulation électrique du nerf optique chez l'embryon de Poulet. *J. Physiol Paris* **63**, 473–496.

Bondy, S. C., Harrington, M. E., and Anderson, C. L. (1978). Effects of prevention of afferentation of the development of the chick optic lobe. *Brain Res. Bull.* **3**, 411–413.

Boydston, W. R., and Sohal, G. S. (1979). Grafting of additional periphery reduces embryonic loss of neurons. *Brain Res.* **178**, 403–410.

Burr, H. S. (1924). Hyperplasia in the brain of amblystoma. *Proc. Soc. Exp. Biol. Med.* **21**, 473–474.

Caddy, K. W. T., and Biscoe, T. J. (1979). Structural and quantitative studies on the normal C3H and lurcher mutant mouse. *Phil. Trans. R. Soc. London* **287**, 167–201.

Cameron, I. L., Pool, M. R. H., and Haage, T. R. (1979). Low level incorporation of tritiated thymidine into the nuclear DNA of Purkinje neurons of adult mice. *Cell Tissue Kinet.* **12**, 445–451.

Carpenter, P., Sefton, A. J., Dreher, B., and Lim, W.-L. (1986). Role of target tissue in regulating the development of retinal ganglion cells in the albino rat: Effects of kainate lesions in the superior colliculus. *J. Comp. Neurol.* **251**, 240–259.

Carr, V. M., and Simpson, S. B. (1978). Proliferative and degenerative events in the early development of chick dorsal root ganglia. *J. Comp. Nourol.* **182**, 727–756.

Carr, V. M., and Simpson, S. B. (1982). Rapid appearance of labeled degenerating cells in the dorsal root ganglia after exposure of chick embryos to triated thymidine. *Dev. Brain Res.* **2**, 157–162.

Catsicas, M., Péquignot, Y., and Clarke, P. G. H. (1992). Rapid onset of neuronal death induced by blockade of either axoplasmic transport or action potentials in afferent fibers during brain development. *J. Neurosci.* **12**, 4642–4650.

Catsicas, S., and Clarke, P. G. H. (1987). Abrupt loss of dependence of retinopetal neurons on their target cells, as shown by intraocular injections of kainate in chick embryos. *J. Comp. Neurol.* **262**, 523–534.

Chen, S., and Hillman, D. E. (1989). Regulation of granule cell number by a predetermined number of Purkinje cells in development. *Dev. Brain Res.* **45**, 137–147.

Chu-Wang, I.-W., and Oppenheim, R. W. (1978a). Cell death of motoneurons in the chick embryo spinal cord. I. A light and electron microscopic study of naturally-occurring and induced cell loss during development. *J. Comp. Neurol.* **177**, 33–58.

Chu-Wang, I.-W., and Oppenheim, R. W. (1978b). Cell death of motoneurons in the chick embryo spinal cord. II. A quantitative and qualitative analysis of degeneration in the ventral root, including evidence for axon outgrowth and limb innervation prior to cell death. *J. Comp. Neurol.* **177**, 59–86.

Chu-Wang, I. W., Oppenheim, R. W., and Farel, P. B. (1981). Ultrastructure of migrating spinal motoneurons in *Anuran* larvae. *Brain Res.* **213**, 307–318.

Clarke, P. G. H. (1982). The genuineness of isthmo-optic neuronal death in chick embryos. *Anat. Embryol.* **165**, 389–404.

Clarke, P. G. H. (1984). Identical populations of phagocytes and dying neurons revealed by intravascularly injected horseradish peroxidase, and by endogenous glutaraldehyde-resistant acid phosphatase, in the brains of chick embryos. *Histochem. J.* **16**, 955–969.

Clarke, P. G. H. (1985). Neuronal death during development in the isthmo-optic nucleus of the chick: Sustaining role of afferents from the tectum. *J. Comp. Neurol.* **234**, 365–379.

Clarke, P. G. H. (1990). Developmental cell death: Morphological diversity and multiple mechanisms. *Anat. Embryol.* **181**, 195–213.

Clarke, P. G. H. (1991). The roles of trophic communication in biological neural networks. *Concepts Neurosci.* **2**, 201–219.

Clarke, P. G. H. (1992). How inaccurate is the Abercrombie correction factor for cell counts. *Trends Neurosci.* **15**, 211–212.

Clarke, P. G. H. (1993). An unbiased correction factor for cell counts in histological sections. *J. Neurosci. Meth.* **49**, 133–140.

Clarke, P. G. H., and Cowan, W. M.(1975). Ectopic neurons and aberrant connections during neural development. *Proc. Natl. Acad. Sci. U.S.A.* **72**, 4455–4458.

Clarke, P. G. H., and Hornung, J. P. (1989). Changes in the nuclei of dying neurons as studied with thymidine autoradiography. *J. Comp. Neurol.* **283**, 438–449.

Coggeshall, R. E. (1992). A consideration of neural counting methods. *Trends Neurosci.* **15**, 9–13.

Coggeshall, R. E., la Forte, R., and Klein, C. M. (1990). Calibration of methods for determining numbers of dorsal root ganglion cells. *J. Neurosci. Meth.* **35**, 187–194.

Coggeshall, R. E., Pober, C. M., Kwiat, G. C., and Fitzgerald, M. (1993). Erythrocyte nuclei resemble dying neurons in embryonic dorsal root ganglia. *Neurosci. Lett.* **157**, 41–44.

Coleman, L.-A., and Beazley, L. D. (1989). Retinal ganglion cell number is unchanged in the remaining eye following early unilateral eye removal in the wallaby *Setonix brachyurus,* quokka. *Dev. Brain Res.* **48,** 293–307.

Cowan, W. M., and Wenger, E. (1967). Cell loss in the trochlear nucleus of the chick during normal development and after radical extirpation of the optic vesicle. *J. Exp. Zool.* **164,** 267–280.

Cowan, W. M., and Wenger, E. (1968). The development of the nucleus of origin of centrifugal fibers to the retina in the chick. *J. Comp. Neurol.* **133,** 207–240.

Crewther, D.P., Nelson, J. E., and Crewther, S. G. (1988). Afferent input for target survival in marsupial visual development. *Neurosci. Lett.* **86,** 147–154.

Dahm, L. M., and Landmesser, L. T. (1991). The regulation of synaptogenesis during normal development and following activity blockade. *J. Neurosci.* **11,** 238–255

Darzynkiewicz, Z., and Crissman, H. A. (eds) (1990). "Flow Cytometry," *Meth. Cell Biol.* **33.**

Decker, R. S. (1978). Retrograde responses of developing lateral motor column neurons. *J. Comp. Neurol.* **180,** 635–660.

Devor, M., and Govrin-Lippmann, R. (1991). Neurogenesis in adult rat dorsal root ganglia: On counting and the count. *Somat. Motor Res.* **8,** 9–12.

Divac, I., and Morgensen, J. (1990). Long-term retrograde labeling of neurons. *Brain Res.* **524,** 339–341.

Dubin, M. W., Stark, L. A., and Archer, S. M. (1986). A role for action-potential activity in the development of neuronal connections in the kitten retinogeniculate pathway. *J. Neurosci.* **6,** 1021–1036.

Dunlop, S. A., and Beazley, L. D. (1987). Cell death in the developing retinal ganglia cell layer of the wallaby *Setonix brachyurus. J. Comp. Neurol.* **264,** 14–23.

Eccles, J. C., Schmidt, R., and Willis, W. D. (1963). Pharmacological studies on presynaptic inhibition. *J. Physiol.* (*London*) **168,** 500–530.

Edmonds, R. H. (1966). Electron microscopy of erythropoiesis in the avian yolk sac. *Anat. Rec.* **154,** 785–806.

Ericson, J., Thor, S., Edlund, T., Jessell, T., and Yamada, T. (1992). Early stages of motor neurons differentiation revealed by expression of homeobox gene *islet-1. Science* **256,** 1555–1560.

Farel, P. B. (1989). Naturally occurring cell death and differentiation of developing spinal motorneurons following axotomy. *J. Neurosci.* **9,** 2103–2113.

Farel, P. B., Wray, S. E., and Meeker, M. L. (1993). Size-related increase in motoneuron number: Evidence for late differentiation. *Dev. Brain Res.* **71,** 169–179.

Feeney, J. F., and Walterson, R. L. (1946). The development of the vascular pattern within the walls of the central nervous systems of the chick embryo. *J. Morphol.* **78,** 231–303.

Ferrer, I., Bernet, E., Soriano, E., DelRio, T., and Fonseca, M. (1990). Naturally occurring cell death in the cerebral cortex of the rat and removal of dead cells by transitory phagocytes. *Neuroscience* **39,** 451–458.

Filogamo, G. (1950). Conseguenze della demolizione dell'abbozzo dell'occhio sullo sviluppo del lobo ottico nell'embrione di pollo. *Riv. Biol.* **42,** 73–79.

Finlay, B. L., Sengelaub, D. R., and Berian, C. A. (1986). Control of cell number in the developing visual system. I. Effects of monocular enucleation. *Dev. Brain Res.* **28,** 1–10.

Fitzgerald, M. (1987). Spontaneous and evoked activity of fetal primary afferents *in vivo. Nature* **326,** 603–605.

Floderus, S. (1944). Untersuchungen über den Bau der menschlichen Hypophyse mit besonderer Berücksichtigung der quantitativen mikromorphologischen Verhältnisse. *Acta Pathol. Microbiol. Scand.* (*Suppl.*) **53,** 1–276.

Friedman, S., and Shatz, C. J. (1990). The effects of prenatal intracranial infusion of tetrodotoxin on naturally occurring retinal ganglion cell death and optic nerve ultrastructure. *Eur. J. Neurosci.* **2,** 243–253.

Furber, S., Oppenheim, R. W., and Prevette, D. (1987). Naturally occurring neuron death in the ciliary ganglia of the chick embryo following removal of preganglionic input: Evidence for the role of afferents in ganglion cell survival. *J. Neurosci.* **7,** 1816–1832.

Gahr, M. (1990). Delineation of a brain nucleus: Comparisons of cytochemical, hodological, and cytoarchitectural views of the song control nucleus HVC of the adult canary. *J. Comp. Neurol.* **294,** 30–36.

Galli-Resta, L., Ensini, M., Fusco, E., Gravina, A., and Morgheritti, B. (1993). Afferent spontaneous electrical activity promotes the survival of target cells in the developing retinotectal system of the rat. *J. Neurosci.* **13,** 243–250.

Giordano, D. L., Murray, M., and Cunningham, T. J. (1980). Naturally occurring neuron death in the optic layers of superior colliculus of the postnatal rat. *J. Neurocytol.* **9,** 603–614.

Glücksmann, A. (1951). Cell death in normal vertebrate ontogeny. *Biol. Rev.* **26,** 59–86.

González-Martín, C., de Diego, I., Crespo, D., and Fairén, A. (1992). Transient c-*fos* expression accompanies naturally occurring cell death in the developing interhemispheric cortex of the rat. *Dev. Brain Res.* **68,** 83–95.

Griffin, W. S. T., Woodward, D. J., and Chanda, R. (1978). Quantification of cell death in developing cerebellum by a ^{14}C tracer method. *Brain Res. Bull.* **3,** 369–372.

Gundersen, H. J. G. (1977). Notes on the estimation of the numerical density of arbitrary profiles: The edge effect. *J. Microsc.* **111,** 219–223.

Gundersen, H. J. G. (1986). Stereology of arbitrary particles: A review of unbiased number and size estimators and the presentation of some new ones. *J. Microsc.* **143,** 3–45.

Gundersen, H. J. G., Bagger, P., Bendtsen, T. F., Evans, S. M., Korbo, L., Marcussen, N., Møller, A., Nielsen, K., Nyengaard, J. R., Pakkenberg, B., Sørensen, F. B., Vesterby, A., and West, M. J. (1988). The new stereological tools: Disector, fractionator, nucleator and point sampled intercepts and their use in pathological research and idagnosis. *Acta Pathol. Microbiol. Immunol. Scand.* **96,** 857–881.

Hamburger, V. (1934). The effects of wing bud extirpation on the development of the control nervous systems in chick embryos. *J. Exp. Zool.* **68,** 449–494.

Hamburger, V. (1958). Regression versus peripheral control of differentiation in motor hypoplasia. *Am. J. Anat.* **102,** 365–410.

Hamburger, V. (1960). "A Manual of Experimental Embryology." University of Chicago Press, Chicago.

Hamburger, V. (1975). Cell death in the development of the lateral motor column of the chick embryo. *J. Comp. Neurol.* **160,** 535–546.

Hamburger, V. (1992). History of the discovery of neuronal death in embryos. *J. Neurobiol.* **23,** 1111–1115.

Hamburger, V., and Levi-Montalcini, R. (1949). Proliferation, differentation and degeneration in the spinal ganglia of the chick embryo under normal and experimental conditions. *J. Exp. Zool.* **111,** 457–501.

Hamburger, V., and Yip, J. W. (1984). Reduction of experimentally induced neuronal death in spinal ganglia of the chick embryo by nerve growth factor. *J. Neurosci.* **4,** 1984.

Hamburger, V., Brunso-Bechtold, J., and Yip, J. W. (1981). Neuronal death in the spinal ganglia of the chick embryo and its reduction by nerve growth factor. *J. Neurosci.* **1,** 60–71.

Hardman, V. J., and Brown, M. C. (1985). Absence of postnatal death among motoneurons supplying the inferior gluteal nerve of the rat. *Dev. Brain Res.* **19,** 1–9.

Harkmark, W. (1956). The influence of the cerebellum on development and maintenance of the inferior olive and the pons. An experimental investigation on chick embryos. *J. Exp. Zool.* **131,** 333–360.

Harman, A. M. (1991). Generation and death of cells in the dorsal lateral geniculate nucleus and superior colliculus of the wallaby. *Setonix brochyurus. J. Comp Neurol.* **313,** 469–478.

Harvey, A. R., Robertson, D., and Cole, K. S. (1990). Direct visualization of death of neurones projecting to specific targets in the developing rat brain. *Exp. Brain Res.* **80,** 213–217.

Heaton, M. B. (1977). A technique for introducing localized long-lasting implants in the chick embryo. *J. Embryol. Exp. Morphol.* **39,** 261–266.

Heaton, M. B., and Kosier, M. E. (1978). Micro-injections in the avian embryo: Technical notes. *Brain Res. Bull.* **3,** 715–717.

Herrup, K., and Sunter, K. (1987). Numerical matching during cerebellar development: Quantitative analysis of granule cell death in staggerer mouse chimeras. *J. Neurosci.* **7,** 829–836.

Hollyday, M., and Hamburger, V. (1976). Reduction of naturally occurring motor neuron loss by enlargement of the periphery. *J. Comp. Neurol.* **170,** 311–320.

Homma, S., Yaginuma, H., and Oppenheim, R. W. (1994). Cell death during the earliest stages of spinal cord development in the chick embryo: A possible means of early phenotype selection. *J. Comp. Neurol.* **345,** 377–395.

Hornung, J. P., Koppel, H., and Clarke, P. G. H. (1989). Endocytosis and autophagy in dying neurons: An ultrastructural study in chick embryos. *J. Comp. Neurol.* **283,** 425–437.

Horsburgh, G. M., and Sefton, J. (1987). Cellular degeneration and synaptogenesis in the developing retina of the rat. *J. Comp. Neurol.* **263,** 553–566.

Howard, V., Reid, S., Baddeley, A., and Boyde, A. (1985). Unbiased estimation of particle density in the tandem scanning reflected light microscope. *J. Microsc.* **138,** 203–212.

Hughes, A. (1961). Cell degeneration in the larval ventral horn of *Xenopus laevis*. *J. Embryol. Exp. Morphol.* **9,** 269–284.

Hyndman, A. G., and Zamenhof, S. (1978). Cell proliferation and cell death in the cerebral hemispheres of developing chick embryos. *Dev. Neurosci.* **1,** 216–225.

Jacobson, M. (1991). "Developmental Neurobiology," 3d Ed.) Plenum Press, New York.

Jeffs, P., Jaques, K., and Osmond, M. (1992). Cell death in cranial neural crest development. *Anat. Embryol.* **185,** 583–588.

Johnson, F., and Bottjer, S. W. (1993). Induced cell death in a thalamic nucleus during a restricted period of zebra finch vocal development. *J. Neurosci.* **13,** 2452–2462.

Kerr, J. F. R., Wyillie, A. H., and Currie, A. R. (1972). Apoptosis: A basic biological phenomenon with wide ranging implications in tissue kinetics. *Br. J. Cancer* **26,** 239–257.

Kirn, J. R., and DeVoogd, T. J. (1989). Genesis and death of vocal control neurons during sexual differentiation in the zebra finch. *J. Neurosci.* **9,** 3176–3187.

Kirn, J. R., Alvarez-Buylla, A., and Nottebohm, F. (1991). Production and survival of projection neurons in a forebrain vocal center of adult male canaries. *J. Neurosci.* **11,** 1756–1762.

Kitamura, T., Nakanishi, K., Watanabe, S., Endo, Y., and Fujita, S. (1987). GFA-protein gene expression on the astroglia in cow and rat brains. *Brain Res.* **423,** 189–195.

Kobayashi, T., Nakamura, H., and Yasuda, M. (1990). Disturbance of refinement of retinotectal projection in chick embryos by tetrodotoxin and grayanotoxin. *Dev. Brain Res.* **57,** 29–35.

Korr, H., and Schultze, B. (1989). Unscheduled DNA synthesis in various types of cells of the mouse brain in vivo. *Exp. Brain Res.* **74,** 573–578.

Korr, H., Wittmann, B., and Schultze, B. (1984). Reutilization of ^{3}H DNA metabolites by proliferating glial and endothelial cells in the brain of the 14-day-old rat. *Acta Histochem. Suppl.* **29,** 159–164.

Korsching, S. (1993). The neurotrophic factor concept: A reexamination. *J. Neurosci.* **13,** 2739–2748.

La Forte, R. A., Melville, S., Chung, K., and Coggeshall, R. E. (1991). Absence of neurogenesis of adult rat dorsal root ganglion cells. *Somat. Motor Res.* **8,** 3–7.

Lance-Jones C. (1982). Motoneuron cell death in the developing lumbar spinal cord of the mouse. *Dev Brain Res.* **4,** 473–479.

Larsell, A. (1929). The effect of experimental excision of one eye on the development of the optic lobe and opticus layer in larvae of the tree-frog (*Hyla regilla*). *J. Comp. Neurol.* **48,** 331–353.

Levi-Montalcini, R. (1950). The origin and development of the visceral system in the spinal cord of the chick embryo. *J. Morphol.* **86,** 253–283.

Linderstrøm-Lang, K., Holter, H., and Soeborg Ohlsen, A. (1934). Studies on enzymatic histochemistry. XIII. The distribution of enzymes in the stomach of pigs as a function of its histological structure. *Comptes-Rendus Trav. Lab. Carlsberg* **20,** 66–127.

Maffei, L., and Galli-Resta, L. (1990). Correlation in the discharges of neighboring rat retinal ganglion cells during prenatal life. *Proc. Natl. Acad. Sci. USA* **87,** 2861–2864.

Mareš, V., Schultze, B., and Maurer, W. (1974). Stability of DNA in Purkinje cell nuclei of the mouse. *J. Cell Biol.* **63,** 665–674.

Martin, D. P., Schmidt, R. E., DiStephano, P. S., Lowry, O. H., Carter, J. G., and Johnson, E. M., Jr. (1988). Inhibitors of protein synthesis prevent neuronal death caused by nerve growth factor deprivation. *J. Cell Biol.* **106,** 829–844.

Martin-Partido, G., and Navascués, J. (1990). Macrophage-like cells in the presumptive optic pathways in the floor of the diencephalon of the chick embryo. *J. Neurocytol.* **19,** 820–832.

Maruyama, Y. (1964). Re-utilization of thymidine during death of a cell. *Nature* **201,** 93–94.

Maurer, H. R. (1981). Potential pitfalls of [³H]thymidine techniques to measure cell proliferation. *Cell Tissue Kinet.* **14,** 111–120.

May, R. M., and Detwiler, S. R. (1925). The relation of transplanted eyes to developing nerve centers. *J. Exp. Zool.* **43,** 83–103.

Mayhew, T. M. (1992). A review of recent advances in stereology for quantifying neural structure. *J. Neurocytol.* **21,** 313–328.

McKay, S. E., and Oppenheim, R. W. (1991). Lack of evidence for cell death among avian spinal cord interneurons during normal development and following removal of targets and afferents. *J. Neurobiol.* **22,** 721–733.

McLoon, S. C., and Lund, R. D. (1982). Transient retinofugal pathways in the developing chick. *Exp. Brain Res.* **45,** 277–284.

Meister, M., Wong, R. D. L., Baylor, D. A., and Shatz, C. J. (1991). Synchronous bursts of action potentials in ganglion cells of the developing mammalian retina. *Science* **252,** 939–943.

Miller, M. W., and Nowakowski, R. S. (1988). Use of bromodeoxyuridine-immunocytochemistry to examine the proliferation, migration and time of origin of cells in the central nervous system. *Brain Res.* **457,** 44–52.

Moody, S. A., and Heaton, M. B. (1981). Morphology of migrating trigeminal motor neuroblasts as revealed by horseradish peroxidase retrograde labeling techniques. *Neuroscience* **6,** 1707–1723.

Morris, V. B., and Taylor, I. W. (1985). Estimation of nonproliferating cells in the neural retinae of embryonic chicks by flow cytometry. *Cytometry* **6,** 375–380.

Naka, K.-I. (1964). Electrophysiology of the fetal spinal cord. I. Action potentials of the motoneuron. *J. Gen. Physiol.* **47,** 1003–1022.

O'Brien, M. K., and Oppenheim, R. W. (1990). Development and survival of thoracic motoneurons and hindlimb musculature following transplantation of the thoracic neural tube to the lumbar region in the chick embryo: Anatomical aspects. *J. Neurobiol.* **21,** 313–340.

O'Connor, T. M., and Wyttenbach, C. R. (1974). Cell death in the embryonic chick spinal cord. *J. Cell Biol.* **60,** 448–459.

Oh, L. J., Kim, G., Yu, J., and Robertson, R. T. (1991). Transneuronal degeneration of thalamic neurons following deafferentation: quantitative studies using [³H]thymidine autoradiography. *Brain Res.* **63,** 191–200.

Okada, A., Furber, S., Okado, N., Homma, S., and Oppenheim, R. W. (1989). Cell death of motoneurons in the chick embryo spinal cord. X. Synapse formation on motoneurons following the reduction of cell death by neuromuscular blockade. *J. Neurobiol.* **20,** 219–233.

Okado, N., and Oppenheim, R. W. (1984). Cell death of motoneurons in the chick embryo spinal cord. IX. The loss of motoneurons following removal of afferent inputs. *J. Neurosci.* **4,** 1639–1652.

O'Leary, D. D. M., and Cowan, W. M. (1982). Further studies on the development of the isthmo-optic nucleus with special reference to the occurrence and fate of ectopic and ipsilaterally projecting neurons. *J. Comp. Neurol.* **212,** 399–416.

O'Leary, D. D. M., Fawcett, J. W., and Cowan, W. M. (1986). Topographic targeting errors in the retinocollicular projection and their elimination by selective ganglion cell death. *J. Neurosci.* **6,** 3692–3705.

Oppenheim, R. W. (1981). Neuronal cell death and some related regressive phenomena during neurogenesis: A selective historical review and progress report. In "Studies in Developmental Neurobiology: Essays in Honor of Viktor Hamburger" (W. M. Cowan, ed.), pp. 74–133. Oxford, New York.

Oppenheim, R. W. (1986). The absence of significant postnatal motoneuron death in the brachial and lumbar spinal cord of the rat. *J. Comp. Neurol.* **246,** 281–286.

Oppenheim, R. W. (1991). Cell death during development of the nervous system. *Annu. Rev. Neurosci.* **14,** 453–501.

Oppenheim, R. W., and Chu-Wang, I. W. (1977). Spontaneous cell death of spinal motoneurons following peripheral innervation in the chick embryo. *Brain Res.* **125,** 154–160.

Oppenheim, R. W., and Chu-Wang, I. W. (1983). Aspects of naturally occurring motoneuron death in the chick spinal cord during embryonic development. *In "Somatic and Autonomic Nerve–Muscle Interactions"* (G. Burnstock and G. Vrbova, eds.), Elsevier, Amsterdam, pp. 57–107.

Oppenheim, R. W., Chu-Wang, I. W., and Foelix, R. (1974). Some aspects of synaptogenesis in the spinal cord of the chick embryo: A quantitative electron microscopic study. *J. Comp. Neurol.* **161,** 383–418.

Oppenheim, R. W., Chu-Wang, I. W., and Maderdrut, J. L. (1978). Cell death of motoneurons in the chick embryo spinal cord. *J. Comp. Neurol.* **177,** 87–112.

Oppenheim, R. W., Maderdrut, J. L., and Wells, D. J. (1982). Cell death of motoneurons in the chick embryo spinal cord. VI. Reduction of naturally occurring cell death in the thoracolumbar column of Terni by nerve growth factor. *J. Comp. Neurol.* **210,** 174–189.

Oppenheim, R. W., Houenou, L., Pincon-Raymond, M., Powell, J. A., Rieger, F., and Standish, L. J. (1986). The development of motoneurons in the embryonic spinal cord of the mouse mutant, muscular dysgenesis (*mdg/mdg*): Survival, morphology and biochemical differentiation. *Dev. Biol.* **114,** 426–436.

Oppenheim, R. W., Cole, T., and Prevette, D. (1989). Early regional variations in motoneuron numbers arise by differential proliferation in the chick embryo spinal cord. *Dev. Biol.* **133,** 468–474.

Oppenheim, R. W., Prevette, D., and Fuller, F. (1992). The lack of effect of basic and acidic fibroblast growth factors on the naturally occurring death of neurons in the chick embryo. *J. Neurosci.* **12,** 1726–2734.

Oppenheim, R. W., Prevette, D., Haverkamp, L. J., Houenou, L., Yin, Q. W., and McManaman, J. (1993). Biological studies of a putative avian muscle-derived neurotrophic factor that prevents naturally occurring motoneuron death *in vivo*. *J. Neurobiol.* **24,** 1065–1079.

Oppenheim, R. W., Prevette, D., Haverkamp, L. J., Houenou, L., Yin, Q. W., and McManaman, J. (1993). Biological studies of a putative avian muscle-derived neurotrophic factor that prevents naturally occurring motoneuron death *in vivo*. *J. Neurobiol.* **24,** 1065–1079.

Osen-Sand, A., Catsicas, M., Staple, J. K., Jones, K. A., Ayala, G., Knowles, J., and Catsicas, S. (1993). Inhibition of axonal growth by SNAP-25 antisense oligonucleotides in vitro and in vivo. *Nature* **364,** 445–448.

Pakkenburg, B., Möller, A., Gundersen, H. J. G., Mouritzen Dam, A., and Pakkenburg, H. (1991). The absolute number of nerve cells in substantia nigra in normal subjects and in patients with Parkinson's disease estimated with an unbiased stereological method. *J. Neurol. Neurosurg. Psychiat.* **54,** 30–33.

Pannese, E. (1974). The histogenesis of the spinal ganglia. *Adv. Anat. Embryol. Cell Biol.* **47,** 1–97.

Pannese, E. (1976). An electron microscopic study of cell degeneration in chick embryo spinal ganglia. *Neuropathol. Appl. Neurobiol.* **2,** 247–267.

Patel, A. J., and Rabié, A. (1980). Thyroid deficiency and cell death in the rat cerebellum during development. *Neuropathol. Appl. Neurobiol.* **6,** 45–49.

Patel, A. J., Bailey, P., and Balazs, R. (1979). Effect of reserpine on cell proliferation and energy stores in the developing rat brain. *Neuroscience* **4,** 139–143.

Péquignot, Y., and Clarke, P. G. H. (1992a). Changes in lamination and neuronal survival in the isthmo-optic nucleus following the intraocular injection of tetrodotoxin in chick embryos. *J. Comp. Neurol.* **321,** 336–350.

Péquignot, Y., and Clarke, P. G. H. (1992b). Maintenance of targeting errors by isthmo-optic axons following the intraocular injection of tetrodotoxin in chick embryos. *J. Comp. Neurol.* **321,** 351–356.

Pilar, G., and Landmesser, L. (1976). Ultrastructural differences during embryonic cell death in normal and peripherally deprived ciliary ganglia. *J. Cell Biol.* **68,** 339–356.

Prestige, M. C., and Wilson, M. A. (1972). Loss of axons from ventral roots during development. *Brain Res.* **41,** 467–470.

Provine, R. R. (1972). Ontogeny of bioelectric activity in the spinal cord of the chick embryo and its behavioral implications. *Brain Res.* **41,** 365–378.

Puchtler, H., Rosenthal, S. L., and Sweat, F. (1964). Revision of the amidoblack stain for hemoglobin. *Arch. Pathol.* **78,** 76–78.

Rager, G. (1976a). Morphogenesis and physiogenesis of the retino-tectal connection in the chicken. I. The retinal ganglion cells and their axons. *Proc. R. Soc. London B* **192,** 331–352.

Rager, G. (1976b). Morphogenesis and physiogenesis of the retino-tectal connection in the chicken. II. The retino-tectal synapses. *Proc. R. Soc. London B* **192,** 353–370.

Rakic, P., and Riley, K. P. (1983). Regulation of axon number in primate optic nerve by prenatal binocular competition. *Nature* **305,** 135–137.

Reiter, H. O., and Stryker, M. P. (1988). Neural plasticity without postsynaptic action potentials: Less-active inputs become dominant when kitten visual cortical cells are pharmacologically inhibited. *Proc. Natl. Acad. Sci. USA* **85,** 3623–3627.

Romanoff, A. (1960). "The Avian Embryo." MacMillan, New York.

Royet, J.-P. (1991). Stereology: A method for analyzing images. *Prog. Neurobiol.* **37,** 433–474.

Sabin, F. R. (1921). The vitally stainable granules as a specific criterion for erythroblasts and the differentiation of the three stains of the white blood cells as seen in the living chick's yolk-sac. *Johns Hopkins Hosp. Bull.* **368,** 314–321.

Saunders, J. W. (1962). Death in embryonic systems. *Science* **154,** 604–612.

Saunders, N. R. (1992). Ontogenetic development of brain barrier mechanisms. *In* "Physiology and Pharmacology of the Blood-Brain Barrier" (M. W. B. Bradbury, ed.), Handbook of Experimental Pharmacology, Vol. 103, pp 327–369. Springer, Berlin.

Scaravilli, F., and Duchen, L. W. (1980). Electron microscopic and quantitative studies of cell necrosis in developing sensory ganglia in normal and sprawling mutant mice. *J. Neurocytol.* **9,** 373–380.

Schelper, R. L., and Adrian, E. K. Jr. (1980). Non-specific esterase activity in reactive cells in injured nervous tissue labeled with ^{3}H-thymidine or ^{125}iododeoxyuridine injected before injury. *J. Comp. Neurol.* **194,** 829–844.

Schmalbruch, H. (1987). The number of neurons in dorsal root ganglia L4-L6 of the rat. *Anat. Rec.* **219,** 315–322.

Schmued, L. C., Beltramino, C., and Slikker, W., Jr. (1993). Intracranial injection of Fluoro-Gold results in the degeneration of local but not retrogradely labeled neurons. *Brain Res.* **626,** 71–77.

Schweichel, J. U., and Merker, H. J. (1973). The morphology of various types of cell death in prenatal tissues. *Teratology* **7,** 253–266.

Sedlácek, J. (1972). Development of the optic afferent system in chick embryos. *In* "Advances in Psychobiology" (G. Newton, and A. H. Riesen, eds.), Vol. 1, pp. 129–170. Wiley, New York.

Sengelaub, D., and Arnold, A. (1989). Hormonal control of neuron number in sexually dimorphic spinal nuclei of the rat. *J. Comp. Neurol.* **280,** 622–629.

Sengelaub, D., and Finlay, B. (1982). Cell death in the mammalian visual system during normal development: I. Retinal ganglion cells. *J. Comp. Neurol.* **204,** 622–629.

Shorey, M. L. (1909). The effect of the destruction of peripheral areas on the differentiation of the neuroblasts. *J. Exp. Zool.* **7,** 25–64.

Silver, J. (1978). Cell death during development of the nervous system. *In* "Handbook of Sensory Physiology" (M. Jacobson, ed.), Vol. 9, pp. 419–436. Springer, New York.

Soriano, E., Antonio del Rio, J., and Auladell, C. (1993). Characterization of the phenotype and birthdates of pyknotic dead cells in the nervous system by a combination of DNA staining and immunohistochemistry for 5'-bromodeoxyuridine and neural antigen. *J. Histochem. Cytochem.* **41,** 819–827.

Spira, A., Hudy, S., and Hannah, R. (1984). Ectopic photoreceptor cells and cell death in the developing rat retina. *Anat. Embryol.* **168,** 293–301.

Sterio, D. C. (1984). The unbiased estimation of number and sizes of arbitrary particles using the disector. *J. Microsc.* **134,** 127–136.

Stuart, G. J., and Redman, S. J. (1992). The role of GABA$_A$ and GABA$_B$ receptors in presynaptic inhibition of 1a EPSPs in cat spinal motoneurons. *J. Physiol (Lond.)* **447,** 675–692.

St. Wecker, P. G. R., and Farel, P. B. (1994). Hindlimb sensory neuron number increases with body size. *J. Comp. Neurol. (in press).*

Thompson, I. D., Morgan, J. E., and Henderson, Z. (1993). The effects of monocular enucleation on ganglion cell number and terminal distribution in the ferret's retinal pathway. *Eur. J. Neurosci.* **5,** 357–367.

Toné, S., Tanaka, S., and Kato Y. (1983). The inhibitory effects of 5-bromodeoxyuridine on the programmed cell death in the chick limb. *Dev. Growth Diff.* **25,** 381–391.

Twitty, V. (1932). Influence of the eye on the growth of its associated structures, studied by means of heteroplastic transplantation. *J. Exp. Zool.* **61,** 333–374.

von Bartheld, C., and Bothwell, M. (1993). Development of the mesencephalic nucleus of the trigeminal nerve in chick embryos: Target innervation, neurotrophic receptors and cell death. *J. Comp. Neurol.* **328,** 185–202.

von Bartheld, C. S., Schecterson, L. C., and Bothwell, M. (1993). Retrograde and anterograde transport of neurotrophins from the eye to the brain in chick embryos. *Soc. Neurosci. Abstr.* **19,** 1101.

Wakai, S., and Hirokawa, N. (1978). Development of the blood–brain barrier to horseradish peroxidase in the chick embryo. *Cell Tiss. Res.* **195,** 195–203.

Waldmeier, P. C., and Baumann, P. A. (1990). Presynaptic GABA receptors. *Ann. N.Y. Acad. Sci.* **604,** 136–151.

West, M. J., and Gundersen, H. J. G. (1990). Unbiased stereological estimation of the number of neurons in the human hippocampus. *J. Comp. Neurol.* **296,** 1–22.

Williams, C. V., and McLoon, S. C. (1991). Elimination of the transient ipsilateral retinotectal projection is not solely achieved by cell death in the developing chick. *J. Neurosci.* **11,** 445–453.

Williams, R. W., and Rakic, P. (1988). Three-dimensional counting: An accurate and direct method to estimate numbers of cells in sectioned material. *J. Comp. Neurol.* **278,** 344–352 and **281,** 355.

Wimer, C. (1977). A method for estimating nuclear diameter to correct for split nuclei in neuronal counts. *Brain Res.* **133,** 376–381.

Wong, R. O. L., and Hughes, A. (1987). Role of cell death in the topogenesis of neural distribution in the developing cat retinal ganglion cell layer. *J. Comp. Neurol.* **262,** 496–511.

Yamada, T., Pfaff, S. L., Edlund, T., and Jessell, T. M. (1993). Control of cell pattern in the neural tube: Motor neuron inducted by diffusible factors from notochord and floor plate. *Cell* **73,** 673–686.

Zamenhof, S. (1976). Final number of Purkinje and other large cells in the chick cerebellum influenced by incubation temperatures during their proliferation. *Brain Res.* **109,** 392–394.

I. Introduction

The nematode *Caenorhabditis elegans* seems so well suited to investigation of genetically specified cell death that one might almost suspect that it was designed for this purpose (see reviews by Ellis *et al.*, 1991a; Driscoll, 1992). The complete cell lineage of this nematode has been recorded (Sulston and Horvitz, 1977; Sulston *et al.*, 1983). Thus, it is well established that of 1090 somatic cells generated during hermaphrodite development, 131 identified cells undergo programmed cell death (PCD) at characteristic times. Because the cuticle of the animal is transparent and individual cells are easily identified, cell deaths can be observed while they occur in the living animal. The ease of observation of specific cells, coupled with the fact that *C. elegans* is readily amenable to genetic analysis, has allowed mutations to be identified that either disrupt the normal pattern of PCD or cause inappropriate cell death. Molecular analysis of the identified genes is facilitated by the fact that the genetic map has been aligned with the physical map of the *C. elegans* genome so each chromosome is now nearly completely represented by an overlapping collection of cosmid or YAC clones that include defined genetic loci (Coulson *et al.*, 1986,1988,1991). In addition, significant progress toward sequencing the genome has been accomplished (Wilson *et al.*, 1994). Finally, it is possible to construct transgenic animals so the activities of engineered genes can be assayed *in vivo* (Fire, 1986; Mello *et al.*, 1992).

Two types of cell death have been studied in *C. elegans*. One type is PCD, which occurs as a component of normal development. Genetic studies have led to the elaboration of a genetic pathway for PCD that includes a negative regulator, *ced-9* (cell death abnormal), that acts to prevent PCD (Hengartner *et al.*, 1992) and two genes, *ced-3* and *ced-4*, that are involved in the execution of the death program (Ellis and Horvitz, 1986). A third group of genes (*ced-1, ced-2, ced-5, ced-6, ced-7, ced-8, ced-10*) is required for efficient removal of corpses of dead cells (Hedgecock *et al.*, 1983; Ellis *et al.*, 1991b). Cells undergoing PCD exhibit stereotypic morphological changes (Robertson and Thomson, 1982; Fig. 1A).

A second type of cell death studied in *C. elegans* resembles necrosis and is characterized by swelling and lysis of specific groups of neurons (see Fig. 1B) (see Chapter 1, this volume). This pathological cell death can be induced by mutations in a family of genes including *deg-1, mec-4,* and *mec-10* (degeneration and mechanosensory abnormal, respectively; Chalfie and Wolinsky, 1990; Driscoll and Chalfie, 1991; Huang and Chalfie, 1994). These genes, called degenerin genes, encode subunits of a newly identified class of ion channels. Degeneration appears to occur as a consequence of inefficient closing of the channel (Hong and Driscoll, 1994).

Interestingly, *C. elegans* genes involved in both normal and degenerative cell death have identified counterparts in mammals (Vaux *et al.*, 1992; Miura *et al.*, 1993; Yuan *et al.*, 1993; Hengartner and Horvitz, 1994; Canessa *et*

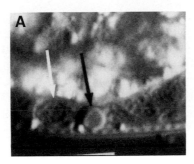
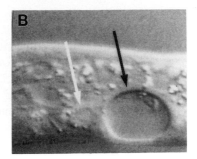

Fig. 1 Characteristic morphologies of cells undergoing cell death. (A) Programmed cell death. The black arrow indicates a "button-like" cell undergoing programmed cell death. The white arrow points to one of many living cells. Window is ~35 μm across. Reprinted with permission from Sulston and Horvitz (1977). (B) Degenerative cell death. The black arrow indicates a swollen touch receptor neuron that is undergoing degeneration in a *mec-4(d)* animal; the white arrow points to one of many normal cells. Window corresponds to ~70 μm. Photograph provided by C. Link.

al., 1993,1994). In several instances, researchers have demonstrated that the activities of these genes have been functionally conserved. For example, human Bcl-2 can partially block PCD in nematodes (Vaux *et al.,* 1992; Hengartner and Horvitz, 1994) and *ced-3,* which exhibits sequence similarity to interleukin-1β-convertase (a cysteine protease) (Yuan *et al.,* 1993), can induce apoptosis in Rat-1 cells (Miura *et al.,* 1993). These experiments support the idea that the mechanism of PCD is conserved and that assays of mammalian genes in the *C. elegans* system can be useful in unraveling the details of the process by which cells die.

This chapter is geared toward investigators who are interested in using the *C. elegans* system to study cell death but have little or no experience in working with this nematode. The emphasis of the included protocols is on assays of the effects of heterologous genes on cell death. Included are a discussion of basic care and maintenance of animals, how to observe and score cell death, how to construct transgenic animals, and how to assay for expressed proteins in the worm. Readers interested in other *C. elegans* methods are referred to Wood (1988) and Epstein and Shakes (1995).

II. Strain Maintenance

The life cycle of *C. elegans* consists of an embryonic period in which the worm develops within an eggshell, four larval stages punctuated by molts, and adulthood. Animals are reared between 15°C (life cycle completed in about 5.5 days) and 25°C (life cycle completed in about 2.5 days). Because *C. elegans* is hermaphroditic, continued mating is not required to maintain a population. One wild-type animal bears about 300 young through self-fertilization.

Caenorhabditis elegans is grown on agar plates that have been overlayed with *Escherichia coli,* the food source. When the food supply is exhausted, young worms enter an alternative L3 larval stage called the dauer larval stage. Dauers can survive for 70 days or longer without feeding. Animals re-enter the reproductive life cycle when placed on a plate containing *E. coli. C. elegans* strains can be stored for years as frozen stocks at −70°C. Useful references for strain maintenance and manipulation are Brenner (1974) and Wood (1988).

A. Media and Methods for Continued Culture

1. Preparation of NGM Agar (Nematode Growth Medium)

a. Stock Solutions

5 mg/ml cholesterol; make solution in 100% ethanol; do not autoclave; sterilization is not necessary

1 M CaCl$_2$; autoclave

1 M MgSO$_4$; autoclave

1 M K-PO$_4$: prepare two solutions: (a) 204.4 g KH$_2$PO$_4$ (monobasic) in a final volume of 1500 ml H$_2$O and (b) 114.12 g K$_2$HPO$_4$ (dibasic) in a final volume of 500 ml H$_2$O; combine the solutions, check pH, and autoclave

100 mg/ml nystatin: suspend 10 g nystatin (Sigma N3503, St. Louis, MO) in 100 ml 70% ethanol; suspension must be shaken thoroughly before it is added; do not autoclave

b. Preparation of NGM Plates

3 g NaCl

2.5 g peptone

0.2 g streptomycin

17 g agar

970 ml distilled H$_2$O

Autoclave 30 min and cool to 55°C. While stirring with a stir bar, add

1 ml 5 mg/ml cholesterol

1 ml 1 M CaCl$_2$

1 ml 1 M MgSO$_4$

1 ml 100 mg/ml nystatin

25 ml 1 M K-PO$_4$, pH 6

Notes. Be careful not to flame the cholesterol and nystatin ethanol solutions when using sterile technique to add stocks. The agar solution should be cloudy. Pour into petri dishes; 60-mm diameter plates are used routinely although

90-mm plates can be used for larger cultures. After allowing plates to dry at room temperature for 1–2 days, store upside-down (agar on top) in sealed bags or in plastic sweater boxes at 4°C. Plates can be stored for a few months.

2. EZ/MYOB Worm Plates (E. Lambie, personal communication)

a. Stock for 100 liters of plates

55 g Tris HCl

24 g Tris Base

460 g Bacto Tryptone

0.8 g cholesterol

200 g NaCl

Place in a labeled plastic container, cover and mix all ingredients throughly. This mix can be stored for use at will.

b. Preparation of EZ plates

7.4 g EZ mix

20 g agar

0.2 g streptomycin sulfate

1000 ml distilled H_2O

Autoclave and cool at 55°. After cooling, 1 ml of a 100 mg/ml (in 70% ethanol) suspension of nystatin can be added to ward off contamination.

Note. This is an easier medium to make, but strains that are very unhealthy as a consequence of certain mutations may not grow well on these plates.

B. Food Source

Worms are fed *E. coli* strain OP50/1, a strain that requires uracil for growth. Since uracil is present in limiting amounts in peptone, the bacteria do not reproduce efficiently and thus do not overgrow the nematodes. OP50/1 is resistant to streptomycin, which is added to the media to restrict growth of contaminants. OP50/1 can be obtained from the *C. elegans* Genetic Stock Center (250 Biological Sciences Center, University of Minnesota, 1445 Gortner Avenue, St. Paul MN 55108-1095; (612) 625-2265).

To prepare a bacterial culture for seeding NGM agar plates, streak out OP50/1 on an LB or NGM plate and incubate overnight at 37°C. On the following day, transfer a single colony (using sterile technique) into a bottle containing about 50 ml LB broth. Incubate the bacteria at 37°C overnight (shaking is not necessary). The resulting sterile culture is used to seed bacterial lawns. This culture can be used for about 1 mo if stored at 4°C.

To spread plates, take up a small amount of OP50/1 suspension in a 1-ml pipet and move the tip along the agar in a streaking pattern while releasing a small volume, about 1/10 ml, to create a thin coating of bacteria over most of the plate. It is important that the bacterial lawn does not touch the edges of the plate since worms will tend to crawl off the agar onto the sides of the dish and desiccate. Also, it is important not to gouge the agar surface, since animals will burrow beneath the agar surface if there are cracks that facilitate entry. Allow the seeded lawn to grow overnight at 37°C before transferring animals. Seeded plates can be stored upside-down at 4°C for 1–2 mo in sealed plastic containers.

C. Transferring Nematodes

1. Transferring Large Numbers of Animals

Many experimental manipulations (DNA injections, antibody staining experiments, etc.) require a plate of well-fed animals. To transfer many animals at once, use a scalpel to cut a chunk of agar from a plate of animals and transfer the agar, worm-side-down, onto a new plate that has been spread with bacteria. Well-fed or starved cultures can be used for this inoculation. The starting plate is usually heavily populated so a small agar chunk (0.5 cm × 0.5 cm) will include many worms. Animals should be transferred using sterile technique. Store the scalpel in a small flask with the blade immersed in 95% ethanol. Flame the blade before and after each transfer.

Another way of transferring large numbers of animals is to use a bacterial loop. Sterilize the loop by heating it to red-hot in a flame. Allow the loop to cool for a moment by touching it to the agar surface. Drag the loop across the agar surface so that a small "peel" of agar and worms is generated. Transfer the peel, worm-side-down, to a fresh plate. Again, usually only a small amount of agar needs to be transferred; the loop must be flamed before and after each use to prevent contamination of different cultures.

Note. The condition of the starting culture is an important determinant of how many animals will crawl out of the transferred agar. Starved dauer animals are recovered with a reasonable frequency for up to 70 days if the plates are not dried out. Plates can dry out rapidly, forming a residue that looks like a potato chip. Animals do not survive as long on such plates. Recovery from dried-out plates can be aided by adding a small amount of H_2O to the agar to soften it and then transferring large chunks of the agar to a new plate. It is essential to flame the scalpel between transfers to prevent cross-contamination of strains.

2. Transferring Individual Animals

To transfer one or only a few animals onto new plates (for example, when selecting transgenic animals), a worm pick is used. The pick is made of a small piece of platinum wire that is inserted into the end of a pasteur pipet (see

Fig. 2). Manipulations are performed using a dissecting microscope at 10–50× magnification. Using a pick is a skill that must be practiced before it becomes routine. In general, an individual can become adept with a few hours of practice.

a. Construction of a Worm Pick (see Fig. 2)

i. Materials

Two pennies (or coins with a flat edge and some flat surface)
about 1″ platinum wire, 32-gauge (VWR #66260-068, Piscataway, NJ)
one 5-3/4″ pasteur pipet
scalpel
tweezers
Bunsen burner
dissecting microscope

ii. Method

1. Flatten the end of a 1″ platinum wire by resting the wire on the smooth surface of a penny and exerting pressure through the edge of a second penny. Check the extent of flattening under the dissecting microscope (50× magnification). Continue pressuring/rolling the penny until the end of the wire becomes flattened.
2. Using a sharp scalpel blade, trim the flattened end to make two linear edges that come to a point.
3. Hold the wire with a pair of tweezers and place the nonflattened edge into the opening of a pasteur pipet. Flame the opening of the pipet until the glass melts to hold the wire.
4. Put a bend in the pick by placing the nonsharpened edge of the scalpel against the wide part of the pick, applying pressure to hold the pick end down, and gently lifting the end of the wire attached to the pipet, making a 120° angle between the very end of the pick and the rest of the wire.
5. Remove any sharp spurs of metal that remain on the pick end by trimming with the scalpel. A pick with rough edges or burrs can kill worms.

b. Picking Individual Worms

1. Heat the pick wire in a flame until the end glows red. Cool the pick momentarily in air.
2. Touch the end of the pick to the bacterial lawn to collect a bit of "sticky" bacteria on the pick; this will make the worms adhere better to the wire.
3. Place the end of the pick gently on the lawn near the animal to be transferred and, approaching the animal from the side, scoop up the animal. Often worms will stick nicely to the bottom of the pick. Animals can also be transferred using the top surface of the pick.

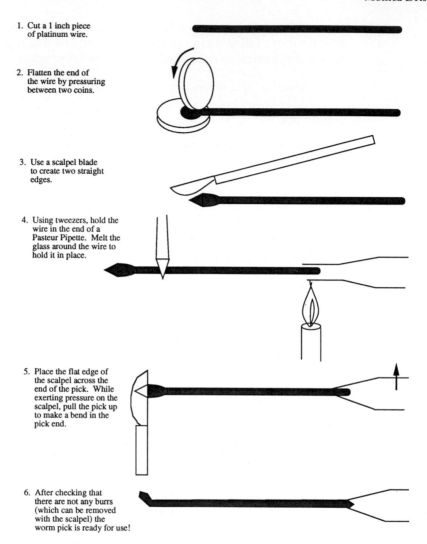

1. Cut a 1 inch piece
   of platinum wire.

2. Flatten the end of
   the wire by pressuring
   between two coins.

3. Use a scalpel blade
   to create two straight
   edges.

4. Using tweezers, hold the
   wire in the end of a
   Pasteur Pipette. Melt the
   glass around the wire to
   hold it in place.

5. Place the flat edge of
   the scalpel across the
   end of the pick. While
   exerting pressure on the
   scalpel, pull the pick up
   to make a bend in the
   pick end.

6. After checking that
   there are not any burrs
   (which can be removed
   with the scalpel) the
   worm pick is ready for use!

Fig. 2 Preparation of a worm pick. Detailed protocol given in Section II,C,2,a,ii.

4. Gently touch the pick to the agar surface of a fresh plate. The animals will usually crawl off voluntarily. They can be encouraged to crawl onto the surface of the new plate by gently moving the pick so the animal is bought in close proximity with the agar surface.

Note. Avoid keeping the animals on the pick for a long period of time since they can dry out. It is good practice to look carefully at the transferred animal to be certain that eggs and other animals have not been inadvertently transferred.

It is also important to avoid gouging the agar surface during transfer so that animals cannot burrow below the surface of the agar. With practice, the dexterity required for transfer without gouging is developed.

D. Storing Nematode Strains

1. Short-Term Storage

Dauer larvae can survive on plates for as long as 70 days in the absence of food, but it is best to maintain stocks by inoculating fresh plates at least once a month. For short term storage, animals can be kept on plates at 15°C. Plates should be sealed with parafilm to prevent excessive drying of the agar.

2. Long-Term Storage

Caenorhabditis elegans strains can be stored for years as glycerol stocks at −70°C. It is good practice to have frozen stocks of all strains in use in the lab.

a. Freezing Materials

i. M9 Buffer (autoclave)

3 g KH_2PO_4
6 g $Na_2 HPO_4$ (or 11.3 g $Na_2HPO_4 \cdot 7H_2O$)
5 g NaCl
1.0 ml 1 M $MgSO_4$
1 liter dH_2O

ii. Freezing Solution

5.85 g NaCl
50 ml 1 M K $\cdot$ PO_4, pH 6.0 (see recipe in Section II,A,1,a)
300 g glycerol (about 200 ml)
dH_2O to 1 liter

Autoclave. After cooling, add 3 ml 0.1 M $MgSO_4$.

Note. Both solutions should be stored at 4°C. It is important that they be cold when used.

b. Freezing Protocol

1. Grow 3 plates (60-mm size) of the strain of interest to a point shortly after the culture has consumed the *E. coli* lawn. The culture should include many L1 and L2 animals since these survive freezing best. It is important that plates be free of bacterial and fungal contamination.

2. Prepare 5 freezer vials for each strain that will be frozen. Each vial should be labeled with the strain name, the allele name, and the reference number for your own strain collection. The vials should be placed on ice shortly before use.

3. Ready a styrofoam box that has slots to hold the vials. This box will be used to hold the vials while they are freezing at −70°C.

4. Harvest the animals. Add 1.5 ml prechilled M9 buffer to one of the plates, swirl the liquid gently so the animals are suspended in the M9, and pour the M9/worm suspension onto the second plate. Swirl and pour the suspension onto the third plate. Finally, transfer the suspension to a disposable 15-ml tube on ice.

5. Add an equal volume (about 1.5 ml) prechilled freezing solution to the worm suspension, mix, and *immediately* aliquot 0.5 ml worms to each vial. Hold vials on ice while working. Transfer vials *immediately* to the styrofoam rack and place in the −70°C freezer.

Note. It is essential to freeze animals shortly after they are introduced into the freezing solution. When freezing several stocks at once, it is best to freeze strains individually or in small groups. After a few days, one tube should be thawed and tested to confirm successful freezing before agar plates of the strain are discarded.

c. Thawing a Strain

1. Label the bottoms of a few *E. coli*-spread NGM plates with the name of the strain to be thawed.

2. Remove one vial of the strain of interest and thaw the contents rapidly by rolling the vial between your hands.

3. As soon as liquid is evident, transfer part of the solution onto a plate. Distribute the liquid around the edges by rotating the plate. The contents of one vial should be distributed to 3–5 60-mm plates.

4. Allow animals to recover overnight and then transfer several to a new plate to inoculate a fresh culture.

3. Cleaning Strains

Worm cultures that become contaminated with mold or foreign bacteria can be disinfected by treating egg-bearing adults with a hypochlorite solution that kills the contaminants and adult worms, but does not kill eggs if exposure time is adequately short.

a. Cleaning Solution—1% Na-hypochlorite, 0.5 M NaOH

2 ml Clorox (alternatively, a 10–20% solution of NaOCl can be purchased from a chemicals supplier)

1 ml 5 *M* NaOH

7 ml H_2O

This solution should be made up fresh.

b. Cleaning on a Small Scale

1. Transfer a few egg-bearing adults from a contaminated culture to a bacteria-free spot on the agar of a new NGM plate.
2. Place a drop of the cleaning solution on the adults. Adults will die and break to liberate eggs. The hypochlorite solution should ultimately be absorbed into the agar and animals should hatch from surviving eggs.

c. Cleaning on a Large Scale

1. Harvest animals from a small plate by adding 1–2 ml M9 buffer to the plate, swirling, and pouring the nematode suspension into a 15-ml tube. The population should include gravid adults.
2. Allow worms to settle to the bottom of the tube. Aspirate off the M9 buffer.
3. Add 1 ml cleaning solution to nematodes. Swirl to keep animals suspended for a few minutes. The condition of the animals should be monitored periodically by examining a drop of the mixture under the microscope. When about 50% of the egg-bearing animals have broken in half, proceed to the next step. Keeping the worms in the hypochlorite solution for too long will kill the eggs.
4. Add 10 ml H_2O to wash out the cleaning solution. Allow worms to settle to the bottom and aspirate off the liquid.
5. Add 10 ml M9 to wash worms.
6. Using a sterile pasteur pipet, remove some of the worm debris and eggs from the bottom of the tube and put a drop onto a fresh plate. Surviving eggs will hatch to start a new, clean culture.

d. Cleaning by Transfer

1. Place a few worms onto a new bacteria-spread plate and let them crawl around for a few hours to release the contents of their guts (which may include contaminants).
2. Transfer animals to a new bacteria-spread plate, letting them crawl for a few hours or overnight. Transfers can be repeated as often as necessary.
3. Once eggs are deposited onto a clean plate, they should be transferred to a new bacteria-spread plate. The population that arises from these worms will often be free of contamination.

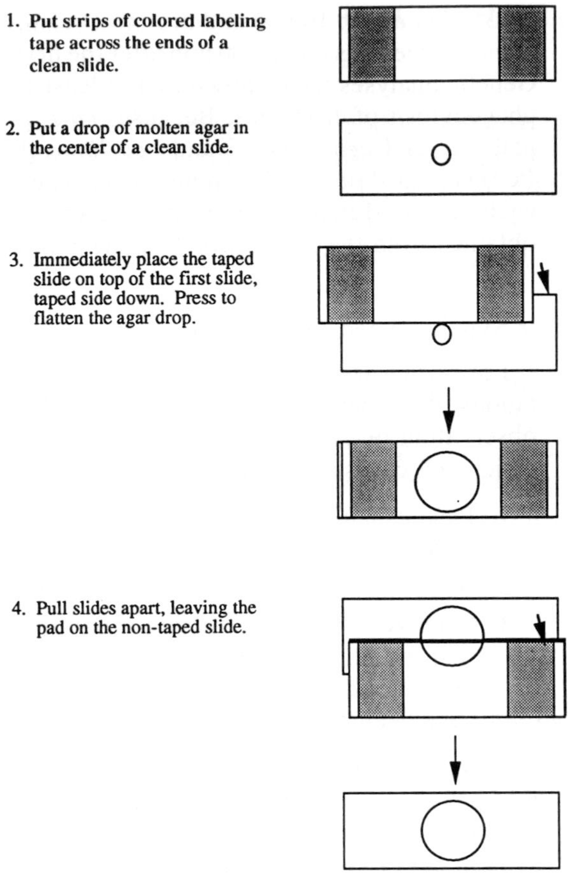

1. Put strips of colored labeling
   tape across the ends of a
   clean slide.

2. Put a drop of molten agar in
   the center of a clean slide.

3. Immediately place the taped
   slide on top of the first slide,
   taped side down. Press to
   flatten the agar drop.

4. Pull slides apart, leaving the
   pad on the non-taped slide.

Fig. 3 Preparation of agar pads for *C. elegans* microscopy. Detailed protocol given in Section III,D,1,b.

b. Preparation of Pads

1. Using the pasteur pipet, place a drop of agar onto the clean slide.
2. Quickly press the taped slide (tape side down) over the slide with the agar drop so that the drop is flattened to a circle about 0.4 mm thick (the thickness of the tape spacers).
3. After the agar solidifies, gently separate the slides so that the agar pad adheres to the nontaped slide. Rest the slide agar-side-up on the bench top.

Note. The agar pad should be prepared just before use so that it does not dry out. Avoid getting bubbles in the agar.

2. Mounting Live Animals

1. Place a 1- to 2-μl drop of M9 containing 50 mM sodium azide (NaN$_3$) onto the center of the agar pad. NaN$_3$ anesthetizes the worms so that they will not move.

2. Transfer the animals to be observed into the drop. Animals can be transferred using a worm pick or an eyebrow hair fastened to a toothpick with wood glue. Getting animals or eggs to stick to the eyebrow hair is a skill acquired with practice. When the pick or hair is moved into the NaN$_3$ drop, animals float off easily. Alternatively, animals can be transferred to agar pads with a "worm pipet." (Construction of a worm pipet is described in Section IV,C,5). If this approach is utilized, the agar pad is best prepared with anesthetic included for a final concentration of 10 mM NaN$_3$ in the agar.

3. Gently lay a coverslip over the animals. Most animals will lie on their sides.

E. Observing Cell Deaths

1. Programmed Cell Death

Using a 100× objective, observe animals for the presence of raised button-like cells or corpses. Figure 1A illustrates dying cells. Note that numbers of viable cells can be scored most easily by counting the number of nuclei in a given region of the body (see in Section V,B).

2. Degenerative Cell Death

With some practice, it is possible to observe and screen for necrotic swelling deaths using the dissecting microscope. To help visualize cells that have swollen, place a drop or a thin film of mineral oil over the animals to be observed. As animals move through the oil, swollen cells pass transmitted light differently than do normal cells and swellings can be recognized as small bubbles. Swellings are difficult to detect at first; it is recommended that the experimenter work with mutant strains such as those that harbor the death-inducing *mec-4(d)* allele to learn to recognize the phenotype. Degenerative deaths are readily observed using Nomarski optics using the protocol described earlier. Figure 1B depicts a cell undergoing degenerative cell death.

IV. *C. elegans* Transformation

A. Special Considerations

The demonstration that the *bcl*-2 gene could partially substitute for the activity of *C. elegans ced-9* in blocking PCD constituted an important advance in the field (Vaux *et al.,* 1992; Hengartner and Horvitz, 1994). As understanding of the biochemical activities of additional factors in mammalian cell death progresses, it will be of interest to test the genes encoding these activities for effects on the *C. elegans* cell death program. In this section, the methods for generating transgenic animals are described.

DNA is introduced into *C. elegans* by injecting a plasmid solution into the gonad of young adult hermaphrodites. Germ cell nuclei in *C. elegans* develop initially in a syncytium; when cell membranes later envelop them, the exogenously added DNA is packaged into oocytes. Transforming DNA does not usually integrate into the chromosomes, but is maintained as a concatamer of introduced sequences. The stability of this extrachromosomal array is variable and depends somewhat on the size of the concatamer; larger arrays are maintained better than smaller ones. The copy number of introduced sequences can be on the order of a few hundred. Thus, the phenotypes of transformants must be interpreted with attention to two facts: (1) many animals are mosaic for the presence of transforming DNA, so in a given animal some cells will have the introduced genes and some cells will not, and (2) transforming DNA is present in high copy number.

B. Recommended References

The protocol outlined here is derived from two critical sources: Fire (1986) and Mello *et al.* (1992).

C. Materials

1. Vectors for Expression of Heterologous Genes in *C. elegans*

To test the effects of the expression of "cell death" genes from other organisms on cell death in *C. elegans,* the gene of interest should lack introns and must be fused to a *C. elegans* promoter. Several plasmids designed to facilitate expression of genes in various cell types have been constructed by A. Fire and colleagues (personal communication, Carnegie Institute of Washington). It is critical that the gene of interest be expressed in the correct cells and at the correct time during development.

Useful promoters for this type of experiment are those from two heat-shock genes that are expressed in distinct but overlapping sets of cells (Stringham *et al.,* 1992) and are included in vectors pPD49.78 and pPD49.83 (Perry *et al.,* 1993). Characterization of the promoters of the *ced-3, ced-4,* and *ced-9* genes has not been reported in detail, although sequences of these genes have been published (Yuan *et al.,* 1993; Yuan and Horvitz, 1992; Hengartner and Horvitz, 1994). Another cell-specific promoter that is useful, especially in studies of degenerative cell death, is the *mec-7* promoter (plasmid pPD52.105; A. Fire, Carnegie Institute of Washington), which drives high level gene expression in the six touch receptor neurons (Hamelin *et al.,* 1992).

2. Equipment for DNA Microinjection

1. *Microscopes.* DNA microinjections are performed using a high-power inverted miscroscope (in which the light source comes from above). Two commonly utilized microscopes are the Zeiss Axiovert 35 and the Nikon Diaphot-200. The microscopes must be fitted with DIC optics and a circular sliding stage; 10× and 40× objectives are utilized.

2. *Micromanipulators.* Several varieties of micromanipulators are available. A popular manipulator is Narishige model number MN-151. Narishige products are marketed in the United States by Kramer Scientific Corporation [(914) 476-8700].

3. *Instrument (needle) holder.* The needle for injections is housed in an instrument holder that is attached to the micromanipulator. The assembly for the instrument holder is Kramer Scientific Corporation part number 520-142. The micro-instrument collar is part number 520-145.

4. *Microinjection control system.* Tritech Research, Inc. markets a microIN-JECTOR™ system that attaches the injection needle to a pressure source (N₂ gas). It includes a foot pedal controller and tubing that attaches the needle assembly to the gas tank via a pressure gauge and a valve that can be manually opened and closed. This is an essential part of the microinjection apparatus.

5. *Pressure source.* DNA is forced through the needle under pressure from a small tank of N₂ gas.

6. *Needle pullers.* Two needle pullers that work well for making nematode injection needles are Narishige PN-3, a horizontal puller controlled by a magnet (settings: heat = 2.5, submagnet 82.9, main magnet 18.3) and Narishige PB-7, a vertical puller that relies on gravity (setting 8.5 for two heavy weights).

3. Preparation of Injection Needles

a. Materials

For the PN-3 horizontal needle puller: 30-30-1 capillary tubing with omega dot fiber for rapid fill, borosil 1.0 mm × 0.5 mm ID/fiber from FMC [Brunswick, ME 04001; (207) 729-1601]

For the PB-7 vertical puller: glass model GD-1, size 1 × 90 mm from Narishige; needles are available from other sources, but the aforementioned ones are the ones that have been used for the recommended puller settings

Small plastic box or large petri dish for needle storage; two rolls of plasticine are set in parallel so that needles can rest on them for storage

b. Notes on Pulling Needles

A good needle is essential for easy injections. The appropriate settings for pulling needles must be determined by trial and error. The previously mentioned settings provide a good starting point. The tip of the needle must be sharp and narrow. In general, needles are melted closed after they are pulled and are opened by gently touching agar debris or the coverslip edge (see Section IV,D,2). The best needles are not perfectly round; they have one edge that is fairly flat. This flat side should be oriented on the bottom when the needle is inserted into the collar of the instrument holder. Store needles by resting across plasticine strips. Be careful not to touch the tip against anything since it can break easily.

4. Preparation of Agarose Pads

1. Make up a solution of 2% agarose in H_2O. Keep agar hot during preparation of pads.
2. Drop 60 μl molten agarose onto a 24 × 50 mm glass coverslip.
3. Immediately place a second coverslip over the first, flattening the agarose drop.
4. When agarose hardens, separate the two coverslips and rest the coverslip holding the pad, agarose side up, at the bottom of a glass box or large petri plate.
5. Bake the agarose pads at 80°C for 30 min.

Note.   Agarose pads can be stored for a few months at room temperature, so it is a good idea to make a fairly large batch at once.

5. Preparation of a Worm Pipet for Manipulation of Injected Animals

1. Heat the center of a 100 μl capillary pipet (for example, Corning #7099S-100) in a flame, pull, and break the glass to create an open, fine-bored end.
2. To use, fasten the pipet to the red-tipped mouth pipetter device that is packaged with the capillary tubes.

6. Solutions for Microinjections

a. 5× Injection Buffer

0.4 g PEG 8000
100 mM K · PO_4, pH 7.5 (0.4 ml 1 M stock made from a solution that is
  67 mM K_2HPO_4, 33 mM KH_2PO_4)
15 mM K Citrate, pH 7.5 (60 μl 1 M stock, pH 7.5 with KOH)
H_2O to 4 ml (about 1.5 ml)

Autoclave and store in 200-μl aliquots.

b. Recovery Buffer

0.1% salmon sperm DNA (1.0 ml 10 mg/ml stock)

4% glucose (2.0 ml 20% stock)

2.4 mM KCl (24 μl 4 M stock)

66 mM NaCl (132 μl 5 M stock)

3 mM MgCl$_2$ (30 μl 1 M stock)

3 mM CaCl$_2$ (30 μl 1 M stock)

3 mM HEPES, pH 7.2 (30 μl 1 M stock)

H$_2$O to 10 ml (6.75 ml)

Note. It may be necessary to add 1 μl (or so) 10 M NaOH to bring pH to 7.2.

7. DNA for Transformation

DNA to be injected can be prepared from rapid mini-preps, but our bias is that tRNA should be removed. Passage of mini-prep DNA over a Sephacryl-S400 (Pharmacia, Piscataway, NJ) column, Quiagen column, or other column is recommended.

DNA is usually co-injected with a plasmid encoding a marker gene that confers a recognizable phenotype. Two commonly utilized markers are *rol-6(su1006),* a dominant allele that causes transformed strains to roll (this is encoded in plasmid pRF4; Kramer *et al.,* 1990), and a construct that synthesizes an antisense complement to a portion of the *unc-22* gene, thereby causing animals to twitch (this marker is encoded on plasmid pPD10.41; Fire *et al.,* 1991). A vector that harbors a suppressor tRNA gene can be used to limit the copy number of introduced DNA since high doses of the suppressor tRNA gene are lethal (see Fire *et al.,* 1990a). This marker must be used in conjunction with a nematode strain that harbors an amber-suppressible allele.

Prepare DNA for microinjection by making a mix of 1× injection buffer, about 50 μg/ml marker DNA, and 50 μg/ml DNA to be assayed. The amounts of DNA introduced can vary between 50 and 200 μg/ml. Only 10 μl total volume is required since needles are filled with very small amounts of DNA.

D. Methods

1. Loading the Injection Needle

1. Make a loading pipet by heating a 100-μl glass micropipet in a Bunsen burner flame and pulling the ends apart as soon as the center melts. The objective is to keep the two ends together, but to create a thin thread of glass between the ends. The ends should be about a foot apart from each other after the pull. Break the glass in half to generate a filling pipet. The "pulled out" region of the filling pipet must be thin enough to fit inside the injection needle and long enough to reach the needle tip.

2. To load the needle, dip the thin tip of the loading pipet into the DNA mix and allow capillary action to fill the pipet end. After a small amount is taken up, place the tip of the pipet into the open end of the injection needle and gently move it toward the needle tip until it touches the end where the taper begins and can move no further. Blow gently into the pipet while slowly pulling the pipet backward, forcing DNA into tip of the needle. A small amount (only a few mm deep) is enough. Be careful not to break the loading pipet inside the needle. Avoid getting an air bubble in the very tip of the needle.

Note. Store the loaded needle on two ridges of plasticine in a small box or petri dish. Be careful not to touch the tip of the needle against anything, since it can break easily. It is a good idea to prepare a few needles for a given DNA, because it is sometimes difficult to get a good open needle.

2. Injection Protocol

a. Materials that Should be Assembled in the Microinjection Area

sterile M9 buffer or recovery buffer

dissecting microscope

worm pipets (see Section IV,C,5) and mouth-tube device

OP50/1 spread plates for injected worms

agarose pads

pipet tips for spreading oil on agarose pads

Halocarbon oil (from Halocarbon Products Corp, 120 Dittman Ct., N. Augusta, SC 29841; (803) 278-3500]

needles loaded with DNA

worm pick (see Section II,C,2,a)

worm brush (eyebrow hair fastened with wood glue to the tip of a toothpick)

worms to be injected

b. Preliminary Set-Up Work

1. Grow a plate of worms that are well supplied with bacteria. Young adults that have no or few eggs inside should be selected for injection. Animals should be free from contamination.
2. Put the DNA-loaded needle into the collar of the instrument holder with the flat side down.
3. Open up the gas line from the N_2 tank to 30–40 psi
4. Position worms on the agarose pad:
   a. Place a drop of halocarbon oil onto an agarose pad.
   b. Using a worm pick, gently transfer 1–6 young adults into the oil.

  c. Using the worm brush, gently press flailing animals to the pad so they stick. It is best to align them in a column so they can be easily injected in a series. Ideally, they should be in a line with their gonads facing in the same direction.

Note. It is best to minimize the time that animals are stuck to the agarose pad. When learning transformation, setting up 1–2 worms at a time is advisable.

C. Injections

1. Transfer slide to the inverted microscope, worm and oil side up. Try to center the worms in the light. They appear as small flecks on the agarose.
2. Bring the animals and the needle into focus.
     a. Using the low power (6 or 10X) objective, bring the animal into focus. Center the target animal.
     b. Look at the tip of the needle without looking through the microscope and use the micromanipulator to position the tip of the needle so it catches the light. This should allow the tip to be lowered close to the animal.
     c. Now, looking at the worm through the low power objective, slowly lower the needle into the oil. Gently lower the needle until the tip comes into focus with the animal.
3. Move the tip a short distance away from the animal and check whether the needle is open. Push on the foot pedal and observe the end of the needle to see if fluid comes out. If it does not, the tip of the needle is probably sealed and must be broken open. The tip can be opened by brushing against a worm, debris in the agar, or the edge of the coverslip. The procedure for breaking open the tip on the coverslip follows.
     a. Gently move the needle back to the edge of the coverslip.
     b. Lower the needle so that the tip is aligned with the edge of the coverslip.
     c. Gently touch the end of the needle to the rough edge of the coverslip. The needle can be moved up and down a bit until it opens. Opening will be evident since DNA will flow out of the needle when the foot pedal is pressed. Be gentle, since touching too hard will cause the needle to break further from the tip and may result in too large an opening in the needle. Learning this procedure requires some trial and error.
4. Once the needle is open, bring the animal into focus, switch to the 40X objective, and align the needle next to the gonad. Bring the large oocytes in the gonad along the edge into focus and then center the bend of the gonad in the working space. The bend in the gonad is the target for insertion of the needle. Move the needle so this region and the needle tip are in

the same focal plane. Good photographs of this are found in Mello *et al.*
(1992).

5. Insert the needle into the gonad either by using the micromanipulator (the needle into worm approach) or by moving the stage slightly with your hands (the worm into needle approach). Press on the foot pedal to force DNA out of the needle into the animal. A sign of a good injection is a ballooning effect seen in the gonad where it looks like liquid is filling the gonad but not the body cavity.

6. Move the worm away from the needle and then release pedal (if you cut off the pressure for flow out of the needle the pressure release can pull material into the end of the needle and clog it). The needle can be used for injection of many worms if it does not become blocked.

Note. As a general practice, we inject 10–25 animals for each construct to be tested. Numbers of injected animals should be higher when one is learning how to microinject. Make sure the gas line is closed after the injection session.

3. Recovery of Injected Worms

1. Return the coverslip to the dissecting microscope and place a small drop of M9 buffer or recovery buffer on top of the worm with the worm pipet. The buffer will sink through the oil and the worm will float up into it.

2. Gently suck the worm into the worm pipet and blow it out onto a plate spread with bacteria.

Note. Be careful not to cross-contaminate plates by reusing the worm pipet. Change to a new glass tip between experiments. Some individuals prefer to recover animals using a worm pick, which has the disadvantage that it is somewhat difficult to get the worm on the pick in the solution, but the advantage that there is less risk of losing the worm (occasionally worms can become stuck in the glass pipet).

4. Isolation of Transformants

1. Incubate animals at 15–25°C, allowing injected animals to lay eggs.

2. Remove transformed animals (rollers or twitchers) to individual fresh plates. These animals are referred to as primary transformants.

3. Examine progeny of the primary transformants for the expression of the transformation marker in the next generation to determine which animals carry introduced DNA in their germ lines. Note that a low percentage of primary transformants will have the introduced array in their germ lines. We find that an average of 15% of primary transformants segregate the marker to their progeny.

4. Score germ line transformants for the percentage of progeny that express the co-transformation marker and select lines that segregate high frequen-

cies of transformed progeny for observation of the effects of the introduced gene.

V. Visualization of Nuclei and Expressed Proteins in Worms

A. Special Considerations

Studies of cell death often require an assay for the presence or absence of cells. Although individual nuclei can be observed visually using Nomarski optics, it is much easier to stain nuclei or to use immunological reagents to identify cells. In this section, techniques for staining nuclei and whole mount immunohistochemistry are described. The immunohistochemistry techniques can also be used to assay for cross-reactivity of antisera of interest and to test for the expression of an introduced fusion gene. Two other techniques for visualizing cells, staining for β-galactosidase activity (Fire *et al.*, 1990b) and monitoring expression of the green fluorescent protein (GFP; Chalfie *et al.*, 1994), can also be used easily after constructing transgenic animals that express these markers. Protocols for the use of these markers are described in detail in the aforementioned references.

B. Staining Nuclei with DAPI

One method for scoring cell death is to count nuclei that have been stained with DAPI (diamidinophenolindole). The wild-type pattern of nuclei is described by Sulston and Horvitz (1977) and Sulston *et al.* (1983). It is advisable to stain wild-type animals of the same stage for easy comparison.

1. Solutions

fixative (Carnoy's solution): 3.0 ml ethanol; 1.5 ml acetic acid; 0.5 ml chloroform

M9 buffer (see Section II,D,2)

DAPI staining solution: 1.0 μg/ml DAPI in M9 buffer (some use up to 1 mg/ml DAPI)

2. Protocol

1. Harvest animals from plates by overlaying with 2 ml M9 and pouring off suspended nematodes into a 15-ml disposable tube. One plate usually yields enough animals for most purposes.
2. Wash to remove bacteria by letting the animals settle, gently aspirating off the M9 buffer, adding 5 ml M9, letting the worms settle again, and aspirating off the M9 buffer.

3. Transfer animals to a microcentrifuge tube. Add 1 ml fixative and fix for at least 1.5 hr. Invert the tube occasionally to suspend worms.

4. Wash 2× with M9 buffer.

5. Stain 30 min or longer in approximately 300 μl DAPI staining solution.

6. Mount animals for observation, as described in Section V,F.

C. Antibody Staining Enzyme-Digested Animals

See Li and Chalfie (1990).

1. Solutions

0.4 M PB stock: 5.3 g $NaH_2PO_4 \cdot H_2O$, 28 g K_2HPO_4, 500 ml H_2O, pH 7.4

4% paraformaldehyde: dissolve 10 g paraformaldehyde in 172 ml H_2O by heating in 0.1 M PB in a waterbath until just steaming and adding 50% w/v NaOH just until the paraformaldehyde goes into solution and the solution clears (about 10 drops); add 62.5 ml 0.4 M PB and dilute to 250 ml; final pH should be about 7.2 (pH can be assayed with pH paper to keep fixative off of pH electrode); store for weeks at 4°C

BTT: 5% β-mercaptoethanol; 1% Triton X-100; 0.1 M Tris base, pH 7.2, at room temperature (pH 6.9 at 37°C)

0.1 M Tris base, pH 7.7

Collagenase buffer: 1 mM $CaCl_2$, 0.1 M Tris base, pH 7.7, at room temperature (pH 7.4 at 37°C)

GPB: 10% goat serum (Gibco, Gaithersburg, MD), 0.5% Triton X-100, 0.2% (by weight) sodium azide (NaN_3), 0.1 M PB

2. Protocol

1. Harvest animals from plates by overlaying with 2 ml M9 and pouring off suspended nematodes into a 15-ml disposable tube. One to several small plates yield enough animals.

2. Wash away bacteria by letting the animals settle, aspirating off the M9 buffer, adding 5 ml M9, letting the worms settle again, and aspirating off the M9 buffer. Note that instead of allowing worms to settle, they can be gently centrifuged at 3000 rpm for 2 min.

3. Fix 12–36 hr in about 1 ml 4% paraformaldehyde in 0.1 M PB, pH 7.4. Fixation time can be decreased if antigen is sensitive to overfixation.

4. Wash animals 6× with 0.1 M PB. Worms can be stored at this stage for several days at 4° C.

5. Incubate animals in BTT for 24–48 hr at 37°C with gentle agitation (e.g., on a labquake or a rotator).

6. Rinse several times with 0.1 *M* PB until the β-mercaptoethanol is no longer detectable. Worms at this stage (in 0.1 *M* PB) can be stored at 4°C for a few days.

7. Rinse once with 0.1 *M* Tris, pH 7.7 (room temperature).

8. Incubate animals at 37°C in 900 units/ml collagenase Type IV (Sigma), in collagenase buffer with gentle agitation. Note that the time of incubation is very dependent on the batch of collagenase. Each batch should be tested for the optimal digestion time. Monitor digestion of a sample withdrawn from the tube by examining the state of the animals under the dissecting microscope. Generally, digestion is sufficient if a few animals are broken and a few eggs have been released.

9. Rinse several times with 0.1 *M* PB. Worms can be stored at this stage for several days.

10. Transfer some or all of the animals into a microcentrifuge tube. Incubate with GBP for at least 30 min at 37°C. Alternatively, incubate at room temperature for a longer time.

11. Incubate with 50–100 μl primary antibody diluted in GBP (initially a 10^{-1}–10^{-4} dilution series of the antibody should be tested) overnight (12–20 hr) at room temperature. Note that some antibodies are sensitive to high concentrations of Triton or to overnight room temperature incubation. It may be necessary to decrease the Triton to 0.1% or to incubate with the primary antibody at 4°C.

12. Wash animals several times with GBP.

13. Incubate with 75–100 μl secondary antibody for 45–60 min at room temperature. (Cappel-, rhodamine-, or fluorescein-conjugated goat anti-rabbit or anti-mouse diluted 1 : 100 in GBP are commonly used secondary antibodies.)

14. Wash animals several times with 0.1 *M* PB.

15. Mount for observation as described in Section V,F.

D. Antibody Staining Homogenized Worms

See Li and Chalfie (1990). This procedure is good for staining the nervous system, muscle, and structures along the body wall. The intestine and gonad are extruded in this preparation.

1. Harvest animals from plates by overlaying with 2 ml M9 and pouring off suspended nematodes into a 15-ml disposable tube. One to several small plates yield enough animals for most purposes.

2. Wash away bacteria by letting the animals settle, aspirating off the M9 buffer, adding 5 ml M9, letting the worms settle again, and aspirating off the M9 buffer.

3. Fix 12–36 hr at 4°C in about 1 ml 4% paraformaldehyde in 0.1 M PB, pH 7.4. Fixation time can be decreased if antigen is sensitive to overfixation.

4. Wash animals 6× with 0.1 M PB.

5. Pipet animals and about 750 μl 0.1 M PB into the bottom of a tissue homogenizer (2 ml Pyrex tissue grinder #7727). Add about 750 μl 0.1 M PB. Dounce animals 4–6×. Monitor breakage of animals under the dissecting microscope by putting a drop of worms on a slide. A good mix is about 50% animal shells and 50% whole or partially broken animals. Animal shells become overly fragmented with too much douncing.

6. Transfer animals to a microcentrifuge tube. The animals can be stored indefinitely at 4°C.

7. Transfer about 10 μl animals into a 0.5-ml microcentrifuge tube. Incubate with GPB for at least 30 min at 37°C. Alternatively, incubate at room temperature for a longer time.

8. Incubate with 50–100 μl primary antibody diluted in GBP (initially a 10^{-1}–10^{-4} dilution series of the antibody should be tested) overnight (12–20 hr) at room temperature. Note that some antibodies are sensitive to high concentrations of Triton or to overnight room temperature incubation. It may be necessary to decrease the Triton to 0.1% or to incubate with the primary antibody at 4°C.

9. Wash animals 6× with GBP.

10. Incubate with 75–100 μl secondary antibody for 45–60 min at room temperature. (Cappel-, rhodamine-, or fluorescein-conjugated goat anti-rabbit or anti-mouse diluted 1 : 100 in GBP are commonly used secondary antibodies.

11. Wash animals several times with 0.1 M PB.

12. Mount for observation as described in Section V,F.

E. Antibody Staining Fixed Worms

(M. Finney, personal communication)

1. Solutions

2× MRWB: 800 μl 2 M KCl (160 mM final), 80 μl 5 M NaCl (40 mM final), 2.0 ml 0.1 M EGTA (20 mM final), 1.0 ml 0.1 M spermidine (10 mM final), 600 μl 0.5 M PIPES, pH 7.4 (15.1 g/100 ml), 5.0 ml 100% MeOH (50% final), ddH$_2$O up to 10 ml

Tris-Triton buffer (TTB): 1.0 ml 1 M Tris HCl, pH 7.4, 100 μl Triton X-100, 20 μl 0.5 M EDTA, ddH$_2$O up to 10 ml

40× BO$_3$ buffer (pH 9.2): 618 mg H$_3$BO$_3$, 5.0 ml 1 M NaOH, ddH$_2$O to 10 ml (pH to 9.2, heat to dissolve)

AbA: 2.0 ml 5× PBS, 5.0 ml 2% BSA, 50 μl Triton X-100, 250 μl 2% NaN_3, 20 μl 0.5 M EDTA, ddH_2O up to 10 ml

AbB: same as AbA, except 500 μl 2% BSA included

5× PBS: 4.0 g NaCl, 100 mg KH_2PO_4, 100 mg KCl, 590 mg Na_2HPO_4, ddH_2O to 100 ml (pH 7.4, store at 4°C)

20% formaldehyde: weigh out about 250 mg paraformaldehyde (but <300 mg) and place in a 15-ml tube; multiply the weight in mg by 4.5 and add that volume in microliters of 5 mM NaOH; heat in 65°C water bath for 30 min with occasional mixing; make fresh

30% H_2O_2: commercially available stock

2. Protocol

1. Harvest worms by washing plates with a few mls H_2O. Pour off nematode suspension into a 15-ml tube and spin worms at 3000 rpm for 5 min, or let settle. Wash again with H_2O, spin, and aspirate off supernatant.

2. Resuspend worms in 500 μl H_2O and transfer to a microcentrifuge tube.

3. Add H_2O to fill the tube; spin at 3000 rpm for 2 min. Gently aspirate off supernatant.

4. Prepare fixative:

    1.0 ml 2× MRWB

    100 μl 20% formaldehyde

    890 μl ddH_2O

5. Add 1.25 ml fixative per tube. Mix by gentle inversion. Do not shake.

6. Immerse the tube in dry ice/ethanol to freeze contents. Watch the tube to make sure it does not open. Frozen samples may be processed immediately or stored indefinitely.

7. Initiate thawing by rubbing the tube between your hands. Place tube on ice to thaw further. Incubate 45 min to overnight on ice. For short incubation times, mix periodically by gentle inversion (approximately every 5 min).

8. Spin worms at 3000 rpm for 2 min. Gently aspirate off supernatant.

9. Resuspend gently in 1 ml 1× TTB. Spin at 3000 rpm for 2 min. Gently aspirate off supernatant.

10. Resuspend gently in 1 ml 1× TTB + 1% β-mercaptoethanol.

11. Incubate at 37°C for 1.5–2 hr with gentle shaking.

12. Spin worms at 3000 rpm for 2 min. Gently aspirate off supernatant.

13. Resuspend in 1 ml 1× BO_3 buffer. Spin down worms and aspirate off supernatant.

14. Resuspend in 1 ml 1× BO_3 + 10 mM DTT. Shake gently for 15 min.

15. Spin worms at 3000 rpm for 2 min. Gently aspirate off supernatant. Resuspend in 1 ml $1\times$ BO_3, spin down worms, and remove supernatant.

16. Resuspend in 1 ml $1\times$ BO_3 + 0.3% H_2O_2. Shake at room temperature for 15 min. Watch tubes carefully since tops may pop open. (Parafilm wrap may be necessary to prevent this.)

17. Spin worms at 3000 rpm for 2 min. Gently aspirate off supernatant.

18. Resuspend worms in 1 ml $1\times$ BO_3 buffer. Spin down worms and remove supernatant.

19. Resuspend worms in 1 ml AbB. Spin down at 3000 rpm for 2 min. Aspirate off supernatant and resuspend in 1 ml AbA.

20. Transfer a 15- to 50-μl aliquot of fixed worms to a microcentrifuge tube for staining.

21. Add an appropriate dilution of antibody (initially dilutions should be 10^{-1}–10^{-4}) in 200–500 μl AbA. Incubate at room temperature for 2 hr or overnight at 4°C. Gentle shaking may be helpful. (Wrap in tin foil if using fluorescently tagged probes such as rhodamine- or fluorescein-labeled primary Abs.)

22. Spin worms at 3000 rpm for 2 min. Gently aspirate off supernatant.

23. Resuspend in AbB and repeat spin. Aspirate off supernatant. Resuspend in AbB and shake gently at room temperature for 2 hr.

24. Spin worms at 3000 rpm for 2 min. Gently aspirate off supernatant.

25. Add an appropriate dilution of fluorescently tagged secondary antibody (1/100 is a good dilution to try first) in 200–500 μl AbA. Incubate at room temperature or 37°C for 2 hr to overnight with gentle shaking. (Keep the tube wrapped in foil.)

26. Spin worms at 3000 rpm for 2 min. Gently aspirate off supernatant.

27. Resuspend worms in AbB and spin at 3000 rpm for 2 min. Aspirate off supernatant. Resuspend in AbB and shake gently at room temperature for 2 hr.

28. Repeat spin, remove supernatant, resuspend in AbA, repeat spin, and remove most supernatant.

29. Mount worms for observation as described in Section V,F.

Note. This protocol works well except on dauer animals and hypochlorite-treated eggs. Hypochlorite treatment and fixation are sufficient to open eggs. Formaldehyde concentration and fixation time should be optimized for each antiserum. A recommended starting condition is 1% formaldehyde for 0.5 hr. More fixation will stabilize some antigens but destroy others. It is reported to be important to spin animals rather than merely let them settle since the mechanical stress appears to increase permeability.

F. Mounting Animals for Microscopy

1. Place 5–10 μl animals in a small drop of 70% glycerol, 30% PB, 5% by weight *N*-propyl gallate and DAPI (1 mg/ml). DAPI is not essential (and can certainly be omitted for mounting animals that have been DAPI stained), but is added so that nuclei can be visualized and cells can be easily identified.
2. Overlay with a coverslip and seal edges with nail polish. Slides can be stored for years in the freezer.

Note. *N*-Propyl gallate slows the bleaching of the fluorochrome. To get *n*-propyl gallate into solution, incubate at 68°C for about 30 min. The solution will turn slightly yellow and can be kept indefinitely at 4°C. Use PB at pH 7.4 for rhodamine, PB at pH 9.5 for fluorescein. When looking at double-labeled samples, optimal conditions for fluoroscein are preferable since fluoroscein fades more quickly than rhodamine.

Acknowledgments

I am indebted to individuals who introduced me to *C. elegans* and provided protocols that have been included, especially M. Chalfie, M. Finney, E. Lambie, C. Li, and D. Miller. I thank J. Sulston and C. Link for providing photographs used in Figure 1, A. Kalgaonkar for help with other figures, and G. Kao, R. Padgett, S. Shaham, and D. Xue for comments on sections of the manuscript.

References

Brenner, S. (1974). The genetics of *Caenorhabditis elegans*. *Genetics* **77**, 71–94.

Canessa, C. M., Horsiberger, J. D., and Rossier, B. C. (1993). Functional cloning of the epithelial sodium channel: Relation with genes involved in neurodegeneration. *Nature* **361**, 467–470.

Canessa, C. M., Schild, L., Buell, G., Thorens, B., Gautschi, I., Horsiberger, J. D., and Rossier, B. C. (1994). Amiloride-sensitive epithelial Na⁺ channel is made up of three homologous subunits. *Nature* **367**, 463–467.

Chalfie, M., and Wolinsky, E. (1990). The identification and suppression of inherited neurodegeneration in *Cuenorhabditis elegans*. *Nature* **345**, 410–416.

Chalfie, M., Yuan, T., Euskirchen, G., Ward, W. W., and Prasher, D. C. (1994). Green fluorescent protein as a marker for gene expression. *Science* **263**, 802–805.

Coulson, A., Sulston, J., Brenner, S., and Karn, J. (1986). Towards a physical map of the genome of the nematode *Caenorhabditis elegans*. *Proc. Natl. Acad. Sci. USA* **83**, 7821–7825.

Coulson, A., Waterston, R., Kiff, J., Sulston, J., and Kohora, Y. (1988). Genome linking with artificial chromosomes. *Nature* **335**, 184–186.

Coulson, A., Kozono, Y., Lutterbach, B., Shownkeen, R., Sulston, J., and Waterston, R. (1991). YACs and the *C. elegans* genome. *BioEssays* **13**, 413–417.

Driscoll, M. (1992). Molecular genetics of cell death in the nematode *Caenorhabditis elegans*. *J. Neurobiol.* **23**, 1327–1351.

Driscoll, M., and Chalfie, M. (1991). The *mec-4* gene is a member of a family of *Caenorhabditis elegans* genes that can mutate to induce neuronal degeneration. *Nature* **349**, 588–593.

Ellis, H. M., and Horvitz, H. R. (1986). Genetic control of programmed cell death in the nematode *Caenorhabditis elegans*. *Cell* **44**, 817–829.

Ellis, R. E., Yuan, J., and Horvitz, H. R. (1991a). Mechanisms and functions of cell death. *Annu. Rev. Cell Biol.* **7,** 663–698.

Ellis, R. E., Jacobson, D. M., and Horvitz, H. R. (1991b). Genes required for the engulfment of cell corpses during programmed cell death in *Caenorhabditis elegans. Genetics* **129,** 79–94.

Epstein, H. F., and Shakes, D. C. (1995). Methods in Cell Biology Vol. 48: *Caenorhabditis elegans:* Modern Biological Analysis of an Organism. Academic Press, San Diego, CA.

Fire, A. (1986). Integrative transformation of *C. elegans. EMBO J.* **5,** 2673–2680.

Fire, A., Kondo, K., and Waterston, R. (1990a). Vectors for low copy transformation of *C. elegans. Nucleic Acids Res.* **18,** 4269–4270.

Fire, A., Harrison, S. W., and Dixon, D. (1990b). A modular set of *lacZ* fusion vectors for studying gene expression in *Caenorhabditis elegans. Gene* **93,** 189–198.

Fire, A., Albertson, D., Harrison, S. W., and Moerman, D. G. (1991). Production of antisense RNA leads to effective and specific inhibition of gene expression in *C. elegans* muscle. *Development* **113,** 503–514.

Hamelin, M., Scott, I. M., Way, J. C., and Culotti, J. G. (1992). The *mec-7* Beta-tubulin gene of *Caenorhabditis elegans* is expressed primarily in the touch receptor neurons. *EMBO J.* **11,** 2885–2893.

Hedgecock, E., Sulston, J. E., and Thomson, J. N. (1983). Mutations affecting programmed cell death in the nematode *Caenorhabditis elegans. Science* **220,** 1277–1280.

Hengartner, M. O., and Horvitz, H. R. (1994). *C. elegans* cell survival gene *ced-9* encodes a functional homolog of the mammalian proto-oncogene *bcl-2. Cell* **76,** 665–676.

Hengartner, M. O., Ellis, R. E., and Horvitz, H. R. (1992). *C. elegans* gene *ced-9* protects cells from programmed cell death. *Nature* **356:**494–499.

Hong, K., and Driscoll, M. (1994). Residues in a transmembrane domain of the *C. elegans* MEC-4 predicted to line a channel pore influence mechanotransduction and neurodegeneration. *Nature* **367,** 470–473.

Huang, M., and Chalfie, M. (1994). Gene interactions affecting mechanosensory transduction in *Caenorhabditis elegans. Nature* **367,** 467–470.

Kramer, J. M., French, R. P., Park, E. C., and Johnson, J. J. (1990). The *Caenorhabditis rol-6* gene, which interacts with the *sqt-1* collagen gene to determine organismal morphology, encodes a collagen. *Mol. Cell Biol.* **10,** 2081–2089.

Li, C., and Chalfie, M. (1990). Organogenesis in *C. elegans:* Positioning of neurons and muscles in the egg-laying system. *Neuron* **4,** 681–695.

Mello, C. C., Kramer, J. M., Stinchcomb, D., and Ambros, V. (1992). Efficient gene transfer in *C. elegans:* Extrachromosomal maintenance and integration of transforming sequences. *EMBO J.* **10,** 3959–3970.

Miura, M., Zhu, H., Rotello, R., Hartweig, E. A., and Yuan, J. (1993). Induction of apoptosis in fibroblasts by IL-1β-converting enzyme, a mammalian homolog of the *C. elegans* cell death gene *ced-3. Cell* **75,** 653–660.

Perry, M. D., Li, W., Trent, C., Robertson, B., Fire, A., Hageman, J. M., and Wood, W. B. (1993). Molecular characterization of the *her-1* gene suggests a direct role in cell signaling during *Caenorhabditis elegans* sex determination. *Genes Dev.* **7,** 216–228.

Robertson, A. M. G., and Thomson, J. N. (1982). Morphology of programmed cell death in the ventral nerve cord of *C. elegans* larvae. *J. Embryol. Exp. Morphol.* **67,** 89–100.

Stringham, E. G., Dixon, D. K., Jones, D., and Candido, E. P. M. (1992). Temporal and spatial expression patterns of the small heat shock (hsp-16) genes in transgenic *Caenorhabditis elegans. Mol. Biol. Cell* **3,** 21–233.

Sulston, J. E., and Horvitz, H. R. (1977). Post-embryonic cell lineages of the nematode *Caenorhabditis elegans. Dev. Biol.* **56,** 110–156.

Sulston, J. E., Schierenberg, E., White, J. G., and Thomson, J. N. (1983). The embryonic cell lineage of the nematode *Caenorhabditis elegans. Dev. Biol.* **100,** 64–119.

Vaux, D. L., Weissman, I. L., and Kim, S. K. (1992). Prevention of programmed cell death in *Caenorhabditis elegans* by human bcl-2. *Science* **258,** 1955–1957.

Yuan, J. Y., and Horvitz, H. R. (1992). The *Caenorhabditis elegans* cell-death gene *ced-4* encodes a novel protein and is expressed during the period of extensive programmed cell death. *Development* **116,** 309–320.

Yuan, J., Shaham, S., Ledoux, S., Ellis, H. M., and Horvitz, H. R. (1993). The *C. elegans* cell death gene *ced-3* encodes a protein similar to mammalian interleukin-1β-converting enzyme. *Cell* **75,** 641–652.

Wilson, R., Alnscough, R., Anderson, K., Baynes, C., Berks, M., Bonfield, J., Burton, J., Connell, M., Copsey, T., Cooper, J., Coulson, A., Craxton, M., Dear, S., Du, Z., Durbin, R., Favello, A., Fraser, A., Fulton, L., Gardner, A., Green, P., Hawkins, T., Hillier, L., Jler, M., Johnston, L., Jones, M., Kershaw, J., Kirsten, J., Laisster, N., Latreille, P., Lightning, J., Lloyd, C., Mortimore, B., O'Callaghan, M., Parsons, J., Percy, C., Rifken, L., Roopra, A., Saunders, D., Shownkeen, R., Sims, M., Smaldon, N., Smith, A., Smith, M., Sonnhammer, F., Staden, R., Sulston, J., Thierry-Mieg, J., Thomas, K., Vaudin, M., Vaughan, K., Waterston, R., Watson, A., Weinstock, L., Wilkinson-Sproat, J., and Wohldman, P. (1994). 2.2 Mb of contiguous nucleotide sequence from chromosome III of *C. elegans*. *Nature* **368,** 32–38.

Wood, W. B., ed. (1988). The Nematode *Caenorhabditis elegans*. "Cold Spring Harbor Laboratory Press, Cold Spring Harbor, New York.

CHAPTER 15

Programmed Cell Death during Mammary Gland Involution

Robert Strange,* Robert R. Friis,† Lynne T. Bemis,* and F. Jon Geske*

* Division of Laboratory Research
AMC Cancer Research Center
Lakewood, Colorado 80214

† Laboratory for Clinical and Experimental Research
University of Bern
CH-3004 Bern, Switzerland

I. Introduction

The death of secretory mammary epithelium that occurs during the postlactational involution of the mammary gland provides a normal physiological setting for studying apoptosis during an important developmental process. As evidenced by morphological changes, oligonucleosomal fragmentation of DNA,

Copyright © 1995 by Academic Press, Inc. All rights of reproduction in any form reserved.

and a pattern of concerted gene expression that includes genes associated with cell death in other systems (SGP-2/clusterin, p53, TGFβ1, c-*myc* and tissue transglutaminase), the bulk of secretory mammary epithelium dies by apoptotic cell death during involution (Walker *et al.*, 1989; Strange *et al.*, 1992). Although initiation of mammary gland involution and the death of secretory epithelium is due to the loss of lactogenic hormone stimulation which occurs after weaning of the young, much less is known about the genes that regulate and execute this process. Mammary involution also involves significant tissue remodeling and renewal in preparation for another cycle of lactation (Strange *et al.*, 1992; Talhouk *et al.*, 1992). This feature complicates the goal of identifying apoptosis-specific genes. Experimental manipulation of remodeling enzyme levels is reported to alter gene expression and the progress of mammary involution (Talhouk *et al.*, 1992). Although this work did not evaluate changes in apoptosis, apoptotic cell death and tissue remodeling appear to be co-regulated or somehow coupled to one another. Certainly *in vivo,* mammary epithelial cells require mesenchymal interactions and an intact basement membrane to maintain differentiated function.

Researchers also have proposed that under the influences of pregnancy and lactogenic hormone stimulus, secretory mammary epithelium becomes terminally differentiated into milk-producing cells and committed to an apoptotic cell death pathway (Ossowski *et al.*, 1979). This hypothesis is supported by experimental manipulations of hormone levels which have shown that after weaning, lactation can be maintained and the progress of involution impeded by injection of hormones (Ossowski *et al.*, 1979; R. Strange and R. Friis, unpublished data). Thus, current understanding supports a role for both the alteration of epithelial–mesenchymal interactions and terminal differentiation in apoptotic death of secretory mammary epithelium. Elements of both are likely to be accurate. This presents an interesting background for detecting and understanding regulation and execution of apoptosis.

A. Mammary Carcinogenesis: Proliferation vs Apoptosis

Mammary tumorigenesis is influenced by the developmental history of the mammary gland. Nulliparity is a significant risk factor, whereas pregnancy and lactation are associated with reduced risk for development of breast cancer (Ewertz *et al.*, 1990; Harris *et al.*, 1992; Russo *et al.*, 1992; Yoo *et al.*, 1992; Yang *et al.*, 1993). Normal development and maintenance of tissue size results from a balance between cell proliferation and cell death. A perturbation in this balance contributes to the development of neoplasia (Williams, 1991; Bursch *et al.*, 1992; Raff, 1992). The balance between proliferation and apoptotic cell death of mammary epithelium may offer an explanation for differences in breast cancer risk. For example, hormonal risk factors for development of breast cancer, such as early menarche and late menopause, promote or permit epithelial cell proliferation which allows accumulation of the mutations requisite for neoplastic development (Cohen and Ellwein, 1990; Preston-Martin *et al.*, 1990). Conditions that limit proliferation or cause cell loss, such as ovariectomy and perhaps apoptotic cell death following pregnancy and lactation, are correlated with reduced risk for developing breast cancer (Ewertz *et al.*, 1990; Harris *et al.*, 1992; Yang *et al.*, 1993). In contrast, an inhibition or a reduction of apoptotic cell death has been associated with mammary tumorigenesis in transgenic mice and with increased risk for breast cancer in humans (Andres *et al.*, 1991; Allan *et al.*, 1992). To the extent that cell proliferation is balanced by cell death, a failure in apoptosis can be viewed as essential for carcinogenesis

II. Induction of Apoptotic Cell Death in Mouse Mammary Gland

In mice, mammary gland involution and the apoptotic cell death that is an integral part of this process are normally induced by the weaning of young at 21 days of age. In fact, toward the end of lactation some of the changes associated with involution and apoptotic cell death have already begun (Walker *et al.*, 1989). Presumably this is due to a reduced demand for milk as the young switch from milk to solid food. This result suggests that the level of lactogenic hormone stimulation is a major controlling factor in maintaining lactation and inhibiting involution and apoptotic cell death. Thus, mammary gland involution and the concomitant apoptotic cell death are initiated by a physiological form of hormone ablation. Indeed, administration of different lactogenic hormones has been found to inhibit mammary gland involution even after the weaning of young (Ossowski *et al.*, 1979; R. Strange and R. Friis, unpublished data). Thus, weaning of young is a simple and physiologically relevant method of inducing apoptotic cell death in the mammary gland.

A. Basic Weaning Protocol

Because the process of mammary gland involution is modulated by many factors (reproductive history, duration of lactation, number of offspring) and is subject to interanimal variation, the window through which one views apoptotic cell death during mammary gland involution can have significant effects on what is seen and, thus, on interpretation. We have chosen the period after the establishment of full lactation as a point for initiating involution. This reduces the interanimal variation found if weaning takes place very early or at 21 days, and allows time for completion of differentiation events that may be important in the commitment to a program of cell death.

The simplest manipulations of involution involve the duration of lactation before weaning of young, either by allowing natural weaning or by enforcing weaning at earlier points in lactation. The process of mammary gland involution can also be manipulated by various forms of hormonal ablation or supplementation. The simplest method for these studies is to inject animals with the hormone of choice (Ossowski *et al.*, 1979). Generally, the lactogenic hormones prolactin (10 U, intramuscular injection, 2 times/day), hydrocortisone (10 mg subcutaneous injection, 1–2 times/day), and oxytocin (0.6 U, intraperitoneal injection, 3 times/day) have been used. This method, although lacking precision, can suggest the influence of individual hormones on the expression of particular genes or proteins. For greater control and reproducibility, a more precisely regulated delivery device such as a time-release pellet or a pump would be preferable (Daniel *et al.*, 1989; Talhouk *et al.*, 1992).

The mammary glands of adult (10 wk of age) virgin, pregnant, and fully regressed mice are useful control tissues in studies of postlactational apoptotic cell death. The mammary glands are removed from virgins and from pregnant mice at day 14–16 of pregnancy. For samples from involuting mammary glands, mice are allowed to deliver their young and litter size is normalized by day 2 of lactation. After lactation is well established (7–8 days), the young are removed. The mice are sacrificed and mammary tissue removed at chosen intervals after weaning (e.g., days 1, 2, 3, 4, 6, 8, and 10 postweaning) and, for full regression, 2 mo after weaning. The mammary tissue is frozen in liquid nitrogen and stored at $-80°C$ for nucleic acid extraction, protein extraction, and preparation of frozen sections. For histological analysis, tissues are fixed using 4% formaldehyde, freshly prepared from paraformaldehyde, in phosphate buffered saline (PBS: 50 mM KH_2PO_4, 150 mM NaCl, pH 7.4) or methacarn (methanol, chloroform, glacial acetic acid; 60 : 30 : 10).

B. Important Considerations in Planning Weaning Protocols

Apoptosis in the involuting mouse mammary gland is rapid. It is completed by 8 days postweaning, but tissue remodeling continues until 10–12 days. Preparation for the next cycle of pregnancy and lactation is complete by 3–4 wk after weaning (Walker *et al.*, 1989; Strange *et al.*, 1992). Many factors must be considered in designing an involution experiment.

1. *Duration of lactation.* Times as short as removal of young immediately postpartum and as long as weaning at 21 days have been used. Involution appears to be more rapid at shorter times (0–2 days of lactation) and slower as one approaches 21 days, the normal length of lactation. Presumably this occurs because (1) full lactation and differentiation of mammary epithelium is not established in short periods of lactation and (2) involution has already begun as the period of lactation approaches normal weaning at 21 days. Full lactation is established at about 7 days.

2. *Number of pups nursed.* Normalization of the litter size is important so the mammary glands of different animals receive about the same level of suckling and, thus, similar lactogenic stimulus. The particular number of pups will vary with the strain of rat or mouse. Normalization should be

carried out early, within the first 2 days after birth, or the mother is likely to reject an introduced pup.

3. *The animal model used.* Compared to the mouse, the rat appears to have a longer period of mammary involution, approaching 14 days in duration, with complete regression delayed until 1 mo or longer (Walker *et al.,* 1989). We have also seen some differences in patterns of gene expression between strains of mice. For example, BALB/c mice show a partial, more locally variable involution than Swiss mice. The period of involution must be empirically determined for each strain or species.

4. *Exclusion of other tissues.* Lymph node and muscle tissue should be excluded from collection with mammary gland tissue. In particular, a band of muscle found between the overlapping second and third mammary fat pads of the thoracic mammary glands should be excluded. The lymph nodes in the fourth mammary fat pad (inguinal mammary gland) should be excised before removing the mammary tissue. For mice, the second, third, and fourth mammary glands provide the bulk of available tissue. For rats, the number five mammary gland is also useful.

5. *Number of animals per time point.* The number of animals used per time point should be sufficient to avoid interanimal variation. These differences can be significant.

6. *Reproductive history.* The reproductive history of the animals may have an influence on involution. The prior exposure of lactogenic hormones and, thus, the differentiation associated with induction of apoptotic cell death and tissue remodeling could alter basal levels of gene expression, the rapidity of onset, and the levels of induced gene expression in a subsequent cycle of lactation and involution. We have observed differences in expression of tissue inhibitor of metalloproteinase (TIMP) between primiparous and multiparous mice during postlactational involution (Strange *et al.,* 1992; Talhouk *et al.,* 1992).

C. Surgical Removal of Mammary Tissue

Pups that are not close to the normal time of weaning, 21 days, will not survive forced weaning and must be either foster nursed or euthanized by inhalation of CO_2 gas. The experimental animals should be anesthetized (Nembutal solution, intraperitoneally; Abbott Laboratories; North Chicago, IL; inhalation of Metofane; Pittman-Moore, Mundeline, IL) or euthanized by cervical dislocation for surgical removal of mammary glands. The animals are attached to cork or styrofoam boards with tape or pins if the animal has been euthanized. If the animal is euthanized prior to removal of tissue or will be afterward, the mammary glands (five pair in the mouse or six pair in the rat) are exposed by making an inverted Y-incision in the abdomen of the animal with the arms of the Y extending into the inguinal area and the base of the Y extending to the neck of the animal (Fig. 1). Survival surgeries require a much smaller incision that is specific to the immediate area of the particular mammary gland being removed (Medina, 1973; Strange *et al.*, 1989). The skin is reflected away from the body. The mammary fat pad can adhere to either the skin or the peritoneum, so care must be taken to avoid losing or damaging the tissue. In both mice and rats, the mammary tissue may be removed with sharp scissors and forceps. Rat mammary glands require more effort to remove than those of mice. Take care not to include skin, muscle, or lymph nodes in the mammary tissue to be used for protein or nucleic acid extractions (Fig. 1). The mouse has two, and

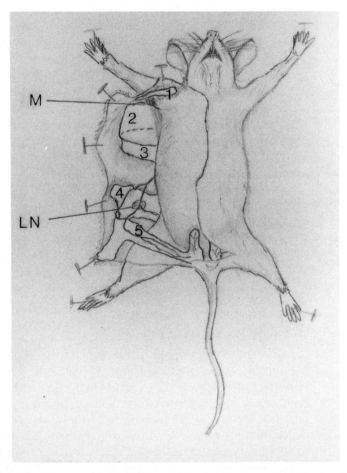

Fig. 1 Drawing of a mouse showing the position of mammary fat pads (1–5). Also shown are the lymph node (LN) in the #4 fat pad (another lymph node proximal to the bodywall is not visible) and the muscle (M) between fat pads 2 and 3 that should be excluded from collection with mammary tissue.

occasionally three, lymph notes in the inguinal (#4) mammary fat pad; the rat has a chain of three in the inguinal (#4) mammary fat pad. Muscle is found running between the layers of the second and third mammary fat pads (second, third, and fourth in the rat). This muscle can be retracted or removed to avoid inclusion. The skin can be included inadvertently with mammary tissue when cutting the mammary fat pad free from the skin, particularly in the rat. The tissue should be handled as little as possible and fixed, quick frozen in liquid nitrogen, or extracted immediately. The #1 and #5 fat pads of the mouse and the #6 fat pad in the rat are generally smaller (except in lactating and early involuting mammary glands) and may be excluded from removal.

Nembutal Solution

Use this solution for mouse anesthesia (Medina, 1973).

30 ml saline

9 ml propylene glycol

5 ml absolute alcohol

5 ml Nembutal solution (Nembutal-Sodium, 50 mg/ml; Abbott Laboratories)

This solution may be stored at room temperature for several months. Inject mice intraperitoneally with 0.01 ml/g body weight of Nembutal solution; reduce total dose by 0.01 ml for mice under 10 g. The anesthesia lasts 30–45 min.

III. Detection and Evaluation of Apoptotic Cell Death in Mammary Epithelium

A. Morphological Criteria

The morphological changes typical of apoptotic cell death are readily observed during postlactational mammary gland involution (see Chapters 1 and 2, this volume). The lactating mammary gland is composed of lobuloalveolar structures that are the synthetic units for milk production. Very few mesenchymal cells are detectable in the mammary gland during lactation. The mammary gland is engorged with milk 1 day after weaning. Although involution is reversible at this point, cells with condensed nuclei consistent with the first morphological changes of cell death are seen. Involution becomes irreversible by 2 days after weaning. At this time, the lobuloalveolar units begin to collapse and secretory epithelium shows the morphology of physiological cell death, including nuclear and cytoplasmic condensation, followed by nuclear fragmentation and formation of apoptotic bodies (Walker et al., 1989; Strange et al., 1992). By days 3 and 4 after weaning, apoptotic cells are seen throughout the collapsed and collapsing lobuloalveolar units. These apoptotic cells decrease and, between days 6 and 8 postweaning, become undetectable. Adipocytes which were difficult to detect during lactation are quite evident by days 3 and 4 after weaning and increase in number until they are the predominant cell component in the fully regressed mammary gland. Finally, mesenchymal changes (folding and thickening of the ECM) occur that are consistent with significant tissue remodeling (Martinez-Hernandez *et al.*, 1976; Warburton *et al.*, 1982; Dickson and Warburton, 1992; Strange *et al.*, 1992). This change is particularly evident between days 6 and 8 postweaning.

B. Evaluation of DNA Integrity

DNA is prepared from mammary tissue quick-frozen in liquid nitrogen and stored at $-80°C$ (see Chapter 3, this volume). Mammary tissue (500 mg/2 ml) is homogenized until connective tissue is well disrupted (about 10 passes) with a dounce tissue homogenizer in DNA extraction buffer (100 mM Tris-HCl, pH 7.5, 10 mM NaCl, 10 mM EDTA, 0.2 mg/ml proteinase K). The homogenate is then transferred to a clean tube (microcentrifuge or larger depending on the amount of tissue), 0.5% SDS is added, and the lysate is incubated at 50°C for 1–2 hr. The lysate is extracted with 2 vol phenol/chloroform/isoamyl alcohol (25 : 24 : 1). After centrifugation, the upper aqueous phase is removed and extracted with 1–2 vol chloroform : isoamyl alcohol (24 : 1). After centrifugation, DNA is precipitated from the aqueous phase with 0.1 vol sodium acetate (2.5 M, pH 5.2) in 2 vol 95% EtOH on ice or at $-20°C$ for 1 hr. The precipitate is pelleted by centrifugation in a microfuge for 20 min, then washed with 70% EtOH. After the second centrifugation, the supernatant is discarded and the pellet is resuspended in TE (pH 7.5). Prior to electrophoresis, samples are treated with DNase-free RNase as needed. Early involution samples have significant amounts of milk protein RNAs. Samples (7.5 μg) are electrophoresed in a 1.5% agarose, TBE gel with 0.25 μg/ml ethidium bromide added to the gel and the running buffer. A 123-bp DNA ladder (GIBCO/BRL, Grand Island, NY) is used as a molecular weight marker. Figure 2 shows the pattern of DNA fragmentation observed during the period of postlactational apoptotic cell death of mammary epithelium. Fragmentation is detectable as early as 1 day after weaning and peaks between days 3 and 4 after weaning.

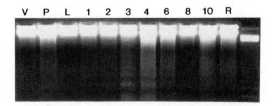

Fig. 2 Electrophoretic analysis of DNA extracted from mouse mammary glands at different development stages: V, virgin; P, pregnant; L, lacation; involution at days 1, 2, 3, 4, 6, 8, and 10 after weaning; R, full regression.

C. Identification of Apoptotic Cells by Terminal Transferase End–Labeling

See Chapter 2 for the general protocol. Tissue samples were fixed for 4 hr 0°C using freshly prepared 4% formaldehyde in PBS. Histological samples were embedded in paraffin and 4-μm sections were obtained, after which hematoxylin–eosin staining was performed. The procedure of Gavrieli *et al.* (1992) was employed with several modifications. Boehringer Mannheim (Heidelberg, Germany) digoxinin-labeled dUTP was employed in the terminal transferase reaction and was detected using alkaline phosphatase-conjugated digoxinin-specific antibodies (Boehringer Mannheim) developed for 30–90 min at room temperature with nitro-blue tetrazolium and 5-bromo 4 chloro-3-indolyl phosphate substrate (Boehringer Mannheim) in the presence levamisole. The major variable to work out in mammary tissue is the length of time required for the protease treatment to allow access to nucleic acids. The length of treatment must be determined empirically, but a reasonable starting point is 15–20 min of treatment with 10–20 μg/ml proteinase K in 20 mM Tris-HCl (pH 8.0) with 2 mM EDTA. It is also important to avoid overfixation. Color Plate 7 shows a 4-μm section from an involuting mouse mammary gland 3.5 days after weaning of young. In this sample, 30–40% of the mammary epithelial cells show staining by terminal transferase end labeling.

D. Gene Expression

Northern analysis of RNA from involuting mammary glands is useful for monitoring changes in apoptotic cell death associated with experimental manipulations. We have detected specific patterns of gene expression. Some of these genes were chosen empirically. Others were detected in differential screening of cDNA libraries from lactating and involuting mammary glands (see Chapter 7, this volume; Li *et al.*, 1994). Figure 3 shows Northern analysis of several genes whose expression is associated with apoptosis. Such putative involution and cell death associated genes include apoptosis-related genes—SGP-2/clusterin, tTG, c-*myc*, p53, and TGFβ1—and possible regulatory genes—24p3, a putative retinoid-binding protein, LZP (a human X-box binding protein related cDNA), and several novel cDNAs (Strange *et al.*, 1992; Li *et al.*, 1994). In normal involuting mouse and rat mammary gland, SGP-2/clusterin, p53, and TGFβ1 are expressed early, particularly SGP-2/clusterin (Fig. 3). In contrast, expression of tissue transglutaminase and c-*myc* is detected relatively late in studies of both mouse and rat mammary involution (Strange *et al.*, 1992; P. Schedin and R. Strange, manuscript in preparation). Control of mammary epithelial cell function, particularly milk protein synthesis, during involution has been hypothesized to be directly influenced by the integrity of, and perhaps regulated by the expression of components of, the extracellular matrix (Streuli *et al.*, 1991; Talhouk *et al.*, 1992). As an indicator for effects of alterations in normal mammary gland differentiated function, expression of remodeling enzymes—stromelysins I and III, plasminogen activators, WDNM1, TIMP,

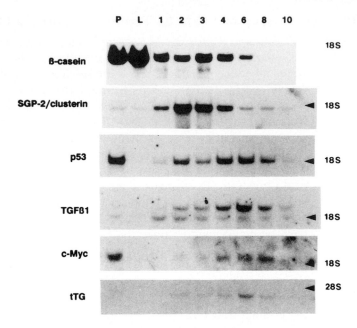

Fig. 3 Northern blot showing expression of genes associated with apoptotic cell death during pregnancy (P), lactation (L), and days 1, 2, 3, 4, 6, 8, and 10 of mammary gland involution, compared with the expression of β-casein, a milk protein gene.

PAI 1—and differentiation markers—glyCAM-1, β-casein, WAP (whey acidic protein), ODC (ornithine decarboxylase), and ADRP (adipose differentiation related protein)—have shown specific regulation during mammary gland involution (Strange *et al.,* 1992; Talhouk *et al.,* 1992; Li *et al.,* 1994). To adjust for equal loading of total RNA samples, we have used acridine orange staining of the gel. For more precise estimates of equal loading and for poly(A)-enriched samples, we have used actin message levels (Strange *et al.,* 1992; Li *et al.,* 1994). Unfortunately, we have not found a gene whose expression is not altered to some degree during mammary gland development. This is particularly true during lactation, when the level of milk protein gene expression may overwhelm other expression. Table I lists the genes that show regulated patterns of expression during the period of apoptotic cell death during mammary gland involution.

IV. Summary

Understanding the cascade of gene expression and subsequent protein interactions that result both in the death of secretory mammary epithelium and the remodeling and renewal of the mammary gland for another cycle of lactation

Table I
Genes Expressed during Mammary Gland Involution

Gene	Putative role(s)	Reference
Adipocyte differentiation related protein (ADRP)	Gene expressed during adipocyte differentiation	Li *et al.* (1994)
α-Lactalbumin	Milk protein	N. L. Shaper, R. Strange, and R.R. Friis, unpublished data
bcl-2	Inhibitor of apoptosis expressed in resting mammary epithelium	Hockenberry et al. (1991)
Caseins (α, β, γ, ε, κ)	Milk proteins	Strange *et al.* (1992); Li *et al.* (1994)
c-*fos*	Gene regulation	Marti *et al.* (1994)
*jun*B	Gene regulation	Marti *et al.* (1994)
*jun*D	Gene regulation	Marti *et al.* (1994)
c-*jun*	Gene regulation	Marti *et al.* (1994)
GlyCAM1	Ligand for L-selectin, expressed in milk	Dowbenko *et al.* (1993); Li *et al.* (1994)
24p3/NGAL	Lipocalin associated with collagenase IV; tissue remodeling	Li *et al.* (1994)
Ornithine decarboxylase	Rate-limiting enzyme in polyamine biosynthesis	Strange *et al.* (1992)
Plasminogen activator inhibitor I	Inhibitor of plasminogen activator	Talhouk *et al.* (1992)
Oct-1	Gene regulation	Marti *et al.* (1994)
p53	Inducer of cell death	Strange *et al.* (1992)
Stromelysin I	Tissue remodeling	Strange *et al.* (1992)
Stromelysin III	Tissue remodeling	Lefebvre *et al.* (1992)
SGP-2/TRPM-2/clusterin	Inhibitor of inflammatory response during apoptosis?	Buttyan *et al.* (1989); Jenne and Tschopp (1989); Strange *et al.* (1992)
Tenascin	ECM protein expressed during estrus, embryogenesis, and in stroma surrounding tumors; role in tissue remodeling	A.-C. Andres, V. Djanov, and R. Strange, unpublished data
TGFβ 1	Inducer of apoptosis; may regulate tissue remodeling	Strange *et al.* (1992)
Tissue plasminogen activator	Protease activator; tissue remodeling	Strange *et al.* (1992); Talhouk *et al.* (1992)
Tissue inhibitor of metalloproteinase I (TIMP)	Metalloproteinase inhibitor; tissue remodeling	Strange *et al.* (1992); Talhouk *et al.* (1992)
Tissue transglutaminase	Cross-linking enzyme; may prevent unregulated release of proteases or DNA	Strange *et al.* (1992)
Urokinase plasminogen activator	Protease activator; tissue remodeling; growth factor activation	Ossowski *et al.* (1979); R. Strange and R. R. Friis, unpublished data
WDNM1	Possible proteinase inhibitor	Li *et al.* (1994)
Whey acidic protein	Major whey milk protein	Strange *et al.* (1992)

poses significant challenges (see Chapters 7 and 8, this volume). The complexity of mammary gland involution warrants caution in sorting through the various potential regulators and executors of apoptotic cell death in the mammary gland. As demonstrated by the number of remodeling enzymes expressed during involution, the relationship between mammary epithelium and its related mesenchyme is important for maintenance of differentiated function (Barcellos-Hoff *et al.*, 1989; Streuli *et al.*, 1991). Components of the extracellular matrix may play the role of survival factors, or may provide a source of factors, as a reserve of matrix-bound growth factors, necessary for survival of the secretory epithelium. Perturbation of this interaction alters mammary-specific differentiation gene expression, for example, production of milk proteins (Parry *et al.*, 1987; Strange *et al.*, 1991; Talhouk *et al.*, 1992). Thus, alteration of the interaction between epithelium and its associated mesenchyme, which is an integral part of mammary involution, may also play a role in epithelial cell death. However, the epithelial–mesenchymal interactions that are the determining features in either mediating or modulating this cell death are just beginning to be defined. Stimuli that alter differentiated function may also induce apoptotic cell death of the epithelium but may have no physiological correlate. They may, however, have significant application in prevention or control of breast neoplasia.

Acknowledgments

This work was supported by grants from the American Cancer Society (RD-370), the Swiss National Science Foundation, the Schweizerische Krebsliga, the Bernische Krebsliga, and the Schweizerische Stiftung für klinische–experimentelle Tumorforschung.

References

Allan, D. J., Howell, A., Poberts, S. A., Williams, G. T., Watson, R. J., Coyne, J. D., Clarke, R. B., Laidlaw, I. J., and Potten, C. S. (1992). Reduction in apoptosis relative to mitosis in histologically normal epithelium accompanies fibrocystic change and carcinoma of the premenopausal human breast. *J. Pathol.* **67**, 25–32.

Andres, A.-C., Bchini, O., Schubar, B., Dolder, B., LeMeur, M., and Gerlinger, P. (1991). H-*ras* induced transformation of mammary epithelium is favored by increased oncogene expression or by inhibition of mammary regression. *Oncogene* **6**, 771–779.

Barcellos-Hoff, M. H., Aggeler, J., Ram, T. G., and Bissell, M. J. (1989). Functional differentiation and alveolar morphogenesis of primary mammary cultures on reconstituted basement membrane. *Development* **105**, 223–235.

Bates, R. C., Buret, A., van Helden, D. F., Horton, M. A., and Burns, G. F. (1994). Apoptosis induced by inhibition of intercellular contact. *J. Cell Biol.* **125**, 403–415.

Bursch, W., Oberhammer, F., and Schulte-Hermann, R. (1992). Cell death by apoptosis and its protective role against disease. *Trends Pharmacol. Sci.* **13**, 245–251.

Buttyan, R., Olsson, C. A., Pintar, J., Chang, C., Bandyk, M., Ng, P-Y., and Sawczuk, I. S. (1989). Induction of the TRPM-2 gene in cells undergoing programmed death. *Mol. Cell. Biol.* **9**, 3473–3481.

Cohen, S. M., and Ellwein, L. B. (1990). Cell proliferation in carcinogenesis. *Science* **249**, 1007–1011.

Daniel, C. W., Silberstein, G. B., Van Horn, K., Strickland, P., and Robinson, S. (1989). TGF-β1-induced inhibition of mouse mammary ductal growth: Developmental specificity and characterization. *Dev. Biol.* **135**, 20–30.

Dickson, S. R., and Warburton, M. J. (1992). Enhanced synthesis of gelatinase and stromelysin by myoepithelial cells during involution in the rat mammary gland. *J. Histochem. Cytochem.* **40**, 697–703.

Dowbenko, D., Kikuta, A., Gillett, N., and Lasky, L. A. (1993). Glycosylation-dependent cell adhesion molecule 1 (GlyCAM 1) mucin is expressed by lactating mammary gland epithelial epithelial cells and is present in milk. *J. Clin. Invest.* **92**, 952–960.

Edwertz, M., Duffy, S. W., Adami, H.-O., Kvåle, G., Lund, E., Meirik, O., Mellemgaard, A., Soini, I., and Tulinius, H. (1990). Age at first birth, parity and risk of breast cancer: A meta-analysis of 8 studies from the Nordic countries. *Int. J. Cancer* **46**, 597–603.

Frisch, S. M., and Francis, H. (1994). Disruption of epithelial cell-matrix interactions induces apoptosis. *J. Cell Biol.* **124**, 619–626.

Gavrieli, Y., Sherman, Y., and Ben-Sasson, S. A. (1992). Identification of programmed cell death in situ via specific labelling of nuclear DNA fragmentation. *J. Cell Biol.* **119**, 493–501.

Guenette, R. S., Corbeil, H. B., Leger, J., Wong, K., Mezl, V., Mooibroek, M., and Tenniswood, M. (1994). Induction of gene expression during involution of the lactating mammary gland of the rat. *J. Mol. Endocrin.* **12**, 47–60.

Harris, J. R., Lippman, M. E., Veronesi, U., and Willet, W. (1992). Breast cancer. *N. Engl. J. Med.* **327**, 319–328.

Hockenberry, D. M., Zutter, M., Hickey, W., Nahm, M., and Korsmeyer, S. J. (1991). BCL2 protein is topographically restricted in tissues characterized by apoptotic cell death. *Proc. Natl. Acad. Sci. USA* **88**, 6961–6965.

Jenne, D. E., and Tschopp, J. (1989). Molecular structure and functional characterization of a human complement cytolysis inhibitor found in blood and seminal plasma: Identity to sulfated glycoprotein 2, a constituent of rat testis fluid. *Proc. Natl. Acad. Sci. USA* **86**, 7123–7127.

Lefebvre, O., Wolf, C., Limacher, J. M., Hutin, P., Wendling, C., LeMeur, M., Basset, P., and Rio, M. C. (1992). The breast cancer-associated stromelysin-3 gene is expressed during mouse mammary gland apoptosis. *J. Cell Biol.* **119**, 997–1002.

Li, F., Bielke, W., Ke, G., Andres, A.-C., Jaggi, R. and Friis, R. R.; Niemann, H.; Bemis, L. T., Geske, F. J., and Strange, R. (1994). Isolation of cDNAs from mammary epithelium undergoing post-lactational apoptotic cell death. Cell Death and Differentiation (*in press*).

Marti, A., Jehn, B., Costello, E., Keon, N., Ke, G., Martin, F., and Jaggi, R. (1994). Protein Kinase A and AP-1 (c-Fos/JunD) are induced during apoptosis of mouse mammary epithelial cells. *Oncogene* **9**, 1213–1223.

Martinez-Hernandez, A., Fink, L. M., and Pierce, G. B. (1976). Removal of basement membrane in the involuting breast. *Lab. Invest.* **34**, 455–462.

Medina, D. (1973). Preneoplastic lesions in mouse mammary tumorigenesis. *Meth. Cancer Res.* **7**, 3–53.

Meredith, J. E., Jr., Fazeli, B., and Schwartz, M. A. (1993). The extracellular matrix as a cell survival factor. *Mol. Biol. Cell* **9**, 953–961.

Ossowski, L., Biegel, D., and Reich, E. (1979). Mammary plasminogen activator: Correlation with involution, hormonal modulation and comparison between normal and neoplastic tissue. *Cell* **16**, 929–940.

Parry, G., Cullen, B., Kaetzel, C. S., Krammer, R., and Moss, L. (1987). Regulation of differentiation and polarized secretion in mammary epithelial cells maintained in culture: Extracellular matrix and membrane polarity influences. *J. Cell Biol.* **105**, 2043–2051.

Preston-Martin, S., Pike, M. C., Ross, R. K., Jones, P. A., and Henderson, B. E. (1990). Increased cell division as a cause of human cancer. *Cancer Res.* **50**, 7415–7421.

Raff, M. C. (1992). Social controls and cell survival and cell death. *Nature* **356**, 397–399.

Russo, J., Rivera, R., and Russo, I. H. (1992). Influence of age and parity on the development of the human breast. *Breast Cancer Res. Treat.* **23**, 211–218.

Strange, R., Aguilar-Cordova, E. C., Young, L. J. T., Billy, H. T., Dandekar, S., and Cardiff, R. D. (1989). Harvey-*ras* mediated neoplastic development in the mouse mammary gland. *Oncogene* **6,** 309–315.

Strange, R., Li, F., Friis, R. R., Reichmann, E., Haenni, B., and Burri, P. H. (1991). Mammary epithelial differentiation *in vitro:* Minimum requirements for a functional response to hormonal stimulation. *Cell Growth Diff.* **2,** 549–559.

Strange, R., Li, F., Saurer, S., Burkhardt, A., and Friis, R. R. (1992). Apoptotic cell death and tissue remodeling during mouse mammary gland involution. *Development* **115,** 49–58.

Streuli, C. H., Bailey, N., and Bissell, M. B. (1991). Control of mammary epithelial differentiation: Basement membrane induces tissue-specific gene expression in the absence of cell-cell interaction and morphological polarity. *J. Cell Biol.* **115,** 1383–1395.

Talhouk, R. S., Bissell, M. J., and Werb, Z. (1992). Coordinated expression of extracellular matrix-degrading proteinases and their inhibitors regulates mammary epithelial function during involution. *J. Cell Biol.* **118,** 1271–1282.

Walker, N. I., Bennett, R. E., and Kerr, J. F. R. (1989). Cell death by apoptosis during involution of the lactating breast in mice and rats. *Am. J. Anat.* **185,** 19–32.

Warburton, M. J., Mitchell, D., Ormerod, E. J., and Rudland, P. (1982). Distribution of myoepithelial cells and basement membrane proteins in the resting, pregnant, lactating, and involuting rat mammary gland. *J. Histochem. Cytochem.* **30,** 667–676.

Williams, G. T. (1991). Programmed cell death: Apoptosis and oncogenesis. *Cell* **65,** 1097–1098.

Yang, C. P., Weiss, N. S., Band, P. R., Gallagher, R. P., White, E., and Daling, J. R. (1993). History of lactation and breast cancer risk. *Am. J. Epidemiol.* **136,** 1050–1056.

Yoo, K.-Y., Tajima, K., Kurioshi, T., Hirose, K., Yoshida, M., Miura, S., and Murai, H. (1992). Independent protective effect of lactation against breast cancer: A case-control study in Japan. *Am. J. Epidemiol.* **135,** 726–733.

CHAPTER 16

Hormonal Control of Apoptosis: The Rat Prostate Gland as a Model System

Marc C. Colombel and Ralph Buttyan

Departments of Urology and Pathology
Columbia University
College of Physicians and Surgeons
New York, New York 10032

I. Introduction

Male sexual accessory tissues grow and develop to their adult size and characteristics under the influence of the male (androgenic) sex steroids, most prominently testosterone. Once at their adult size (postpuberty), any reductions in the systemic level of androgens will result in the loss of function as well as the drastic shrinkage of many male sexual accessory tissues in a process termed involution. Involution occurs when cells of these tissues are deleted by apoptosis. Therefore, this process defines cell types that are dependent on the presence of androgenic steroids for their survival. Classically, the mammalian prostate gland has been the most preferred tissue in which to study this

Copyright © 1995 by Academic Press, Inc. All rights of reproduction in any form reserved.

androgen-regulated response (see Chapter 15). This choice is strongly motivated by the fact that, in humans, the prostate gland is the site for an inordinate number of debilitating diseases associated with aging, most prominently prostate cancer and a condition of benign enlargement known as benign prostatic hyper-trophy. Moreover, both benign and malignant prostate growth diseases can be treated by androgen withdrawal strategies. Treatment of these diseases, therefore, depends on the ability to induce apoptosis in this gland.

Found only in males, the prostate gland is located at the base of the urinary bladder (see McNeal, 1983). Embryologically, it originates as a series of bilateral invaginations of the urogenital sinus and lower Wolffian ducts that bud out into the surrounding mesenchyme. These buds develop into a relatively simple ductal system that, when mature, will direct exocrine secretions containing complex biochemicals such as citrates and polyamines in addition to protein-aceous substances into the urethra during ejaculation. The value of this particu-lar contribution to male fertility is not especially clear; prostatic fluid might aid in nutrition of sperm as well as provide some antibacterial protection. The adult human prostate gland has a compact structure that, with scrutiny, can be subdivided into three "zones" that differ in cellular composition and seem to be distinguished by the site at which the ducts originate from the urethra.

In the rat, the most studied laboratory animal model for prostate development, growth and response, the prostatic tissue is more conspicuously divided into three visually distinct lobes known as the dorsal, lateral, and ventral prostate glands. Like the zones of the human prostate gland, the rat lobes differ in size, cellular composition, secretory components, and site at which they originate from the urethra. The most prominent of the rat prostatic lobes is the ventral prostate which can weigh up to 0.8 g (for both lobes) in the mature male. Additionally, the ventral prostate appears to be the most sensitive to androgen withdrawal strategies, making it the most commonly studies prostatic lobe with respect to apoptosis. This sensitivity can be explained by the remarkable cellular composition of the ventral prostate which consists overwhelmingly (up to 80–85% cell population) of differentiated tall columnar secretory epithelium, the prostatic cell type most dependent on androgenic steroids (see Color Plate 8).

By removing the testis, the major source of androgenic steroids, all these cells will be deleted from the gland by apoptosis within 2 wk (Fig. 1; Color Plate 8). Another remarkable trait of this tissue is the rapidity with which the involuted ventral prostate gland can be completely regenerated by returning testosterone to the long-term castrated rat (Fig. 1). The regenerated tissue remains as sensitive to androgen withdrawal as the original adult gland; remov-ing the source of androgens will once again initiate apoptosis of the epithelial cells and result in re-involution of the prostate (Sanford et al., 1984). These features make the ventral prostate gland a useful model for studying androgen action in proliferative stimulation and differentiation as well as in apoptotic stimulation.

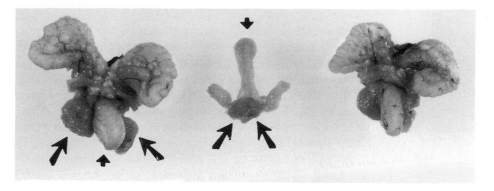

Fig. 1 Regression and regrowth of the rat ventral prostate gland in response to androgen manipulation. Tissue on the left displays a portion of the urogenital tract of the mature male Sprague–Dawley rat. Short arrow identifies the bladder and large arrows identify individual lobes of the ventral prostate gland. The two unmarked large lobes at the top of the tissue are the rat seminal vesicles and coagulating glands. Same tissue obtained from a 2-wk castrated rat is shown in the middle of the panel. The large arrows mark the remnants of the ventral prostate gland. Tissue on the right was obtained from a male rat that had been castrated for 2 wk and then received testosterone supplementation (25 mg time-release subcutaneous pellet) for 10 days. At this time, the male sexual accessory tissues have fully regenerated.

II. Initiation of Apoptosis in the Rat Ventral Prostate Gland

The prostate cell's response to androgens is regulated by three important effector molecules. The first is the male steroid hormone component or androgen. Testosterone is the primary male steroid produced by the Leydig cells of the testis. Without further metabolism, this particular hormone is unable to support adult prostate development. Lesser amounts of other types of androgenic steroids are produced by the adrenals, but at physiological levels these substances also appear to be insignificant in supporting male sexual accessory tissue development. Second, circulating testosterone must be metabolized to its more active form, dihydrotestosterone, by a prostatic membrane-bound enzyme referred to as 5-α reductase. This enzyme, now known to have multiple forms, converts the hydroxyl group of testosterone to a ketone. This conversion is critically important for adult prostate gland development and maintenance; men genetically lacking this enzyme do not develop adult prostate tissue. The importance of this conversion for prostate growth and development is the rational for the development of newer generation drugs that suppress prostate growth. The third important component of the androgenic response is the intracellular androgen receptor protein itself. Like other classic steroid receptors, the androgen receptor is believed to be a transcription activating protein when engaged by an androgenic steroid. Unlike the prototypic receptors for estradiol or glucocorticoids however, the DNA binding domain for this receptor remains poorly defined at this time.

To initiate apoptosis of the ventral prostate epithelial cells, androgen withdrawal is most effectively accomplished by the surgical removal of the testis. For the laboratory rat, castration is performed through a simple scrotal incision. The scrotal skin of the anesthetized male rat is swabbed with a dilute iodine disinfectant, followed by 70% ethanol, and a short incision is made through the skin and the tunica vaginalis with sterile scissors. Individual testes can be squeezed through this opening and pulled out of the scrotal cavity to expose the testicular cord above the epididymis. The cord and gubernaculum are then tightly ligated with sterile surgical silk thread; subsequently the cord is cut below the ligation so the testis can be removed. Once both testes are removed, the scrotal incision can be closed with sterile surgical clips. Done with appropriate care, this is a minor operation: infection control subsequent to the surgery is not necessary, survival is 100%, and rats show little sign of distress other than wound-licking behavior.

Although surgical castration immediately removes the source of testosterone, at least 6 hr are required for the circulating testosterone levels to be metabolized out of the system. Prostatic dihydrotestosterone levels reach the permanent castrated state within 12 hr (Kyprianou and Isaacs, 1988). Within 24 hr of castration, a slight increase in the apoptotic parameters of the prostate gland is already seen, as will be discussed later. Generally, cell loss and the detection of apoptotic bodies in the rat prostate gland are most apparent on the third day after castration (Sanford et al., 1984; Berges et al., 1993). On this day alone, the gland loses over 20% of its cells (Berges et al., 1993). By 96 hr following castration, cell loss by apoptosis is already in decline and counts of apoptotic bodies fall to levels comparable with those in uncastrated tissues after the first week. By 2 wk after castration, the gland has lost up to 85% of its cellular composition and its remarkable small size (approximately 50 mg) makes it difficult to distinguish in the fat pad that remains.

To mimic recently developed anti-androgen therapies for human prostate diseases, drugs such as cyproterone acetate or flutamide (anti-androgenic ligands) or inhibitors of 5-α reductase, the enzyme that converts circulating testosterone to the active ligand dihydrotestosterone, have also been used to initiate involution of the rat prostate gland. Cyproterone acetate can be delivered in an oil-based vehicle by subcutaneous injection (Shao et al., 1993). Flutamide is designed to be delivered orally or by subcutaneous injection and is subsequently metabolized to the active form hydroxyflutamide (Labrie, 1993). Although the studies that were performed with these drugs reflect expected results, that is, any drug that inhibits androgenic steroid action initiates prostate shrinkage via apoptosis of the prostate epithelial cells, most drug regimens do not appear to be as effective as surgical castration (Shao et al., 1994). However, these regimens have the benefit that they are readily reversible on termination of drug treatment. Intriguingly, a recent study found that the 5-α reductase inhibitor finasteride, currently being used to treat human benign prostate overgrowth, was capable of shrinking the rat ventral prostate gland without apparently inducing apoptosis (Rittmaster *et al.*, 1991). This could occur if the drug were simply suppressing the secretory functions of the gland, thereby leading to a loss in wet weight.

III. Criteria Identifying Apoptosis in the Rat Ventral Prostate Gland

Like most other forms of apoptosis, androgen-regulated apoptosis of the rat ventral prostate epithelial cell can be defined by (1) a number of distinct morphological changes initiated by castration, (2) the appearance of internucleosomal DNA fragmentation, and (3) the induced synthesis of certain gene products.

A. Morphological Characteristics

The adult rat ventral prostate gland is a ductal system that can be described as having a tree-like structure in that a major duct commencing at the prostatic urethra gradually branches into a succession of smaller ducts as one moves distally away from the urethra. Ultimately, these ducts terminate in a series of tubular glands in which the prostatic secretions originate. The ductal system is lined by a single layer of epithelium attached to the basement membrane. The characteristics of the epithelial cells vary from a rather inactive cuboidal cell type lining the ducts near the urethra to a distinct columnar secretory epithelial cell in the distal branches and the glands. As early as 1973, researchers recognized that the prostatic epithelium was the site for extensive apoptosis following castration (Kerr and Searle, 1973). In a 3-day castrated rat, these investigators identified the characteristic morphological criterion of apoptotic cells (the appearance of apoptotic bodies) associated with this early stage of prostate involution. These bodies are easily viewed by conventional microscopy of thin sections made from formalin-fixed and paraffin-embedded tissues subsequent to standard histochemical staining with hematoxylin or hematoxylin combined with eosin (see Chapters 1 and 2). As shown in Color Plate 8, these apoptotic bodies frequently appear in vacuoles of the prostatic epithelium. Obviously, the apoptotic cells have detached from the basement membrane and have broken their tight junctions with neighboring epithelial cells. These cells are extensively shrunken relative to neighboring epithelial cells and generally present in a spherical shape. The cytoplasm of the apoptotic body is eosinophilic when compared with surrounding epithelial cells. The nucleus is small, fragmented, and darkly stained by hematoxylin, demonstrating the remarkable condensation and pyknosis associated with apoptosis. This characteristic morphology observed after hematoxylin and eosin (H&E) staining enables the quantification of apoptosis in prostate tissue as a function of time after castration. By counting overt apoptotic bodies subsequent to H&E staining of regressing rat prostate tissues, researchers showed early on that apoptosis increases within 24 hr after castration and peaks at 3 days following castration (Sanford *et al.*, 1984). Although quantitative analysis of the prostatic stroma has shown that approximately 15% of this cell population is deleted by castration (English *et al.*, 1985), no description of apoptotic fibroblasts in the rat prostate gland has been published.

Within the epithelium, some of the apoptotic bodies are actively phagocytized. This phagocytic activity was originally attributed to the presence of intraepithelial macrophages (Helminen and Ericsson, 1972). However, our immunohistochemical analysis of regressing rat prostate tissue for macrophages or for any cell of the rat monocyte lineage using monoclonal antibodies directed against rat surface proteins of these cell types never identified any of these blood-derived cell types on the luminal side of the basement membrane. This result suggests that the phagocytosis of apoptotic bodies is mostly being performed by cells outside the monocyte lineage, potentially by neighboring epithelial cells that are (transiently) converted into phagocytes. The effectors of this epithelium-to-phagocyte conversion are poorly understand and provide an intriguing point of further study in apoptotic systems. Macrophages (monocytes) can, however, be recognized within the stroma of the normal rat prostate and their numbers seem to increase slightly after castration. These cells may be involved in the removal of the small number of interstitial cells that are also lost following castration.

The ultimate fate for many of the epithelial apoptotic bodies seems to be simple extrusion out into the ductal lumens. This is illustrated by the occasional observation of clusters containing large numbers of apoptotic bodies in prostatic lumens. This phenomenon of extrusion of apoptotic bodies into luminal spaces can also be observed in the developing rat kidney, suggesting that extrusion may represent a basic strategy for removing dead cell bodies from ductal organ systems with defined exit ports from the body.

B. Biochemical Characteristics

As are most other forms of apoptosis, castration-induced apoptosis in the rat ventral prostate gland is accompanied by extensive degradation of the chromosomal DNA, particularly within the internucleosomal linker region (Kyprianou and Issacs, 1988) (see Chapter 3). Using a simple DNA extraction and agarose gel electrophoretic analysis method on postcastration rat ventral prostate tissue, the canonical 180-bp ladder generated by breaks in the internucleosomal region of the DNA is readily visible by UV irradiation (Fig. 2). Visualization of the apoptotic DNA ladder can be further enhanced by a Southern blotting/hybridization procedure (Santarosa *et al.*, 1994). In this procedure, the electrophoresed DNA is briefly denatured and transferred to a nylon or nitrocellulose filter by standard Southern blotting techniques. This Southern blot can be hybridized with radiolabeled total rat DNA that has been fragmented by double restriction endonuclease digestion. This technique is also adaptable to the newer, chemiluminescent-based probe systems.

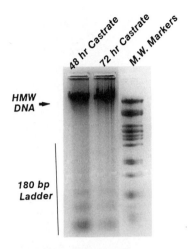

Fig. 2 DNA fragmentation associated with rat ventral prostate gland regression. High molecular weight (HMW) DNA was extracted from liquid nitrogen frozen, pulverized ventral prostate tissue taken from a 48- or 72-hr castrated rat, as indicated. DNA extraction was performed with the HMW DNA isolation reagents supplied by Boehringer Mannheim, Inc. (Indianapolis, Indiana) using the protocol provided by the manufacturer. DNA from 0.5 g frozen tissue was resuspended in 100 μl 10 mM Tris-HCl, pH 8.0, 1 mM EDTA, and was quantified by spectrophotometry at 260 nm. Aliquots of DNA containing 5 μg were electrophoresed on a 1.5% agarose gel. The gel was stained with an ethidium bromide solution and DNA bands were photographed by UV transillumination. HMW indicates the position of the high molecular weight DNA that is observed when DNA is extracted from the normal rat ventral prostate gland; a series of DNA fragments in multiples of 180 bp is also found in these samples from regressing ventral prostate glands. MW identifies the lane containing the DNA molecular weight markers coelectrophoresed adjacent to the tissue DNA specimens.

   Further enzymatic manipulations of DNA extracted from regressing rat ventral prostate tissue suggest that the degradation of nuclear DNA may also entail extensive single-strand damage in addition to the double-stranded breaks associated with DNA ladder formation. Figure 3 shows that mung bean nuclease treatment (a single strand-specific nuclease) of DNA extracted from 48-hr castrated rat ventral prostate dramatically increases the amount of DNA appearing in the lower molecular weight regions of the DNA ladder (Fig. 3). This kind of analysis implies that single-stranded digestions in the internucleosomal DNA may actually precede the double-stranded breaks that lead to the 180-bp DNA ladder. The existence of extensive single-stranded regions in such DNA is also supported by the ability of the Klenow DNA polymerase to act on DNA from the regressing ventral prostate. This activity is demonstrated by the results shown in Fig. 4; in this study digoxigenin-modified nucleotides were shown to be readily incorporated into DNA from regressing prostate DNA but not into DNA from normal prostate tissues.

   One can take advantage of this nuclear DNA fragmentation to label enzymatically and subsequently identify apoptotic cells *in situ* (see Chapter 2: Gavrieli

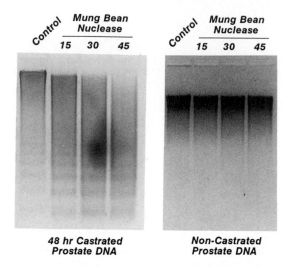

Fig. 3 Susceptibility of DNA from the regressing prostate gland to the action of mung bean nuclease indicates the presence of single-stranded gaps. (*Left*) A series of reactions in which a 5-μg aliquot of high molecular weight DNA extracted from a 48-hr castrated rat ventral prostate gland was incubated at 37°C in a buffer containing 30 mM sodium acetate, pH 4.5, 50 mM NaCl, 1 mM ZnCl$_2$ without mung bean nuclease for 60 min (control) or with mung bean nuclease (10 units/ml) for 15, 30, or 45 min as indicated. The shift of high molecular weight DNA into the 180-bp ladder DNA indicates the presence of single-stranded gap regions in regressing prostate DNA that can be digested by a single-strand-specific nuclease to give multimers of 180 bp. (*Right*) The same experiment done using high molecular weight DNA extracted from the normal rat ventral prostate gland. Note the lack of digestion by mung bean nuclease, indicating the absence of single-stranded gaps.

et al., 1992). Thin sections from formaldehyde-fixed and paraffin-embedded prostate tissues can be treated directly with Klenow polymerase in the presence of modified nucleotides (biotin or digoxigenin). Incorporation of such modified nucleotides into nuclear DNA marks cells with fragmented genomes (Santarosa *et al.*, 1994). This incorporation can be visualized by an immunohistochemical procedure employing monoclonal antibodies against digoxigenin or biotin. When this procedure is applied to normal prostate tissues, very few nuclei stain. In contrast, tissues recovered following castration stain extensively (Color Plate 9). It is remarkable in this tissue that the stained nuclei are almost exclusively identified in apoptotic bodies. Therefore, by this parameter, extensive DNA degradation would appear to be a late event in apoptosis associated with the morphological transition to the apoptotic body.

C. Molecular Characteristics

Comparable to certain other systems that are used to study apoptosis, castration-induced regression of the rat ventral prostate gland is inhibited by

Fig. 4 DNA fragmentation in the regressing prostate gland makes the DNA a substrate for the action of DNA polymerase. High molecular weight DNA was extracted from a mature adult ventral prostate gland or from a 48-hr regressing ventral prostate gland. Aliquots of 5 μg from each specimen were incubated in a buffer containing 25 mM Tris-HCl, pH 7.5, 5 mM MgCl$_2$, 1 μM dATP and dCTP, 10 μM dGTP, 0.1 nM digoxigenin-11-2'-deoxyuridine triphosphate (dig-dUTP) and 2 mM 2-mercaptoethanol. Then 5 U Klenow DNA polymerase was added and the mixture was incubated at 37°C for 30 min. DNA was precipitated by the addition of sodium acetate, pH 5.5, to 0.3 M and 3 volumes of ethanol, and was incubated at −20°C for more than 20 min. The precipitate was washed with 70% ethanol, dried, and resuspended in 25 μl TE buffer, pH 8.0. Treated DNA specimens were coelectrophoresed on a 1.2% agarose gel and were subsequently transferred to a nylon filter by standard Southern blot techniques. The nylon filter was treated with blocking solution provided in the Boehringer Mannheim digoxigenin immunodetection kit and was subsequently incubated for 1 hr in blocking solution containing alkaline phosphatase-labeled mouse monoclonal anti-digoxigenin antibody (Boehringer Mannheim). After extensive washing in 0.1 M sodium maleate, pH 7.4, the blot was washed in a buffer containing Tris-HCl, pH 9.5, 50 mM NaCl, and 10 mM MgCl$_2$ and was subsequently incubated in this same buffer containing the colorimetric alkaline phosphatase substrates BCIP and NBT. This Western blot procedure reveals extensive digoxigenin-labeled nucleotide incorporation into DNA from regressing ventral prostate but not from normal ventral prostate tissue.

treatments with RNA or protein synthesis inhibitors (see Section IV,A). This finding supports the concept that androgen withdrawal activates a genetic pathway within the prostate epithelial cell and has driven the search for new gene products synthesized in conjunction with castration. Molecular examination of bulk regressing prostate tissue by various types of RNA or protein analytical techniques confirms that the gene activity of the rat ventral prostate gland changes rapidly after castration. Some gene products such as the major secretory protein (prostatein) decline drastically and rapidly at both the RNA and the protein level (Page and Parker, 1982). In contrast, a large number of gene products are actively induced in this tissue by castration. The plethora of new gene activities associated with the first few days after castration often seems paradoxical. These induced gene products, however, plausibly appear to sort into three different catagories: (1) gene products such as sulfated glycoprotein-2 (clusterin) (Bandyk *et al.*, 1990), transforming growth factor -β (Kyprianou and Isaacs, 1989), and glutathione S-transferase (Chang *et al.*, 1987) with potential protective function (for the dying cell or the surrounding tissue); (2) gene products such as cathepsin D (Sensibar *et al.*, 1990), plasminogen activator (Rennie *et al.*, 1984), and transglutaminase, with potential degradative and apoptotic body removal function; and (3) gene products associated with cell cycle transition, for example, c-*fos*, c-*myc*, cyclin D1, and p53 (Buttyan *et al.*, 1988; Colombel *et al.*, 1992; Zhang *et al.*, 1994). The purpose of this chapter is not to discuss the relevance of these categories to the mechanism by which androgens control apoptosis in the prostate cell. It is important to establish, however, that these changes in gene activity can be viewed by several different types of molecular technology and ultimately can be localized to the cells that are undergoing apoptosis or to the cells that will survive this process.

Because nucleic acid isolation and identification procedures are now so standard and gene specific, many approaches to investigating gene activity in the regressing rat ventral prostate gland involve analyses of messenger RNAs. RNAs are easily extracted from liquid nitrogen-frozen and pulverized rat prostate tissues using the Chomczynski and Sacchi method [(1987; using RNazol B (Cinna/Biotech, Houston, TX) according to the manufacturers recommendations)]. Simple Northern blotting and comparative hybridization analysis of RNAs extracted from ventral prostate tissues at different times after castration have been utilized for many studies of prostate gland regression (Fig. 5). More recently, RNase protection assays, often considered more quantitative than Northern blot assays, have been adapted to this analysis and generally confirms data visualized by Northern blot techniques (Zhang *et al.*, 1994). Both kinds of RNA analyses determine whether the concentration of any given species of mRNA (in the total mRNA population) changes as a function of time after castration. Interpretation of these assays also depends on consecutive analysis of same RNA specimens for the concentration of "control" mRNA molecules that might not change after castration. Clearly, so-called "housekeeping" genes such as α-actin and β-tubulin are not appropriate control markers; these mRNAs are rapidly induced after castration as well as following testosterone restimulation of prostatic growth. Messenger RNAs for the cKi-*ras* gene and glyceraldehyde 3'-phosphate dehydrogenase show minimal fluctuations over the period of 1 wk after castration (Buttyan *et al.*, 1988; Zhang *et al.*, 1994). Equal RNA loading for Northern blot assays can also be ensured by ethidium bromide

Time After Castration

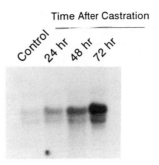

Fig. 5   Northern blot assay demonstrates highly induced expression of transforming growth factor β_1 (TGF-β_1) during regression of the rat ventral prostate gland. RNA was extracted from adult rat ventral prostate glands and from the same tissue at 24, 48, or 72 hr after castration by the RNazol B method (Cinna/Bioteck Inc., Houston, Texas). Aliquots containing 25 μg total RNA were coelectrophoresed in adjacent lanes of a 1.2% formaldehyde–agarose gel. The RNAs in the gel were transferred to a charge-modified nylon filter by capillary blotting. The RNA was fixed to the filter by UV irradiation; then the blot was hybridized to a ^{32}P-labeled probe for mouse TGF-β_1. Autoradiograph of hybridized blot shows highly induced expression of 2.4-kb transcript associated with this growth factor.

fluorescence of residual ribosomal RNA bands or by hybridization of the Northern blot with a probe for ribosomal RNA.

Interpretations of specific mRNA changes detected by these RNA analytical techniques must be made with the understanding that: (1) general mRNA production and metabolism in the rat prostate gland is rapidly and drastically altered by castration; (2) the cellular composition of the rat prostate gland changes after castration; and (3) cells undergoing apoptosis degrade their mRNAs (e.g., for β-actin) which then become undetectable in the apoptotic body. With regards to issue 1, overall RNA synthesis in the rat prostate gland is apparently reduced after castration (Lee, 1981). This reduction is at least partially reflected by the drastic cessation of the expression of the major rat prostatic secretory proteins (referred to as C1, C2, and C3) which cumulatively make up as much as 25% of the mRNA of the differentiated prostatic epithelial cell (Page and Parker, 1982). Studies of these proteins have shown that the decline in these mRNAs occurs because of a coordinate decrease in the specific transcription rate in addition to an increased rate of degradation of the mature mRNAs for these species. Of interest, a study of radiolabeled uridine incorporation into regressing rate prostate tissues *in vitro* identified an enhanced rate of uridine incorporated into regressing tissue RNA, compared with RNAs from sham-castrated tissues (Lee, 1981). Selective loss of RNA was attributed in this study to an increase in the activity of prostatic RNase. Concerning item 2; epithelial cell loss by apoptosis begins to occur in the rat ventral prostate gland by 24 hr after castration. This loss reaches a maximum at 72 hr after castration and by 10 days postcastration, cellular homeostasis is again achieved in the tissue. If alterations in bulk gene expression of the prostate gland were the result of the selective loss of a major cellular compartment (the secretory epithelial cells in this case), one would expect to detect a small increase in mRNA over the first few days, a rapid increase for the next 3 days, and an ultimate plateau in mRNA after the first week of castration. This expression pattern has already been observed for some specific gene products, for example, expression of bcl-2 mRNA or the cytokeratin protein (Hsieh et al., 1992; McDonnell et al., 1992) in the regressing rat prostate gland. In contrast, however, many of the mRNA gene products that are discussed in association with rat prostate regression are induced during a transient burst of gene expression that peaks within 3–5 days after castration and declines to control (noncastrate) levels thereafter (Montpetit et al., 1987; Buttyan et al., 1988). This coordinated pattern of gene expression in conjunction with the period of maximum apoptosis adds support to the hypothesis that these gene products might play a role in the cell death process. More interesting is the ability to discern subpatterns during this peak expression period. For example, in the initial studies of proto-oncogene activation associated with rat prostate regression, researchers found that expression of the c-fos gene (peaking approximately 36 hr after castration) preceded the peak induction period for c-myc (at 48–60 hr after castration) (Buttyan et al., 1988). Finally, with respect to issue 3, some studies have tried to correlate changes in mRNA expression with cellular content of the regressing prostate gland by devising a scale of mRNAs/cell (Berges et al., 1993). This scale supposedly coordinates mRNA levels (through densitometry) to cellular content of the regressing prostate gland (by measurement of glandular DNA content). However, this index fails to take into account the concept of the apoptotic body, which contains DNA but does not appear to display residual mature mRNA transcripts. Measurements of mRNA/cellular DNA content would therefore underestimate induction levels for any given RNA product, especially on days 3 and 4 after castration when up to 20% of the cells are found to be apoptotic.

One additional advantage of RNA analysis as a means of identifying changes in gene expression associated with apoptosis is the ability to localize gene activity to a given cell type subsequently by the procedure of in situ hybridization. This technique works best for gene products showing large fluctuations in expression during the period of prostatic regression and has been applied to analysis of SGP-2 (clusterin) (Buttyan et al., 1989) and c-myc (Quarmby et al., 1987) to date. This type of analysis has allowed the assignment of these postcastration gene activities to the prostatic epithelial cells. A major drawback of this technique is the inability to identify any of these mRNA molecules by in situ hybridization within apoptotic bodies. Apparently, this inability is related to a generalized degradation of mRNA in the remnants of dead cells.

Comparing the patterns of gene expression in normal and regressing rat prostate glands has allowed the isolation of some gene products that are very specific for the period of regression. For example, the original cloning of the SGP-2 (formerly TRPM-2) (clusterin) gene product which is highly associated with prostate cell apoptosis, was facilitated because this transcript is relatively rare in normal prostate RNA but accumulates to very high levels during prostate gland regression (Montpetit *et al.*, 1987). This mRNA species becomes so abundant after castration that is was isolated by simple sucrose gradient sedimentation procedures and was cloned directly. The same was true for TRPM-1, now known to be the Yb_1 subunit of rat glutathione S-transferase (Chang *et al.*, 1987). More recently, comparative cDNA subtraction hybridization methods have been used to isolate additional castration-induced gene products that accumulate to 3- to 8-fold (at the mRNA level) in the regressing prostate gland (Briehl and Miesfeld, 1991).

Obviously, increased mRNA transcript abundance is only one step toward the increased synthesis of a protein, a prerequisite if that protein is to play a role in the cell death process. Although mRNA analytical techniques are often much easier to perform because of the general availability of specific probes, it is important to perform quantitative studies of protein expression as well. To identify and quantify changes in specific protein levels requires the availability of appropriate antibodies. Given the species specificity of many antibodies, it has proven difficult to extend some of the studies of gene expression to the protein level, especially when using the rat as a model system. In the regressing rat prostate gland, increased expression of SGP-2 (clusterin) (Bandyk *et al.*, 1989), heat-shock 70K protein (see Fig. 6) cytokeratin (Hsieh *et al.*, 1992), and the

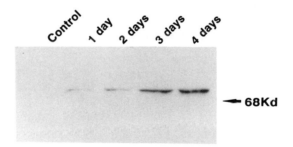

Fig. 6 Immunoprecipitation and Western blot assay demonstrates highly induced expression of hsp-70-related protein (hsp-72) in the regressing rat ventral prostate gland. Ventral prostate tissue from adult male or at 1, 2, 3, or 4 days after castration was homogenized in RIPA buffer containing 50 m*M* Tris, pH 8.0, 150 m*M* NaCl, 1% NP-40, 0.5% sodium deoxycholate, and 0.1% SDS. Aliquots containing 30 μg protein were electrophoresed by standard Laemmli denaturing polyacrylamide gel procedures. Proteins were transferred to a nitrocellulose filter by electroblotting. The filter was blocked with 10% powdered milk in TBS (Tris-buffered saline), and then incubated with monoclonal anti-human hsp-72 (Amersham, Inc., Arlington Heights, Illinois). After extensive washing with TBS, antibody binding was detected with the Chemiluminiscent Immunodetection Kit (Amersham) and exposure to film for autoradiography. This antibody identifies the highly induced synthesis of the rat hsp-70-related protein in the rat ventral prostate gland after castration in coordination with the previously described induction of the mRNA encoding this protein (Buttyan *et al.*, 1988).

tumor suppressor protein p53 (Zhang *et al.*, 1994) have been confirmed by Western blotting techniques as well as by *in situ* immunohistochemistry. Other proteins (cathepsin D and plasminogen activator) have been shown to be induced based on enzymatic activity or direct immunohistochemical analysis (Rennie *et al.*, 1984; Sensibar *et al.*, 1990).

IV. Systemic Modulators of Rat Ventral Prostate Apoptosis

To date, at least five different drugs have proven to be modulators (specifically suppressors) of prostatic regression when given systemically to the rat. Two of the drugs are gene expression blockers (RNA and protein synthesis inhibitors) whereas the other three drugs affect intracellular calcium ion homeostasis. The effectiveness of both classes of drugs has interesting implications for the mechanism involved in castration-induced apoptosis of prostate cells.

A. RNA/Protein Synthesis Inhibitors

In many but not all well-defined systems of apoptosis, protein and RNA synthesis inhibitors have proven to be modulators of this process. When both protein and RNA synthesis inhibitors act to inhibit or delay the onset of apoptosis, researchers have considered the results evidence for a requirement of "new" gene expression and an indication that apoptosis is a genetically active process. Castration-induced apoptosis of rat ventral prostate epithelial cells may be an example of a system in which apoptosis is dependent on new RNA and protein synthesis. In early studies (Lee, 1981), researchers showed that treatment with an RNA synthesis inhibitor (actinomycin D; daily subcutaneous injections of 166 μg/kg in saline) or a protein synthesis inhibitor (cycloheximide; daily subcutaneous injections of 1 mg/kg in saline), can delay the loss of wet weight of the adult rat ventral prostate gland after castration, presumably because of a delay in the loss of cells. These experiments are complicated by the metabolism, as well as systemic toxicity, of the drugs when administered to living animals and the uncertainty of establishing a dose–response effect at drug concentrations beyond the level of systemic toxicity. Moreover, more recent techniques to quantify the occurrence of apoptosis will be required to validate the extent to which these substances are actually blocking apoptosis in the androgen-deprived adult male rat prostate gland.

B. Calcium Metabolism Effectors

Among the classic cellular physiological changes thought to be associated with the onset of apoptosis is a drastic disturbance of intracellular calcium ion homeostasis. Therefore, it is intriguing that three drugs with the ability to affect Ca^{2+} ion homeostasis (verapamil, nifedipine, and chloroquin) have proven effective in supressing castration-induced regression of the rat ventral prostate gland (Lee, 1981; Connor *et al.*, 1988; Kyprianou *et al.*, 1988). All three of these drugs are designed for oral delivery and have effects when added to the water supply. Moreover, both verapamil (100 mg) and nifedipine (50 mg) can be given as subcutaneous time release pellets (Innovative Research, Inc., Toledo, OH) to mature male rates. Treatment with these reagents not only slowed the loss of prostatic wet weight and DNA content, but also preserved histological appearance of the prostate tissue and suppressed the activation of apoptosis-associated gene products such as c-*fos* and SGP-2 (clusterin) following castration (Connor *et al.*, 1988). Although these two drugs have proven to suppress castration-induced regression of the rat ventral prostate gland, no evidence yet demonstrates the presence of the classical voltage-gated calcium ion channels on the surface of prostatic epithelial cells. Therefore, whether the physiological mode of action of these agents is in blocking uptake of extracellular Ca^{2+} or in preventing its release from intracellular stores remains unclear.

V. Summary

Because of the large proportion of cells that undergoes apoptosis in response to castration and because of the predictable time in which apoptosis occurs subsequent to castration of an adult male, the rat ventral prostate gland provides a superior model system in which to study the process of apoptosis *in vivo*. This model system has already proven to be one of the more fertile systems for the identification of specific gene products that have the potential to effect apoptosis. Unfortunately, this *in vivo* system has limited usefulness for the types of genetic manipulations required to prove the role of any given gene product in the onset and procession of apoptosis. Direct genetic manipulation of a living tissue remains a goal of molecular biology-based therapies, especially for peripheral tissues such as the prostate gland. Appropriate *in vitro* (cell culture) models in which to study androgen-regulated apoptosis of prostate cells are currently unavailable because prostate epithelial cells, once established in culture, are no longer dependent on androgenic steroids. In the future, genetic approaches involving gene targeting through transgenic mouse technology may provide the kind of information needed to evaluate the role of individual gene products in prostate cell apoptosis.

References

Buttyn, M. G., Sawczuk, I. S., Olsson, C. A., Katz, A. E., and Buttyan, R. (1990). Characterization of the products of a gene expressed during androgen-programmed cell death and their potential use as a marker of urogenital injury. *J. Urol.* **143,** 407–412.
Berges, R. R., Furuya, Y., Remington, L., English, H. F., Jacks, T., and Isaacs, J. T. (1993). Cell proliferation, DNA repair and p53 function are not required for programmed death of prostatic glandular cells induced by androgen ablation. *Proc. Natl. Acad. Sci. USA* **90,** 8910–8914.

Briehl, M. M., and Miesfeld, R. L. (1991). Isolation and characterization of transcripts induced by androgen withdrawal and apoptotic cell death in the rat ventral prostate. *Mol. Endocrinol.* **5,** 1381–1388.

Buttyan, R., Zakeri, Z., Lockshin, R., and Wolgemuth, D. (1988). Cascade induction of the c-fos, c-myc and hsp-70K transcripts in the regressing rat ventral prostate gland. *Mol. Endocrinol.* **2,** 650–657.

Buttyan, R., Olsson, C. A., Pintar, J., Chang, C., Bandyk, M., Ng, P.-Y., and Sawczuk, I. S. (1989). Induction of the TRPM-2 gene in cells undergoing programmed death. *Mol. Cell. Biol.* **9,** 3473–3481.

Chang, C., Saltzman, A. G., Sorensen, N. S., Hiipakka, R. A., and Liao, S. (1987). Identification of glutathione S-transferase Yb_1 mRNA as the androgen-repressed mRNA by cDNA cloning and sequence analysis. *J. Biol. Chem.* **262,** 11901–11903.

Chomczynski, P., and Sacchi, N. (1987). Single step method of RNA isolation by acid guanidinium thiocyanate-phenol-chloroform extraction. *Anal. Biochem.* **162,** 156–159.

Colombel, M., Olsson, C. A., Ng, P.-Y., and Buttyan, R. (1992). Hormore-regulated apoptosis results from reentry of differentiated prostate cells onto a defective cell cycle. *Cancer Res.* **52,** 4313–4319.

Connor, J., Buttyan, R., Olsson, C. A., O'Toole, K., Ng, P.-Y., and Sawczuk, I. S. (1989). Calcium channel antagonists delay regression of androgen-dependent tissues and suppress gene activity associated with cell death. *Prostate* **13,** 119–130.

English, H. G., Drago, J. R., and Santen, R. J. (1985). Cellular response to androgen depletion and repletion in the rat ventral prostate: Autoradiographic and morphometric analysis. *Prostate* **7,** 41–51.

Gaurieli Y., Sherman, Y., and Ben-Sasson, S. A. (1992). Identification of programmed cell death in situ via specific labeling of nuclear DNA fragmentation. *J. Cell Biol.* **119,** 492–510.

Helminen, H. J., and Ericsson, J. L. E. (1972). Ultrastructural studies on prostatic involution in the rat. Evidence for focal irreversible damage to epithelium and heterophagic digestion in macrophages. *J. Ultrastruct. Res.* **39,** 443–455.

Hsieh, J. T., Zhau, H. E., Wang, X. H., Liew, C. C., and Chung, L. W. (1992). Regulation of basal and luminal cell-specific cytokeratin expression in rat accessory sex organs. Evidence for a new class of androgen-repressed genes and insight into their pairwise control. *J. Biol. Chem.* **267,** 2303–2310.

Kerr, J. F. R., and Searle, J. (1973). Deletion of cells by apoptosis during castration-induced involution of the rat prostate. *Virchows Arch. Zellpathol.* **13,** 87–102.

Kyprianou, N., and Isaacs, J. T. (1988). Activation of programmed cell death in the rat ventral prostate after castration. *Endocrin.* **122,** 552–562.

Kyprianou, N., and Isaacs, J. T. (1989). Expression of transforming growth factor-β in the rat ventral prostate during castration-induced programmed cell death. *Mol. Endocrinol.* **3,** 1515–1522.

Kyprianou, N., English, H. F., and Isaacs, J. T. (1988). Activation of a Ca^{++}-Mg^{++}-dependent endonuclease as an early event in castration-induced prostate cell death. *Prostate* **13,** 103–118.

Labrie, F. (1993). Mechanism of action and pure antiandrogenic properties of flutamide. *Cancer* **72 (Suppl.),** 3816–3827.

Lee, C. (1981). Physiology of castration-induced regression in rat prostate. *Prog. Clin. Biol Res.* **75A,** 145–159.

McDonnell, T. J., Troncoso, P., Brisbay, S. M., Logothesis, C., Chung, L. W. K., Hsieh, J.-T., Tu, S.-M., and Campbell, M. L. (1992). Expression of the proto-oncogene bcl-2 in the prostate and its association with the emergence of androgen-independent prostate cancer. *Cancer Res.* **52,** 6940–6944.

McNeal, J. E. (1983). Relationship of the origin of benign prostatic hypertrophy to prostate structure of man and other mammals. *In* "Benign Prostatic Hypertrophy" (F. Hinman, Jr., ed.), pp. 152–166. Springer-Verlag Press, New York.

Montpetit, M. L., Lawless, K. R., and Tenniswood, M. P. (1987). Androgen-repressed messages in the rat ventral prostate. *Prostate* **8,** 25–36.

Page, M. J., and Parker, M. G. (1982). Effects of androgen on the transcription of rat prostatic binding genes. *Mol. Cell. Endocrinol.* **27,** 343–355.

Quarmby, B. E., Beckman, W. C., Wilson, E. M., and French, F. S. (1987). Androgen regulation of c-myc messenger ribonucleic acid levels in rat ventral prostate. *Mol. Endocrinol.* **1,** 865–824.

Rennie, P. S., Bouffard, R., Bruchovsky, N., and Cheng, H. (1984). Increased activity of plasminogen activators during involution of the rat ventral prostate. *Biochem. J.* **221,** 171–178.

Rittmaster, R. S., Magor, K. E., Manning, A. P., Norman, R. W., and Lazier, C. B. (1991). Differential effect of 5 alpha-reductase inhibition and castration on androgen-regulated gene expression in rat prostate. *Mol. Endocrinol.* **5,** 1023–1029.

Sanford, M. L., Searle, J. E., and Kerr, J. F. R. (1984). Successive waves of apoptosis in the rat prostate after repeated withdrawal of testosterone stimulation. *Pathology* **16,** 406–410.

Santarosa, R., Colombel, M. C., Kaplan, S., Munson, F., Levin, R. M., and Buttyan, R. (1994). Hyperplasia and apoptosis: Opposing cellular processes that regulate the response of the rabbit bladder to transient outlet obstruction. *Lab. Invest.* **70,** 503–510.

Sensibar, J. A., Liu, X., Patai, B., Alger, B., and Lee, C. (1990). Characterization of castration-induced cell death in the rat prostate by immunohistochemical localization of cathepsin D. *Prostate* **16,** 263–276.

Shao, T. C., Kong, A., and Cunningham, G. R. (1993). Effects of 4-MAPC, a 5 alpha-reductase inhibitor, and cyproterone acetate on regrowth of the rat ventral prostate. *Prostate* **24,** 212–220.

Zhang, X., Colombel, M., Raffo, A., and Buttyan, R. (1994). Enhanced expression of p53 mRNA and protein in the regressing rat ventral prostate gland. *Biochem. Biophys. Res. Commun.* **198,** 1189–1194.

CHAPTER 17

Genetic Approaches for Studying Programmed Cell Death during Development of the Laboratory Mouse

Electra C. Coucouvanis, * **Gail R. Martin,** * **and Joseph H. Nadeau†**

* Department of Anatomy
University of California at San Francisco
San Francisco, California 94143

† Department of Human Genetics
Montreal General Hospital
Montreal, Quebec, Canada H3G1A4

I. Introduction

Predictable, physiologically appropriate, and naturally occurring cell death (usually termed programmed cell death, PCD) occurs in organisms ranging from worms to humans via a genetic pathway with multiple regulatory and effector components. Classically, such multistep pathways are studied using a genetic approach, which has already proven very successful in understanding the control of cell death in relatively simple organisms, most notably the nematode *Caenorhabditis elegans* (see Chapter 8, this volume). In vertebrates, the genetic pathway by which cell death is controlled is much less well understood. The purpose of this chapter is to discuss the use of genetically based methods and resources such as mice with genetic aberrations (whether identified initially as random mutations or deliberately engineered using transgenic technology) to study PCD during development. Although the incidence of PCD during embryonic development and mouse mutants exhibiting aberrant patterns of PCD have been studied for more than 60 years, recent technical advances allowing the generation of transgenic mice—essentially custom or designer mutants—has greatly increased the power of the genetic approach in unraveling complex biological pathways in mammals.

In Section I, we provide a comprehensive overview of the tissues in which and the times at which PCD has been observed in normal mouse development. Several examples are discussed in detail to illustrate key functions of PCD.

In Section II, we discuss techniques currently being used to detect cell death *in situ* in embryonic tissues or in isolated populations of cells, and we identify the feature of the dying cells on which each technique is based. Because these techniques have been described in detail elsewhere (many in this volume), we refer the reader to the appropriate reference rather than detailing the protocols here.

In Section III we compile a list of genes that are known or are strongly suspected to play a role in cell death *in vitro* or *in vivo,* followed by a brief discussion of the specific roles some of these genes may play in cell death. Such genes are excellent candidates for functional analysis by means of misexpression, overexpression, or knockout (nonexpression) experiments in transgenic mice; in fact, some are currently being studied in this manner.

Section IV is devoted to the identification of mouse mutants currently available in which cell death has been formally identified as aberrant or is suspected of being responsible for the mutant phenotype. Techniques for identifying the genes affected in such mutants are discussed. A list of resources that are available for genetic studies of PCD is included in an appendix at the end of this chapter. We also discuss transgenic mice that have been characterized as having cell death-related abnormalities.

Finally, in Section V, we discuss strategies utilizing the mutant and transgenic mice described in Section IV and the cell death genes described in Section III to begin to dissect and elucidate the pathways that control the regulation and execution of PCD during mouse development.

II. Cell Death during Normal Development

Mutations can affect PCD in two ways: cells that normally live may die or cells that normally die may live. The first important tool in analyzing cell death mutants is a detailed knowledge of normally occurring cell death during development. This allows the classification of embryonic cell death observed in a mutant or genetically manipulated mouse as normal or abnormal. In addition, an abnormality observed in a given tissue or structure can be analyzed with respect to the known pattern of physiological cell death.

A. Definition of Programmed Cell Death: Relation to Apoptosis

Prior to describing the patterns of cell death that have been observed in normal developing embryos, it is important to define "programmed cell death" and to address the terminology used in the field.

Programmed cell death refers to normally occurring, physiologically appropriate death that occurs in a stereotypic manner at specific times during development and also throughout life. Another simplistic way to define PCD is death that is beneficial rather than harmful, that is, death that serves a purpose within the organism.

Apoptosis is a morphological characterization of a particular type of cell death that involves membrane blebbling, nuclear and cytoplasmic shrinkage, and chromatin condensation. The term apoptosis refers to the seasonal fall of leaves from trees and was coined to distinguish a naturally occurring type of cell death distinct from necrosis. The causes and morphological features of necrosis are very different from those of apoptosis; necrotic death is characterized by cell swelling and lysis, and results from injury to a cell.

Researchers have long debated whether the terms PCD and apoptosis are interchangeable. Some incidences of cell death in invertebrates are clearly "programmed," for example, loss of intersegmental muscles in the moth *Manduca sexta* at the end of metamorphosis, but are not characterized by the morphological features found in mammalian cells undergoing apoptosis (Schwartz *et al.*, 1993). In contrast, nearly all vertebrate cells undergoing PCD appear to exhibit the general morphological features of apoptosis (but see

Clarke, 1990, for an expanded discussion of this topic). Confusion arises when apoptosis is used as a term to describe the *mechanism* by which death occurs. Many, and perhaps most, apoptotic deaths are characterized by intranucleosomal digestion of the cell's DNA by an endonuclease, but apoptotic morphology has been reported in the absence of this phenomenon and vice versa (see Section II). If apoptosis is used solely as a morphological term, as it will be in this chapter, much confusion is avoided.

B. Incidence of Programmed Cell Death during Development

Glucksmann (1951) classified embryonic cell deaths into three categories based on their presumed function in development. (1) Morphogenetic cell deaths are those that participate in the generation of form, for example, by creating lumens in or canals through tissues, or by facilitating the detachment of an invaginated tubular structure from a sheet of cells, as in the formation of the neural tube. (2) Histogenetic deaths are those associated with cell differentiation, and can be involved in changes in the cell types composing an organ or tissue as, for example, in the developing kidney (see subsequent discussion). Finally, (3) phylogenetic deaths function to eliminate transient or vestigial structures or cells, such as the tail in human embryos. Table I lists instances of cell death that have been reported during development and is organized according to these classifications, although clearly there can be overlap among the categories. Most examples listed in Table I are from the mouse, although when indicated other organisms including chick, rat, and human are included. Many of the examples of cell death listed in Table I have not been re-examined since their initial description 40–50 years ago. Of those that have, the majority display the features of apoptosis. Many instances of cell death remain that could be better defined with the improved methods currently available for detecting PCD (see Section II); doubtless, many have gone undetected to date.

Several systems in which cell death occurs have been analyzed in detail and are discussed subsequently. By studying examples in which the control of cell death is at least partially understood, it may be possible to gain insight into the general mechanisms that may be functioning in other instances of PCD (see Chapters 13, 15, and 16, this volume).

1. Cell Death in Limb Morphogenesis

Cell death serves a morphogenetic function during development of the vertebrate limb, where it is primarily involved in shaping and pattern formation. A comparison of the incidence of cell death in the mouse and chick limb provides insight into the role played by PCD in limb morphogenesis.

In the developing chick limb beginning at stage 23–24, cell death occurs in the two regions of mesenchyme that underly the ends of the apical ectodermal ridge (AER) (reviewed by Hinchliffe, 1981). These areas are historically termed the anterior and posterior necrotic zones (ANZ and PNZ, respectively). The use of the term necrotic is misleading, however, since the dying cells display the morphological features of apoptosis. Grafting experiments performed by Saunders *et al.* (1962) demonstrated that as early as stage 17, the cells in the PNZ have received a signal instructing them to die at stage 24. Even if cells from this region are transferred to an ectopic site such as flank mesenchyme, they die at stage 24, just as they would have in the posterior part of the limb. However, the death program can be altered if these cells are transplanted to central wing mesenchyme instead, in which case they live beyond stage 24. In contrast, if cells from the PNZ are grafted to central wing mesenchyme after the embryo reaches stage 22, the cells die on schedule at stage 24. These elegant experiments demonstrate that a pathway involving at least three steps functions during cell death in the PNZ: the first step is a reversible decision to die (occurring at stage 17), step two is an irreversible commitment to die (occurring at stage 22), and step three is the actual execution of cell death. Of course, it is possible that appropriate manipulations of post-stage-22 PNZ cells could also prevent death at stage 24, allowing further dissection of the control and programming of this example of PCD.

The ANZ and PNZ appear to limit the length of the AER, which is dependent for its survival on the underlying mesenchyme. In turn, the AER determines the width of the subsequent digit field (reviewed by Hinchliffe, 1981). After the death of the underlying ANZ and PNZ, the overlying region of the AER regresses (also displaying an apoptotic morphology). Support for this model comes from the fact that normal mouse limbs lack a PNZ and contain a smaller ANZ than chick limbs. As a result, mouse limb buds are wider than those of the chick and form five digits rather than the three formed in a normal chick wing. In *talpid*[3] mutant chicks, which lack both ANZ and PNZ, an abnormally long AER results in the formation of supernumerary digits (Hinchliffe and Ede, 1967). In a related mutant, *talpid*[2], similar abnormalities are due to a defect in the mesenchyme, since mutant ectoderm combined with normal mesenchyme produces a normal limb whereas the reverse combination (mutant mesenchyme with normal ectoderm) produces a polydactylous limb (Goetinck and Abbott, 1964). Similarly, mice carrying the mutation *Dominant hemimelia (Dh)* lack an ANZ altogether, and commonly display 7–8 digits per paw; the extra digits are always located anteriorly (Knudsen and Kochhar, 1981).

Programmed cell death also occurs in the interdigital tissues of vertebrate limbs, where it functions to separate the digits. The mouse *Polysyndactyly (Ps)* mutation (Johnson, 1969) and the chick *talpid*[3] mutation (Hinchliffe and Thorogood, 1974) result in inhibition of this death and the digits remain joined by webs of mesenchymal tissue.

In these examples, mutant phenotypes are caused by too little cell death. Excessive cell death can also produce limb defects, such as that caused by the *wingless (ws)* mutation of the chick. In severe cases, early and excessive cell death in the ANZ eliminates the wing bud entirely (Hinchliffe and Ede, 1973). Likewise, the chick *rumpless* mutation *(Rp)* results in degeneration of the

Table I
Cell Death during Development

Incidence	Reference
Morphogenetic cell death	
Formation of neural grooves	Glucksmann (1951; review)
Invagination, closure, and detachment of neural tube from ectoderm	Glucksmann (1951; review)
Closure of ventral body wall	Glucksmann (1951; review)
Invagination of optic cup	Glucksmann (1951; review)
Invagination of lens and detachment from ectoderm	Glucksmann (1951; review)
Formation of otocyst and separation from ectoderm	Glucksmann (1951; review)
Invagination of olfactory pit and detachment from palate	Glucksmann (1951; review)
Regression of epithelial seam formed during closure of palate	Ferguson (1988; review)
Regression of epidermal seams in temporary closure of lids, lips, urethral orifice, vulva (human)	Glucksmann (1951; review)
Endodermal cells in formation of thymus, thyroid, parathyroid	Glucksmann (1951; review)
Lumen formation in the salivary glands	Glucksmann (1951; review)
Hollowing out of epithelial occlusions of intestinal tract (human)	Glucksmann (1951; review)
Regression of pharyngeal and cloacal membranes	Glucksmann (1951; review)
Separation of tracheal and esophageal analgen	Glucksmann (1951; review)
Bifurcation of aorta (chick)	Glucksmann (1951; review)
Partial regression of notochord	Glucksmann (1951; review)
Regression of apical ectodermal ridge of limbs	Hinchliffe (1981; review)
Posterior and anterior necrotic zones in limb (chick)	Hinchliffe (1981; review)
Opaque zone or patch in limb (chick)	Hinchliffe (1981; review)
Interdigital zone in limbs	Hinchliffe (1981; review)
Histogenetic (differentiation-associated) cell death	
Blastocyst at 70- to 89-cell stage, both polar trophectoderm and ICM	El Shershaby and Hinchliffe (1974); Copp (1978)
Primitive streak and node (chick)	Glucksmann (1951; review)
Innervation of olfactory epithelium	Glucksmann (1951; review)
Cranial and caudal somites	Glucksmann (1951); Jeffs and Osmond (1992) (chick)
Sclerotome prior to cartilage formation	Glucksmann (1951; review)

Myotome, dermatome	Glucksmann (1951; review)
Early differentiation of myoblasts	Glucksmann (1951; review)
Precartilaginous mesenchyme	Glucksmann (1951; review)
Cartilage prior to ossification	Glucksmann (1951; review)
Regression of Mullerian ducts in males	Glucksmann (1951; review)
Regression of Wolffian ducts in females	Glucksmann (1951; review)
Prespermatocytes (hamster)	Miething (1992)
Prenatal oogonia, spermatogonia oocytes, and spermatocytes	Coucouvanis et al. (1993)
Metanephric mesenchyme during formation of epithelial structures of nephron	Koseki et al. (1992); Coles et al. (1993)
Intestinal epithelium (rat)	Harmon et al. (1984)
Motoneurons; spinal, oculomotor facial (trochlear, trigeminal—chick)	Oppenheim (1991; review)
Ganglia: trigeminal, spinal, sympathetic (pineal, ciliary, cochlear, vestibular, nodose—chick)	Oppenheim (1991; review)
Granule cells of cerebellum	Wood et al. (1993)
Forebrain—cerebral hemisphere (chick)	Oppenheim (1991; review)
Cochlear nuclei	Oppenheim (1991; review)
Inferior olive	Oppenheim (1991; review)
Inferior colliculus	Oppenheim (1991; review)
Parabigeminal nucleus	Oppenheim (1991; review)
Lateral genicula e nucleus	Oppenheim (1991; review)
Retina	Young (1984)
Hippocampus	Oppenheim (1991; review)
Corpus striatum	Oppenheim (1991; review)
Olfactory cortex	Oppenheim (1991; review)
Cerebral cortex	Oppenheim (1991; review)
Oligodendrocytes	Barres et al. (1992)
Phylogenetic cell death	
Undifferentiated mesenchyme of branchial region	Glucksmann (1951; review)
Regression of pronephros and mesonephros	Glucksmann (1951; review)
Regression of tail (human)	Fallon and Simandl (1978)

393

PCD—the presence of multiple single- or double-strand breaks in DNA (see Chapter 3, this volume). Among the methods used to detect PCD, it is still unclear which, if any, identify universal features of physiologically occurring cell death. Indeed, growing evidence suggests that some of the candidate markers, including presence of DNA fragmentation and up-regulation of genes encoding molecules such as ubiquitin or Sgp2 are not detectable concomitant with the appearance of the morphological hallmarks of death in some cells (for example, see Zakeri *et al.*, 1993; Martin *et al.*, 1992). Clearly, the most likely universal markers would be gene products directly involved in cell killing, namely, cell death effector proteins. The recent demonstration that the *C. elegans ced-3* gene encodes a cysteine protease related to a known mammalian protein, the interleukin-1β-converting enzyme (ICE) (Yuan *et al.*, 1993) and that overexpression of ICE causes PCD in cultured rat fibroblasts (Miura *et al.*, 1993) makes it possible to test whether this or any related protease may serve as a reliable marker of cell death.

The methods described in this section will be divided into two categories. The first includes methods used to identify the classic morphological features of PCD including condensed DNA, smaller cell and nuclear size, cell membrane blebbing or cell breakup into apoptotic bodies, or engulfment of a dying cell by a neighboring cell. The second category includes methods based on the few biochemical characteristics that are frequently associated with PCD.

A. Methods Based on Morphological Features of Apoptotic Cell Death

1. Electron Microscopy

The highly condensed chromatin in apoptotic cells is readily apparent in tissue sections stained with uranyl or lead acetate and examined by transmission electron microscopy (TEM) (see Chapter 1). Normal euchromatin is easily distinguished from the much darker, usually solid black mass of chromatin in a shrunken apoptotic nucleus (Fig. 1). Other ultrastructural features including increased density of the cytoplasm, the presence of blebs in the cellular membrane, or the presence of an engulfing phagocytic cell are also indicators of cell death visible by TEM.

Scanning EM has been utilized successfully to detect surface blebbing seen in apoptotic thymocytes. However, in the developing embryo this method is less useful when examining material in which blebbing may be constrained by the close proximity of neighboring cells; phagocytic cells may engulf the blebs immediately, and the external contours of the cells may not be visible. Scanning EM may be useful for examining embryonic cell populations that can be easily dissociated or that do not have tight connections to other cells at some point in development, for example, primordial germ cells and cells of the hematopoietic lineage.

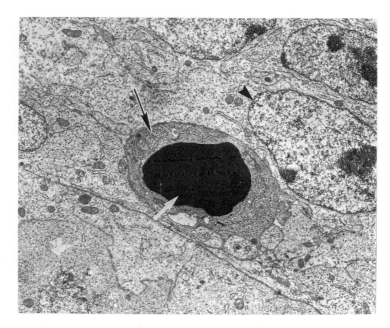

Fig. 1 Detection of apoptotic morphology by electron microscopy. Transmission electron micrograph of an apoptotic primordial germ cell in an embryonic day 13 male gonad. Note the shrunken cytoplasm (black arrow) and highly condensed nuclear chromatin (white arrow). Compare the apoptotic nucleus to a normal nucleus (black arrowhead). Magnification: 4400×. Reprinted with permission from Coucouvanis *et al.*, (1993).

2. Light Microscopy Using Dyes that Bind to DNA

So-called supravital staining (staining with or without fixation, using standard dyes such as toluidine blue or neutral red) has historically been used to detect PCD in sections of tissues or embryos. Pyknotic nuclei can be detected in these preparations, but determination of cell death is often equivocal. A simple improvement of this method that consistently works well in identifying condensed or apoptotic nuclei is to stain thin sections with fluorescent DNA binding dyes such as DAPI, propidium iodide, acridine orange, chromomycin, and others (see Chapters 9 and 10, this volume). These dyes may be used in conjunction with a light counterstain, or alternate sections may be stained with another dye, antibody, or reagent of interest. When examined using the appropriate excitation lamps and detection filters, specific staining of the nuclei allows clear visualization of the difference between a viable cell with a normal sized, diffusely stained nucleus and an apoptotic cell, with smaller solid spheres or blobs or condensed chromatin (Fig. 2, large white arrow). Thin sections are important when using this method because superimposed nuclei in a section can cause local areas of brightness, obscure the black space that typically surrounds an

B. Methods Based on Biochemical Properties of Cells Undergoing PCD

1. Use of Marker Genes to Identify PCD *in Situ*

An ideal assay for PCD *in situ* would be the detection of expression of a gene or genes known to regulate or implement PCD in all cell types. Although several genes and gene products are up-regulated or otherwise specifically expressed during PCD in several tissues, it is presently unclear whether any represent universal markers of PCD. (See Section III for a list of genes that have been shown or are suspected to be involved in the regulation or execution of PCD.) Genes such as *Clu* (also known as SGP-2 and TRPM-2) and *myc*, which are expressed at elevated levels in many cell types undergoing apoptosis, are not induced in association with neuronal apoptosis (Martin *et al.*, 1992; D'Mello and Galli, 1993). Researchers anticipate that suitable marker genes may eventually be identified. In the meantime, *in situ* techniques for detecting biochemical features of PCD are limited to those that detect DNA strand breakage.

2. *In Situ* Detection of DNA Digestion

A biochemical property of cells that is often considered a diagnostic feature of PCD is endonucleolytic digestion of a cell's DNA. When such digested DNA is isolated and subjected to gel electrophoresis, a characteristic "ladder" pattern is observed. However, the role DNA degradation plays in cell death and whether it is necessarily associated with apoptotic morphology is currently unclear. In the nematode *C. elegans,* activation of a Ca^{2+}/Mg^{2+}-independent endonuclease is one of the last steps in the PCD pathway. The nuclease activity is not required for cell death and functions to degrade the DNA of dead cells during the final phase of corpse removal (reviewed by Ellis *et al.*, 1991). In contrast, in mammalian cells, a Ca^{2+}/Mg^{2+}-dependent endonuclease functions during earlier stages of cell death, at least in some cells (reviewed by Arends and Wyllie, 1991). The observation that treating cultured mammalian cells with micrococcal nuclease produces the condensed nuclear morphology characteristic of apoptosis as well as a ladder of DNA fragments suggests that an endogenous nuclease might act similarly to produce both apoptotic morphology and DNA cleavage in cells undergoing PCD (Arends *et al.*, 1990). However, in other studies the presence of a DNA ladder has not been detected in cells with classical apoptotic morphology (Zakeri *et al.*, 1993). In at least one type of cell death, the endonuclease activity is a dispensable feature (Ucker *et al.*, 1992). Therefore, endonucleolytic cleavage of DNA may be necessary for some or even most forms of PCD, but not all.

Isolation and electrophoretic separation of DNA for the purpose of detecting endonuclease-cleaved fragments and hence PCD is a technique that is useful for analyzing cell populations in which most or all of the cells die simultaneously, for example, in response to a specific signal. For the purpose of detecting

PCD in most tissues of the mouse embryo, where cell death may be far less synchronous, *in situ* detection of DNA strand breaks is quite useful. As early as 1972, terminal deoxyribonucleotidyl transferase (TdT)-catalyzed addition of labeled deoxyribonucleotides was used to identify free DNA ends (i.e., DNA digestion) in individual nuclei in tissue sections (Modak and Bollum, 1972). More recent versions of this technique make use of biotinylated dNTPs and different enzymes to catalyze their addition, but the basic principle is the same.

The TUNEL (TdT-mediated dUTP–biotin nick end labeling) method described by Gavrieli *et al.* (1992; see also Chapter 2) involves protease treatment of paraffin sections followed by incubation with TdT in the presence of biotin-labeled dUTP. Avidin conjugated peroxidase is then used to detect the labeled nucleotides. Pretreatment with DNase I, which results in the staining of all cells in the treated section, serves as a positive control.

Two other recent papers describe similar methods of *in situ* end labeling (ISEL), which differ from TUNEL because *Escherichia coli* polymerase I (pol I) or the Klenow fragment rather than TdT is used to label DNA ends with biotin-conjugated nucleotides in protease-treated, paraffin-embedded tissue sections Ansari *et al.*, 1993; Wijsman *et al.*, 1993). Both these references note the importance of carefully optimizing the length of the protease digestion period, the concentration of the polymerase, and the incubation time to achieve the highest sensitivity and to avoid nonspecific staining. Finally, both papers caution researchers that ISEL is not specific to cells undergoing PCD, since necrotic cells may also contain free DNA ends due to digestion by lysosomal nucleases. Although only minimal data substantiate this assertion, it seems logical that a DNA strand need not have been cleaved by a PCD-associated endonuclease to serve as a substrate for the labeling reaction used in any of the protocols mentioned here.

Another variation of the basic *in situ* labeling technique has been introduced by Wood *et al.* (1993). In this method, T7 DNA polymerase is employed rather than pol I or Klenow, frozen sections are used, and there is no protease treatment of the tissue sections. The omission of protease treatment eliminates one of the possible causes of nonspecific staining, and also helps preserve cell structure, facilitating localization of epitopes of interest in nucleotide-labeled cells. In addition, the use of T7 polymerase may increase the sensitivity of the technique. Like Klenow, T7 DNA polymerase has both a $5' \rightarrow 3'$ polymerase activity and a $3' \rightarrow 5'$ exonuclease. However, T7 differs because it is the most processive of all known DNA polymerases and its exonuclease activity is 1000 times that of the Klenow fragment (Sambrook *et al.*, 1989). Interestingly, Wood *et al.* (1993) noted that the exonuclease activity was required for the labeling reaction because without it, reduced labeling was observed on a nuclease-treated control sample.

As a general caution regarding these *in situ* techniques, note that very limited data are available pertaining to the question of sensitivity. In an attempt to obtain some measure of the minimum number of strand breaks detectable by their method, Wood *et al.* (1993) irradiated 3T3 fibroblasts with 5000–20,000 cGy and fixed the cells immediately to prevent DNA repair. When these cells were then stained with T7 and biotin–dUTP, no labeled cells were detected. Thus, although the irradiated cells presumably sustained a substantial amount of DNA breakage, the level was still too low to be detectable by the T7 assay, implying that a high level of DNA digestion is necessary to produce a positive signal with this assay in putative examples of PCD.

The extent to which these techniques have been successful in detecting dying cell *in situ* has been variable. Using the TUNEL method, Gavrieli *et al.* (1992) saw staining in nuclei in the uppermost layer of the stratified epithelium in the epidermis and in lymphatic tissues including thymus and spleen. All these are tissues in which cell death with apoptotic morphology and DNA fragmentation has been shown to occur by other methods such as TEM or detection of a DNA ladder. Staining was also present in rat, mouse, and human intestinal epithelia, in particular in cells at the tips of villi. Interestingly, in these cells, labeling with TdT was not associated with characteristic apoptotic morphology. Because the staining was nevertheless specific and because apoptotic cell death has been shown by EM to occur in this tissue (Harmon *et al.*, 1984), the authors suggest that the cells are indeed dying and are detected after some DNA fragmentation has occurred, but prior to the appearance of classical apoptotic morphology.

However, the TUNEL method failed to detect DNA fragmentation in postlactation breast tissue undergoing involution in rats and mice, despite the fact that this process has been shown to involve PCD by electron microscopy (Walker *et al.*, 1989), light microscopy, and detection of DNA laddering (Strange *et al.*, 1992). Because the presence of a nucleosomal ladder was clearly demonstrated (Strange *et al.*, 1992), the lack of detection of strand breaks in mammary gland tissue by the TUNEL method (Gavrieli *et al.*, 1992) is a serious concern and may indicate that the technique has limited sensitivity (see Chapter 15). Although one must bear these concerns about sensitivity and the discrepancies in results among techniques in mind, the TUNEL method has been used to make an observation that is quite intriguing: cells that show only weak labeling by TUNEL often display localized staining specifically at the periphery of the nucleus. One interpretation of this result is that DNA digestion, at least in some instances, may initiate the nuclear boundary and then spread inward (Gavrieli *et al.*, 1992).

Variability was also found when the ISEL technique was used to examine PCD *in situ*. In some cases such as involuting prostate after castration (Wijsman *et al.*, 1993) and in interdigital tissue of rat limb buds (Ansari *et al.*, 1993), ISEL staining was found to correlate very well with apoptotic morphology. In other cases this correlation was not observed. For example, in agreement with the results obtained using the TUNEL method, labeling was detected in cells appearing morphologically normal at the tips of intestinal villi (Ansari *et al.*, 1993). Conversely, in cultured rat thymocytes induced to undergo PCD by methylprednisolone, many cells that could be identified morphologically as apoptotic based on acridine orange staining were negative by the ISEL technique (Ansari *et al.*, 1993).

Wood *et al.* (1993) applied the T7 *in situ* labeling technique to the detection of DNA fragmentation in the developing mouse cerebellum, where they investigated two distinct episodes of cell death. By this method too, there was no correlation between *in situ* labeling and apoptotic morphology. A major period of cell death is known to occur among granule cells during the 3rd–5th postnatal weeks and has been previously identified on the basis of pyknotic nuclei. However, in the experiments by Wood and colleagues, even those granule cells displaying apoptotic morphology at this time were not labeled in the T7 assay and thus do not appear to contain very large amounts of fragmented DNA. However, Wood *et al.* (1993) were able to identify a period during which many granule cells were positively labeled by the T7 method. During the period of histogenesis of the cerebellum, in the first 2 weeks of postnatal development, numerous granule cells were found to be stained positively for DNA breaks. Interestingly, many of the labeled cells did not display apoptotic morphology. Although granule cell death had not been previously identified at this time during development, the authors assume that the high level of labeling reflects a sufficiently high number of strand breaks (more than in lethally irradiated cells, at least) to prevent cell recovery. They speculate that the lack of pyknotic morphology precluded the previous identification of these dying cells. In light of these two observations, Wood *et al.*, (1993) conclude that it is possible for different mechanisms of cell death to occur within the same population of cells during development.

Unfortunately, it is difficult to make meaningful comparisons among the various methods available to detect DNA strand breaks *in situ*. Based solely on the known characteristics of each enzyme, T7 DNA polymerase might be expected to be more sensitive than pol I, Klenow, or TdT, but a direct comparison under identical conditions on a common template is needed. Treatments such as irradiation or controlled nuclease digestion, which could be used to produce a "standard curve" of degree of DNA fragmentation, may prove useful for this purpose. From the results obtained using the methods described here, it seems clear that *in situ* labeling of DNA strand breakage and apoptotic morphology are not always coincident. Indeed, apparently specific ISEL staining in cells without apoptotic morphology was seen using all the protocols discussed earlier. One interpretation of these data is that multiple mechanisms of PCD occur—some involving DNA fragmentation, some involving nuclear condensation, and some involving both. Alternatively, topological constraints on DNA in the nuclei of certain cell types may prevent even extensively cleaved DNA from condensing to produce nuclear morphology typical of apoptosis. Whatever the eventual resoluton of these issues, for the time being, rather than being considered a universal marker of all types of PCD, detection of DNA fragmentation should be used in conjunction with other methods of identifying cell death. Characterization of cells in terms of as many putative markers of cell death as possible will produce a clearer picture of the mechanism(s) of PCD occurring *in vivo*.

IV. Genes Implicated in Programmed Cell Death

As indicated in Section II, the ideal marker of PCD would be a gene or genes, the expression of which could be considered diagnostic for PCD. Such a gene is not currently in hand, but many candidate genes are being actively investigated (see Chapters 7 and 8, this volume).

In any attempt to determine whether a particular gene is involved in a biological process, in this case PCD, several levels of investigation are undertaken. The first is to ask whether or not a gene of interest is expressed at the time at which and in the cells in which PCD is occurring. However, because regulation of cell death can be mediated by protective proteins like Bcl-2, which must be absent or functionally inactivated for PCD to proceed, the observation that a particular gene product is absent may therefore be just as informative as its presence. An additional consideration is that alterations in gene expression in dying cells may simply be a reflection of loss of transcriptional integrity or breakdown of the cell's internal order. Therefore, genes identified at this first level only should be viewed as candidates for involvement in PCD.

Further experimentation can demonstrate a function for candidate genes compatible with a role in PCD *in vivo*. One can ask whether the gene in question is able to effect or prevent cell death, for example. The addition of growth factors and cytokines to trophic factor-dependent cultured cell lines is an assay frequently used to address this question. Numerous studies demonstrate that diverse factors can rescue a single cell type from PCD, and also that a single factor may induce apoptosis in one cell type but protect another from the same fate. Such experiments show that the molecules assayed are capable of modulating cell death, but do not address the question of whether they actually do so *in vivo*.

The ultimate test of a gene's role in a biological process is to remove it and determine whether it is necessary for the process to occur. *In vitro,* this can be accomplished by inhibiting gene function using a variety of methods such as treatment with antibodies or antisense oligonucleotides. Again, the nature of PCD makes misexpression a very informative approach as well. Although *in vitro* experiments are valuable, the removal of a gene product is potentially most informative *in vivo*. The production of transgenic mice that misexpress or no longer express a gene of interest is a powerful way to study the functional roles of genes *in vivo*. Transgenic mice in which cell death has been affected by altering the expression of specific genes are discussed in Section IV.

In compiling a list of genes that may play a role in cell death (Table II), we chose very few genes solely on the basis of an interesting expression pattern in the absence of functional data; those that were included on the basis of their expression patterns were chosen because they were found to be expressed specifically in multiple cell types undergoing cell death. We intend Table II to serve not as a definitive listing of cell death genes, but as a roster of good possibilities for further investigation. For purposes of clarity, we have organized Table II according to a model for the sequence of events during cell death. The process can be broken down into the following discrete steps, which are also depicted shematically in Fig. 4:

1. An initial signal is received by the cell, telling it to die. This process may be mediated by the presence or absence of growth or trophic factors, hormones, cytokines and their respective receptors, or other cell-signaling molecules.

2. The signal is transferred to the nucleus and the cell then becomes committed to die. This process is presumably mediated by signal transducers, transcription factors, and proteins that control the cell cycle. These gene products could be the same as those involved in other cellular processes such as division or differentiation, or they may be specific to cell death.

3. The cell dies or is killed. This event appears to involve the activity of nucleases, proteases, and other proteins able to modify cell structure.

4. The dead cell is removed and its remnants disposed of by breakup into apoptotic bodies and phagocytosis or autophagy, which may be mediated by recognition molecules on the surface of the dead cell or the phagocytic cell. Degradation of the phagocytosed cellular material may involve additional nucleases and proteases, ubiquitins, or other catabolic proteins.

In addition to genes involved in these four basic steps are several gene products that are known to protect cells from PCD, although it is not yet clear at what point in the death pathway they act. These genes include, among others, bcl-2 and bcl-x. Interestingly, some genes in this family can also act to accelerate cell death.

Because space does not permit a detailed analysis of the evidence for every putative cell death gene listed in Table II, we will limit ourselves to a discussion of several examples of genes that act at each step in the process.

A. Extracellular Signals: Growth Factors, Cytokines, Hormones, and Receptors

Several members of the TGF-β superfamily have been implicated in the control of PCD in vitro and in vivo. In vitro, TGF-β1 inhibits cell proliferation and increases death by apoptosis in primary cultures of rabbit uterine epithelial cells (Rotello et al., 1991) and also causes apoptosis in cultured hepatocytes as well as regressing liver in vivo (Oberhammer et al., 1992). Mullerian inhibiting substance (MIS; also known as anti-Mullerian hormone) is responsible for the regression of the Mullerian ducts during normal male development, a process accomplished by PCD (reviewed by Hinchliffe, 1981).

Table II
Known or Candidate Cell Death Genes

Gene symbol	Chromosome location[a]	Gene name	Reference
Signalling molecules—cytokines, growth factors, and hormones			
Actva	?	Activin a	Nishihara et al. (1993)
Amh	10	Anti-Mullerian hormone (= MIS)	Cate et al. (1986)
Bdnf	2	Brain-derived neurotrophic factor	Oppenheim et al. (1992)
Cd30l	4	CD30 ligand	Smith et al. (1993)
Cd30r	?	CD30 receptor	Smith et al. (1993)
Cd40	2	CD40 antigen	Liu et al. (1991)
Cntf	19	Ciliary neurotrophic factor	Kessler et al. (1993); Louis et al. (1993)
Csfg	11	Colony-stimulating factor, granulocyte	Williams et al. (1990)
Csfgm	11	Colony-stimulating factor, granulocyte–macrophage	Williams et al. (1990)
Csfm	3	Colony stimulating factor, macrophage	Lotem and Sachs (1992)
Egf	3	Epidermal growth factor	Rawson et al. (1991)
Epo	5	Erthyropoietin	Muta et al. (1993)
Epor	9	Erythropoietin receptor	Nakamura et al. (1992)
Grl1	18	Glucocorticoid receptor 1	Dieken and Miesfeld (1992)
Igf1	10	Insulin-like growth factor-1	Barres et al. (1992)
Igf2	7	Insulin-like growth factor-2	Barres et al. (1992)
Il1a	2	Interleukin 1 alpha	Hogquist et al. (1991); Mangan et al. (1991)
Il1b	2	Interleukin 1 beta	Hogquist et al. (1991); Mangan et al. (1991)
Il2	3	Interleukin 2	Cohen (1985)
Il3	11	Interleukin 3	Mekori et al. (1993)
Il4	11	Interleukin 4	Migliorate et al. (1993)
Il6	5	Interleukin 6	Sabourin and Hawley (1990); Yonish-Rouach et al. (1991)
Il7	3	Interleukin 7	Matsue et al. (1993)
Ifg	10	Interferon gamma	Mangan and Wahl (1991)
Lif	11	Leukemia inhibitory factor	Kessler et al. (1993); Pesce et al. (1993)
Mgf	10	Mast cell growth factor (= stem cell factor)	Mekori et al. (1993); Muta et al. (1993); Pesce et al. (1993)
Ngfr	11	Nerve growth factor receptor, low affinity	Rabizadeh et al. (1993)
Ntrk1	3	Nerve growth factor receptor 1, high affinity	Klein et al. (1991)
—	?	Nur 77	Liu et al. (1994); Woronicz et al. (1994)
Pdgfa	5	Platelet-derived growth factor a	Barres et al. (1992)
Pdgfb	15	Platelet-derived growth factor b	Barres et al. (1992)
Tnfa	17	Tumor necrosis factor alpha	Laster et al. (1988); Mangan et al. (1991); Kawase et al. (1994)
Tnfb	17	Tumor necrosis factor beta	Laster et al. (1988); Mangan et al. (1991)
Tnfr1	6	Tumor necrosis factor receptor 1	Tartaglia et al. (1993)
Tnfr2	4	Tumor necrosis factor receptor 2	Heller et al. (1992)

Tgfb1	7	Transforming growth factor beta 1	Schwartz *et al.* (1990a); Rotello *et al.* (1991)
Tgfb2	1	Transforming growth factor beta 2	Schwartz *et al.* (1990a)
Tgfb3	12	Transforming growth factor beta 3	Schwartz *et al.* (1990a)
—	?	Leukalexin	Liu *et al.* (1987)

Processors of the cell death signals—signal transduction molecules, oncogenes, and cell cycle control proteins

Abl	2	Abelson murine leukemia oncogene	Evans *et al.* (1993)
Fos	12	FBJ osteosarcoma oncogene	Buttyan *et al.* (1988)
Jund2	8	Jun D2 oncogene	Colotta *et al.* (1992)
Kit	5	Kit oncogene	Mekori *et al.* (1993)
Max	12	Max protein	Amati *et al.* (1993)
Myc	15	Myelocytomatosis oncogene	Evan *et al.* (1992)
Piml	17	Proviral integration site 1	Moroy *et al.* (1993)
Raf1	6	Murine sarcoma 3611 oncogene	Troppmair *et al.* (1992)
Rel	11	Reticuloendotheliosis oncogene	Abbadie *et al.* (1993)
Rras	7	Harvey rat sarcoma oncogene R (= p23)	Fernandez-Sarabia and Bischoff (1993)
Tr-53	11	Transformation related protein 53	Yonish-Rouach *et al.* (1991); Clarke *et al.* (1993); Lowe *et al.* (1993)
—	?	Cyclin D	Freeman *et al.* (1992)
—	?	p34cdc2	Shi *et al.* (1994)

Effectors of cell death—proteases

Ctla1	14	Cytotoxic T-lymphocyte antigen 1 (granzyme B)	Brunet *et al.* (1991)
Ctla3	13	Cytotoxic T-lymphocyte antigen 3 (granzyme A)	Gershenfeld and Weissman (1986)
Il1bc	9	Interleukin 1 beta convertase	Miura *et al.* (1993); Yuan *et al.* (1993)
—	?	Fragmentin 2	Shi *et al.* (1994)
—	?	Nedd 2	Kumar *et al.* (1992); Kumar *et al.* (1994)

Proteins involved in recognition and degradation of dead cells

Ubb	?	Ubiquitin B	Manetto *et al.* (1988)
Ubc	?	Ubiquitin C	Manetto *et al.* (1988)
—	?	Transglutaminase	Gentile *et al.* (1992); Fesus *et al.* (1987)
—	?	Vitronectin receptor	Savill *et al.* (1990)

Molecules that protect cells from programmed cell death

Bax	?	Bcl-2 associated x	Oltvai *et al.* (1993)
Bcl2	1	B-cell lymphoma-2	
Bclx	?	B-cell lymphoma x	Boise *et al.* (1993)
Mcll	?	Myeloic cell leukemia 1	Kozopas *et al.* (1993)

Miscellaneous

Apt4	?	Apoptosis 4	Schwartz and Osborne (1993)
Apt5	?	Apoptosis 5	Schwartz and Osborne (1993)
Egr1	18	Early growth response 1	Schwartz and Osborne (1993)
Fas	19	Fas antigen (= Apo1)	Watanabe-Fukunaga *et al.* (1992); Watson *et al.* (1992)

(continues)

Table II (*continued*)

Gene symbol	Chromosome location[a]	Gene name	Reference
Fasl	?	Fas antigen ligand	Suda *et al.* (1993)
Hsp70	17	Heat shock protein 70	Buttyan *et al.* (1988)
Pdeb	5	Phosphodiesterase beta	Chang *et al.* (1993)
Plp	X	Myelin proteolipid protein	Dautigny *et al.* (1986)
Ppl1r	15	Placental protein-11-related gene (= Tcl-30)	Kingsmore *et al.* (1993)
Prph	15	Photoreceptor peripherin	Chang *et al.* (1993)
Rho	6	Rhodopsin	Chang *et al.* (1993)
Sgp2	14	Sulfated glycoprotein-2	Buttyan *et al.* (1988)

[a] A ? indicates that the chromosomal location has not been established.

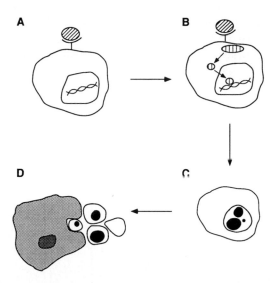

Fig. 4 A schematic diagram illustrating the steps leading to PCD. (A) An extracellular signal occurs, for example, the presence (or absence) of some cytokine or growth factor (cross-hatched oval). (B) Transduction of the death signal to the nucleus. (C) Shrinkage of the cell and condensation of chromatin. (D) Breakup of the cell into apoptotic fragments and phagocytosis by a macrophage or other neighboring cell type (shaded cell). Although it is not illustrated in the diagram, another step—the prevention of cell death—also plays a role in many or most incidences of PCD. However, the point or points in the process at which such intervention acts is not known.

The tumor necrosis factor (TNF) family has multiple members capable of inducing apoptosis. TNFα, TNFβ and the related molecule leukalexin can induce apoptosis in a limited number of tumor cell types (reviewed by Cohen *et al.*, 1992). Another molecule recently identified as a member of the TNF family is the ligand for the Fas receptor (Suda *et al.*, 1993). The Fas receptor itself (also known as Apo-1) belongs to another family of death-associated molecules. This receptor was first identified by the ability of anti-Fas monoclonal antibodies to cause rapid cell death by apoptosis in cell lines expressing the receptor. The function of the Fas receptor–ligand system *in vivo* is not presently understood, but it may eliminate autoreactive T cells or limit the amplitude and duration of the immune response (reviewed by Cory, 1994). Other members of the Fas/TNF receptor family include the two TNF receptors and the low affinity NGF receptor, p75NGFR. This interesting molecule appears to cause apoptosis when it is expressed but not bound by NGF (Rabizadeh *et al.*, 1993).

Leukemia inhibitory factor (LIF) inhibits apoptotic death in primordial germ cells in culture (Pesce *et al.*, 1993), but is also able to *induce* apoptosis in cultured sympathetic neurons (Kessler *et al.*, 1993). Apoptosis is also suppressed by related molecules such as granulocyte colony stimulating factor (G-CSF) and ciliary neurotrophic factor (CNTF). G-CSF promotes survival of hematopoietic precursor cells (Williams *et al.*, 1990) whereas CNTF acts on oligodendrocytes (Louis *et al.*, 1993).

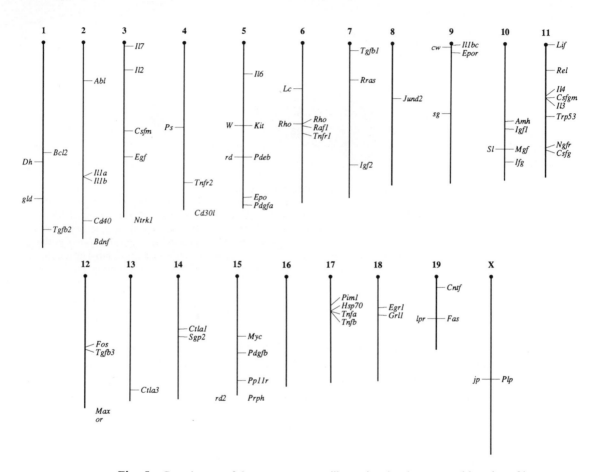

Fig. 5 Genetic map of the mouse genome illustrating the chromosomal location of known or suspected cell death genes and mutant genes for which evidence suggests that aberrant cell death patterns are a primary defect. Known or susptected cell death genes are illustrated on the right side of each chromosome and mutated genes are illustrated on the left. Genes that have been assigned to a chromosome but whose location on the chromosome has not yet been determined are illustrated without a tick mark at the bottom of the chromosome. When the gene altered in a particular mutant has been identified, the tick marks are coincident (e.g., W and Kit on chromosome 5). Additional information about these genes and mutants, as well as references, are provided in Tables II and III. Map locations were obtained from the Chromosome Committee Reports (Chapman *et al.*, 1993). Corrections and additions to the map should be addressed to J. H. Nadeau.

B. Intracellular Mediators: Signal Transducers and Transcription Factors

Little is known about the signal transduction mechanisms that mediate PCD, although Ca^{2+} ions, cAMP, and PKC have all been implicated in various systems (see Chapters 3 and 10, this volume; reviewed by Lee *et al.*, 1993). The human Ras-related protein R-*ras* p23 has recently been shown to bind to Bcl-2, raising the possibility that this molecule may be a component of a signal transduction pathway involved in the regulation of apoptosis (Fernandez-Sarabia and Bischoff, 1993).

Evan *et al.* (1992) convincingly demonstrated that deregulated *myc* expression induces apoptosis when combined with a block to cell proliferation. Rat fibroblasts constitutively expressing *myc* survive in culture when serum growth factors are present. However, after serum starvation the inability of these cells to down-regulate *myc* expression as they normally do when deprived of growth factors results in cell death. Additional support for a role for *myc* in the control of apoptosis comes from studies in T-cell hybridomas in which *myc* antisense oligonucleotides were shown to block expression of Myc protein and prevent normal activation-induced apoptosis in these cells (Shi *et al.*, 1992).

Another nuclear transcription factor, p53, mediates apoptosis triggered by DNA damaging agents such as radiation, but not other types of PCD occurring as a normal part of development (Lane, 1992). Although it is not currently understood how p53 induces PCD, its ability to remove cells with mutations in growth control genes (by causing their death by apoptosis) makes it easy to imagine how lack of p53 function leads to malignancy.

C. Direct Effectors of Cell Death

Researchers have known for some time that several classes of cells in the immune system are able, by a receptor-mediated process, to kill cells by a mechanism that produces the features of apoptosis (reviewed by Cohen *et al.*, 1992). Cytotoxic T lymphocytes (CTLs), for example, release granule serine proteases such as granzyme B (Shi *et al.*, 1992) which trigger apoptotic PCD in target cells.

Interleukin-1β convertase (ICE) is homologous to the protein product of *ced-3*, a cysteine protease responsible for most of the normally occurring cell death in *C. elegans* (Yuan *et al.*, 1993). Remarkably, overexpression of ICE in rat fibroblasts causes death characterized by the morphological features of apoptosis, and induction of death was shown to be dependent on the cysteine protease activity of ICE. Additionally, death initiated in this way can be suppressed by Bcl-2 (Miura *et al.*, 1993). It has recently been shown that the product of the *Nedd2* gene is a protease similar to CED-3 and ICE, which likewise can induce apoptosis when overexpressed *in vitro* (Kumar *et al.*, 1994). Thus, ICE and NEDD2 appear to fulfill several of the criteria for genes responsible for PCD in mammals. It remains to be determined whether one or both of these proteins typically function in cell death *in vivo*.

The polyubiquitin gene is up-regulated in moth muscles soon after these cells receive the hormonal signal to die but before cell death is morphologically evident. Thus this gene could facilitate protein degradation necessary for cell death to occur (Schwartz *et al.*, 1990b). In mammalian neurons induced to die by deprivation of NGF, however, no up-regulation of polyubiquitin gene expression is detected (Martin *et al.*, 1992).

D. Breakup and Removal of Dead Cells

Intracellular transglutaminases are enzymes that catalyze the cross-linking of proteins, for example, during the formation of a cornified cell envelope (Polakowska and Goldsmith, 1991). Tissue transglutaminase is an intracellular protein found in many cells and tissues that is thought to be involved in apoptotic cell death (Fesus *et al.*, 1987). Gentile *et al.* (1992) showed that the overexpression of tissue transglutaminase in 3T3 fibroblasts causes blebbing and cellular fragmentation that is very reminiscent of that observed during apoptosis.

The vitronectin receptor, a member of the $\beta 3$ family of integrins, appears to play an important role in the removal of some dead cells. Using competition experiments in which peptides and antibodies known to bind the vitronectin receptor specifically inhibited phagocytosis by macrophages, Savill *et al.* (1990) showed that macrophage recognition and phagocytosis of apoptotic lymphocytes and neutrophils is mediated by the vitronectin receptor.

E. Prevention of Cell Death: The *bcl*-2 Family

Bcl-2 is a membrane-associated protein capable of preventing cell death in multiple contexts, including growth factor-deprived cells (Nunez *et al.*, 1990; Garcia *et al.*, 1992), irradiated cells, and lymphoid cells treated with glucocorticoids (Sentman *et al.*, 1991). In contrast, ciliary neurons deprived of CNTF are not rescued by *bcl*-2 expression (Allsopp *et al.*, 1993). Evidence from Hockenbery *et al.* (1993) suggests that Bcl-2 may prevent cell death by functioning in an antioxidant pathway.

The recent identification of genes structurally related to the *bcl*-2 gene, such as *bcl*-x, *bax*, and *mcl*-1, indicates an additional level of complexity in the control of cell death by *bcl*-2. For example, Bcl-x can also function as a regulator of cell death via two alternatively spliced forms (Boise *et al.*, 1993). Bcl-x_L, the larger form of the protein, acts similarly to Bcl-2 and prevents death after growth factor withdrawal in factor-dependent cells. In contrast, the smaller form, Bcl-x_S, interferes with the ability of Bcl-2 to prevent death under the same circumstances (Boise *et al.*, 1993).

Bcl-2 can also interact with another related protein designated Bax (Oltvai *et al.*, 1993). Bax forms homodimers and can also heterodimerize with Bcl-2. Overexpression of Bax both accelerates induced apoptotic cell death and interferes with the ability of Bcl-2 to inhibit cell death.

Finally, another family member, Mcl-1, was identified in a human myeloid cell line undergoing differentiation (Kozopas *et al.*, 1993). Although it remains to be seen whether this protein also functions to prevent cell death, the authors note parallels between the role Bcl-2 plays in lymphoid development and a possible role for Mcl-1 in myeloid differentiation.

V. Genetic Resources for Identifying and Characterizing Cell Death Genes

One of the most informative methods of studying PCD is the analysis of mutations that affect the process. Mouse mutations provide a means for identifying genes that play a role in PCD and for studying the function of those genes *in vivo*. These mutant mice can also be used to define the pathways leading to cell death (see Section VI). The value of a genetic approach is illustrated by the elegant work done in *C. elegans* and *Drosophila,* in which mutant genes have been used to begin defining the hierarchies of gene action that lead ultimately to PCD (Ellis *et al.*, 1991; Abrams *et al.*, 1993).

The classical genetic approach in mice is to study spontaneous and mutagen-induced mutations that affect a process such as PCD. Such mutations can be detected because of their overt phenotypic consequences. The effects of mutation can be very different depending on which kind of gene is affected and whether the mutation causes loss of function or gain of function. For example, a loss-of-function mutation in a gene that causes PCD would result in survival of cells that would normally die, whereas the same type of mutation in a gene that suppresses PCD would cause the death of cells that normally live.

For most mutants, the underlying molecular alteration responsible for the abnormal phenotype is unknown. Most are probably loss-of-function alleles that result from deletions, simple base substitutions in the coding region, or other sequence changes affecting the gene product, including splicing defects. Much less common are gain-of-function mutations in which the gene product is expressed or functions abnormally, disrupting the normal physiological process in which it is involved. Although the molecular nature of mutations cannot be controlled when they arise spontaneously, the types of molecular alterations that are obtained can be controlled by treatment of mice with chemical mutagens such as ethylnitrosourea (base substitutions) (Favor, 1990) or chlorambucil (deletions) (Russell *et al.*, 1990),

By identifying the gene that is mutated and thus responsible for certain cell death abnormal phenotypes, one can identify genes that play a role in PCD. One method for identifying the gene responsible for a given spontaneous or mutagen-induced mutation, which is often called "candidate gene analysis," is to identify cloned (candidate) genes that are closely linked to the mutation of interest, and then to explore the possibility that one of these candidate genes is altered by the mutation. Numerous resources are available for mapping cloned genes in the mouse and for establishing their close linkage to the mutation of interest (see Appendix 1). In some cases, the phenotype of the mutation of interest is consistent with an alteration in the candidate gene, providing a strong impetus to pursue their potential allelic relationship. In other instances it is not obvious how a mutation in the candidate gene could result in the phenotype displayed by the mutant animals. Once a candidate gene is identified, the next step is usually to determine whether it can be separated from the mutation of interest by recombination. If it can, this is strong evidence for nonidentity. In contrast, absence of recombination is strong evidence for identity, but not definitive proof; additional molecular studies are usually required to demonstrate that the candidate gene is altered in the mutant.

If the gene affected by a mutation cannot be identified by the candidate gene approach, another more laborious approach known as "positional cloning" is available. This approach is based on knowledge of the precise location of the mutation of interest on genetic maps, the isolation of a large genomic segment (from 100 kb to several megabases in length) that is likely to contain the gene of interest, and the construction of a detailed physical map of the region. Access to large-insert clone libraries is essential for this approach (see Appendix 2). The candidate genomic segment is then analyzed to identify genes contained within it. These genes are then evaluated to determine which has been altered by the mutation of interest. This can be a daunting task if the molecular defect is difficult to locate, as a single base change would be. This approach has the advantage of identifying genes without preconceptions about their function. As a result, novel genes and gene families are often discovered. However, because of the technical and laborious nature of this approach, few genes have been positionally cloned in mice and none appear to involve a PCD gene (Copeland *et al.*,1993). Many positional cloning projects are under way and considerable progress can be expected in the foreseeable future.

In many cases a cloned gene rather than a particular mutation is the focus of interest; the aim is to obtain mice that carry mutant alleles of that gene. Again, the "candidate" approach can be very productive; in this instance, a mutant is the candidate. Once the map location of the gene of interest is established, one can identify a closely linked candidate mutation, determine whether the gene of interest can be separated from it by recombination, and, if appropriate, perform a molecular analysis of the gene of interest in the mutant mice. An excellent example of the success of this approach is the identification of the Fas antigen-encoding gene as the locus that is mutated in *lpr* and *lpr^cg* mice (see subsequent discussion).

If animals carrying mutations in the gene of interest are unavailable, or if particular types of mutations are desired, it is now possible to create specific mutations in genes of interest through transgenic technology. A method known as gene targeting, which involves altering the gene of interest by homologous recombination in embryonic stem cells cultured *in vitro* (reviewed by Capecchi, 1989), has enabled investigators to produce mice carrying loss-of-function mutations in known genes and to explore the phenotypic consequence of these mutations *in vivo*. This approach has recently been used to target the mouse *bcl-2* gene (Veis *et al.*, 1993; see subsequent discussion).

Finally, transgenic technology can be used to produce mice in which expression of a particular gene is altered by modifying or replacing expression control regions. This approach can be used to test whether changing the time or location of gene expression activates or suppresses PCD pathways, for example. If an unusual phenotype results in mice mis-expressing a gene, the result is formally equivalent to a spontaneous gain-of-function mutation.

The mouse biomedical research community is blessed with an exceptional collection of resources that facilitate the genetic approaches just outlined as well as with a long tradition of sharing these resources with colleagues, often before publication. In addition to the resources available for candidate gene analysis and positional cloning described in Appendices 1 and 2, numerous other types of genetic resources exist. Appendix 3 describes readily accessible mouse mutant collections and repositories.

The volume and diversity of information discovery is outstripping the capacity of print publications. Just as it was necessary to store and access DNA sequences electronically, it is becoming imperative to manage and distribute other kinds of biological knowledge in electronic information resources. These resources provide more timely updates and more experimental detail than is usually available through other mechanisms. The electronic resources that are of particular utility for genetic analysis in the mouse are described in Appendix 4.

A. Mouse Mutations that Result in Abnormal Patterns of Programmed Cell Death

In this section, we provide a brief description of existing mouse mutations that appear to affect genes that play a role in PCD. For some of the mutations discussed, the evidence is compelling that the primary effect is on PCD; for others the evidence is circumstantial. We divided putative PCD genes and mutants into two classes depending on the nature of their effect. The first class includes those that result in survival of cells that would normally die, and the second class includes those that result in death of cells that would normally live. Although this categorization is somewhat arbitrary, it provides a framework for thinking about the effect of PCD genes on mouse development.

Hundreds, if not thousands, of spontaneous mutations have been identified and systematically studied in mice. Compelling evidence that a spontaneous mutation disrupts PCD is available for remarkably few genes (see Table III). The number of examples is so modest for two principal reasons. First, until recently few rigorous methods were available for identifying cells undergoing PCD. Second, because the importance of PCD during development was not generally recognized, even these limited assays were not commonly used. Many mutations other than those listed in Table III probably affect PCD. For example, although syndactylism and polydactyly, which can result from alterations in

Table III
Examples of the Various Kinds of Mutations in Which the Normal Patterns of PCD Are Altered

Gene symbol	Chromosomal location	Gene name	Phenotypic consequences	Reference
Mutations that cause cells to live that would normally die				
Dh	1	Dominant hemimelia	Hemimelia, no spleen	Rooze (1977)
Fas^{lpr}	19	Fas antigen	Lymphadenopathy	Watanabe-Fukunaga et al. (1992); Watson et al. (1992)
gld	1	Generalized lymphoproliferative disease	Indistinguishable from lpr	Roths et al. (1984); Takahashi et al. (1994)
or	12	Ocular retardation	Retardation of the retinal anlage	Theiler et al. (1976)
Ps	4	Polysyndactyly	Polysyndactyly	Johnson (1969)
cw^{thd}	9	Curly whiskers, tail hair depletion	Altered inflammatory response	Les and Roths (1975); Roths (1978)
Mutations that cause cells to die that would normally live				
Lc	6	Lurcher	Purkinje and granule cell loss	Heintz (1993)
$Pdeb^{rd}$	5	cGMP-phosphodiesterase, beta subunit	Photoreceptor death	Chang et al. (1993)
$Prph^{rd2}$	15	Photoreceptor peripherin	Photoreceptor death	Chang et al. (1993)
Plp^{jp}	X	Myelin proteolipid protein	Myelin deficiency	Dautigny et al. (1986)
Rho	6	Rhodopsin	Photoreceptor death	Chang et al. (1993)
sg	9	Staggerer	Granule and Purkinje cell loss	Vogel et al. (1989)
Kit^{W}	5	Kit oncogene	Altered hematopoiesis, pigmentation, and fertility	Geissler et al. (1988)
Mgf^{Sl}	10	Mast cell factor	Altered hematopoiesis, pigmentation, and fertility	Copeland et al. (1990); Zsebo et al. (1990)

the normal patterns of PCD, are common phenotypes, evidence for involvement of cell death is available for only two mutations (see subsequent text). With the development of new methods for assessing cell death, as discussed in Section II, these and other mouse mutants should be re-evaluated.

1. Mutations that Allow Cells to Live that Would Normally Die

a. Fas^{lpr} and Fas^{lpr-cg} *(formerly known as lpr and lpr^{cg})*

These spontaneous mutations provide the first conclusive demonstration that an abnormal phenotype results from a mutation in a gene involved in PCD. The Fas antigen, which is defined by an antibody recognizing a cell surface molecule, is a member of the TNF receptor family. Ligands for these receptors have been implicated in PCD. Expression of the human Fas antigen in mouse cells leads to antibody-induced apoptotic cell death (Itoh *et al.*, 1991); expression in the thymus suggests a role in negative selection of autoreactive T cells through apoptosis.

A crucial step in the analysis was the discovery from routine mapping studies that the gene encoding the Fas antigen was closely linked to *lpr* on chromosome 18 (Watanabe-Fukunaga *et al.*, 1992; Watson *et al.*, 1992), thus making it a candidate for the gene mutated in *lpr* mice. The phenotype of mice homozygous for the *lpr* mutation consists of an autoimmune disease resembling systemic lupus erythematosus (Morse *et al.*, 1985). Lymphadenopathy and glomerulonephritis are common in older mice. The abnormal accumulation of T cells results from a defect in thymic T-cell selection rather than from overproliferation of a normal cell type. An obvious explanation for this unusual phenotype and important disease model was that a gene encoding a molecule responsible for PCD (i.e., Fas) was defective.

Molecular studies were undertaken to determine whether *fas* was mutated in *lpr* and *lpr^{cg}* mice. These studies revealed a 1.4-kb deletion in the *fas* gene in mice homozygous for the *lpr* mutation. This deletion covers an intronic region and does not detectably affect Fas antigen expression levels, although it presumably affects protein function (Watanabe-Fukunaga *et al.*, 1992; Watson *et al.*, 1992). Molecular analysis of a second *lpr* mutation, *lpr^{cg}*, revealed a T to A transition at nucleotide 786, causing an amino acid replacement in the cytoplasmic portion of the Fas molecule. (Because the mutant alleles have now been shown to be caused by a mutation in the gene encoding the Fas antigen, they have been renamed *Fas^{lpr}* and *Fas^{lpr-cg}*.) The discovery that lupus-like diseases result from mutations in a PCD gene illustrates the benefits of mapping genes responsible for cell death.

b. gld

This spontaneous mutation, which is located on chromosome 1, causes a generalized lymphoproliferative disease that in many respects is indistinguishable from the disease in *lpr* mice (Davidson *et al.*, 1986). The discovery that the gene mutated in *lpr* mice is a cell surface receptor immediately led to the prediction that the ligand for this receptor is mutated in *gld* mice. This hypothesis has recently been confirmed (Takahashi *et al.*, 1994).

Several other mutations probably result from a failure in PCD but the mutated gene has not yet been identified. These mutations include *Dominant hemimelia* (*Dh;* chromosome 1) and *Polysyndactyly* (*Ps;* chromosome 4), both of which show digit abnormalities probably resulting from inappropriate survival of cells in the interdigital zone. Another important mutation in this class is *ocular retardation* (*or;* chromosome 12), which is associated with a general failure of PCD during development of the eye, resulting in congenital blindness.

c. Dh

Heterozygous mice show numerous abnormalities of the axial and apendicular skeleton including reduced numbers of ribs, preaxial polydactyly, and tibial hemimelia (Searle, 1964). Homozygotes die near birth with more severe abnormalities than those found in heterozygotes (Searle, 1964). As discussed in Section I, cell death patterns in *Dh* mice are disrupted during limb development, showing less PCD than expected (Rooze, 1977). These mice lack an anterior necrotic zone in the developing limb bud (Knudsen and Kochhar, 1981), a defect that could readily explain the observed limb phenotype.

d. Ps

Heterozygous mice show soft tissue syndactylism and occasionally an extra digit between digits III and IV on the hindfeet or a duplicated digit V on the forefeet (Johnson, 1969). *Ps* homozygotes die at or near birth and show edematous limbs and abnormal apical ectodermal ridge morphology characterized by hypertrophy and a wavy appearance. Normal patterns of PCD are disrupted; on E14, cell death fails to occur in the interdigital zone of mutant mice as it does in normal mice (Johnson, 1969). *Ps* is thus another example of a mutation that results in the survival of cells that would normally die. The apical ectodermal ridge is known to be an important source of signals during limb morphogenesis. A defect in the function of the ridge may lead, among other effects, to a failure to activate PCD in the interdigital zone.

e. or

Cell death plays an important role in shaping the eye during development (Silver and Hughes, 1973; see Table I). In *or* mice, a general failure of cell death occurs during development of the eye. The eye develops normally until about E10–E11 when the frequency of cell death is markedly reduced. The principal result of this defect is a closed optic stalk which blocks the path of the central artery to the eye and the path of optic nerve fibers to the brain, eventually resulting in blindness (Theiler *et al.*, 1976).

f. cw^{thd}

Interest in this mutation, *curly whiskers*, is based on the findings that it maps near *Ilbc* (Kingsley, 1992), the mouse homolog of the *C. elegans* cell death gene *ced-3* (Yuan *et al.*, 1993), which encodes a cysteine protease called ICE that activates the β-chain of IL-1 and can also induce PCD in mammalian cells *in vitro* (Miura *et al.*, 1993). The phenotype of *cw^{thd}* (Les and Roths, 1975; Roths, 1978) could potentially result from an alteration in ICE expression, thus providing an impetus to study whether *cw^{thd}* is a mutant allele of *Ilbc*.

2. Mutations that Cause Cells to Die that Would Normally Live

Many mutants display varying degrees of degeneration of cells or tissues, but few have been characterized in terms of the nature of the death. Although the following examples involve alterations in genes not usually thought of as cell death genes, the identification of the mechanism of cell degeneration in these mutants as PCD suggests that we may need to broaden our ideas about the types of genes that can play a role in cell death.

a. Pdeb^{rd}

Mice with this mutation (originally designated *rd*, retinal degeneration) display degeneration of rod photoreceptor cells in the retina that begins at approximately 8 days after birth and continues until 4 wk of age, by which time no photoreceptors remain (Bowes *et al.*, 1990). The observation that the degeneration is preceded by a build-up of cGMP in the affected cells led to the discovery that this mutation is caused by a viral insertion into the gene encoding the cGMP phosphodiesterase β subunit (Bowes *et al.*, 1990). Researchers showed that the mechanism of degeneration in the photoreceptors of *Pdeb^{rd}* mice is PCD, since the dying cells display both apoptotic morphology and a DNA ladder (Chang *et al.*, 1993). Investigators have known for some time that PCD is normally associated with histogenesis of the retina in the mouse (Young, 1984) and a comparison of PCD in normal vs mutant mice showed that in the mutants, the timing of this developmental death is the same but the amount of death is greater (Chang *et al.*, 1993).

b. Prph^{rds}

This mutation was initially designated *rds*, retinal degeneration slow. Although the phenotype is similar to that of *Pdeb^{rd}* mice, a different locus encoding the peripherin protein is affected. Peripherin is a neuron specific cytoskeletal protein that may function to maintain photoreceptor stability. As in *Pdeb^{rd}* mice, the cell degeneration in *Prph^{rds}* mice was also demonstrated to involve DNA fragmentation and the dying cells display apoptotic morphology. However, in *Prph^{rds}* mice, the degeneration occurs after the normal period of developmental PCD in the retina has ceased (Chang *et al.*, 1993).

In addition to the mice with retinal degeneration phenotypes, at least three other neurological mutations—lurcher (*Lc*) (Heintz, 1993), staggerer (*sg*) (Vogel *et al.*, 1989), and jimpy (*Plp^{jp}*) (Knapp *et al.*, 1986)—exist in which PCD has been implicated in cell loss and abnormal behavioral phenotypes. The evidence in these three examples is primarily based on histological analyses.

c. *Mgf^{Sl}* and *Kit^W*

Mutations at both the *Steel* (*Sl*) and the *Dominant white spotting* (*W*) loci produce mice with defects in the germ cell, hematopoietic and melanocyte lineages resulting in sterility, anemia, and characteristic lack of coat pigmentation (Russell, 1979). Researchers now know that *Sl* and *W* phenotypes result from mutations in the genes for mast cell growth factor (MGF; also called Steel factor) and its receptor, the c-*kit* proto-oncogene, respectively (Chabot *et al.*, 1988; Geissler *et al.*, 1988; Copeland *et al.*, 1990; Huang *et al.*, 1990; Zsebo *et al.*, 1990). The *Sl* and *W* genes have consequently been renamed *Mgf^{Sl}* and *Kit^W*. Although it is not known whether the reduction in numbers of the affected cell types seen in the mutants involves PCD, the observation that MGF promotes the survival of mouse primordial germ cells (PGCs) in culture (Godin *et al.*, 1991), combined with the fact that some PGCs die by PCD both *in vitro* in the absence of trophic factors (Pesce *et al.*, 1993) and *in vivo* as a normal part of their development (Coucouvanis *et al.*, 1993), suggests that these mutations may produce their effects by causing excessive PCD in the cell types affected.

B. Transgenic Mice Involving Cell Death Genes

Another approach to the characterization of cell death-associated genes *in vivo* is the use of reverse genetics: start with a gene of interest and determine its function by specifically altering it in mice. Although this strategy is time consuming and expensive, the rewards include the ability to make very specific changes in gene expression such as ablation of expression, selective overexpression, or targeted misexpression.

1. Targeted Disruption of the *bcl-2* Gene

The gene targeting approach via homologous recombination in ES cells has been used to produce mice deficient for a number of cell death-associated genes. *bcl-2*-deficient mice, for example, are characterized by massive apoptotic involution of the thymus and spleen, hypopigmented hair, and polycystic kidney disease (Veis *et al.*, 1993). Hematopoiesis, in contrast, is unaffected. The apparently normal prenatal development of these mice is surprising considering that Bcl-2 is widely expressed in the developing embryo. For example, Bcl-2 is expressed in digital zones of the wild-type developing limb, but not in the interdigital regions, where cell death occurs (Veis *et al.*, 1993). If Bcl-2 alone were responsible for preventing the death of digital cells, the *bcl-2* −/− embryos would be expected to lack digits. The fact that bcl-2 −/− embryos develop normal digits may be explained by the existence of several bcl-2-related genes that may be able to substitute for bcl-2 functionally during development. Nevertheless, bcl-2-deficient mice allow processes more and less dependent on bcl-2-mediated prevention of cell death to be distinguished—an important step in understanding normal control of cell death.

2. Selective Overexpression of the Gene Encoding Mullerian Inhibiting Substance

Experiments involving misexpression of putative and established cell death genes have demonstrated that it is possible to perturb PCD in a highly specific fashion. Mullerian inhibiting substance (MIS), a member of the TGF-β superfamily, is normally secreted by Sertoli cells in the male reproductive system. Its normal expression in males results in regression of the Mullerian ducts (precursors of the oviducts, uterus, and upper vagina) by initiating a localized program of cell death via a receptor-mediated event (reviewed by Catlin et al., 1993). Behringer et al., (1990) produced transgenic mice that constitutively express MIS under the control of the mouse metallothionein-1 promoter. In females, this expression produced adult mice lacking uteri, oviducts, and ovaries. The majority of transgenic males developed normally. However, some individuals from lines with the highest levels of MIS expression exhibited aberrant sexual development such as undescended testes and impaired Wolffian duct development. This result may be due to down-regulation of the MIS receptor in these individuals or some other function of MIS in testicular development (Behringer et al., 1990).

These female MT–MIS mice display a phenotype with cell death specifically induced in response to misexpression of a single gene, strongly suggesting that MIS is directly or indirectly responsible for normal, PCD-mediated Mullerian duct regression. These mice could be used to ask additional questions about the control of cell death in sexual development. For example, crossing MT–MIS mice with mice specifically overexpressing Bcl-2 in the Mullerian duct might show whether Bcl-2 can inhibit this example of developmental cell death in vivo.

3. Targeted Cell Ablation of a Macrophage Subset

Another example of a transgenic mouse model displaying a highly specific cell death phenotype was produced by the ablation of a macrophage subset with diphtheria toxin (Lang and Bishop, 1993). By placing diphtheria toxin expression under the control of a genomic DNA fragment known to drive gene expression in the ocular and serosal subsets of mouse macrophages, Lang and Bishop (1993) ablated these two cellular subsets. The resulting mice displayed an interesting eye phenotype characterized by the persistence of normally transient eye tissues, the hyaloid vasculature and the pupillary membrane. Apparently, these two tissues are normally removed by the missing subset of ocular macrophages. Tissue regression of the hyaloid capillary network in normal animals had been previously reported, although the mechanism of death was not established (Szirmai and Balazs, 1958). Since the ocular macrophages are normally closely associated with the hyaloid vessels and pupillary membrane during their phases of regression, the authors interpret the transgenic phenotype to mean that the ocular macrophage is required for developmentally programmed tissue remodeling in the eye, and may provide an extracellular signal that initiates cell death (Lang and Bishop, 1993). Although it is unclear whether cell and chromatin condensation or other features indicative of apoptosis are normally present in these regressing cell populations, phagocytosis is a key feature consistent with the PCD scheme. Alternatively, macrophage-mediated regression may be a distinct mode of developmental cell death.

4. Gain-of-Function Transgenic Mice: The Rhodopsin Gene

A spontaneous mutation in the rhodopsin gene in humans causes the disease retinitis pigmentosa (RP), a degenerative disorder of the retina. As discussed earlier, mutations in two other mouse loci, in the genes encoding peripherin (*Prph^{rds}*) and the cGMP phosphodiesterase β subunit (*Pdebrd*), also produce this phenotype. To test the hypothesis that an altered rhodopsin gene in mice might also produce retinal degeneration similar to that seen in the human syndrome, transgenic mice expressing a mutant rhodopsin gene were generated (Chang *et al.*, 1993). These transgenic mice display essentially the same phenotype as the *Prph^{rds}* mice, including retinal degeneration occurring after normal developmental death in the retina has ended. Furthermore, the cell degeneration in these mice was shown to involve apoptotic morphology and the presence of a DNA ladder (Chang *et al.*, 1993). Why alterations in rhodopsin function cause PCD remains to be determined.

Several other cell death-associated genes have been manipulated in transgenic experiments. However, these results have proven more difficult to interpret with respect to the precise role that the gene in question may play in cell death. The genes include (but are not limited to) *tgfb*1 (Shull *et al.*, 1992), *ngfr* (Lee *et al.*, 1992), *lif* (Escary *et al.*, 1993), *csfgm* (Lang *et al.*, 1987), *il*2 (Ishida *et al.*, 1989), *jun* (Hillberg *et al.*, 1993), and *fos* (Wang *et al.*, 1992).

VI. Experimental Approaches for Dissecting the PCD Pathway in Mice

The phenomenon of programmed cell death is universal because it occurs in myriad cell types and contexts. However, as in other universal processes such as cell differentiation and proliferation, many context-dependent variations are likely to operate, despite the fact that the physiological end result is the same.

Indeed, the sheer number and diversity of gene products able to modulate PCD (Table II) confirms that at least some of the signals and events outlined in Section III are likely to be accomplished by different molecules in different cell types.

In contrast, some genes involved in PCD appear to function in a less context-dependent manner. For example, Bcl-2 expression is able to prevent the PCD that occurs in response to many different signals (see Section III). Likewise, endonuclease activity has been observed in many cell types undergoing PCD. Thus, the challenge for the investigator is to unravel the hierarchy of genes involved in PCD, and to determine which components are cell type specific and which (if any) are universal.

In the preceding sections of this chapter, we provided background information pertinent to future study of the mechanisms and control of PCD, especially during development. The summary of the incidence of PCD in the embryo (Table I), the list of genes that have been identified as potential mediators of the process (Table II), and the examples of mice with random or engineered mutations affecting PCD (Table III) are all intended to stimulate further progress by facilitating the genetic approach to understanding PCD. A discussion follows of how knowledge of normal cell death patterns, candidate genes, and mice with PCD-altering mutations can be used to learn more about the genetic pathway controlling PCD.

A. *In Vitro* Analysis

Once a gene has been identified that is able to cause or prevent PCD, further investigation *in vitro* can lead to an understanding of its role in a PCD pathway. For example, one can test whether the expression of the gene is responsive to the levels of Bcl-2 or to the growth factors that can keep the affected cell types alive. Rapid and specific up-regulation of transcription of the gene in question in response to treatment of cells with trophic factors, combined with an observed ability of that gene to rescue cells from PCD, might suggest that the gene acts downstream from the trophic factor in a pathway that results in prevention of PCD.

To investigate the PCD pathway in which transcription factors such as Myc act, gel shift assays could be used to identify genes activated by the transcription factor (see Chapter 11, this volume). In turn, the genes activated by Myc during PCD may themselves encode new or previously known cell death molecules.

Although *in vitro* studies can be informative, mutant or transgenic mice are needed to extend studies of PCD pathways to the whole organism.

B. Downstream Gene Expression in Mutant/Transgenic Mice

Mice with altered patterns of cell death can be examined by *in situ* methods to determine whether the expression patterns of genes other than the one altered by the mutation are affected. For example, a mutation or knockout of a trophic factor receptor in a specific population of cells may result in the initiation of a cascade of abnormal expression of other genes leading to cell death. The identification of genes that are expressed abnormally immediately prior to the onset of inappropriate cell death could identify likely participants in the wild-type sequence of events leading to PCD.

Likewise, a comparison of gene expression in mutant or transgenic mice lacking a normal episode of cell death may point to genes that play a role in PCD. One could then supply the products of such genes in *trans,* either *in vitro* or by mating the altered mice with transgenic mice carrying the putative downstream gene of interest driven by the appropriate tissue-specific promoter. Restoration of the wild-type pattern of cell death would establish a role for the gene in question in PCD and would suggest the possible order of the genes in a cell death pathway.

The identification of proteases ICE and NEDD2, which are capable of producing apoptotic cell death *in vitro,* has made it possible to analyze the relative dominance of controller and effector genes. To assess the role of these putative death effector genes *in vivo,* transgenic mice with the gene in question under the control of a tissue specific promoter could be used. A comparison of the effects of expressing the putative death gene early and late during a differentiation sequence could address whether the timing of death gene expression influences the outcome of a cell's life-or-death decision.

C. Double Transgenics

The existence of multimember gene families in which family members can functionally substitute for one another *in vivo* is adding a new wrinkle to the transgenic approach to determination of gene function. The phenotype of *bcl-2*-deficient mice may be a good example of this phenomenon. The fact that development is normal in such mice despite widespread expression of *bcl-2* during embryogenesis suggests that another related gene (or genes) can substitute when *bcl-2* is absent. Examples of family members that could substitute for *bcl-2* include the mouse version of *mcl-1, bcl-x,* or other related genes not yet cloned. It may be necessary to produce transgenic mice deficient in two members of a multigene family with potentially overlapping functions to demonstrate a phenotypic abnormality.

Double transgenic mice may also be useful for determining the interaction between two genes thought to be involved in the same process. Double mutants in *Drosophila* and *C. elegans,* for example, have helped to elucidate the order and interaction of genes in multistep pathways (see Chapter 14, this volume). Although mammalian systems are less amenable to the straightforward analysis of double mutant phenotypes, confining the effects to a single tissue or cell lineage via transgenes with expression limited by specific promoters may simplify interpretation of results.

One of the potential problems with the study of cell death during mouse development is that expression of a death gene by traditional transgenic means could lead to early lethality of the affected embryos. This premature death would preclude examination of the function of the gene in cell death at later stages of gestation or postnatally. An exciting new avenue for transgenic research and one that provides a solution to this potential problem is the binary transgenic strategy (for review, see Kilby *et al.*, 1993). This approach can also be used to produce gene knockouts that are tissue specific.

VII. Concluding Remarks

Unfortunately, the current status of our knowledge of the genetic pathways controlling PCD in vertebrates leaves us with many more questions than answers. It is unclear, for example, whether any of the events in PCD are common to all cell types. Likewise, we do not know how many different pathways lead to PCD. Although in Section III we outlined a series of steps that can be involved in the PCD process, it is likely that in some examples of cell death, one or more of these steps is by-passed. For example, if RBC differentiation is in fact a specialized kind of PCD, it should be considered an example of this phenomenon, because the steps of phagocytosis and breakup of the cell into fragments are either prevented or delayed in these cells. It is remarkable that the precise control of different aspects of PCD might be used to produce such highly specialized cells. The relative resistance of a cell type to PCD is another interesting question. Can some cell types be killed more easily than others? How is that feature a reflection of the genetic status of such cells?

The opportunities are exceptional for answering these and other questions about the genetic pathways of cell death using the laboratory mouse. There are many new assays for PCD, an extraordinary collection of mouse mutants that are largely uncharacterized with respect to PCD, and a large number of genes that remain to be identified, mapped, and characterized. Utilization of these rich resources is likely to produce considerable progress in understanding the genetic control of cell death in the mouse in the near future.

Appendix 1. Resources for Mapping Genes in the Mouse

A. Special Strains of Mice and Panels of Genomic DNA from Linkage Testing Crosses

These strains and panels are exceptional resources because they make it possible to determine the map position of a gene of interest without having to perform a new genetic cross. Instead, by typing a given panel for the gene of interest, its map location can be readily determined by simply comparing those typing results with results for numerous other loci that were previously obtained for the panel by other investigators.

1. Recombinant Inbred Strains

These strains are derived through inbreeding mice derived from intercrosses of hybrids between two inbred strains (Taylor, 1981). Their value is that each strain captures a unique combination of recombination events. These strains can be used for gene mapping and for mapping traits that require *in vivo* testing. Some panels such as the BXD strain set (the progenitors are C57BL/6J and DBA/2J) have been typed for more than 1000 loci. Taylor (1995) has summarized the status and availability of the commonly used recombinant inbred strain sets. Most of these strains can be obtained by contacting B. A. Taylor, Jackson Laboratory [tel. (207)288-3371; e-mail bat @ aretha.jax.org]. Genomic DNA from most of these strains can be obtained by contacting the Mouse DNA Resource, Jackson Laboratory [Marie Ivey; tel. (207)288-3371].

2. Interspecific Mapping Panels

Gene mapping underwent a dramatic revolution with the introduction of interspecific crosses (Avner *et al.,* 1988). These crosses involve conventional strains of inbred mice such as C57BL/6J and strains derived from different subspecies such as *Mus musculus castaneus* (e.g., CAST/Ei) and *Mus musculus molossinus* (e.g., MOLG/Ei), and strains derived from different species such as *Mus spretus* (SPRET/Ei and SPE/Pas). Because of substantial evolutionary divergence, the frequency of genetic variants among these strains is high. As a result, it is possible to type large numbers of loci in a single cross, thereby resolving many of the problems of inferring gene order and of preparing comprehensive multilocus maps. Mapping panels that have been prepared from backcrosses involving these mice have already made extraordinary contributions to gene mapping (Copeland *et al.,* 1993). The following is a listing of the principal panels that are available for public use. Some are available only through collaboration; others are freely available.

1. The Copeland–Jenkins panel. This panel is composed of ~300 progeny derived from backcrosses (C57BL/6J x *M. spretus*)F_1 hybrid females to C57BL/6J males (Copeland and Jenkins, 1991). This panel has been typed for more than 1500 loci. Information about access to the panel can be obtained by contacting N. G. Copeland or N. A. Jenkins, Frederick Cancer Research Facility, Frederick, Maryland [tel. (301)846-1260; fax (301)846-6666].

2. EUCIB—the European Community Interspecific Backcross panels. These panels are composed of ~500 progeny derived from backcrosses of (C57BL/6J x *M. spretus*)F_1 hybrid females to C57BL/6J males and another ~500 progeny derived from backcrosses of the same hybrids to *M. spretus* males. These panels have been typed for more than 500 loci. Information about access to these panels can be obtained by contacting S. D. M. Brown, Biochemistry Department, St. Mary's Hospital, London (e-mail s. brown@sm.ic.ac.uk).

3. The Jackson Laboratory panels. These panels are composed of ~100 progeny derived from backcrosses of (C57BL/6J x *M. spretus)* F_1 hybrid females to C57BL/6J males and another ~100 progeny derived from backcrosses of the same hybrids to *M. spretus* males (Rowe *et al.,* 1994). The panel derived from the backcross to *M. spretus* has been typed for more than 700 loci and the

panel derived from the backcross to C57BL/6J has been typed for ~300 loci. Information about access to these panels can be obtained by contacting Lucy Rowe, Jackson Laboratory (207)288-3371; fax (207)288-5079; email lbr@aretha.jax.org).

4. The Johnson–Davisson panels. These panels are composed of mice from two crosses. The first panel of 144 progeny is derived from a backcross of (C57BL/6J x CAST/Ei)F$_1$ hybrid females to C57BL/6J males. The second panel, which is also composed of 144 progeny, is derived from a backcross of (C57BL/6J x SPRET/Ei)F$_1$ hybrid females to SPRET/Ei males. Approximately 250 loci have been typed in each panel. Information about access to these panels can be obtained by contacting K. R. Johnson, Jackson Laboratory [tel. (207)288-3371; fax (207)288-5079].

5. The Kozak panels. These two panels are composed of ~500 progeny derived from backcrosses of (NFS/N x *M. musculus musculus*)F$_1$ hybrids to *M. musculus musculus* and from backcrosses of (NFS/N x *M. spretus*)F$_1$ hybrids to C58/J and to *M. spretus*. These panels have been typed for more than 600 loci. Information about access to these panels can be obtained by contacting C. A. Kozak, Department of Molecular Microbiology, National Institute of Allergy and Infectious Diseases, National Institutes of Health, Building 4, Rm 324, Bethesda, Maryland [tel. (301)496-0972; fax (301)480-2808, e-mail christine_kozak@d4.niaid.nih.gov).

6. The Seldin panel. This panel is composed of ~500 progeny derived from backcrosses of (C3H/HeJ-*gld* x *M. spretus*)F$_1$ gld/+ hybrid females to C3H/HeJ-*gld*/+ females. This panel has been typed for more than 500 loci. Information about access to the panel can be obtained by contacting M. Seldin, Duke University Medical Center, Durham, North Carolina [tel. (919)684-6152; fax (919)681-6070; e-mail seldi001@mc.duke.edu).

7. Other panels. Many other private or semi-private mapping panels exist. Information about their existence and availability can be obtained either through review of the primary literature or by contacting members of the Mouse Chromosome Committees.

B. Simple Sequence Repeat Polymorphisms

The ease and power of linkage mapping was substantially enhanced with the introduction of large numbers of highly variable loci defined by simple sequence repeat polymorphisms (Dietrich *et al.*, 1992). These loci are dinucleotide repeat tracts, usually, CA$_n$, that are flanked by unique DNA sequences that serve as primer binding sites. Primer pairs for many of these loci are available from Research Genetics, Inc. (Huntsville, Alabama). Information about these primer pairs is available from the MIT Center for Genome Research through e-mail (send the message "help" to genome_database @ genome.wi.mit.edu) or from the Encyclopedia of the Mouse Genome (see Appendix 4,B).

C. Genomic DNA

For many mapping experiments, a small sample of genomic DNA rather than a whole mouse is needed. An example of a possible use of this DNA includes identification of a restriction fragment length variant that can be used for gene mapping. Another example involves comparing the restriction maps and genomic DNA sequences for a gene in normal and mutant mice as a test for whether a mutation in a cell death gene is responsible for the mutant phenotype. For these and other applications, it is often more economical to purchase a small quantity of high quality genomic DNA than to purchase a mouse. For a modest fee, the Jackson Laboratory Mouse DNA Resource provides aliquots of genomic DNA from most inbred, congenic, recombinant inbred, and mutant mouse strains. Information about DNA availability can be obtained by contacting Marie Ivey, Jackson Laboratory [tel. (207)288-3371].

D. cDNA Libraries

In many cases, studies aimed at mapping the gene of interest are hindered by the availability of only a short sequence tag, or because one is searching for cross-hybridizing genes. Access to appropriate cDNA libraries that could be used to isolate full-length cDNA sequences of the gene of interest would greatly facilitate the work. Relatively few noncommercial cDNA libraries are available. However, at least four "public" libraries are available including an 8.5-day whole embryo library (Farnher *et al.*, 1987), an equalized (normalized clone representation) library composed of cDNAs prepared from embryos from each of the 20 days of embryonic development (Ko, 1990), several cDNA libraries from pre-implantation stage embryos (Rothstein *et al.*, 1993), and several other embryonic cDNA libraries (see Zehetner and Lehrach, 1994).

Appendix 2. Resources for Positional Cloning in the Mouse: Large Insert Clone Libraries

Large insert clones are essential for positional cloning of disease genes and for physical mapping of large genomic regions. Construction of a large library should not be undertaken casually. Transformation frequencies are low and other technical problems are commonly encountered (Burke *et al.*, 1991; Pierce *et al.*, 1992; Strauss *et al.*, 1992; Kusumi *et al.*, 1993). Considerable time, resources, and expertise are required. Common problems, which appear to be an intrinsic feature of all large insert libraries, include a substantial frequency of chimeric clones, clone instability, and internal deletions or rearrangements. Despite these problems and technical difficulties, at least five large insert clone libraries are available.

1. Dupont P1 library. An important library was made using a bacteriophage P1 vector (Pierce *et al.*, 1992). This vector–host system provides easy clone handling and should provide reduced rates of clone chimerism. The genomic

DNA sources were C127 mouse fibroblasts and 129/Sv-Ter mice. Average insert size is 70 kb (range 50–95 kb) and genome redundancy is ~3x. The library is available from Genome Systems, Inc. [7166 Manchester Rd., St. Louis, MO; tel. 1-800-248-7609; fax (314)647-4134].

2. ICRF YAC library. This was the first mouse YAC library reported (Larin *et al.*, 1991). The genomic DNA sources were C3H and C57BL/6J mice. Average insert size is 700 kb (range 150–1300 kb) and genome redundancy is ~3x. Access to these and other libraries is described by Zehetner and Lehrach (1994).

3. MIT YAC library. The genomic DNA source was C57BL/6J female mice (Kusumi *et al.*, 1993). Average insert size is 480 kb for part A of the library and 680 kb for part B; genome redundancy is ~4.3x. The library is available from Research Genetics, which also provides library screening services.

4. Princeton YAC library. The genomic DNA source was C57BL/6J female mice (Burke *et al.*, 1991). Average insert size is 265 kb (range 250–275 kb) and genome redundancy is ~2.2x. The library is available from Research Genetics, which also provides library screening services.

5. St. Mary's YAC library. To test whether recombination in the yeast host contributed to YAC chimerism, a YAC library was constructed using a Rad52 mutant yeast host (Chartier *et al.*, 1992). The genomic DNA source was C57BL/10J female mice. Average insert size is 240 kb and genome redundancy is ~3x. Access to the library can be obtained by contacting S. D. M. Brown, Biochemistry Department, St. Mary's Hospital, London (e-mail s.brown-@sm.ic.ac.uk) or by using the Reference Library System (Zehetner and Lehrach, 1994).

Appendix 3. Access to Mutant Mice

A. Mouse Repositories

Several institutions are major repositories of mutant mice. These include Institut Pasteur (Paris, France), National Institute of Genetics (Mishima, Japan), Jackson Laboratory (Bar Harbor, Maine), MRC Radiobiology Unit (Chilton, UK), and Oak Ridge National Laboratory (Oak Ridge, Tennessee). Many other institutions and investigators have important collections. Listings of mutant mice available from these and other sources can be found in the journal *Mouse Genome* (Oxford University Press), in the book "Genetic Variants and Strains of the Laboratory Mouse" (Green, 1981; Lyon and Searle, 1989; Lyon *et al.*, 1995), and in an informal publication called "List of Mutations and Mutant Alleles" (formerly known as the "Lane List," Jackson Laboratory).

B. Induced Mutant Resource

A national resource for targeted mutants and selected transgenic mice was recently established at the Jackson Laboratory (Culliton, 1993). Targeted mutants and transgenic mice that are relevant to research on programmed cell death are discussed in Section IV. Many of these are available from the Induced Mutant Resource at the Jackson Laboratory. Information about current holdings and availability can be obtained by calling 1-800-422-MICE.

C. Mouse Locus Catalog

A summary of the phenotypic characteristics of mutant mice is provided in the Mouse Locus Catalog, which has been published in "Genetic Strains and Variants of the Laboratory Mouse" (Green, 1981; Lyon and Searle, 1989). An update will appear in the next edition of this book (Lyon *et al.*, 1995). Access to electronic versions, which are regularly updated, is available through GBASE, gopher, and the World Wide Web (see Appendix 4).

Appendix 4. Electronic Information Resources

Several criteria were used to select resources for description here. First, the resources must be idiosyncratic for the laboratory mouse. For example, GenBank was not included because its utility is generally known. In contrast, the Mouse Locus Catalog, which is a text description of genes and phenotypic mutants in the mouse, is much less widely known. Second, the information in these resources must be important for research on programmed cell death. For example, gene mapping databases are crucial for evaluating mapping results for new cell death genes. Similarly, descriptions of known mutant mice are essential for evaluating whether a cell death gene might be mutated in a closely linked gene responsible for a mutant phenotype.

A. Chromosome Committee Reports

The mouse genetics community annually publishes comprehensive reports of the genetic, cytogenetic, and physical maps for each mouse chromosome. These reports are prepared by experts on particular chromosomes and chromosome segments. They are perhaps the best status report of the maps for each chromosome. These reports are published in *Mammalian Genome* (Chapman *et al.*, 1993) and are available electronically (see "Encyclopedia of the Mouse Genome" and other electronic resources available from the Mouse Genome Informatics Project at the Jackson Laboratory).

B. Mouse Genome Informatics Project

Several databases and reports are available from this project at the Jackson Laboratory. This project is responsible for maintaining several important databases concerning the biology and genetics of the laboratory mouse and for providing easy to learn and use, electronic access to these resources. These databases and reports include GBASE, a genomic database for the mouse;

SSLP data from the MIT Center for Genome Research; Comparative Mapping Database for more than 40 mammalian species; Probes and Clones; MLDP, a database of raw mapping information; and MATRIX, a database of genetic variation among inbred strains. Inquiries concerning access can be directed to User Support, Mouse Genome Informatics Project, Jackson Laboratory, Bar Harbor, Maine or to the e-mail address mgi-help@informatics.jax.org. Many of these databases are also available through the "Encyclopedia of the Mouse Genome," gopher, and the World Wide Web (http://www.informatics.jax.org/). Finally, an electronic Community Bulletin Board is available to the mouse genetics community. This bulletin board can be used for sharing results and requesting information. To gain access to the bulletin board, send the message "subscribe mgi-list <your name>" to the e-mail address listserver-@informatics.jax.org. For additional information, send an e-mail message to the Mouse Genome Informatics Project at the Jackson Laboratory (mgi-help@informatics.jax.org).

C. TBASE

This database documents the characteristics of transgenic mice and targeted mutations (Woychik *et al.,* 1993). The database is available through the Johns Hopkins University Computational Biology Gopher Server. Inquiries can also be directed via e-mail to gopher://merlot.gdb.org/11/Database-local/mcuse/tbase or /World Wide Web (http://www.gdb.org/Dan/tbase/tbase.html).

References

Abbadie, C., Kabrun, N., Bouali, F., Smardova, J., Stehelin, D., Vandenbunder, B., and Enrietto, P. J. (1993). High levels of c-*rel* expression are associated with programmed cell death in the developing avian embroyo and in bone marrow cells in vitro. *Cell* **75**, 899–912.

Abrams, J. M., White, K., Fessler, L. I., and Steller, H. (1993). Programmed cell death during *Drosophila* embryogenesis. *Development* **117**, 29–43.

Allsopp, T. E., Wyatt, S., Paterson, H. F., and Davies, A. M. (1993). The proto-oncogene *bcl*-2 can selectively rescue neurotrophic factor-dependent neurons from apoptosis. *Cell* **73**, 295–307.

Amati, B., Littlewood, T. D., Evan, G. I., and Land, H. (1993). The c-Myc protein induces cell cycle progression and apoptosis through dimerization with Max. *EMBO J.* **12**, 5083–5087.

Ansari, B., Coates, P. J., Greenstein, B. D., and Hall, P. A. (1993). In situ end-labelling detects DNA strand breaks in apoptosis and other physiological and pathological states. *J. Pathol.* **170**, 1–8.

Arends, M. J., and Wyllie, A. H. (1991). Apoptosis: mechanisms and roles in pathology. *Int. Rev. Exp. Pathol.* **32**, 223–254.

Arends, M. J., Morris, R. G., and Wyllie, A. H. (1990). Apoptosis. The role of the endonuclease. *Am. J. Pathol.* **136**, 593–608.

Avner, P., Amar, I., Dandolo, L., and Guenet, J. L. (1988). Genetic analysis of the mouse using interspecific crosses. *Trends Genet.* **8**, 18–23.

Barres, B. A., Hart, I. K., Coles, H. S., Burne, J. F., Voyvodic, J. T., Richardson, W. D., and Raff, M. C. (1992). Cell death and control of cell survival in the oligodendrocyte lineage. *Cell* **70**, 31–46.

Behringer, R. R., Cate, R. L., Froelick, G. J., Palmiter, R. D., and Brinster, R. L. (1990). Abnormal sexual development in transgenic mice chronically expressing mullerian inhibiting substance [see comments]. *Nature* **345,** 167–170.

Boise, L. H., Gonzalez-Garcia, M., Postema, C. E., Ding, L., Lindsten, T., Turka, L. A., Mao, X., Nunez, G., and Thompson, C. B. (1993). bcl-x, a bcl-2-related gene that functions as a dominant regulator of apoptotic cell death. *Cell* **74,** 597–608.

Bowes, C., Li, T., Danciger, M., Baxter, L. C., Applebury, M. L., and Farber, D. B. (1990). Retinal degeneration in the rd mouse is caused by a defect in the beta subunit of rod cGMP-phosphodiesterase [see comments]. *Nature* **347,** 677–680.

Brunet, J. F., Dosetto, M., Denizot, F., Mattei, M. G., Clark, W. R., Haqqi, T. M., Perrier, P., Nabholz, M., Schmitt-Verhulst, A. M., Luciani, M. F., and Golstein, P. (1991). The inducible cytotoxic T-lymphocyte associated gene transcript CTLA-1 sequence and gene localization in the mouse. *Nature* **322,** 268–271.

Burke, D. T., Rossi, J. M., Leung, J., Koos, D. S., and Tilghman, S. M. (1991). A mouse genomic library of yeast artificial chromosome clones. *Mamm. Gen.* **1,** 65.

Buttyan, R., Zakeri, Z., Lockshin, R., and Wolgemuth, D. (1988). Cascade induction of c-fos, c-myc, and heat shock 70K transcripts during regression of the rat ventral prostate gland. *Mol. Endocrinol.* **2,** 650–657.

Capecchi, M. R. (1989). Altering the genome by homologous recombination. *Science* **244,** 1288–1292.

Cate, R. L., Mattaliano, R. J., Hession, C., Tizard, R., Farber, N. M., Cheung, A., Ninfa, E. G., Frey, A. Z., Gash, D. J., Chow, E. P., Fisher, R. A., Bertonis, J. M., Torres, G., Wallner, B. P., Ramachandran, K. L., Ragin, R. C., Manganaro, T. F., MacLaughlin, D. T., and Donahoe, P. K. (1986). Isolation of the bovine and human genes for Mullerian inhibiting substance and expression of the human gene in animal cells. *Cell* **45,** 685–698.

Catlin, E. A., MacLaughlin, D. T., and Donahoe, P. K. (1993). Mullerian inhibiting substance: New perspectives and future directions. *Microsc. Res. Tech.* **25,** 121–133.

Chabot, B., Stephenson, D. A., Chapman, V. M., Besmer, P., and Bernstein, A. (1988). The proto-oncogene c-kit encoding a transmembrane tyrosine kinase receptor maps to the mouse W locus. *Nature* **335,** 88–89.

Chang, G. Q., Hao, Y., and Wong, F. (1993). Apoptosis: Final common pathway of photoreceptor death in rd, rds, and rhodopsin mutant mice. *Neuron* **11,** 595–605.

Chapman, V. M., Goodfellow, P., Nadeau, J. H., and Silver, L. M. (1993). Encyclopedia of the mouse genome III. *Mamm. Gen.* **4,** S1-S283.

Chartier, F. L., Keer, J. T., Sutcliffe, M. J., Henriques, D. A., Mileham, P., and Brown, S. D. (1992). Construction of a mouse yeast artificial chromosome library in a recombination-deficient strain of yeast. *Nat. Genet.* **1,** 132–136.

Clarke, A. R., Purdie, C. A., Harrison, D. J., Morris, R. G., Bird, C. C., Hooper, M. L., and Wyllie, A. H. (1993). Thymocyte apoptosis induced by p53-dependent and independent pathways [see comments]. *Nature* **362,** 849–852.

Clarke, P. G. (1990). Developmental cell death: Morphological diversity and multiple mechanisms. *Anat. Embryol. (Berl.)* **181,** 195–213.

Cohen, J. J., Duke, R. C., Chervenak, R., Sellins, K. S., and Olson, L. K. (1985). DNA fragmentation in targets of CTL: An example of programmed cell death in the immune system. *Adv. Exp. Med. Biol.* **184,** 493–508.

Cohen, J. J., Duke, R. C., Fadock, V. A., and Sellins, K. S. (1992). Apoptosis and programmed cell death in immunity. *Annu. Rev. Immunol.* **10,** 267–293.

Coles, H., Burne, J., and Raff, M. (1993). Large-scale normal cell death in the developing rat kidney and its reduction by epidermal growth factor. *Development* **118,** 777–784.

Colotta, F., Polentarutti, N., Sironi, M., and Mantovani, A. (1992). Expression and involvement of c-fos and c-jun protooncogenes in programmed cell death induced by growth factor deprivation in lymphoid cell lines. *J. Biol. Chem.* **267,** 18278–18283.

Copeland, N. G., and Jenkins, N. A. (1991). Development and applications of a molecular genetic linkage map of the mouse genome. *Trends Genet.* **7,** 113–118.

Copeland, N. G., Gilbert, D. J., Cho, B. C., Donovan, P. J., Jenkins, N. A., Cosman, D., Anderson, D., Lyman, S. D., and Williams, D. E. (1990). Mast cell growth factor maps near the steel locus on mouse chromosome 10 and is deleted in a number of steel alleles. *Cell* **63**, 175–183.

Copeland, N. G., and Jenkins, N. A., Gilbert, D. J., Eppig, J. T., Maltais, L. J., Miller, J. C., Dietrich, W. F., Weaver, A., Lincoln, S. E., Steen, R. G., Stein, L. D., Nadeau, J. H., and Lander, E. S. (1993). A genetic linkage map of the mouse: Current applications and future prospects [see comments]. *Science* **262**, 57–66.

Copp, A. J. (1978). Interaction between inner cell mass and trophectoderm of the mouse blastocyst. I. A study of cellular proliferation. *J. Embryol. Exp. Morphol.* **48**, 109–125.

Cory, S. (1994). Fascinating death factor [news]. *Nature* **367**, 317–318.

Coucouvanis, E. C., Sherwood, S. W., Carswell-Crumpton, C., Spack, E. G., and Jones, P. P. (1993). Evidence that the mechanism of prenatal germ cell death in the mouse is apoptosis. *Exp. Cell Res.* **209**, 238–247.

Culliton, B. J. (1993). A home for the mighty mouse [news]. *Nature* **364**, 755.

Dautigny, A., Mattei, M. G., Morello, D., Alliel, P. M., Pham-Dinh, D., Amar, L., Arnaud, D., Simon, D., Mattei, J. F., Guenet, J. L., Jolles, P., and Avner, P. (1986). The structural gene coding for myelin-associated proteolipid protein is mutated in jimpy mice. *Nature* **321**, 867–869.

Davidson, W. F., Dumont, F. J., Bedigian, H. G., Fowlkes, B. J., and Morse, H. C. (1986). Phenotypic, functional, and molecular genetic comparisons of the abnormal lymphoid cells of C3H-lpr/lpr and C3H-gld/gld mice. *J. Immunol.* **136**, 4075–4084.

Dieken, E. S., and Miesfeld, R. L. (1992). Transcriptional transactivation functions localized to the glucocorticoid receptor N terminus are necessary for steroid induction of lymphocyte apoptosis. *Mol. Cell. Biol.* **12**, 589–597.

Dietrich, W., Katz, H., Lincoln, S. E., Shin, H. S., Friedman, J., Dracopoli, N. C., and Lander, E. S. (1992). A genetic map of the mouse suitable for typing intraspecific crosses. *Genetics* **131**, 423–447.

D'Mello, S. R., and Galli, C. (1993). SGP2, ubiquitin, 14K lectin and RP8 mRNAs are not induced in neuronal apoptosis. *Neuroreport* **4**, 355–358.

Ellis, R. E., Yuan, J. Y., and Horvitz, H. R. (1991). Mechanisms and functions of cell death. *Annu. Rev. Cell Biol.* **7**, 663–698.

El-Shershaby, A. M., and Hinchliffe, J. R. (1974). Cell redundancy in the zona-intact preimplantation mouse blastocyst: A light and electron microscope study of dead cells and their fate. *J. Embryol. Exp. Morphol.* **31**, 643–654.

Escary, J. L., Perreau, J., Dumenil, D., Ezine, S., and Brulet, P. (1993). Leukaemia inhibitory factor is necessary for maintenance of haematopoietic stem cells and thymocyte stimulation. *Nature* **363**, 361–364.

Evan, G. I., Wyllie, A. H., Gilbert, C. S., Littlewood, T. D., Land, H., Brooks, M., Waters, C. M., Penn, L. Z., and Hancock, D. C. (1992). Induction of apoptosis in fibroblasts by c-myc protein. *Cell* **69**, 119–128.

Evans, C. A., Owen-Lynch, P. J., Whetton, A. D., and Dive, C. (1993). Activation of the Abelson tyrosine kinase activity is associated with suppression of apoptosis in hemopoietic cells. *Cancer Res.* **53**, 1735–1738.

Fahrner, K., Hogan, B. L., and Flavell, R. A. (1987). Transcription of H-2 and Qa genes in embryonic and adult mice. *EMBO J.* **6**, 1265–1271.

Fallon, J. F., and Simandl, B. K. (1978). Evidence of a role for cell death in the disappearance of the embryonic human tail. *Am. J. Anat.* **152**, 111–129.

Favor, J. (1990). Toward an understanding of the mechanisms of mutation induction ethylnitrosourea in germ cells of the mouse. *In* "Biology of Mammalian Germ Cell Mutagenesis" (J. W. Allen, B. A. Bridges, M. F. Lyon, M. J. Moses, and L. B. Russell, eds.), Banbury Report 34, pp. 221–236. Cold Spring Harbor Laboratory Press, Cold Spring Harbor, New York.

Ferguson, M. W. (1988). Palate development. *Development* **103**(Supplement), 41–60.

Fernandez-Sarabia, M. J., and Bischoff, J. R. (1993). Bcl-2 associates with the ras-related protein R-ras p23. *Nature* **366**, 274–275.

regulation of apoptosis and cell proliferation by transforming growth factor beta 1 in cultured uterine epithelial cells. *Proc. Natl. Acad. Sci. USA* **88**, 3412–3415.

Roths, J. B. (1978). Personal communication. *Mouse Newslett.* **58**, 50.

Roths, J. B., Murphy, E. D., and Eicher, E. M. (1984). A new mutation, gld, that produces lymphoproliferation and autoimmunity in C3H/HeJ mice. *J. Exp. Med.* **159**, 1–20.

Rothstein, J. L., Johnson, D., Jessee, J., Skowronski, J., DeLoia, J. A., Solter, D., and Knowles, B. B. (1993). Construction of primary and subtracted cDNA libraries from early embryos. *Meth. Enzymol.* **225**, 587–610.

Rowe, L. B., Nadeau, J. H., Turner, R., Frankel, W. N., Letts, V. A., Eppig, J. T., Ko, M. S., Thurston, S. J., and Birkenmeier, E. H. (1994). Maps from two interspecific backcross DNA panels available as a community genetic mapping resource. *Mamm. Gen.* **5**, 253–274.

Russell, E. S. (1979). Hereditary anemias of the mouse: A review for geneticists. *Adv. Genet.* **20**, 357–459.

Russell, L. B., Russell, W. L., Rinchik, E. M., and Hunsicker, P. R. (1990). Factors affecting the nature of induced mutations. *In* ''Biology of Mammalian Germ Cell Mutagenesis'' (J. W. Allen, B. A. Bridges, M. F. Lyon, M. J. Moses, and L. B. Russell, eds.), Banbury Report 34, pp. 271–292. Cold Spring Harbor Laboratory Press, Cold Spring Harbor, New York.

Sabourin, L. A., and Hawley, R. G. (1990). Suppression of programmed death and G1 arrest in B-cell hybridomas by interleukin-6 is not accompanied by altered expression of immediate early response genes. *J. Cell Physiol.* **145**, 564–574.

Sambrook, J., Fritsch, E. F., and Maniatis, T. (1989). ''Molecular Cloning: A Laboratory Manual.'' Cold Spring Harbor Laboratory Press, Cold Spring Harbor, New York.

Sanwal, M., Muel, A. S., Chaudun, E., Courtois, Y., and Counis, M. F. (1986). Chromatin condensation and terminal differentiation process in embryonic chicken lens in vivo and in vitro. *Exp. Cell Res.* **167**, 429–439.

Saunders, J. W., Gasseling, M. T., and Saunders, L. C. (1962). Cellular death in morphogenesis of the avian wing. *Dev. Biol.* **5**, 147–178.

Savill, J., Dransfield, I., Hogg, N., and Haslett, C. (1990). Vitronectin receptor-mediated phagocytosis of cells undergoing apoptosis. *Nature* **343**, 170–173.

Schwartz, L. M., and Osborne, B. A. (1993). Programmed cell death, apoptosis and killer genes. *Immunol. Today* **14**, 582–590.

Schwartz, L. M., Kosz, L., and Kay, B. K. (1990a). Gene activation is required for developmentally programmed cell death. *Proc. Natl. Acad. Sci. USA* **87**, 6594–6598.

Schwartz, L. M., Myer, A., Kosz, L., Engelstein, M., and Maier, C. (1990b). Activation of polyubiquitin gene expression during developmentally programmed cell death. *Neuron* **5**, 411–419.

Schwartz, L. M., Smith, S. W., Jones, M. E., and Osborne, B. A. (1993). Do all programmed cell deaths occur via apoptosis? *Proc. Natl. Acad. Sci. USA* **90**, 980–984.

Searle, A. G. (1964). The genetics and morphology of two luxoid mutants in the house mouse. *Genet. Res.* **5**, 171–197.

Sentman, C. L., Shutter, J. R., Hockenbery, D., Kanagawa, O., and Korsmeyer, S. J. (1991). bcl-2 inhibits multiple forms of apoptosis but not negative selction in thymocytes. *Cell* **67**, 879–888.

Shi, L., Kraut, R. P., Aebersold, R., and Greenberg, A. H. (1992). A natural killer cell granule protein that induces DNA fragmentation and apoptosis. *J. Exp. Med.* **175**, 553–566.

Shi, L., Nishioka, W. K., Th'ng, J., Bradbury, E. M., Litchfield, D. W., and Greenberg, A. H. (1994). Premature p34cdc2 activation required for apoptosis. *Science* **263**, 1143–1145.

Shi, Y., Glynn, J. M., Guilbert, L. J., Cotter, T. G., Bissonnette, R. P., and Green, D. R. (1992). Role for c-myc in activation-induced apoptotic cell death in T cell hybridomas. *Science* **257**, 212–214.

Shull, M. M., Ormsby, I., Kier, A. B., Pawlowski, S., Diebold, R. J., Yin, M., Allen, R., Sidman, C., Proetzel, G., Calvin, D., Annunziata, N., and Doetschman, T. (1992). Targeted disruption of the mouse transforming growth factor-beta 1 gene results in multifocal inflammatory disease. *Nature* **359**, 693–699.

Silver, J., and Hughes, A. F. (1973). The role of cell death during morphogenesis of the mammalian eye. *J. Morphol.* **140,** 159–170.

Smith, C. A., Gruss, H. J., Davis, T., Anderson, D., Farrah, T., Baker, E., Sutherland, G. R., Brannan, C. I., Copeland, N. G., Jenkins, N. A., Grabstein, K. H., Gliniak, B., McAlister, I. B., Fanslow, W., Alderson, M., Falk, B., Gimpel, S., Gillis, S., Din, W. S., Goodwin, R. G., and Armitage, R. J. (1993). CD30 antigen, a marker for Hodgkin's lymphoma, is a receptor whose ligand defines an emerging family of cytokines with homology to TNF. *Cell* **73,** 1349–1360.

Strange, R., Li, F., Saurer, S., Burkhardt, A., and Friis, R. R. (1992). Apoptotic cell death and tissue remodeling during mouse mammary gland involution. *Development* **115,** 49–58.

Strauss, W. M., Jaenisch, E., and Jaenisch, R. (1992). A strategy for rapid production and screening of yeast artificial chromosome libraries. *Mamm. Gen.* **2,** 150–157.

Suda, T., Takahashi, T., Golstein, P., and Nagata, S. (1993). Molecular cloning and expression of the Fas ligand, a novel member of the tumor necrosis factor family. *Cell* **75,** 1169–1178.

Szirmai, J. A., and Balazs, E. A. (1958). Studies on the structure of the vitreous body. *Arch. Ophthal.* **59,** 34–47.

Takahashi, T., Tanaka, M., Brannan, C. I., Jenkins, N. A., Copeland, N. G., Suda, T., and Nagata, S. (1994). Generalized lymphoproliferative disease in mice, caused by a point mutation in the Fas ligand. *Cell* **76,** 969–976.

Tartaglia, L. A., Ayres, T. M., Wong, G. H., and Goeddel, D. V. (1993). A novel domain within the 55 kd TNF receptor signals cell death. *Cell* **74,** 845–853.

Taylor, B. A. (1981). Recombinant inbred strains: Use in gene mapping. *In* "Origins of Inbred Mice" (H. C. Morse, ed.), pp. 423–438. Academic Press, New York.

Taylor, B. A. (1995). Recombinant inbred strains. *In* "Genetic Variants and Strains of the Laboratory Mouse" (M. F. Lyon, S. Rastan and S. D. M. Brown, eds.). Oxford University Press, Oxford (*in press*).

Theiler, K., Varnum, D. S., Nadeau, J. H., Stevens, L. C., and Cagianut, B. (1976). A new allele of ocular retardation: Early development and morphogenetic cell death. *Anat. Embryol.* **150,** 85–97.

Troppmair, J., Cleveland, J. L., Askew, D. S., and Rapp, U. R. (1992). v-Raf/v-Myc synergism in abrogation of IL-3 dependence: v-Raf suppresses apoptosis. *Curr. Top. Microbiol. Immunol.* **182,** 453–460.

Ucker, D. S., Obermiller, P. S., Eckhart, W., Apgar, J. R., Berger, N. A., and Meyers, J. (1992). Genome digestion is a dispensable consequence of physiological cell death mediated by cytotoxic T lymphocytes. *Mol. Cell Biol.* **12,** 3060–3069.

Veis, D. J., Sorenson, C. M., Shutter, J. R., and Korsmeyer, S. J. (1993). Bcl-2-deficient mice demonstrate fulminant lymphoid apoptosis, polycystic kidneys, and hypopigmented hair. *Cell* **75,** 229–240.

Vogel, M. W., Sunter, K., and Herrup, K. (1989). Numerical matching between granule and Purkinje cells in lurcher chimeric mice: A hypothesis for the trophic rescue of granule cells from target-related cell death. *J. Neurosci.* **9,** 3454–3462.

Walker, N. I., Bennett, R. E., and Kerr, J. F. (1989). Cell death by apoptosis during involution of the lactating breast in mice and rats. *Am. J. Anat.* **185,** 19–32.

Wang, Z. Q., Ovitt, C., Grigoriadis, A. E., Mohle-Steinlein, U., Ruther, U., and Wagner, E. F. (1992). Bone and haematopoietic defects in mice lacking c-fos. *Nature* **360,** 741–745.

Watanabe-Fukunaga, R., Brannan, C. I., Copeland, N. G., Jenkins, N. A., and Nagata, S. (1992). Lymphorproliferation disorder in mice explained by defects in Fas antigen that mediates apoptosis. *Nature* **356,** 314–317.

Watson, M. L., Rao, J. K., Gilkeson, G. S., Ruiz, P., Eicher, E. M., Pisetsky, D. S., Matsuzawa, A., Rochelle, J. M., and Seldin, M. F. (1992). Genetic analysis of MRL-lpr mice: Relationship of the Fas apoptosis gene to disease manifestations and renal disease-modifying loci. *J. Exp. Med.* **176,** 1645–1656.

Wier, K. A., Fukuyama, K., and Epstein, W. L. (1971). Nuclear changes during keratinization of normal human epidermis. *J. Ultrastruct. Res.* **37,** 138–145.

Wijsman, J. H., Jonker, R. R., Keijzer, R., Van De Velde, C. J., Cornelisse, C. J., and Van Dierendonck, J. (1993). A new method to detect apoptosis in paraffin sections: in situ end-labeling of fragmented DNA. *J. Histochem. Cytochem.* **41**, 7–12.

Williams, G. T., Smith, C. A., Spooncer, E., Dexter, T. M., and Taylor, D. R. (1990). Haemopoietic colony stimulating factors promote cell survival by suppressing apoptosis. *Nature* **343**, 76–79.

Wood, K. A., Dipasquale, B., and Youle, R. J. (1993). In situ labeling of granule cells for apoptosis-associated DNA fragmentation reveals different mechanisms of cell loss in developing cerebellum. *Neuron* **11**, 621–632.

Woronicz, J. D., Calnan, B., Ngo, V., and Winoto, A. (1994). Requirement for the orphan steroid receptor Nur77 in apoptosis of T-cell hybridomas. *Nature* **367**, 277–281.

Woychik, R. P., Wassom, J. S., Kingsbury, D., and Jacobson, D. A. (1993). TBASE: A computerized database for transgenic animals and targeted mutations [published erratum appears in *Nature* (1993) **363(6430)**, 656]. *Nature* **363**, 375–376.

Yonish-Rouach, E., Resnitzky, D., Lotem, J., Sachs, L., Kimchi, A., and Oren, M. (1991). Wild-type p53 induces apoptosis of myeloid leukaemic cells that is inhibited by interleukin-6. *Nature* **352**, 345–347.

Young, R. W. (1984). Cell death during differentiation of the retina in the mouse. *J. Comp. Neurol.* **229**, 362–373.

Yuan, J., Shaham, S., Ledoux, S., Ellis, H. M., and Horvitz, H. R. (1993). The *C. elegans* cell death gene ced-3 encodes a protein similar to mammalian interleukin-1 beta-converting enzyme. *Cell* **75**, 641–652.

Zakeri, Z. F., Quaglino, D., Latham, T., and Lockshin, R. A. (1993). Delayed internucleosomal DNA fragmentation in programmed cell death. *FASEB J.* **7**, 470–478.

Zehetner, G., and Lehrach, H. (1994). The reference library system—Sharing biological material and experimental data. *Nature* **367**, 489–491.

Zsebo, K. M., Williams, D. A., Geissler, E. N., Broudy, V. C., Martin, F. H., Atkins, H. L., Hsu, R. Y., Birkett, N. C., Okino, K. H., Murdock, D. C., Jacobsen, F. W., Langley, K. E., Smith, K. A., Takeishi, T., Gattanach, B. M., Galli, S. J., and Suggs, S. V. (1990). Stem cell factor is encoded at the Sl locus of the mouse and is the ligand for the c-kit tyrosine kinase receptor. *Cell* **63**, 213–224.

Zwilling, E. (1942). The development of dominant rumplessness in chick embryos. *Genetics* **27**, 641–662.

INDEX

VOLUMES IN SERIES

Founding Series Editor
DAVID M. PRESCOTT

Volume 1 (1964)
Methods in Cell Physiology
Edited by David M. Prescott

Volume 2 (1966)
Methods in Cell Physiology
Edited by David M. Prescott

Volume 3 (1968)
Methods in Cell Physiology
Edited by David M. Prescott

Volume 4 (1970)
Methods in Cell Physiology
Edited by David M. Prescott

Volume 5 (1972)
Methods in Cell Physiology
Edited by David M. Prescott

Volume 6 (1973)
Methods in Cell Physiology
Edited by David M. Prescott

Volume 7 (1973)
Methods in Cell Biology
Edited by David M. Prescott

Volume 8 (1974)
Methods in Cell Biology
Edited by David M. Prescott

Volume 9 (1975)
Methods in Cell Biology
Edited by David M. Prescott

Volume 10 (1975)
Methods in Cell Biology
Edited by David M. Prescott

Volume 11 (1975)
Yeast Cells
Edited by David M. Prescott

Volume 12 (1975)
Yeast Cells
Edited by David M. Prescott

Volume 13 (1976)
Methods in Cell Biology
Edited by David M. Prescott

Volume 14 (1976)
Methods in Cell Biology
Edited by David M. Prescott

Volume 15 (1977)
Methods in Cell Biology
Edited by David M. Prescott

Volume 16 (1977)
Chromatin and Chromosomal Protein Research I
Edited by Gary Stein, Janet Stein, and Lewis J. Kleinsmith

Volume 17 (1978)
Chromatin and Chromosomal Protein Research II
Edited by Gary Stein, Janet Stein, and Lewis J. Kleinsmith

Volume 18 (1978)
Chromatin and Chromosomal Protein Research III
Edited by Gary Stein, Janet Stein, and Lewis J. Kleinsmith

Volume 19 (1978)
Chromatin and Chromosomal Protein Research IV
Edited by Gary Stein, Janet Stein, and Lewis J. Kleinsmith

Volume 20 (1978)
Methods in Cell Biology
Edited by David M. Prescott

Advisory Board Chairman
KEITH R. PORTER

Volume 21A (1980)
**Normal Human Tissue and Cell Culture, Part A: Respiratory,
    Cardiovascular, and Integumentary Systems**
Edited by Curtis C. Harris, Benjamin F. Trump, and Gary D. Stoner

Volume 21B (1980)
Normal Human Tissue and Cell Culture, Part B: Endocrine, Urogenital, and Gastrointestinal Systems
Edited by Curtis C. Harris, Benjamin F. Trump, and Gary D. Stoner

Volume 22 (1981)
Three-Dimensional Ultrastructure in Biology
Edited by James N. Turner

Volume 23 (1981)
Basic Mechanisms of Cellular Secretion
Edited by Arthur R. Hand and Constance Oliver

Volume 24 (1982)
The Cytoskeleton, Part A: Cytoskeletal Proteins, Isolation and Characterization
Edited by Leslie Wilson

Volume 25 (1982)
The Cytoskeleton, Part B: Biological Systems and *in Vitro* Models
Edited by Leslie Wilson

Volume 26 (1982)
Prenatal Diagnosis: Cell Biological Approaches
Edited by Samuel A. Latt and Gretchen J. Darlington

Series Editor
LESLIE WILSON

Volume 27 (1986)
Echinoderm Gametes and Embryos
Edited by Thomas E. Schroeder

Volume 28 (1987)
***Dictyostelium discoideum:* Molecular Approaches to Cell Biology**
Edited by James A. Spudich

Volume 29 (1989)
Fluorescence Microscopy of Living Cells in Culture, Part A: Fluorescent Analogs, Labeling Cells, and Basic Microscopy
Edited by Yu-Li Wang and D. Lansing Taylor

Volume 30 (1989)
Fluorescence Microscopy of Living Cells in Culture, Part B: Quantitative Fluorescence Microscopy—Imaging and Spectroscopy
Edited by D. Lansing Taylor and Yu-Li Wang

Volume 31 (1989)
Vesicular Transport, Part A
Edited by Alan M. Tartakoff

Volume 32 (1989)
Vesicular Transport, Part B
Edited by Alan M. Tartakoff

Volume 33 (1990)
Flow Cytometry
Edited by Zbigniew Darzynkiewicz and Harry A. Crissman

Volume 34 (1991)
Vectorial Transport of Proteins into and across Membranes
Edited by Alan M. Tartakoff

Selected from Volumes 31, 32, and 34 (1991)
Laboratory Methods for Vesicular and Vectorial Transport
Edited by Alan M. Tartakoff

Volume 35 (1991)
Functional Organization of the Nucleus: A Laboratory Guide
Edited by Barbara A. Hamkalo and Sarah C. R. Elgin

Volume 36 (1991)
***Xenopus laevis:* Practical Uses in Cell and Molecular Biology**
Edited by Brian K. Kay and H. Benjamin Peng

Series Editors
LESLIE WILSON AND PAUL MATSUDAIRA

Volume 37 (1993)
Antibodies in Cell Biology
Edited by David J. Asai

Volume 38 (1993)
Cell Biological Applications of Confocal Microscopy
Edited by Brian Matsumoto

Volume 39 (1993)
Motility Assays for Motor Proteins
Edited by Jonathan M. Scholey

Volume 40 (1994)
A Practical Guide to the Study of Calcium in Living Cells
Edited by Richard Nuccitelli